S0-BLS-439

The Hall Effect and Its Applications

The Hall Effect and Its Applications

Edited by

C. L. Chien

and

C. R. Westgate

The Johns Hopkins University
Baltimore, Maryland

PLENUM PRESS • NEW YORK AND LONDON

Library of Congress Cataloging in Publication Data

Symposium on Hall Effect and Its Applications, Johns Hopkins University, 1979.
The Hall effect and its applications.

Includes index.
1. Hall effect–Congresses. I. Chien, Chia Ling. II. Westgate, C.R. III. Title.
QC612.H3S97 1979 537.6′22 80-18566
ISBN 0-306-40556-3

Proceedings of the Commemorative Symposium on the Hall Effect and its Applications, held at The Johns Hopkins University, Baltimore, Maryland, November 13, 1979.

A Division of Plenum Publishing Corporation
227 West 17th Street, New York, N.Y. 10011

Printed in the United States of America

PREFACE

In 1879, while a graduate student under Henry Rowland at the Physics Department of The Johns Hopkins University, Edwin Herbert Hall discovered what is now universally known as the Hall effect. A symposium was held at The Johns Hopkins University on November 13, 1979 to commemorate the 100th anniversary of the discovery. Over 170 participants attended the symposium which included eleven invited lectures and three speeches during the luncheon.

During the past one hundred years, we have witnessed ever expanding activities in the field of the Hall effect. The Hall effect is now an indispensable tool in the studies of many branches of condensed matter physics, especially in metals, semiconductors, and magnetic solids. Various components (over 200 million!) that utilize the Hall effect have been successfully incorporated into such devices as keyboards, automobile ignitions, gaussmeters, and satellites.

This volume attempts to capture the important aspects of the Hall effect and its applications. It includes the papers presented at the symposium and eleven other invited papers. Detailed coverage of the Hall effect in amorphous and crystalline metals and alloys, in magnetic materials, in liquid metals, and in semiconductors is provided. Applications of the Hall effect in space technology and in studies of the aurora enrich the discussions of the Hall effect's utility in sensors and switches. The design and packaging of Hall elements in integrated circuit forms are illustrated. Some historical notes, most notably portions of the hitherto unpublished notebook of E. H. Hall, provide the background and the course of the discovery.

This commemorative symposium was jointly sponsored by The Johns Hopkins University and the Micro Switch Division of Honeywell. We are indebted to members of the organizing committee, numerous colleagues from Johns Hopkins and Honeywell for their suggestions, assistance and cooperation. We would also like to thank E. M. Pugh of Carnegie-Mellon University, H. H. Hall of the University of New Hampshire and S. J. Nelson of Honeywell, Inc. for their illuminating speeches during the luncheon; C. Hurd of the National Research Council of Canada, L. Berger of Carnegie-Mellon University and R. O'Handley of the IBM Yorktown Heights Research Center for their helpful suggestions; and F. Kafka of Micro Switch for his inexhaustible assistance. Special thanks are due to Mrs. E. Carroll for her meticulous typing and retyping of the manuscripts of this volume. Finally, to all the lecturers, chairmen, contributors and participants, many thanks.

C. L. Chien and C. R. Westgate

CONTRIBUTORS

C. J. Adkins	Cavendish Laboratories Cambridge
R. S. Allgaier	Naval Surface Weapons Center
L. E. Ballentine	Department of Physics Simon Fraser University
A. C. Beer	Battelle Columbus Laboratories
L. Berger	Department of Physics Carnegie-Mellon University
G. Bergmann	Institut für Festkörperforschung Kernforschungsanlage, Jülich
R. V. Coleman	Department of Physics University of Virginia
D. Emin	Sandia National Laboratories
A. Fert	Laboratoire de Physique des Solids Université Paris-Sud, Orsay
R. J. Gambino	IBM Thomas J. Watson Research Center
M. L. Geske	Solid State Electronics Center Honeywell
H. J. Güntherodt	Institut für Physik Universität Basel
A. Hamzić	Institute of Physics of the University Zagreb
O. Hannaway	History of Science Department The Johns Hopkins University

C. M. Hurd — Division of Chemistry, National Research Council of Canada

B. R. Judd — Department of Physics, The Johns Hopkins University

R. W. Klaffky — Department of Physics, University of Virginia

H. U. Künzi — Institut für Physik, Universität Basel

D. Long — Honeywell Corporate Material Sciences Center

W. H. Lowrey — Department of Physics, University of Virginia

J. T. Maupin — Micro Switch Division, Honeywell

T. R. McGuire — IBM Thomas J. Watson Research Center

F. F. Mobley — Applied Physics Laboratories, The Johns Hopkins University

P. Nagels — Materials Science Department, S. C. K./C. E. N.

R. C. O'Handley — IBM Thomas J. Watson Research Center

T. A. Potemra — Applied Physics Laboratories, The Johns Hopkins University

L. G. Rubin — Francis Bitter National Magnet Laboratory, M.I.T.

H. H. Sample — Department of Physics, Tufts University

K. R. Sopka — Harvard University

O. N. Tufte — Honeywell Corporate Material Sciences Center

E. A. Vorthmann — Micro Switch Division, Honeywell

CONTENTS

"PRESSING ELECTRICITY": A HUNDRED YEARS OF HALL EFFECT IN CRYSTALLINE METALS AND ALLOYS

Colin M. Hurd

Division of Chemistry
National Research Council of Canada
Ottawa, Ontario, Canada, K1A OR9

1. INTRODUCTION

The quotation in the title comes from Hall's first paper (Hall 1879) where he proposed that a magnetic field applied to a current-carrying wire would generate "...a state of stress in the conductor, the electricity pressing...toward one side of the wire". Although Hall was writing some years before the electron was identified, his intuitive view of "pressed" electricity was appropriate. The current in a conductor in a magnetic field is indeed under stress, for the electrons are forced against an inside face of the sample. The resulting concentration gradient of charge is the source of the electrostatic field that Hall and Rowland (see Miller 1976) observed.

Phenomenologically, Hall found that the transverse electric field he had discovered (E_H) was proportional to the product of the primary current density I^D and the component of magnetic flux density $\underline{B}$ normal to the plane containing I^D and E_H. For a set of rectangular axes with E_H along the y-direction, we can write:

$$E_H = RI_x^D B_z, \tag{1.1}$$

where R is the Hall coefficient. Hall's original experimental arrangement of a thin foil mounted perpendicular to the applied field is still in common use, and the practical form of Eq. (1.1) is:

$$R = V_H t / IB_z. \tag{1.2}$$

Here V_H is the transverse voltage measured in say the y-direction,

I is the total current in the x-direction and t is the sample's thickness in the z-direction.

If the electrons acquire a uniform and constant drift velocity v due to the applied electric field, in the steady state the transverse force qE_H acting on any electron due to the piled up charge will equal the Lorentz force qv_xB_z. Equating these, and substituting for the current density $I_x^D = nqv_x$ gives $E_H = (I_x^D/nq)B_z$, where n is the density of carriers. Thus:

$$R = 1/nq \tag{1.3}$$

$q = -1.6 \times 10^{-19}$C for an electron.

This simple expression was the starting point for the modern studies of the Hall effect in metals and alloys. It implies R is temperature independent and that E_H varies linearly with applied field strength. Neither Eq. (1.3) nor its implications are obeyed well by any solid metal (Ziman 1961) and contemporary studies are essentially attempts to explain these discrepancies. It used to be in vogue to make $n = n^*$ a variable parameter and to consider the behaviour of n^* versus n, where n is now the mean density of electrons calculated from the valency (and the concentration of solute, for an alloy). There were numerous attempts to interpret the variation of n^* versus n for different metallic systems. Initially, these emphasized that the charge carriers in a real metal are not independent, free electrons, as assumed in Eq. (1.3), but quasiparticles having variable dynamical properties that depend upon the interaction with the lattice. The itinerant electron population comprises particles showing different dynamical responses to the applied fields so that no single value of the transverse field can exactly balance the Lorentz force on each constituent; hence the discrepancy between n^* and n. Subsequently, it became clear that this explanation in terms of the electrons' interactions with the lattice is not sufficient to explain the behaviour of n^* in alloys. Coles (1965) showed that when scattering by solute atoms is important the Hall effect may also reflect the microscopic details of their scattering mechanisms.

Since the 1950's these views have crystallized into well-defined conditions depending on the dominant influence on the electrons' motion. In the high-field condition, which is defined in Sec. 2, the Lorentz force completely dominates the electrons' response, so the effect of collisions is a negligible perturbation by comparison. (For all practical purposes, this requires measurements at low temperatures on very pure samples.) In the limit of this condition the galvanomagnetic properties are independent of the dominant scattering process and are determined solely by the metal's electronic structure as expressed in the topology of the Fermi surface (Fawcett 1964). Thus the high-field Hall effect reflects global properties of the Fermi surface such as its connectivity, the volume of occupied phase space, and the proximity in energy of its different sheets. At the other extreme, in the low-field condition, collisions completely dominate the electrons' motion so that the Lorentz force is a negligible perturbation by comparison. The global properties of the Fermi surface then do not manifest themselves in the galvanomagnetic effects. The Hall effect reflects instead the electrons' dynamical properties as expressed by local properties of the Fermi surface, such as its curvature at each point and the anisotropy over it of the relaxation time. The low-field Hall effect thus depends on microscopic details of the dominant scattering process, although these cannot always be deduced unambiguously from the results.

Nowadays the preoccupation with the interpretation of n^* has passed. Instead, the first question to ask of Hall effect measurements in metals and alloys is: do they refer to the high- or low-field condition? In other words, do they reflect the metal's electronic structure or the effect of the dominant scatterer? Obviously, the interpretation branches depending on the answer. But frequently there is no clear answer because the experimentalist, by varying field strength, temperature, or sample purity, has put the experiment into the intermediate-field region where vestiges of both high- and low-field behaviour persist, so complicating considerably the interpretation of the results. The measurement of the low-field Hall effect in pure metals has particularly been a problem because of the need to apply an adequately large field to produce measurable signals (Barnard 1977).

It is convenient to arrange the paper according to the above divisions (Fig. 1.1). After describing the field conditions and computational models applicable to them in Secs. 2 and 3, respectively, we consider in Secs. 4-6 representative results and interpretations progressing successively from the high- to the low-field conditions. Each of Secs. 4-6 comprises different topics falling within the scope of its field condition. Section 4 discusses magnetic breakdown and intersheet scattering in Mg and Cd, as examples of prominent effects in the high-field condition. Section 5 considers the effect of the high-field/low-field transition in Cu and Al, as examples of the intermediate-field condition. Finally, Section 6 considers several topics relating to the effect of electron scattering on the low-field Hall effect. This section is divided according to whether the scattering centre has a localized magnetic moment (Fig. 1.1). In a sense we shall then have come a full circle because the final topic discussed in Sec. 5 is the sign of the Hall effect in the magnetic 3d transition metals, and one of Hall's first observations was that the sign of the effect in Fe at room temperature is opposite to that in Au. We show that

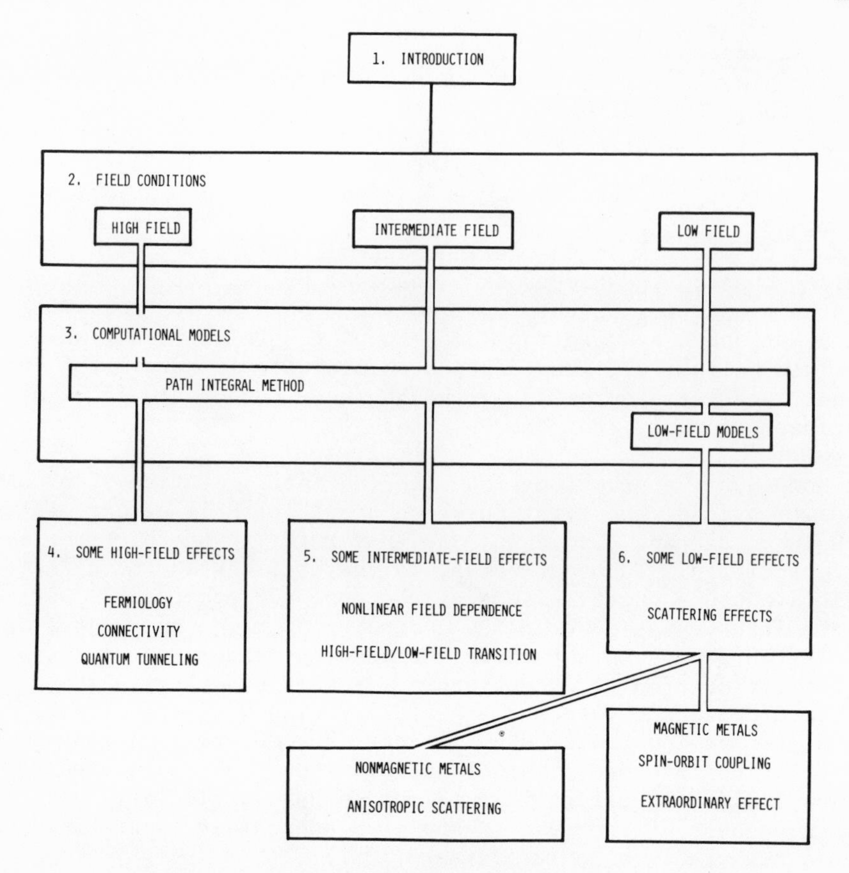

Fig. 1.1 The layout of the paper.

the explanation of these opposite signs is still a contentious point.

2. FIELD CONDITIONS AND THEIR MANIFESTATIONS

2.1 The Ideal Metal

Chambers (1960) has distinguished five regimes for galvanomagnetic measurements; three classical and two quantal. We consider just the classical ones, since they cover the majority of the Hall effect results reported for metals and alloys, and for reasons of space we shall not discuss results pertaining to the quantum oscillatory or quantum high-field limits.

If ω is the cyclotron frequency of a representative electron moving in a flux B, and τ is its average relaxation time in the cyclotron orbit, the classical field conditions introduced in Sec. 1 are formally defined for an orbit by:

(1) $\omega\tau \ll 1$ gives the low-field condition. The representative point completes an infinitesimal segment of the total cyclotron orbit during the electron's lifetime between collisions so that topological features of the orbit are not manifested. When all orbits for a given orientation of $\underline{B}$ are in the low-field condition, the Hall effect is independent of the global features of the Fermi surface and depends only upon the electrons' effective mass, velocity and anisotropic relaxation time over the Fermi surface.

(2) $\omega\tau \approx 1$ gives the intermediate-field condition. Experimentally this is a very important region in which many studies of alloys and less-pure metals have been made.

(3) $\omega\tau \gg 1$ gives the high-field condition. The electron's lifetime between collisions is very long compared with the time necessary to complete the cyclotron orbit. When all orbits for a given $\underline{B}$ are in the high-field condition the Hall effect depends only upon the topology of the Fermi surface; electron scattering is generally not manifested, except where it affects the topology (Sec. 4.1).

Figure 2.1 represents schematically an electron's response in the three field conditions. In the high-field limit the unrestricted effect of the Lorentz force prevents the electron from any forward progress along the direction of the applied electric field $\underline{E}$. Such progress is made only through collisions, as in the intermediate- and low-field cases (Garcia-Moliner 1968); a collision disrupts the orbit, and the electron restarts its trochoidal path but from a displaced starting point. In the low-field condition the electron's trajectory can be visualized as a series of slightly curved segments joining points at which the electron's velocity is randomized. For the ideal metal, in which all itinerant electrons have identical dynamical properties, Fig. 2.1 depicts the transient behaviour that exists from the moment $\underline{B}$ is applied until the build up of electron concentration due to the sample's finite dimensions prevents further transverse drift. Then longitudinal current flow is re-established, although with a transverse concentration gradient. But in a real metal with $\underline{B}$ applied there is a range of drift velocities and dynamical properties so that some electrons are inevitably acted on by unbalanced transverse forces—whence the deviation from free electron behaviour—and Fig. 2.1 illustrates their fate under different field conditions in the steady state.

Returning to the ideal metal, we have from standard theory for the steady state (Pippard 1965) an expression for the

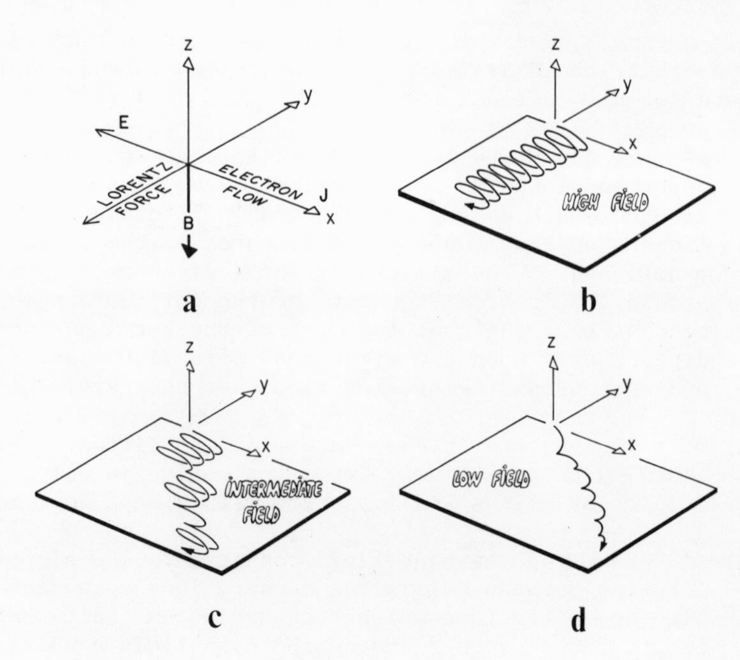

Fig. 2.1 Showing schematically the behaviour of an electron contributing to the Hall voltage for each of the field conditions discussed. In an ideal metal it represents the transient state for all electrons; in a real metal it represents the fate of some electrons even in the steady state. (Hurd 1975)

magnetoconductivity tensor σ_{ij} defined by $J_i = \sigma_{ij}E_j$:

$$\sigma_{ij} = \frac{e^2 n\tau}{m} \times \begin{vmatrix} \gamma & -\omega\tau\gamma & 0 \\ \omega\tau\gamma & \gamma & 0 \\ 0 & 0 & 1 \end{vmatrix} \tag{2.1}$$

where m is the electron's mass and $\gamma \equiv (1+\omega^2\tau^2)^{-1}$. To obtain the the corresponding experimentally measured quantities we must invert Eq. (2.1) to give the magnetoresistivity tensor:

$$\rho_{ij} = \frac{m}{e^2 n\tau} \times \begin{vmatrix} 1 & \omega\tau & 0 \\ -\omega\tau & 1 & 0 \\ 0 & 0 & 1 \end{vmatrix} \tag{2.2}$$

where ρ_{21} is the Hall resistivity and ρ_{11} gives the transverse magnetoresistance $\Delta\rho/\rho_o = (\rho_{11}(B)-\rho_o)/\rho_o$.

Hence the well-known result that in the ideal metal there is a magnetoconductivity effect—σ_{11} decreases with increasing field in the high-field condition because of the electrons' increasingly reduced forward progress (Fig. 2.1)—but the behaviour of σ_{21} is such that there is no transverse magnetoresistivity ρ_{11}. This is because in the ideal metal the transverse forces on each electron are exactly balanced in the steady state, so the current flows along $\underline{E}$, as when B = 0. The Hall resistivity is independent of τ and linear in the applied field, since ω~B.

2.2 Hole-like Response and Quantum Tunneling Effects

The ideal metal is too simplified to be much use in interpreting the Hall effect in real metals. A more realistic model includes two complications: 1) the itinerant electron's interaction with the periodic potential of the lattice, making it effectively a quasiparticle with variable dynamical properties, including a negative effective mass; 2) the possibility that an itinerant electron may modify its cyclotron orbit by quantum mechanical tunneling to appropriate states lying nearby in momentum space. This tunneling may be enhanced by electron scattering (the so-called "hot spot scattering" discussed in Sec. 4.1) or by the applied magnetic field ("magnetic breakdown" mentioned in Sec. 4.3). We shall illustrate here the effect of the periodic potential and leave tunneling until Sec. 4.

Since electrons are fermions they establish in metals an energy hierarchy that puts a small proportion into higher energy states having a strong interaction with the periodic lattice potential. (The lighter alkali metals are possible exceptions.) These electrons are close to the Bragg reflection condition, as explained in any standard textbook. When the interaction is strong enough the electron appears to have a negative effective mass in the direction perpendicular to the Bragg plane because it is accelerated by an applied electrostatic field in the direction opposite to that expected for a free electron. This may be regarded as an elastic Bragg reflection but it must not be thought to arise abruptly at a critical momentum: the electron's motion is affected progressively as it approaches the critical momentum and so the particle can show negative effective mass along a segment of its cyclotron orbit. Since the electron energy states with strong interaction are generally at the top of a nearly filled energy band, it is frequently convenient to consider the empty rather than the full spaces in the band. (Just as the bubble in a spirit level is more useful to the user than the liquid.) An empty state in a nearly filled band behaves like a positively charged particle so it is a matter of taste whether we regard an electron close to the Bragg reflection condition as quasi-particle having an effective positive charge or a negative mass. Both views lead to correct predictions for the sign of the Hall effect, but the latter view is arguably closer to reality (Cross 1978). It should be emphasized that the itinerant microscopic body is always the negatively charged electron; its variable dynamical properties are characteristic of its path viewed macroscopically. It is generally only possible to assess an electron's response to external fields after its trajectory has been completed.

2.3 Some Typical Magnetoconductivity Behaviours...

Figure 2.2 shows cyclotron orbits to illustrate the above. The free electron orbit (a) supports only electron-like response, so its macroscopic effect is electron-like no matter what segment is traced out during the electron's lifetime between collisions. But rearranging the free electron segments as in (b), where a Bragg reflection is implied at each cusp, gives an orbit with a response that depends on the segment traced out. A cusp contributes a hole-like response while a circular segment gives an electron-like one. In the high-field condition the orbit's net response is hole-like because the electron goes around in the opposite sense to that expected for a free electron. But in the intermediate- or low-field condition the net response is not so obvious for it depends sensitively upon the balance between the competitive hole-like contributions from a cusp and electron-like ones from a circular segment (Banik and Overhauser 1978). A further complication, which we shall neglect until Sec. 5.1.1, is that in a real metal the relaxation time is generally not constant around an orbit and may emphasize the contribution of some feature, say a cusp over a circular segment.

Another manifestation of the electrons' interaction with the lattice is the so-called open orbit. This is illustrated in Fig. 2.2(d) where a circular orbit with four points at which a Bragg reflection may occur (Fig. 2.2c) is reassembled into an open orbit and a closed "lens" orbit. In real space the electron in this open orbit is channelled by repeated Bragg reflections at yz-lattice planes. Such an electron has unrestricted motion along the y-direction but makes no progress along the x-direction in the high-field limit.

Figures 2.3-2.5 show quantitatively the behaviours of the magnetoconductivities and magnetoresistivities for the above examples as $\omega\tau$ goes from the low-field limit to well into the high-field condition. The results refer throughout to rectangular axes with the primary electron flow in the x-direction and the applied magnetic field along the z-direction; τ is isotropic. (Falicov

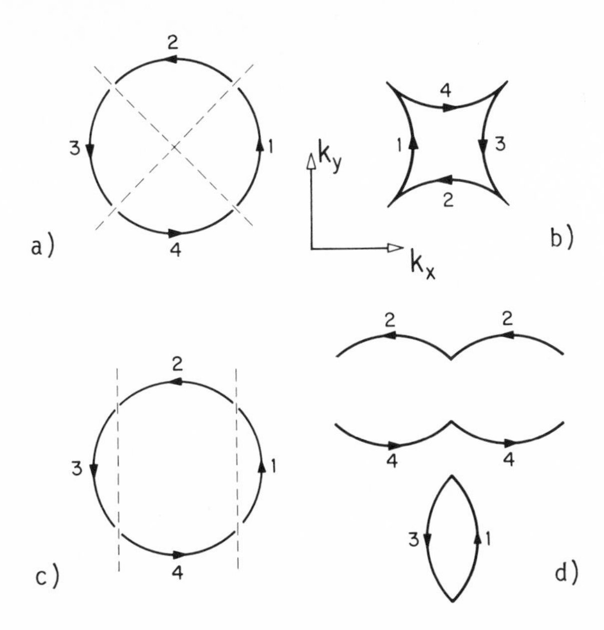

Fig. 2.2 Cyclotron orbits used to illustrate two deviations from free electron behaviour due to the electron's interaction with the periodic potential of the ionic lattice. (a) and (c) represent free electron orbits divided into segments by Bragg planes; (a) reassembles into the hole orbit (b), and (c) into the open-orbit and "lens" combination (d). The field dependences of the magnetoconductivities and magnetoresistivities of these orbits are shown in Figs. 2.3-2.5.

and Sievert 1965, have considered similar cases in which the shape of the orbit is also allowed to vary with $\omega\tau$.) The magnetoconductivities are shown in Figs. 2.3 and 2.4(a). As Eq. (2.1) shows σ_{21} ($= \sigma_{12}$) increases linearly from zero with increasing $\omega\tau$ in the low-field condition and switches to an $(\omega\tau)^{-1}$ dependence in the high-field limit. The changeover gives a local maximum in the intermediate-field range at $\omega\tau \approx 1$. σ_{11} ($= \sigma_{22}$) is independent of $\omega\tau$ in the low-field limit and switches to an $(\omega\tau)^{-2}$ dependence at higher $\omega\tau$.

The behaviour for the hole orbit of Fig. 2.2(b) is shown in Fig. 2.3. (The same arbitrary units are used throughout Figs. 2.3-2.5, as the scales imply.) σ_{11} ($= \sigma_{22}$) shows qualitatively the same behaviour as the electron orbit but σ_{21} ($= -\sigma_{12}$) changes sign from positive to negative with increasing $\omega\tau$. The change occurs at $\omega\tau \approx 0.6$ (see also Fig. 7 of Banik and Overhauser, 1978). This is the change from electron-like to hole-like behaviour as $\omega\tau$ in-

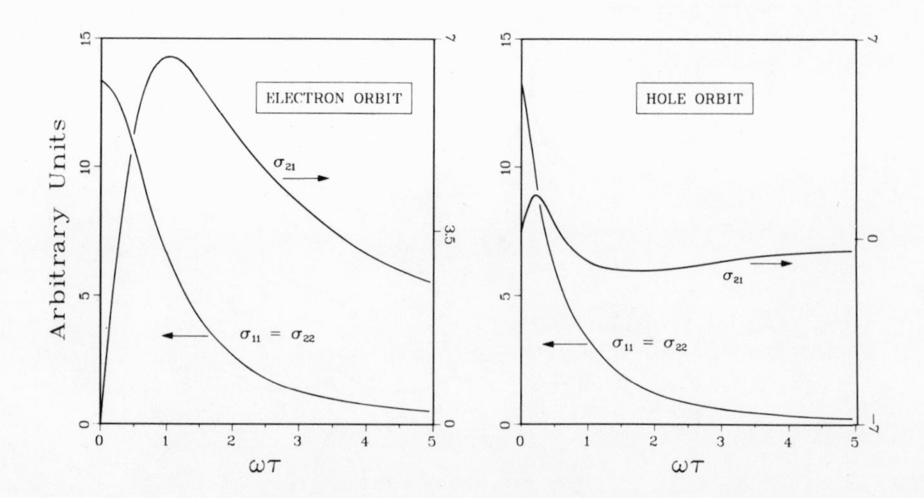

Fig. 2.3 Showing the behaviour of elements of the magnetoconductivity tensor for the orbits of Fig. 2.2(a) and (b) as $\omega\tau$ varies between the low- to the high-field conditions. The primary electron flow is along k_x and the magnetic flux is along k_z (Fig. 2.2).

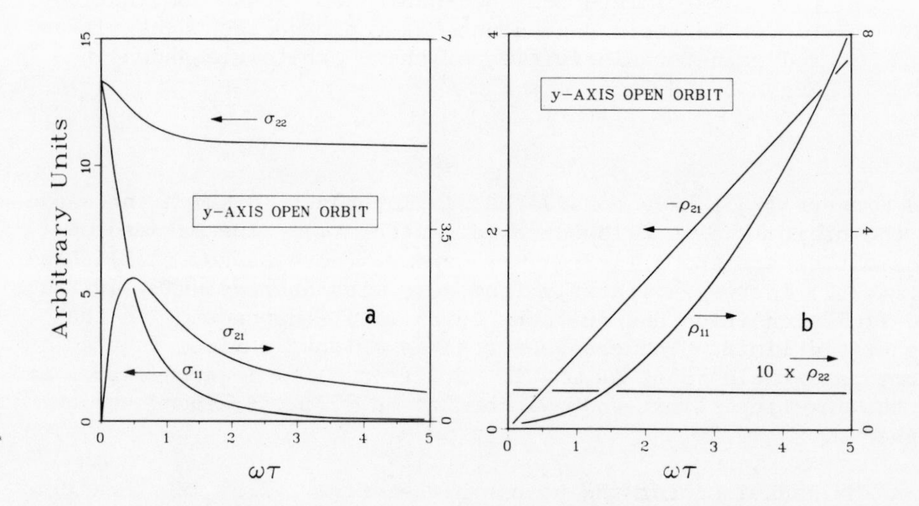

Fig. 2.4 As in Fig. 2.3 except here the magnetoconductivity and magnetoresistivity elements for the orbit of Fig. 2.2(d) are compared.

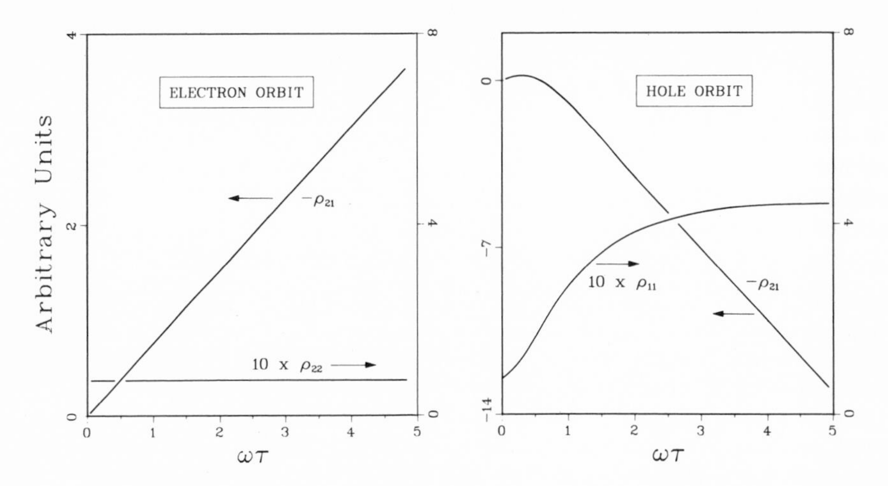

Fig. 2.5 As in Fig. 2.3 except here for the magnetoresistivity elements.

creases and the segment of the orbit completed during the electron's lifetime includes more of the cusps' contributions.

The behaviour for the y-axis open orbit and lens combination of Fig. 2.2(d) is shown in Fig. 2.4(a). σ_{21} (= $-\sigma_{12}$) has qualitatively the same behaviour as the electron orbit but reduced in magnitude. Calculation shows this reduction arises because σ_{21} comes predominantly from the closed lens part of the orbit with hardly any contribution from the open orbit sections. σ_{21} from a closed orbit in the high-field condition is proportional to the area of momentum space enclosed by the orbit. Hence the different magnitudes of $\sigma_{21}(\omega\tau)$ for the electron- and open-orbit cases. The effect of the electron channelling along the y-axis is seen in the σ_{11} behaviour. In the low-field limit $\sigma_{11} = \sigma_{22}$, but with increasing field σ_{22} does not follow the free electron-like behaviour of σ_{11}; it tends to a nonzero value that is independent of field because in the high-field condition electrons in the open orbit cannot respond to the Lorentz force as they are channelled between lattice planes. Thus the conductivity along the y-direction does not go to zero in the high-field limit. If instead the open orbit is along the x-axis—that is, along the primary current direction—then σ_{11} and σ_{22} in Fig. 2.4 are interchanged, while σ_{21} remains unchanged. The same physical arguments apply, of course, except that now as the field is increased electrons are channelled along the primary current direction and so are not deflected into cycloidal paths by the Lorentz force; σ_{21} is not affected, however, because it arises mainly from the closed part of the orbit, which is unchanged.

2.4 ...and the Corresponding Magnetoresistivities

The preceding shows how the combination of the Lorentz force and lattice potential effects give behaviours of $\sigma(\omega\tau)$ characteristic of orbits found typically in many metals. To be useful in the interpretation of the Hall effect in real metals, however, two further complications must be recognized. Firstly, the experimentalist normally measures the Hall resistivity ρ_{21}, which is related to the conductivities by tensor inversion:

$$\sigma_{21} = -\sigma_{21}/(\sigma_{11}\sigma_{22} - \sigma_{21}\sigma_{12}). \quad (2.3)$$

Secondly, normally more than one type of orbit contributes to the magnetoconductivity tensor in a real metal. For a given orientation of $\underline{B}$ with respect to a Fermi surface, there may be a hole-, electron- and open-orbits contributing. To obtain then the Hall resistivity for a given $\omega\tau$, the $\sigma_{ij}(\omega\tau)$ for each orbit must be added to give the total $\underset{\sim}{\sigma}$ for the orientation, then inverted to obtain the element $\sigma_{21}(\omega\tau)$. The situation can be complicated but fortunately there are frequently one or two types of orbit that dominate the conduction for a given orientation. By neglecting the less important orbits it is sometimes possible to relate features seen in $\rho_{21}(\omega\tau)$ to the microscopic motion of the electrons. We describe in Sec. 5.2.1 how this can be done for Cu.

When interpreting experimental $\rho_{21}(\omega\tau)$, one normally does not know the corresponding $\sigma_{ij}(\omega\tau)$. This is a handicap because the σ_{ij} are easier to relate to microscopic origins. To emphasize this we show in Figs. 2.4(b) and 2.5 the $\rho_{ij}(\omega\tau)$ for the cases considered above. (Again, the same arbitrary units are used in each figure, as the scales imply.) For the electron orbit, ρ_{21} is negative and linear in $\omega\tau$, as explained in connection with Eq. (2.2). ρ_{22} (= ρ_{11}) is independent of $\omega\tau$, for there is no magnetoresistance. For the hole orbit (Fig. 2.5), ρ_{21} is negative in the low-field condition but changes sign at $\omega\tau \approx 0.7$—due to the sign change of σ_{21}—and tends at large $\omega\tau$ values towards σ_{21}^{-1}. Note that there is a magnetoresistance in this case for ρ_{11} increases with increasing $\omega\tau$ from its low-field limit of $1/\sigma_{11}$ to approach the fixed value at large $\omega\tau$. A magnetoresistance arises here but not in the free electron case because $\rho_{21} \approx 0$ for the hole orbit, unlike the case of the electron orbit. Thus from the expression:

$$\rho_{11} = \sigma_{22}/(\sigma_{11}\sigma_{22} - \sigma_{21}\sigma_{12}) \quad (2.4)$$

we find that ρ_{11} increases roughly as σ_{11}^{-1} in the range $\omega\tau < \sim 3$. For larger $\omega\tau$ values $|\sigma_{11}| << |\sigma_{21}|$ holds and $\sigma_{11}\sigma_{22}$ becomes negligible compared to σ_{21}^{2}. Hence ρ_{11} tends to the ratio $\sigma_{22}/\sigma_{21}^{2}$, which is constant.

For the y-axis open orbit (Fig. 2.4), ρ_{21} is linear in $\omega\tau$, just as in the free electron case. The reason, which follows from Eq. (2.3), is that $\sigma_{11}\sigma_{22}$ has the same behaviour in both cases even though the $\sigma_{22}(\omega\tau)$ behaviours are different. $\rho_{22}(\omega\tau)$ is also identical to the free electron behaviour—there is no magnetoresistance along the y-direction—because $\sigma_{11}(\omega\tau)$ is the same in both cases. But there is a magnetoresistance in the x-direction, for $\rho_{11}(\omega\tau)$ increases without limit. This increase in the range $\omega\tau <\sim 1$ is because σ_{22} remains large while $|\sigma_{11}|$ and $|\sigma_{21}|$ are approximately equal and small relative to σ_{11}. Thus $\sigma_{11}\sigma_{22}$ is the dominant term in the denominator of Eq. (2.4) and so ρ_{11} varies roughly as σ_{11}^{-1}. For larger $\omega\tau$ the increase in ρ_{11} with increasing $\omega\tau$ is maintained because the terms of $\sigma_{11}\sigma_{22}$ and σ_{21}^2 in Eq. (2.4) are about the same order of magnitude so σ_{21}^2 in Eq. (2.4) can be replaced by a constant times $(\sigma_{11}\sigma_{22})$, and we again find that $\rho_{11} \sim \sigma_{11}^{-1}$. It is worth noting that with open orbits along both the x- and y-directions both $\sigma_{11}(\omega\tau)$ and $\sigma_{22}(\omega\tau)$ tend to nonzero values as $\omega\tau$ increases. The term $\sigma_{11}\sigma_{22}$ does not then tend to zero in the high-field limit and hence $|\rho_{21}|$ is reduced from its free electron behaviour (Eq. 2.3). We shall see exactly this situation in Sec. 5.2 for the case of Cu with $\underline{B}$ applied along a [211] axis.

These examples cover some of the principal types of cyclotron orbit encountered in metals. The results are useful in following sections in the qualitative interpretation of features seen in $\rho_{21}(\omega\tau)$. They also serve as a warning that the inversion between $\underset{\sim}{\rho}_{21}$ and $\underset{\sim}{\sigma}_{21}$ is a stumbling block that can mislead the interpretation of experimental $\rho_{21}(\omega\tau)$ behaviours. It is risky to attempt to explain a measured $\rho_{21}(\omega\tau)$ directly in terms of the microscopic details of cyclotron motion without the guidance of the corresponding $\sigma_{ij}(\omega\tau)$ and $\sigma_{ii}(\omega\tau)$ behaviours.

3. COMPUTATIONAL MODELS

Two models used to make quantitative interpretations of the galvanomagnetic effects in metals are described. Both derive from prescriptions for solving the Boltzmann equation but have different emphasis for different applications. The first (Sec. 3.1) is the so-called path integral method based on Chamber's (1952) integral solution of the Boltzmann equation with developments mainly by Falicov and coworkers (Falicov and Sievert 1965, Falicov 1973) and Datars and coworkers (Douglas and Datars 1973, 1974, 1975). It is particularly useful for incorporating into the calculation high-field effects such as magnetic breakdown, quantum tunneling, open- and extended-orbits. The method is not limited to the high-field condition, however, and it has the marked advantage of being valid for all values of $\omega\tau$ outside the quantum limit. It has proved useful in disentangling the origins of features found in the behaviours of $\rho_{21}(\omega\tau)$ and $\rho_{11}(\omega\tau)$ for several pure metals in the intermediate-field condition (Douglas and Datars 1974, 1975, Hurd et al. 1976, 1977 and Hurd and McAlister 1977). Some examples are considered in Secs. 4 and 5.

The second model (Sec. 3.2), which is limited to the low-field condition, has been developed by Böning and coworkers (Böning 1970, Pfänder et al. 1979, Böning et al. 1979) for cubic metals. It emphasizes the interpretation of the anisotropic relaxation time arising from electron scattering, and has been applied particularly to the study of the Hall effect from scattering by lattice imperfections, induced defects, and foreign ions. Examples are considered in Sec. 6.1. Although the path integral method could be used to interpret results from the low-field condition, a specifically low-field model is still preferred by some authors, so we include Böning's model as an alternative approach to the path integral method. A description of the model is also useful because it covers other more restricted expressions for the low-field Hall coefficient that have appeared previously (Tsuji 1958, Cooper et al. 1965, Ziman 1964).

3.1 The Path Integral Method

3.1.1 The conductivity equation. The standard procedure is used to find the statistical distribution function to specify the probability of an electron having its momentum and position within small ranges. The rate of change of this distribution under the influences of applied fields and dissipative scattering is expressed by the Boltzmann equation. An electron can only contribute to the distribution function $f(\underline{k},\underline{r},t)$ if it is at A (Fig. 3.1) at time t. To arrive there it must have followed a simple trajectory T since its last collision, which will have occurred at say $B=(\underline{k}',\underline{r}')$ at the time t'. Since during its mean free path B→A the electron acquires an energy $\Delta\varepsilon$ from the external field, its energy at A is ε only if it was $\varepsilon-\Delta\varepsilon$ at B. We obtain $f(\underline{k},\underline{r},t)$ by adding together, for the whole elapsed time up to t, the number of electrons that are scattered onto T at previous points like B and that survive on T up to A. Note that we assume "catastrophic scattering" processes, so that an electron's velocity is randomized at each collision. At a point B it is returned to the unperturbed distribution and brings to T no memory of any excess energy acquired previously from applied fields.

It is assumed that the dominant scattering process at any point $(\underline{k},\underline{r})$ can be described by a relaxation time $\tau(\underline{k},\underline{r})$ such that $1/\tau(\underline{k},\underline{r})$ is the average scattering rate at that point. Thus an electron that has survived without collision up to some time u has a probability of collision in the next small interval Δu of approximately $\Delta u/\tau$.

By considering the gain and loss of electrons in equivalent elementary volumes of phase space centered on A and B of Fig. 3.1, it is possible to derive the distribution function (Chambers 1969):

$$f(\underline{k},\underline{r},t) = \int_{-\infty}^{t} \frac{dt'}{\tau(t')_{\underline{k}',\underline{r}'}} \cdot f_o(\varepsilon-\Delta\varepsilon)\cdot\exp\left[-\int_{t'}^{t} du/\tau(u)_{\underline{k},\underline{r}}\right] \tag{3.1}$$

Descriptively, the terms in the integrand represent respectively from left to right: the probability of scattering during the time interval dt', the probability of having the correct energy at the instant t, and the probability of surviving until t without scattering.

Expanding the integrand to first order in E by writing $f_o(\varepsilon-\Delta\varepsilon) = f_o(\varepsilon) - \Delta\varepsilon \frac{\partial f_o}{\partial\varepsilon}$, and integrating gives:

$$f(\underline{k},\underline{r},t)=f_o-\frac{\partial f_o}{\partial\varepsilon}\int_{-\infty}^{t} \Delta\varepsilon \exp\left[-\int_{t'}^{t} du/\tau(u)_{\underline{k},\underline{r}}\right]\frac{dt'}{\tau(t')_{\underline{k}',\underline{r}'}} \tag{3.2}$$

Budd (1962) has shown that Eq. (3.1) is an exact solution of the Boltzmann equation whereas Eq. (3.2) solves only the linearized equation.

Integrating Eq. (3.2) by parts gives:

$$f-f_o= -\frac{\partial f_o}{\partial\varepsilon}\cdot\Delta\varepsilon+\frac{\partial f_o}{\partial\varepsilon}\int_{-\infty}^{t}\frac{d(\Delta\varepsilon)}{dt'}\exp\left[-\int_{t'}^{t} du/\tau(u)_{\underline{k},\underline{r}}\right]dt'$$

and

$$f-f_o(\varepsilon-\Delta\varepsilon)=\frac{\partial f_o}{\partial\varepsilon}\int_{-\infty}^{t}\frac{d(\Delta\varepsilon)}{dt'}\exp\left[-\int_{t'}^{t} du/\tau(u)_{\underline{k},\underline{r}}\right]dt' \tag{3.3}$$

When τ is independent of time the exponential factor in the integrand becomes $\exp[-(t-t')/\tau_{\underline{k},\underline{r}}]$ throughout.

No energy is gained by the electron from the magnetic field but its rate of gain from $\underline{E}$ in going from B→A is:

$$\Delta\varepsilon(\underline{r},t') = q\int_{t'}^{t} \underline{E}(\underline{r},s)\cdot\underline{v}(\underline{r},s)\,ds \tag{3.4}$$

where s is a dummy time variable. The rate of increase of energy is:

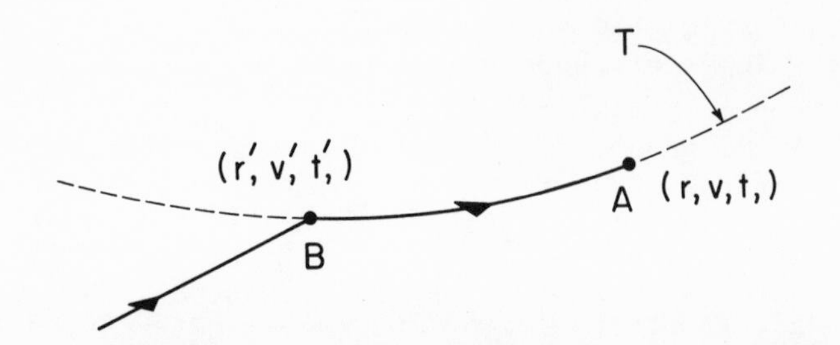

Fig. 3.1 A trajectory T in phase space for an electron under the combined influences of external electric and magnetic fields.

$$\frac{d\Delta\varepsilon}{dt'} = -q\ \underline{E}(\underline{r},t')\cdot\underline{v}(\underline{r},t') \tag{3.5}$$

[The negative sign arises because the variable t' is the lower limit in Eq. (3.4).] Substituting Eq. (3.5) into (3.3) gives Chamber's path integral expression for $g(\underline{k},\underline{r},t) = f(\underline{k},\underline{r},t) - f_o(\underline{k},\underline{r},t)$:

$$g(\underline{k},\underline{r},t) = -q\ \frac{\partial f_o}{\partial\varepsilon}\int_{-\infty}^{t} dt'\ \underline{E}(t')\cdot v(t')_{\underline{k}} \exp\left[-\int_{t'}^{t} du/\tau(u)_{\underline{k}}\right] \tag{3.6}$$

The current density $\underline{J}$ due to the perturbation g is given by the standard result

$$\underline{J} = \frac{q}{4\pi^3}\int_{\underline{k}} g(\underline{k},t)\underline{v}_{\underline{k}}\ d^3k \tag{3.7}$$

so the conductivity tensor $(\sigma_{ij} = J_i/E_j)$ can be written:

$$\sigma_{ij} = \frac{-q^2}{4\pi^3}\sum_{\sigma}\int_{\underline{k}} \frac{\partial f_o}{\partial\varepsilon}\ v^i_{\sigma n}\ d^3k\int_{-\infty}^{t} dt' v^j_{\sigma n}(t')\exp\left[-\int_{t'}^{t} du/\tau(u)_{\underline{k}}\right] \tag{3.8}$$

where the band n and spin σ indices have been reintroduced.

Equation (3.8) is the basic path integral equation for the conductivity. f_o is the Fermi factor, ε the energy, v_i and v_j are the components of the velocity $\underline{v} = \hbar^{-1}\nabla_k\varepsilon$, u and t' are dummy time variables, and d^3k is a volume element of $\underline{k}$-space. The intuitive interpretation of Eq. (3.8) (Visscher and Falicov 1972) is that the current produced in the i-direction due to the perturbation of electrons in the n-th band is the integral over the whole Fermi surface (all $\underline{k}$) of the velocity v_i times a factor that represents the number of extra electrons present because of the applied field. This factor is an integral (going backwards in time from the present instant t) of the velocity v_j in the direction of the applied field multiplied by the exponential coherence factor that accounts for the probability that an electron is scattered before it is able to contribute to the current at the time t'.

3.1.2 Phase variable form of conductivity equation. A more convenient form of Eq. (3.8) is obtained by incorporating variables which take the geometry of the electron's path more naturally into account (Chambers 1956, Falicov and Sievert 1965, Douglas and Datars 1973). We replace the volume element d^3k by $(qB/\hbar^2)\, d\varepsilon dk_z dt$, and introduce a new phase variable θ which measures the angular position reached at a given instant around the path (Chambers 1956). Between two instants t' and t, this parameter for an orbit α changes by:

$$\theta_t - \theta_{t'} = (\omega_\alpha \hbar/qB) \int_{t'}^{t} \frac{dk_{\parallel}}{v_{\perp}} \tag{3.9}$$

ω_α is the angular frequency for the orbit and the subscripts refer to mutually perpendicular directions tangential and normal to the electron's path. Taking the value of the phase variable at the instants t' and t to be φ and θ, respectively, leads to the expression for the conductivity tensor for a single spin index:

$$\sigma_{ij} = \frac{-q^2}{4\pi^3} \int_0^\infty \frac{df_o}{d\varepsilon}\, d\varepsilon \int_{-\infty}^{\infty} dk_z \int_0^{2\pi} v_i(\theta,k_z)\, \frac{m}{\hbar^2}\, d\theta \int_{-\infty}^{\theta} \frac{v_j(\phi,k_z)}{\omega}\, e^{(\phi-\theta)/\omega\tau_{\underline{k}}}\, d\theta \tag{3.10}$$

The simpler form of the exponential coherence factor is assumed here, appropriate to the case when $\tau_{\underline{k}}$ is independent of time. A standard result for a metal shows $-(df_o/d\varepsilon)$ is essentially non-zero only in a very narrow range of energies close to the Fermi energy ε_f and that for semiclassical effects considered at low temperatures it is a good approximation to replace it by the negative delta function $-\delta(\varepsilon_{\underline{k}} - \varepsilon_f)$. Consequently the first integral in Eq. (3.10) can be replaced by -1 to this approximation, and we are left with:

$$\sigma_{ij} = \frac{q^2}{4\pi^3\hbar^2} \int_{-\infty}^{\infty} dk_z \sum_{\substack{\text{ALL}\\ \text{ORBITS}}} \frac{m}{\omega} \int_0^{2\pi} v_i(\theta,k_z)\,d\theta \int_{-\infty}^{\theta} d\phi\; v_j(\phi,k_z) \times \exp\,[(\phi-\theta)/\omega\tau_{\underline{k}}] \tag{3.11}$$

3.1.3 Computation form of σ_{ij}: arcs, pieces, and orbits. A more convenient form for computation, which incorporates the possibility that an electron may jump from one point on a cyclotron orbit to another, is the matrix method introduced by Falicov and Sievert (1965) and developed by Douglas and Datars (1973, 1974, 1975). It permits the simulation of the effects of changing contours of cyclotron orbits due to such as magnetic breakdown or localized scattering on the Fermi surface.

We distinguish between an arc, a piece and an orbit. An orbit is the macroscopic path followed by the quasiparticle under the influence of the applied electric and magnetic fields. A piece is a segment of orbit between points at which transfer is possible to another piece on a different orbit as a result of intersheet scattering or magnetic breakdown. A piece is formed from contiguous, circular arcs. Thus in Fig. 3.2(a) we show a circular orbit made up of 6 pieces (AB, BC, etc.) each in turn made up of three arcs (numbered in the figure). An orbit is divided into pieces so that interpiece scattering can be accommodated. A piece is divided into arcs so that we can vary over a piece the arc-dependent parameters, particularly the electron's relaxation time. The arcs are circular so that a number of useful analytical integrals can be used.

The computation form of Eq. (3.11), and the steps in the calculation have been outlined in Hurd et al. (1976). Briefly, model orbits are chosen and divided into pieces and arcs appropriate to the simulated Fermi surface. Each arc is specified by six parameters: initial and final azimuth angles, components of the Fermi velocity parallel and perpendicular to $\underline{B}$, a relaxation time τ characteristic of the orbit, and an anisotropy parameter b_p such that $b_p\tau$ is the relaxation time on the p-th arc. The differential form of Eq. (3.11):

$$\frac{d\sigma_{ij}(k_z)}{dk_z} = \frac{q^2 m}{4\pi^3\hbar^2\omega} \sum_{\substack{\text{ALL}\\ \text{ORBITS}}} \int_0^{2\pi} v_i(\theta)\,d\theta \int_{-\infty}^{\theta} d\phi\, v_j(\phi)\, \exp[(\phi-\theta)/\omega\tau_{\underline{k}}] \tag{3.12}$$

is evaluated piece by piece from analytical equations for the contributions from each constituent arc. The summation is made for all pieces on an orbit to give the orbit's contribution to the nine components of the differential conductivity. If there are other contributing orbits in the same plane k_z = constant, then the process must be repeated and the summation made over the whole plane. Finally, the process must be repeated and summed over all appropriate planes k_z = constant in the Brillouin zone. The accumulated magnetoconductivity tensor is then inverted to give the magnetoresistive tensor. Interpiece transitions are incorporated by the

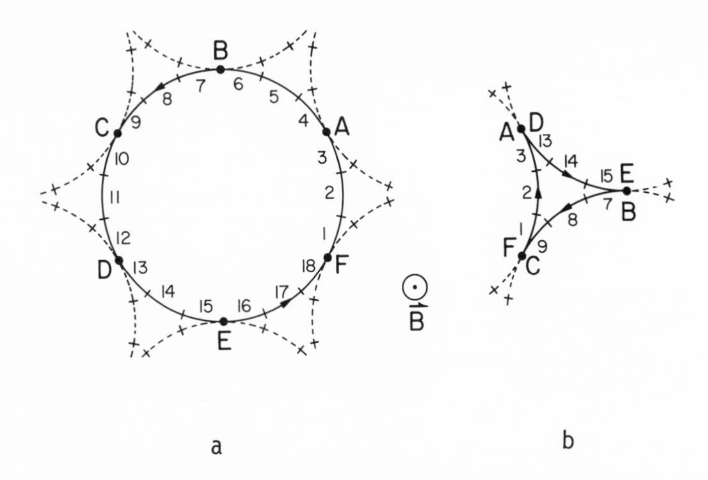

Fig. 3.2 A plan view of typical cyclotron orbits used in a path integral calculation. (a) The solid lines ABC... represent an orbit. The segments AB, BC, etc., are pieces made up from the numbered, circular arcs. Interpiece transitions are permitted only at the ends of pieces. (b) An example of a new orbit formed when the transitions are between diametrically opposed points on (a). These orbits are used to simulate electron motion in the basal plane of Cd (Sec. 4.2).

matrix formulation introduced by Falicov and Sievert (1965) and Douglas and Datars (1973). Examples of the application of the method are described in Secs. 4 and 5.

3.2 Böning's Low-field Model for Cubic Symmetry

3.2.1 Current density in the low-field condition. This starts from the assumption, supported by various electron transport studies (Springford 1971), that the electron relaxation time $\tau_{\underline{k}}$ varies over the Fermi surface (Böning, 1970, Böning et al. 1979). An orthogonal set of axes is defined in $\underline{k}$-space with the magnetic field $\underline{B}$ along $\underline{k}_z$ and the electric field $\underline{E}$ along $\underline{k}_x$ in the transverse arrangement. The crystalline axes $(\underline{X},\underline{Y},\underline{Z})$ of the cubic metal may have any orientation relative to $(\underline{k}_x,\underline{k}_y,\underline{k}_z)$. An electron represented at a point $\underline{k}$ on an arbitrary Fermi surface has the equation of motion:

$$\hbar\dot{\underline{k}} = q(\underline{E} + \underline{v}\wedge\underline{B}) \tag{3.13}$$

Let $\underline{dS}$ be an incremental area centered on $\underline{k}$ and specified as shown

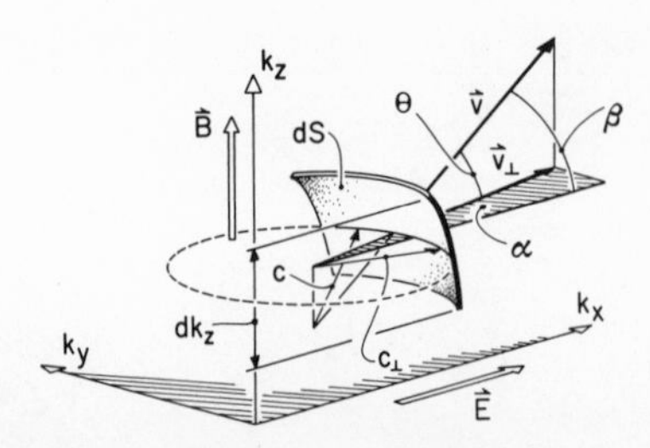

Fig. 3.3 The arrangement of parameters connected with the arbitrary incremental area $\underline{dS}$ of Fermi surface discussed in the text. c is the curvature of $\underline{dS}$ in the plane defined by $\underline{v}$ and $\underline{v}\wedge\underline{B}$, $v_\perp$ and $c_\perp$ are, respectively, the components of $\underline{v}$ and $\underline{c}$ in the k_xk_y-plane.

in Fig. 3.3. Pippard's (1965, 1966) artifice is used to calculate the effect of $\underline{E}$ and $\underline{B}$ upon the conductivity: first, $\underline{E}$ is applied for a time dt that is much shorter than either the collision relaxation time or the cyclotron period; subsequently the effect of $\underline{B}$ is added. During the impulse $\underline{E}$ dt the shift of $\underline{dS}$ along the direction of $\underline{E}$ leads to a current density;

$$\begin{aligned}\delta\underline{J}(0) &= (q/4\pi^3)\ \underline{v}\ (\underline{dS}\cdot\underline{dk_E}) \\ &= (q/4\pi^3)\ \underline{v}\ dSdk_E\ \cos\beta \\ &= (q^2E/4\pi^3\hbar)\ \underline{v}c_\perp(v/v_\perp)\cos\beta\ d\alpha dt dk_z \qquad (3.14)\end{aligned}$$

The total current density at the end of the impulse is given by the integral of Eq. (3.14) over the Fermi surface, and will lie along the x-axis in a cubic metal. Thus:

$$\begin{aligned}dJ_x(0) &= (q^2E/4\pi^3\hbar)\ dt \int_{FS} c_\perp\ v\ \cos\alpha\ \cos\beta\ d\alpha\ dk_z \\ &= (q^2E/4\pi^3\hbar)\ dt \int_{FS} v\ \cos^2\beta\ dS \qquad (3.15)\end{aligned}$$

and

$$dJ_y(0) = 0 \qquad (3.16)$$

Each point $\underline{k}$ on the Fermi surface is given an angular velocity $\omega_{\underline{k}}$ as the result of $\underline{B}$ during the impulse so α in Fig. 3.3 is time dependent and $\underline{v}_\perp$ sweeps through $\omega_{\underline{k}} t$ during an elapsed time t. Consequently, at some time t after the electric field is switched off at t = 0 the current density dJ(t) is no longer along the k_x-direction, although still confined to the $k_x k_y$-plane. Taking into account that after the pulse is switched off the electron population reverts to equilibrium and the current dies away exponentially, the steady values of the components of the current density can be expressed as a complex sum (Pippard 1965):

$$J_x + iJ_y = \int_0^\infty dJ_x(0) \exp\,[i\omega_{\underline{k}} - 1/\tau_{\underline{k}}]\; dt \tag{3.17}$$

In the low-field condition with $\omega_{\underline{k}}\tau_{\underline{k}} << 1$ the integration with respect to time can be carried out before that over the Fermi surface, whence:

$$J_x + iJ_y = dJ_x(0)\; \tau_{\underline{k}}(1 - i\omega_{\underline{k}}\tau_{\underline{k}}) \tag{3.18}$$

Isolating the real and imaginary parts and substituting from Eq. (3.15) gives eventually:

$$J_x = Q \int_{FS} \ell \cos^3\beta \; dS = Q \int_{FS} (v_x^2/v)_{\underline{k}} \; dS \tag{3.19}$$

$$J_y = Q(qB/\hbar) \int_{FS} \ell^2 \cos^2\beta (1/c) \; dS \tag{3.20}$$

$\ell \equiv v\tau_{\underline{k}}$ and $Q \equiv q^2 E/4\pi^3\hbar$. Substituting for dS and replacing cosβ by $(v_\perp/v)\cos\alpha$ gives alternatively $(\ell_\perp = \tau_{\underline{k}} v_\perp)$:

$$J_y = Q(qB/\hbar) \int_{FS} \ell_\perp^2 \cos^2\alpha \; d\alpha \; dk_z \tag{3.21}$$

3.2.2 Expression for the low-field Hall effect. In the conventional experimental arrangement with applied fields aligned as in Sec. 3.2.1 the sample's geometry confines $\underline{J}$ to the y-direction. Consequently $\underline{E}$ is no longer colinear with the x-axis but has longitudinal ($E_{||}$) and transverse ($E_\perp$) components relative to the sample. Their ratio defines the Hall angle θ_H: $\tan\theta_H = E_\perp/E_{||}$. An expression for the low-field θ_H thus follows directly from Eqs. (3.19) and (3.21) (Cooper et al. 1965):

$$\tan\theta_H = (qB/\hbar) \int_{FS} \ell_\perp^2 \cos^2\alpha \, d\alpha \, dk_z \Big/ \int_{FS} \ell \cos^2\beta \, dS$$

We see also from the ratio of Eqs. (3.20) and (3.19) that if $\tau_k = \tau$ is independent of $\underline{k}$, the low-field Hall angle is given by $\tan\theta_H = \omega\tau$.

The Hall coefficient [Eq. (1.1)] is defined by $R = E_\perp/J_{||}B$, which in the low-field condition is approximately:

$$R(LF) = (E_{||} \tan\theta_H)/B\, J_x = (E/B)(J_y/J_x^2) \tag{3.22}$$

in terms of the current densities relative to the axes of Fig. 3.3. Direct substitution from Eqs. (3.19) and (3.21) gives the expression used by Van der Mark et al. (1969):

$$R(LF) = (4\pi^3/q) \int_{FS} \ell_\perp^2 \cos^2\alpha \, d\alpha \, dk_z \times \left[\int_{FS} \ell \cos^2\beta \, dS \right]^{-2} \tag{3.23}$$

Böning (1970) has shown that the current density expressions (3.19)-(3.21) can be reduced further for cubic symmetry, leading to other expressions for the low-field Hall coefficient. These involve the principal radii of curvature for $\underline{dS}$ defined in Fig. 3.4. After some algebra to calculate the mean of $\overline{\cos^2\beta}$, $\cos^2\alpha$ and $\cos^2\phi\cos^2\beta$ over all orientations (Figs. 3.3 and 3.4), the following symmetry-reduced forms of Eqs. (3.19)-(3.21) are obtained:

$$J_x = Q/3 \int_{FS} \ell \, dS \tag{3.24}$$

$$J_y = (Q/2)(qB/\hbar) \int_{FS} \ell_\perp^2 \, d\alpha \, dk_z \tag{3.25}$$

$$J_y = \frac{1}{3} Q(qB/\hbar) \int_{FS} \ell^2 \, (\overline{\tfrac{1}{c}}) \, dS \tag{3.26}$$

where $(\overline{1/c}) = \frac{1}{2}(1/c_1 + 1/c_2)$. Expressions for the Hall coefficient follow directly from substitution into Eq. (3.22). Thus Eqs. (3.24) and (3.26) give (Van der Mark et al. 1969):

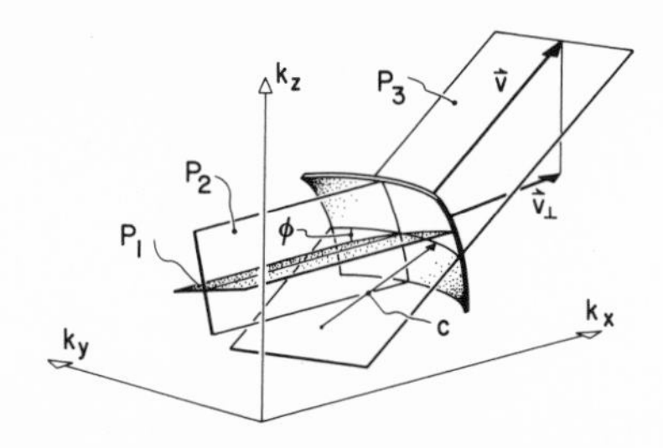

Fig. 3.4 Showing the relationship between the various principal planes and curvatures for the surface $\underline{dS}$ of Fig. 3.3. P_1 and P_2 are the principal sections of $\underline{dS}$ through the point $\underline{k}$. They are orthogonal and contain the extreme values of the principal radii of curvature c_1 and c_2, respectively. P_3 is the normal plane of $\underline{k}$ containing $\underline{v}$ and $\underline{c}$.

$$R(LF) = (18\pi^3/q) \frac{\int_{FS} \ell_\perp^2 \, d\alpha \, dk_z}{[\int_{FS} \ell \, dS]^2} \tag{3.27}$$

Equations (3.24) and (3.26) give:

$$R(LF) = (12\pi^3/q) \frac{\int_{FS} \ell^2 \, (\frac{\bar{1}}{c}) \, dS}{[\int_{FS} \ell \, dS]^2} \tag{3.28}$$

This is Tsuji's (1958) expression for the low-field Hall coefficient in a cubic metal with an anisotropic relaxation $\tau_{\underline{k}}$. When $\tau_{\underline{k}}$ is isotropic it cancels from Eq. (3.28) to give:

$$R(LF) = (12\pi^3/q) \frac{\int_{FS} v^2 \, (\frac{\bar{1}}{c}) \, dS}{[\int_{FS} v \, dS]^2} \tag{3.29}$$

One or another of Tsuji's formulas is commonly used to interpret the behaviour of R(LF), for example: Dugdale and Firth 1969, Shiozaki 1975, Böning et al. 1975, Bergmann 1975, Kesternich et al. 1976, Barnard 1977, Yonemitsu et al. 1978.

For a free electron metal v is constant over the Fermi surface and Eq. (3.29) becomes:

$$R(LF) = (12\pi^3/q)(1/cS)$$

$$= (12\pi^3/q)(1/3V) \tag{3.30}$$

where $V = (4/3)\pi c^3$ is the volume of phase space within the Fermi sphere. Since the density of electrons n in phase space is $1/4\pi^3$, Eq. (3.29) becomes:

$$R(LF) = 1/nq, \tag{3.31}$$

which brings us full circle to Eq. (1.3),

4. SOME HIGH FIELD EFFECTS

Two examples are discussed of mechanisms that are effective only in the high-field condition and lead to distinctive behaviours in $\rho_{21}(\underline{B},T)$: intersheet scattering and magnetic breakdown.

4.1 Intersheet Scattering in Cd

In exceptional circumstances Umklapp scattering can modify the topology of a Fermi surface leading to pronounced features in the behaviour of the Hall voltage in the high-field condition. This occurs for electrons moving parallel to the basal plane in Cd. The Fermi surface of Cd has sheets in three Brillouin zones: hole sheets in the first and second zones (known as the "caps" and the "monster", respectively) and the electron "lens" in the third zone. Throughout we are concerned only with $\underline{B}$ along [0001]. Since the Hall voltage is positive over the whole temperature range from the melting point down to ~4K, it follows that in the low- and intermediate-field conditions the conduction is dominated by electrons represented on hole sheets. (Those on the lens have their velocity vectors along the hexagonal axis and therefore are not major contributors to the basal-plane conduction.) The volume of phase space occupied by the second-zone caps is much less than that occupied by the monster, so we conclude that conduction is dominated by electrons on the monster surface.

Below ~10K some dramatic changes are seen (Fig. 4.1) in the high-field Hall voltage (Katyal and Gerritsen 1969, Lilly and Gerritsen 1974, Hurd et al. 1976). The explanation centres on a type of quantum tunneling between parts of the monster surface adjacent in phase space, called "intersheet" scattering (Young et al. 1969). In the repeated zone scheme the monster forms a trifoliate surface (Fig. 4.2) of "clover leaf" section at its closest points

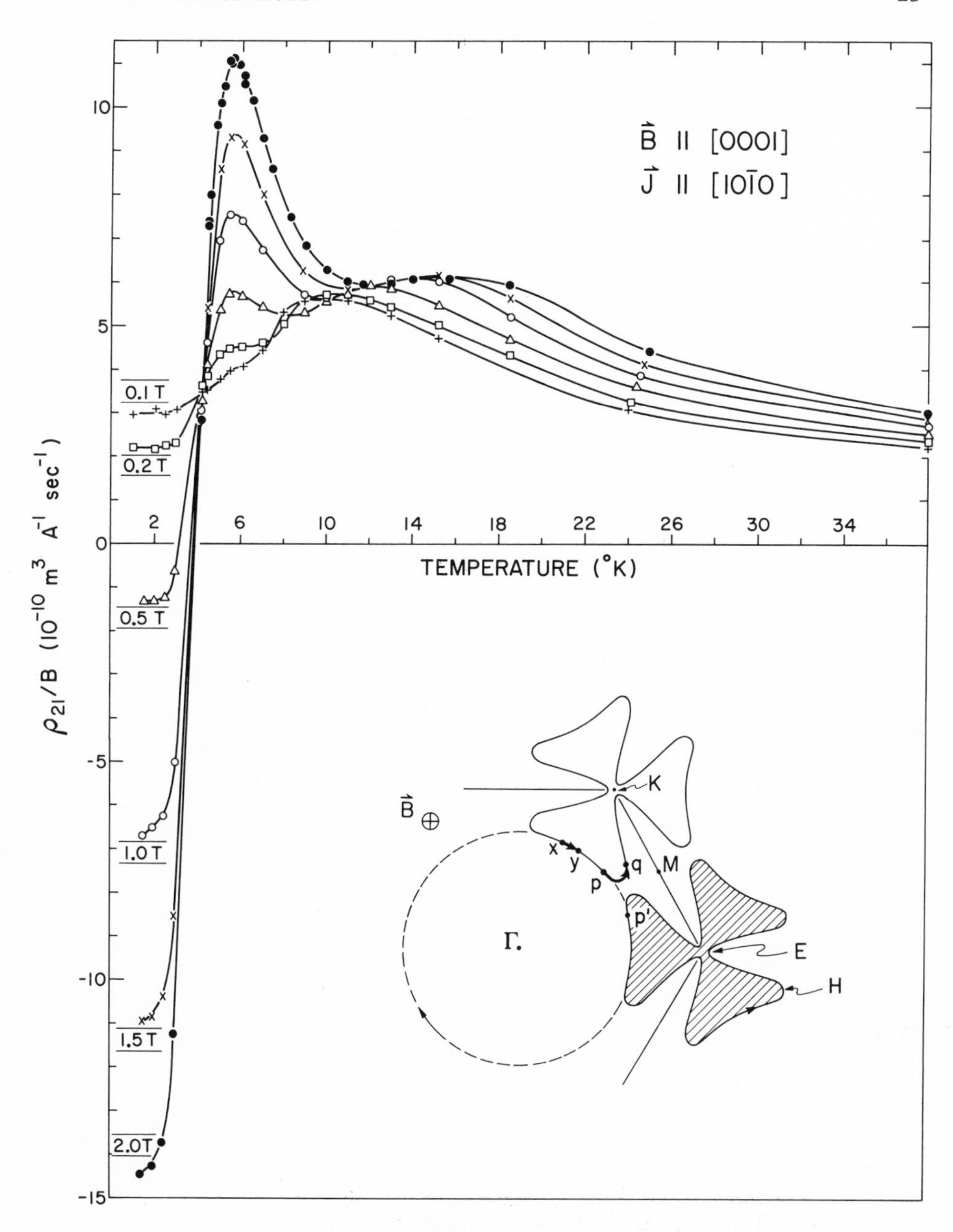

Fig. 4.1 Showing the temperature dependence of the Hall resistivity in a Cd monocrystal with residual resistance ratio (RRR) = 25,600 in different applied fields (tesla). (Hurd et al. 1976.) The insert shows a section through the basal plane of phase space containing the clover-leaf section of the monster's trifoliate surfaces (Fig. 4.2).

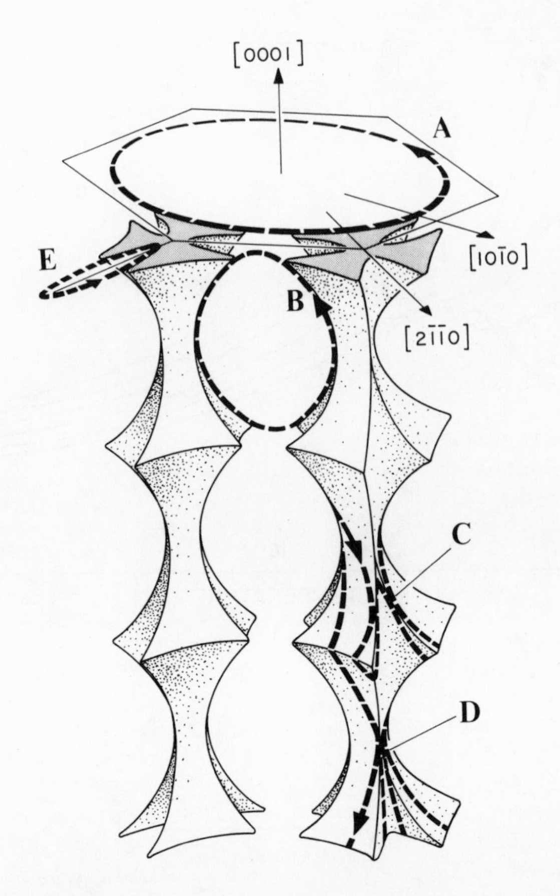

Fig. 4.2 A sketch of the trifoliate surfaces of the second-zone monster in the repeated zone scheme. In a complete diagram the surface would be repeated at the vertices of a space-filling array of hexagons. A and E are electron orbits produced in the basal plane by intersheet scattering. Other orbits are labelled but not discussed in this context. (Alderson et al. 1976.)

of approach (insert to Fig. 4.1). Tunneling occurs by the following mechanism between the tips of adjacent clover leaf orbits.

Consider the hole orbit of Fig. 2.2(b) shown in Fig. 4.3. (To be more realistic the cusps are rounded.) A Bragg reflection is implied at A, B, C, D, and the magnetic flux is into the plane of the page. The dashed lines show the sense of a free electron orbit. However low the temperature, there are always some low energy phonons present that normally are energetic enough to give only small-angle scattering between adjacent points on the orbit. But exceptionally, in regions close to the Brillouin zone, they can

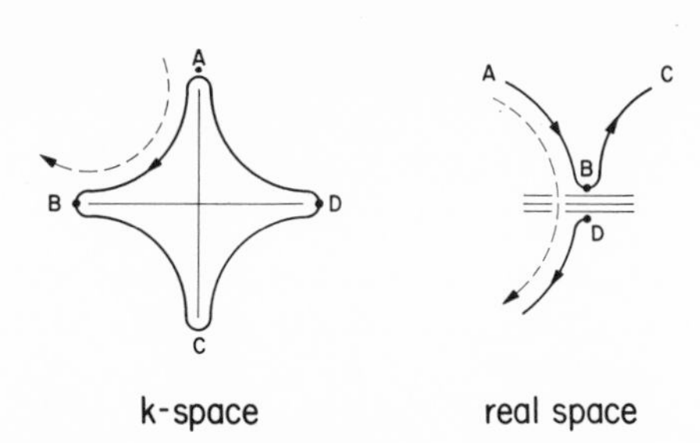

Fig. 4.3 This illustrates the effect of intersheet scattering from a 'hot spot' region B (Young et al. 1969). With no scattering present, the electron suffers a Bragg reflection at B and continues toward C on the hole orbit. But if the electron undergoes an Umklapp scattering at B, and so reappears at D in the Brillouin zone, its path in real space is effectively electron-like. Hot spot scattering is believed to affect the Hall voltage in Cd and W.

give large-angle scattering through an Umklapp process; as in scattering from B→D, or A→C, in an Umklapp event. Regions of the Fermi surface where the probability of such events is appreciable are called "hot spots" (Young 1968, Young et al. 1969). A point moving along the orbit in k-space from say A→B can thus suffer only small angle scattering on the main, electron-like segment away from A or B, but near B it has the choice of two channels. Either it is Bragg reflected, and so continues on the segment B→C with an overall hole-like response, or it undergoes an Umklapp scattering to continue on the segment D→A. The corresponding trajectories in real space are sketched in Fig. 4.3, with the lattice planes at B shown symbolically. The electron's velocity parallel to these planes is reversed in the Umklapp process so that effectively it follows the free electron trajectory, whereas the Bragg reflected electron does not. The character of an orbit can thus change from hole-like to electron-like in the high-field condition. The latter requirement follows because the scattering changes the response of the complete orbit, which is only manifested when $\omega\tau$ is large.

The difference between intersheet scattering and magnetic breakdown, considered below, is that tunneling in the latter is produced when the perturbation of the electron's motion by the applied field dominates that from the periodic potential; effectively, the electron does not 'see' the potential barrier separating the pieces of Fermi surface and so follows a more free-electron-like path. But in intersheet scattering the Bragg reflection is central through its role in the Umklapp event. Magnetic breakdown increases with increasing applied magnetic field whereas intersheet scattering decreases. According to Young et al. (1969), this is because with increasing cyclotron frequency the representative points spend less time in the hot spot regions.

4.2 Effect of Intersheet Scattering on $\rho_{21}(\underline{B},T)$

Given that intersheet scattering between the tips of the clover leaf orbits can produce orbits like A and E of Fig. 4.2, it can be seen qualitatively how the $\rho_{21}(\underline{B},T)$ behaviours might arise. The clover leaf orbit is analogous to that of Fig. 2.2(b), for in the low- and intermediate-field conditions it can support both electron- and hole-like responses. A segment $x \rightarrow y$ (refer to Fig. 4.1) traced out during an electron's lifetime makes an electron-like contribution, whereas $p \rightarrow q$ makes a hole-like one. The sign of the Hall effect at higher temperatures—ensuring the low-field condition—shows that hole-like responses outweigh electron-like ones, although the details of this balance are not understood. (Possibly an isotropic $\tau_{\underline{k}}$ around the orbit affects the balance.) In the high-field condition the orbit's contribution can only be hole-like, so as temperature is reduced sufficiently in a fixed field, the clover leaf's response changes from a mixture of electron- and hole-respon-responses to a purely hole-like one. Hence the increase of $\rho_{21}(T)$ towards more positive values that is seen in Fig. 4.1 as the temperature is reduced below ~10K in the higher applied fields. But at ~6K this tendency is suddenly reversed, the more so the higher the applied field. This reversal is attributed to electrons on orbits like A and E of Fig. 4.2 (A is shown dashed in Fig. 4.1) through intersheet scattering. When $\omega\tau_{\underline{k}}$ for these orbits approaches unity, their contribution to the Hall effect is equivalent to that from a fictitious electron sheet. The negative behaviour of $\rho_{21}(T)$ increases as temperature falls as long as $\omega\tau_{\underline{k}}$ increases corresondingly, but eventually at the lowest temperatures the scattering by inevitable impurities limits further increase in $\omega\tau_{\underline{k}}$ and $\rho_{21}(T)$ tends to a temperature-independent value. This qualitative interpretation of $\rho_{21}(\underline{B},T)$ has been supported by a path integral calculation using a scale model Fermi surface as described in Sec. 3 (Hurd et al. 1976, Alderson et al. 1977). A cylindrical sheet of Fermi surface is used to simulate the clover leaf orbit with the cross-section shown in Fig. 3.2. Interpiece scattering (of variable probability) is permitted at the six points A, B...F, corresponding to a transition like $p \rightarrow p'$ in Fig. 4.1. Although the model is a crude imitation of the orbits involved, it contains the essential features for a successful simulation of the observed $\rho_{21}(\underline{B},T)$ and confirms the qualitative interpretation outlined above. Other experiments and model calculations have been done to study the effect on $\rho_{21}(\underline{B},T)$ of different solutes in Cd (Lilly and Gerritsen 1974, Gerritsen 1974, 1975, Hurd et al. 1976) but are not considered here.

Intersheet scattering may also be important in W and Mo

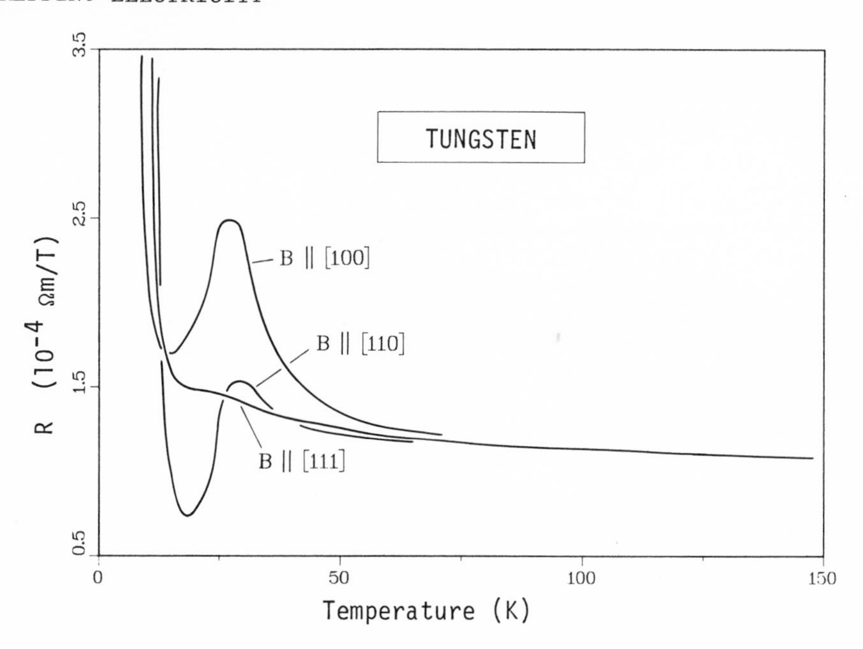

Fig. 4.4 Tungsten is believed to be another metal in which Umklapp scattering from "hot spots" (Fig. 4.2) affects the Hall voltage. These different temperature dependences of the Hall coefficient for given orientations of B are cited as evidence of intersheet scattering (Volkenshteyn et al. 1978).

(Volkenshteyn et al. 1978). Figure 4.4 shows the isomagnetic temperature dependence of the Hall coefficient of W measured for three orientations of B in monocrystalline samples. The pronounced tendency to positive values seen for all orientations at the lowest temperatures is attributed to the establishment of the high-field condition, but the local maximum seen at ~27K for two orientations has been interpreted (Volkenshteyn et al. 1978) in terms of intersheet scattering. This is proposed to occur between hot spots on the electron "jack" and the hole octahedron sheets of the Fermi surface (Cracknell 1971). The results fit qualitatively this conjecture because it would be expected from the geometry of the Fermi surface that intersheet scattering would be prominent for B || <110> or <100> but much less evident for B || <111>. Similar behaviour is reported for Mo (Volkenshteyn et al. 1978), which has an analogous electronic structure.

4.3 Magnetic Breakdown in Mg

Cadmium and Mg have similar crystallographic and electronic

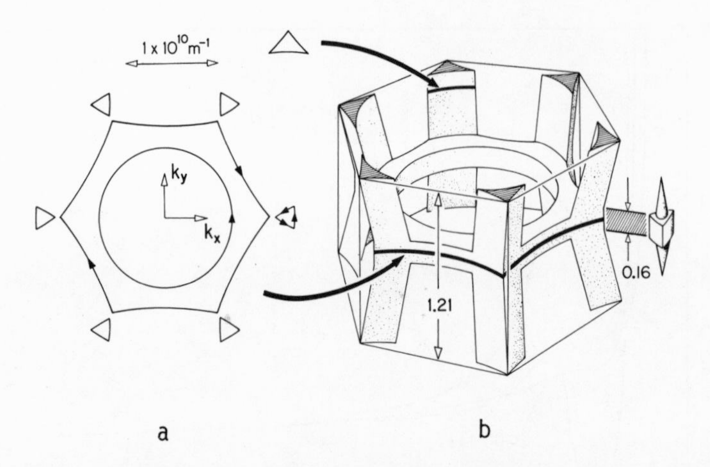

Fig. 4.5 A sketch of the "monster" and "cigar" surfaces of the Fermi surface of Mg. Magnetic breakdown is permitted across the shaded region in (b). The numbers refer to a scale model surface used by McAlister et al. (1977) in a path integral simulation of $\rho_{21}(\underline{B},T)$.

structures but they have important differences. The lattice potential has less pronounced effects upon the electron's motion in Mg, which is therefore a more free-electron-like metal. This is reflected in the Hall voltage, which is negative at room temperature in Mg but positive in Cd. The Fermi surface of Mg differs from that of Cd principally in the monster, the arms of which in Mg are joined (Cracknell 1971), so the tips of the clover leaf orbits in Fig. 4.1 are merged in Mg and there is no possibility of intersheet scattering (Sec. 4.1). Magnetic breakdown can occur, however, for electrons moving parallel to the basal plane, between the second-zone monster and elongated third-zone electron surfaces known as "cigars" (Falicov et al. 1966, Falicov 1973), as sketched in Fig. 4.5. With no breakdown, the cyclotron orbits contributing from the basal plane are the circular inner electron orbit (Fig. 4.5a) and the six-segmented hole orbit from the monster, and the three-segmented electron orbits from the cigars. When the probability of breakdown is unity, these reduce to the inner electron orbit and a large free electron orbit that circumscribes the cigars (Falicov et al. 1966; the "outer electron orbit" of Fig. 8 in Hurd et al. 1977). A tendency toward more electron-like behaviour is thus expected as the magnetic breakdown increases. This is evident in the results shown in Fig. 4.6. Compared to Cd (Fig. 4.1), the swing to more negative values at lower temperatures, in the high-field region, is markedly dependent on $\underline{B}$. This reflects the different origins of the behaviours: both intersheet scattering and magnetic breakdown affect the topology of the Fermi surface and so

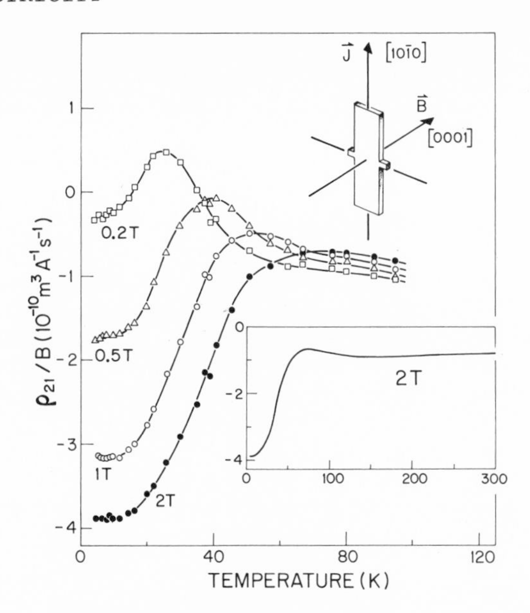

Fig. 4.6 Showing the temperature dependence of the Hall resistivity (here divided by flux density for easier graphing) for a Mg monocrystal with RRR = 425 (McAlister et al. 1977). The field-dependent swing to negative values at low temperatures is produced by magnetic breakdown in the basal plane.

give more prominent effects as $\omega\tau$ increases into the high-field region, but the probabilities of the two effects vary oppositely with $\underline{B}$. The probability of breakdown is given by $P = \exp[-B_o/B]$, where B_o is related to the energy barrier and other electronic parameters (Cohen and Falivoc 1961, Blount 1962). Hence breakdown increases with applied field strength. A more quantitative interpretation of Fig. 4.6 has been made by McAlister et al. (1977) using a path integral calculation and a scale model Fermi surface (Sec. 3), of which Fig. 4.5 shows part. Using the matrix formulation mentioned (Sec. 3.1), the calculation incorporates all the geometrical possibilities arising from magnetic breakdown in the basal plane. Briefly, it shows how breakdown ensures that the monster's contribution to σ_{21} is strongly positive and dominant at lower temperatures. Hence ρ_{21} is negative (Eq. 2.3). The approach to the high-field limit also reduces progressively σ_{11}, by the mechanism described in Sec. 2.3, so $|\rho_{21}|$ increases. Whence the negative tendency of $\rho_{21}(\underline{B},T)$ due to increasing breakdown.

Although we take the example of Mg with $\underline{B} \,||\, [0001]$ to illustrate the effect of magnetic breakdown on $\rho_{21}(\underline{B},T)$ other cases could have been used. Douglas and Datars (1975) describe breakdown parallel to the hexagonal axis of Cd; Hurd et al. (1977) for Zn in both basal- and axial-plane orientations; and McAlister et al. (1977) for Mg in the axial plane.

5. SOME INTERMEDIATE-FIELD EFFECTS

The Hall effect in the intermediate-field condition arises from a mixture of high- and low-field magnetoconductivity contributions that is complicated and potentially less informative than a result from a clearly defined limit. This condition should be shunned by all, yet experimentally it is important because so many studies of $\rho_{21}(\underline{B},T)$ in pure metals and dilute alloys at cryogenic temperatures encompass it, intentionally or not. A variation of temperature, effective purity, or applied magnetic field strength can change the sample's field condition between predominantly high- and low-field situations so that either the Lorentz force or the scattering processes becomes the dominant perturbation. This transition—called the "high-field/low-field transition"—can produce distinctive behaviours in $\rho_{21}(\underline{B},T)$.

5.1 Non-linear $\rho_{21}(\underline{B})$

According to simple expectations (Sec. 2.1) the Hall voltage measured isothermally should vary linearly with the applied field strength, but this is frequently not the case. Two principal origins of non-linearity with field can be distinguished. The first is from an extra voltage that appears in the direction of the Hall voltage when the sample has a net magnetization $\underline{M}$. It is called the extraordinary or anomalous Hall effect and is the subject of Sec. 6.2.

A second source of non-linear field dependence is the high-field/low-field transition, of which an example for an alloy is shown in Fig. 5.1. The transition can also be evident in pure metals when a prominent topological feature on the Fermi surface manifests itself as $\omega\tau$ increases. Figure 5.2 shows an example that relates directly to Sec. 4.3. With $\underline{B} \parallel [0001]$ in Mg, the Hall voltage is initially positive in the low-field condition, but increasing $\underline{B}$ increases both the magnetic breakdown probability and $\omega\tau$ so that its effects are evident. The positive tendency of the low-field condition is reversed and eventually changes the sign of the Hall voltage (Papastaikoudis and Kesternich 1975, McAlister et al. 1977).

The Hall voltage also changes sign due to high-field/low-field effects in Ru, from positive to negative with increasing applied

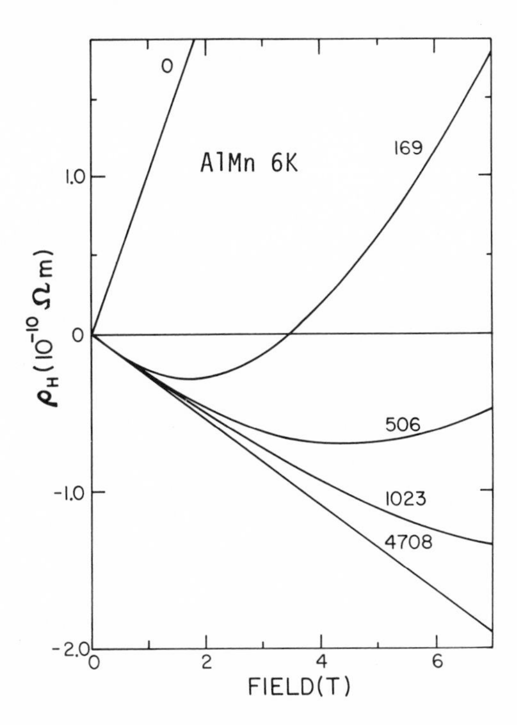

Fig. 5.1 A cause of the nonlinearity in the field dependence of the Hall voltage is the high-field/low-field transition, here in Al containing small amounts of Mn, a nonmagnetic solute. (Solute concentration is shown in at.ppm). $\rho_{21}(\underline{B})$ is linear in the high-field condition (pure Al) and in the low-field condition (Al + 4708 at.ppm Mn) but is nonlinear in the intermediate-field range (McAlister et al. 1979).

field (Volkenshteyn et al. 1974). This hexagonal metal has a Fermi surface including a monster hole sheet similar to that sketched in Fig. 4.5(b). The sign change in this case is believed to arise from the influence of the inner electron orbits, equivalent to the circular orbit in Fig. 4.5(a) (Volkenshteyn et al. 1974). These orbits form a wide band on the Ru Fermi surface and, in the high-field condition, their electron contribution is believed to dominate that from the hole monster. Hence the negative Hall voltage at higher fields.

<u>5.1.1 Sign change of $\rho_{21}(\underline{B})$ in Al.</u> The change of sign of the Hall voltage in Mg and Ru discussed above arises because the topology of a limited part of the Fermi surface changes in the high-field/low-field transition. Prominent nonlinearity of $\rho_{21}(\underline{B})$,

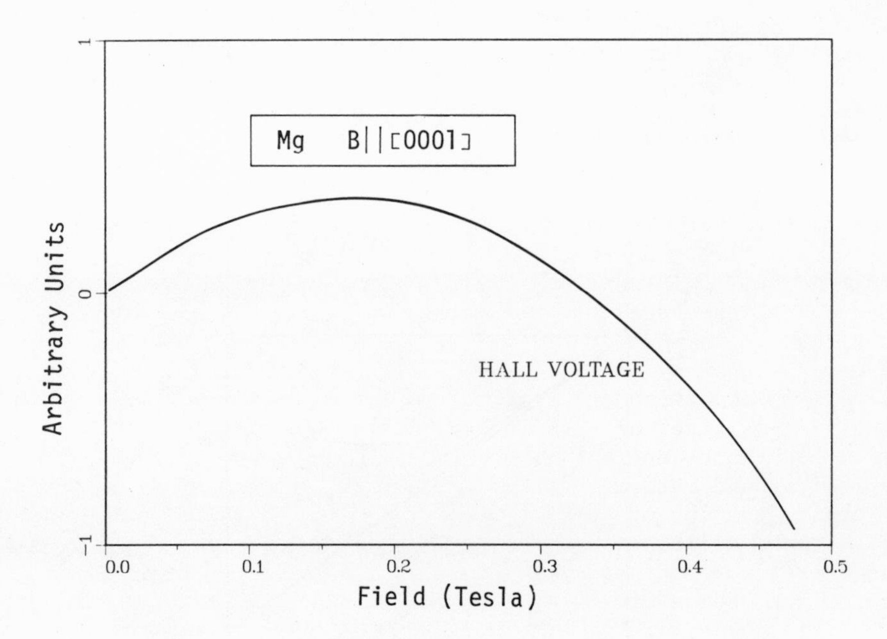

Fig. 5.2 The nonlinear field dependence of the Hall voltage can include a change of sign. This example shows the Hall voltage from electrons moving in the basal plane of Mg. In higher flux densities, magnetic breakdown (Sec. 3.2) dominates the Hall effect and eventually produces a positive Hall voltage. (Papastaikoudis and Kesternich 1975.)

including a change of sign, can also occur when the response of a whole sheet of Fermi surface changes as in Al and In (Forsvoll and Holwech 1964, Garland and Bowers 1969, Douglas and Datars 1973, 1974, Kesternich et al. 1976, Ozimek et al. 1978). We choose Al for illustration because recent results have modified the previously established explanation of the $\rho_{21}(B)$ behaviour; corresponding calculations for In have not yet been made.

The Fermi surface of Al (Cracknell 1971) consists of two major sheets: a hole sheet in the second Brillouin zone and an electron sheet in the third zone formed from segments of the free electron sphere which run close to the edges of the zone. This third-zone sheet comprises about 7% of the free electron surface. Calculation shows (McAlister et al. 1979) that with the currently accepted anisotropy of $\tau_{\underline{k}}$ on the second-zone surface, the third-zone surface makes an insignificant contribution to the magnetoconductivity. This is an important point because several quantitative explanations of the sign change of $\rho_{21}(\omega\tau)$ (for example: Ashcroft 1969,

Douglas and Datars 1973, Kesternich et al. 1976, Sato et al. 1978) invoke the competition between contributions from the second- and third-zone sheets.

A path integral calculation (Sec. 3) shows that the situation in Al is similar to that encountered in Sec. 2.3 (McAlister et al. 1979). The second-zone hole surface supports orbits like that of Fig. 2.2(b), but with the important complication of an anisotropic relaxation time. Sorbello (1974) has shown that states on regions near the Brillouin zone boundaries have distinct quantum symmetry, those along the edges of the second-zone sheet have p-like symmetry, so electrons in them are scattered more by interstitially localized potentials than by potentials located at lattice sites. In the high-field condition the surface's response is hole-like and hence R(HF) is positive, as observed experimentally. But decreasing $\omega\tau$ gives both hole- and electron-like contributions, depending on the segment traced out, as described in Sec. 2.3. R(LF) is determined by the anisotropy of the dominant scattering process, which might be phonons, dislocations, a residual impurity, and so forth. If the anisotropy increases the relative importance of electrons in states near edges of the Fermi surface, then R(LF) shifts to more positive values. Earlier measurements found R(LF) to be close to its free electron value. We think this is fortuitous (McAlister et al. 1979), but others (Frosvoll and Holwech 1964) associate it with a more fundamental cause.

5.2 High-field/Low-field Effects in $\rho_{21}(T)$

A moderately pure metal in a field of a few tesla may undergo a gradual transition to the high-field condition as the temperature is reduced sufficiently. (Indeed, some high-field behaviour in pure metals has been seen in various metals up to ~100K. See Ref. 23 of Hurd et al. 1976.) Strikingly different behaviours for $\rho_{21}(T)$ can be found with $\underline{B}$ along different crystallographic directions in a monocrystal. In a polycrystal the effects are smeared out and $\rho_{21}(T)$ is averaged over the orientations of the crystallites.

Figure 5.3 shows an example of the high-field/low-field transition in $\rho_{21}(T)$ for various orientations of field in Au and Cu—an homologous pair—together with corresponding results for polycrystalline samples. The high-field condition is first manifested at ~ 50K (B of Fig. 5.3) where topological features in the {100} and {110} planes show their effects. At ~30K (A of Fig. 5.3) the difference between the behaviours for other orientations is distinguishable. The local maximum in $|\rho_{21}(T)|$ of polycrystalline Cu at ~30K (Fig. 5.3) has been the source of controversy. A similar behaviour has been seen in polycrystalline Ag (Dugdale and Firth 1969, Alderson et al. 1968) and, to a lesser extent, in Au

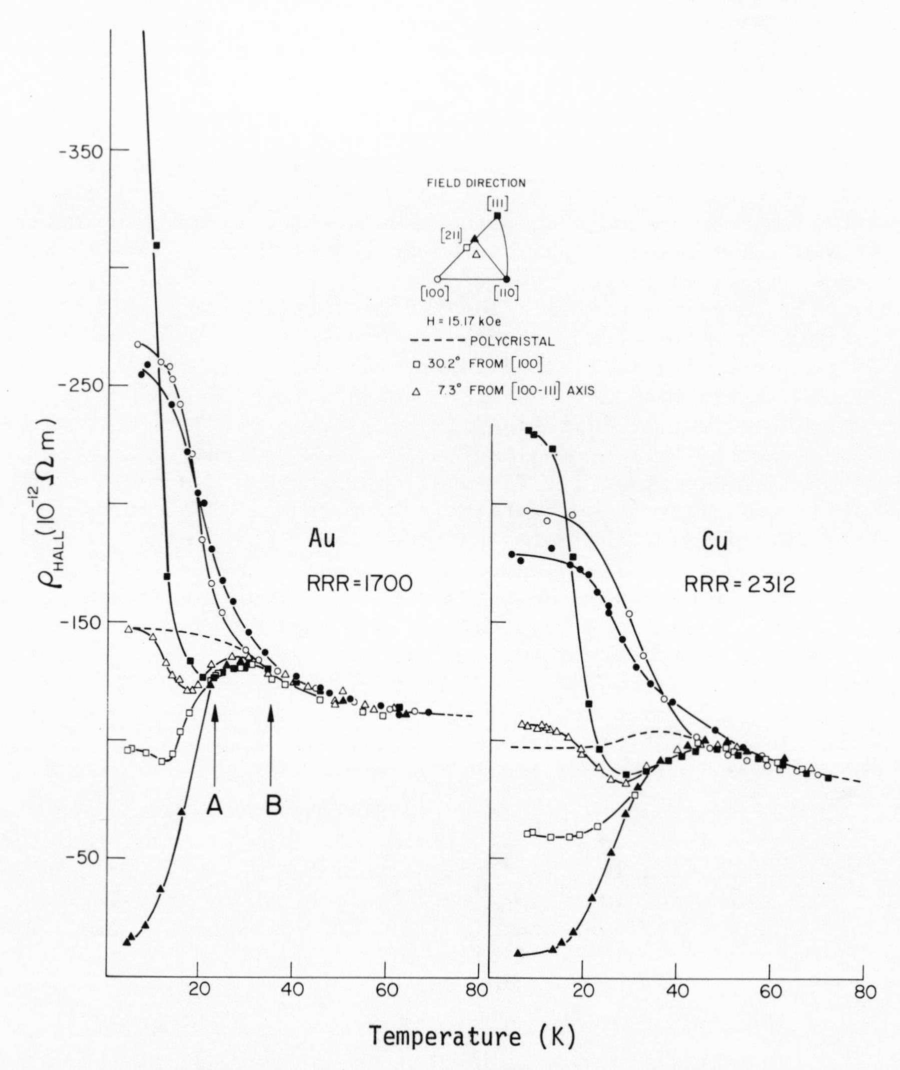

Fig. 5.3 Showing the isomagnetic temperature dependences of the Hall resistivity in monocrystals of Au and Cu through the high-field/low-field transition. (Alderson and Hurd 1973.)

(Alderson et al. 1968, and Fig. 5.3). Numerous explanations of this maximum have been proposed including: phonon drag (Alderson et al. 1970), suppression of Umklapp processes from localized regions of the Fermi surface (Barnard and Sumner 1969), transition between phonon-dominated and impurity-dominated scattering ranges (Dugdale and Firth 1969), and different anisotropies in the phonon scattering involving normal and Umklapp processes (Kimura and Honda 1971). Barnard (1977), working with polycrystals of Cu and Ag, showed the maximum in $|\rho_{21}(T)|$ exists even in applied fields of 0.0085T, sufficient to ensure low-field conditions throughout the experiment. The high-field/low-field transition therefore cannot be the origin of the behaviour in polycrystals and it seems much more likely to be due to the transition between phonon- and impurity or defect-dominated scattering ranges.

There have been two attempts to explain quantitatively the behaviour shown for Cu in Fig. 5.3. Both essentially solve a Boltzmann equation to derive the magnetoconductivity and both involve adjustable parameters to describe the electron-phonon interaction. That by Mann and Schmidt (1975) and Schmidt and Mann (1979) gives qualitative agreement with observations except for certain prominent maxima in $\rho_{21}(T)$ when $\underline{B} \parallel [111]$ or [211]. The disadvantage of this treatment is that it does not transparently show the physical origins of the various $\rho_{21}(\underline{B},T)$ that it produces. The second approach (Hurd and McAlister 1977) is a path integral calculation (Sec. 3) and is more flexible since it uses a scale model Fermi surface on which various prominent topological features can be manipulated to show their role in the origins of the $\rho_{21}(\underline{B},T)$. It gives a quantitative explanation of practically all the features seen in Fig. 5.3. Consider, for example, the case with $\underline{B} \parallel [211]$, since this has two orthogonal bands of open orbits (A conducting primarily along [110] and B along [111]) like the case encountered in Sec. 2.4. In the calculation, the high-field condition for these orbits is reached at temperatures below ~20K, but σ_{11} and σ_{22} do not go to zero in the high-field limit in this case (see left-hand side of Fig. 5.4) for the reasons explained in Sec. 2.4: the electrons are channeled between lattice planes by successive Bragg reflections and cannot respond to the Lorentz force. The product $\sigma_{11}\sigma_{22}$ dominates the denominator of Eq. (2.3) and gives directly the observed reduction of $\rho_{21}(T)$ at low temperatures. The right-hand side of Fig. 5.4 shows the fictitious cases with the two types of open orbits removed separately. It reflects the behaviour seen in Fig. 2.4. With the primary current's direction k_x along [111], orbits A enhance σ_{22} at low temperatures, while B enhance σ_{11}. A similar interpretation of $\rho_{21}(\underline{B},T)$ is obtained (Hurd and McAlister 1977) for all orientations considered in Fig. 5.3, but the big handicap in extending this approach to other metals—or, indeed, to making it more secure for the group-1B metals—is the lack of knowledge of the anisotropy of τ_k over the Fermi surface and its temperature variation. In every application we have described, of the path integral calculation to a model

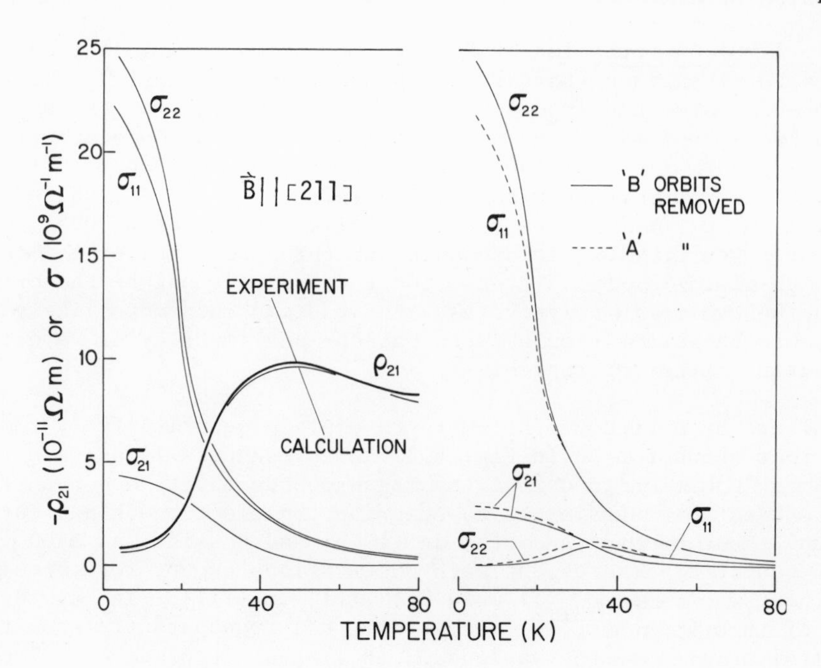

Fig. 5.4 (Left) A comparison of the calculated and experimental (Fig. 5.3) $\rho_{21}(T)$ with $\underline{B}\,||\,[211]$, including the corresponding calculated $\sigma_{ij}(T)$. (Right) The behaviour of elements in the magnetoconductivity when different orbits are eliminated in the calculation. (Hurd and McAlister 1977.)

Fermi surface, $\tau_{\underline{k}}(T)$ is a parameter that has to be chosen either by heuristic or arbitrary methods.

6. SOME LOW-FIELD EFFECTS

Section 3.2 explains how the Hall voltage, which is a scalar in a cubic metal when $\omega\tau \ll 1$ for all cyclotron orbits, is then determined by the effective mass, electron velocity and anisotropic relaxation times over the Fermi surface. If the metal is magnetic or contains a magnetic solute, however, the situation is complicated by spin-orbit coupling between the localized moments and the itinerant electrons. This produces an extra voltage in the same orientation as that from the Lorentz force, leading to the "extraordinary" or "anomalous" Hall effect. It is convenient to consider separately the low-field behaviour in nonmagnetic and magnetic systems.

6.1 Diamagnetic Systems

6.1.1 Establishing the low field condition. Figure 6.1 is a composite sketch comprising the concentration dependence of the low-field Hall coefficient of binary Al-based alloys, but its features are generally applicable to other systems. The solute is a typical point scatterer, such as a substitutional nonmagnetic species. Since the R(LF) reflects the anisotropy of the relaxation time $\tau_{\underline{k}}$ produced by the dominant scatterer, it changes rapidly upon first addition of solute from a value controlled by the anisotropy of residual defects, impurities or phonons, to a concentration-independent value that is dominated by the solute's scattering (as in Fig. 6.1a). Clearly, R(LF) for the pure solvent metal is experimentally unattainable because a residual scattering from phonons will always dominate when the impurities and imperfections are reduced sufficiently. It is meaningless to talk of the low-field Hall coefficient of a pure metal because a binary system is always involved; R(LF) is characteristic of a given scatterer in a given solute. In Fig. 6.1, R(LF) appropriate to the solute should ideally be deduced from the extrapolation to zero concentration of the concentration-independent plateau for results obtained at absolute zero, but in practice results from ~4K have to suffice (as with the dashed line in Fig. 6.1. The caption refers to examples of this use). With increasing solute concentration, R becomes concentration-dependent (as at 4.2K in Fig. 6.1c) due to changes in the scattering anisotropy produced by solute-solute interaction or to changes in the Fermi surface. At higher temperatures, say 77K, a concentration-independent R can also be obtained but it is shifted from that at lower temperatures due to the anisotropy of the extra phonon scattering introduced. The effect of the phonon's anisotropy may be even more pronounced at higher temperatures, as indicated in Fig. 6.1(c), but since these results relate to AlCu alloys the behaviour may also reflect changes in the Fermi surface. Figure 6.1 emphasizes that to obtain an R(LF) representative of a given dominant scatterer, the experimental data should satisfy at least two criteria:

(1) R should be field-independent in the range of fields up to that used in its determination, thus ensuring the low-field condition;

(2) R should be independent of concentration of the presumed dominant scatterer for a reasonable range about that of the sample, thus ensuring that the dominant scatterer is correctly identified.

6.1.2 R(LF) and anisotropic scattering in some binary systems. The group-1B metals and Al are frequently solvents used to study the anisotropy of scattering by different solutes. Their popularity is partly metallurgical but mainly because much is known about the variation over their Fermi surfaces of $\tau_{\underline{k}}$ and the

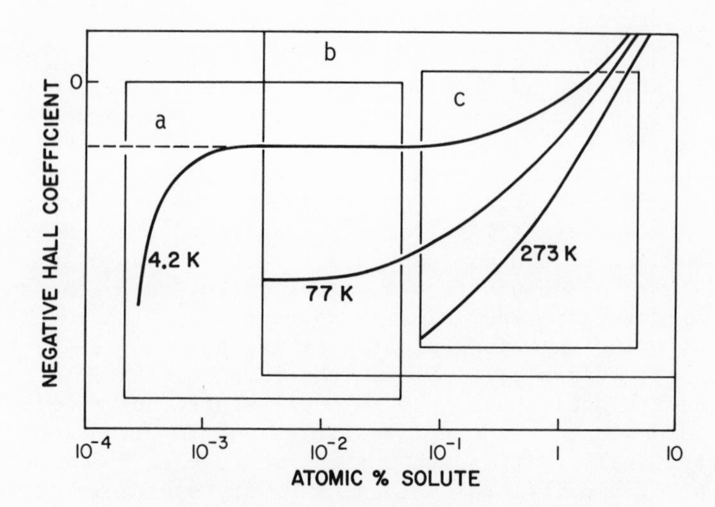

Fig. 6.1 A schematic and composite sketch of different circumstances seen in the literature for the typical behaviour of the low-field Hall coefficient of a binary alloy. It is based upon results for Al containing an s-type scatterer, such as Zn, Cu or Si. (a) is typical of the range covered by Böning et al. (1975) or Papastaikoudis et al. (1976) for various point scatterers in Al; (b) is typical of the range covered by Osamura et al. (1973) and (c) is typical of the range covered by Bradley and Stringer (1974).

symmetry of the wave functions (Springford 1971, Sorbello 1974). Briefly, states near "neck" regions on the Fermi surface of Cu have p-type symmetry (defined by analogy with the nomenclature for the atomic case) with wave functions of greatest amplitude between the ion sites. Electrons in such states are particularly susceptible to scattering by interstitially-localized scattering potentials. States on the "belly", however, have generally s-type symmetry and electrons in these states are affected most by scattering potentials localized on the ion sites. In the case of Al, we saw in Sec. 5.1.1 that states near the edges of the second-zone sheet have p-type symmetry while those on its main body are s-type. Theory also predicts an anisotropy on the third-zone sheet (Sorbello 1974, Meador and Lawrence 1977) but we explained in Sec. 5.1.1 that electrons in such states probably do not contribute significantly to the magnetoconductivity.

The simplification is made that states on the Fermi surface of Cu, or the second-zone sheet of Al, can be divided into two groups with distinct relaxation times: "neck" and "belly" states in Cu, and "edge" and "body" states in Al. (The assignment of these areas is arbitrary and is a great weakness in the method.) A relationship between R(LF) for the solvent metal and the ratio of the

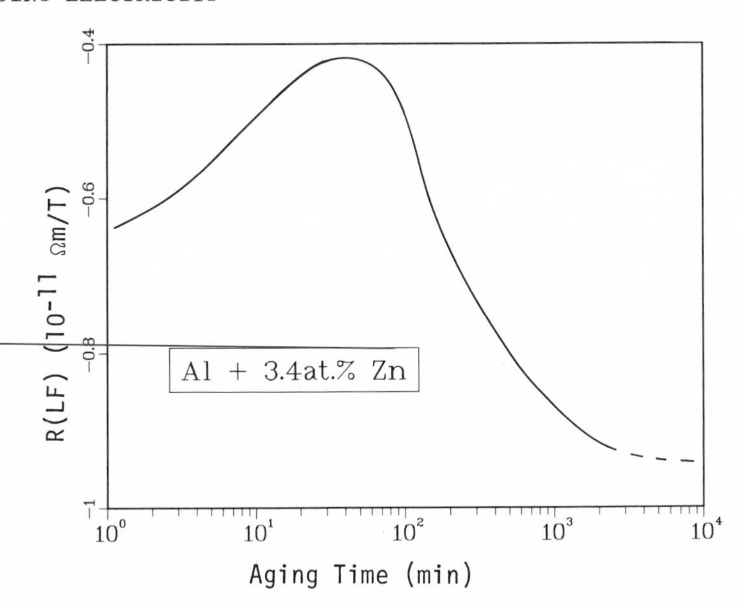

Fig. 6.2 The low-field Hall coefficient versus aging time at 263K for an alloy containing spherical precipitates (Smugeresky et al. 1969). A shift of R(LF) towards negative values implies a greater scattering potential in the interstices of the Al lattice.

relaxation times for the two groups is then obtained using a low-field expression (Sec. 3.2.2) or its equivalent, together with a model Fermi surface. An R(LF) measured for a dominant scatterer thus yields a corresponding ratio of relaxation times, and this is interpreted as the anisotropy in τ_k due to the scatterer. One of the first applications of this approach was by Dugdale and Firth (1969) for Cu- and Ag-based alloys. More recently it has been used by McAlister et al. (1979) for Al-based systems. These measurements give information about the degree of interstitial influence of the scatterer, and the anisotropy in τ_k it produces, but they are susceptible to misinterpretation because of metallurgical effects. For example, the scattering centres may not be isolated or uniformly distributed but may cluster together giving an interstitial influence that is atypical of the isolated centre.

Figure 6.2 shows R(LF) measured at 77K for an Al + 3.4 at.% Zn alloy after different aging times at 263K (Smugeresky et al. 1969). The aging promotes the growth of spherical clusters of solute through migration, until ultimately about 2% of the solute ions are involved. Details of the kinetics are still the subject of study, but the Hall results imply, for AlZn at least, a two-stage process. The first clusters to form shift R(LF) to more positive values, equivalent to increasing the amount of scattering

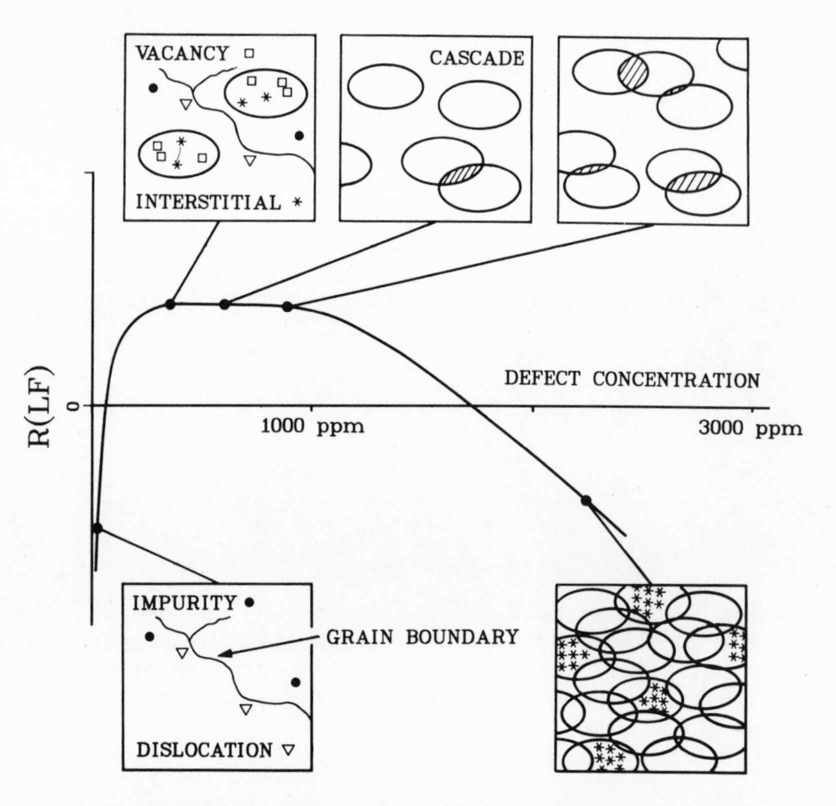

Fig. 6.3 The low-field Hall coefficient of Al shows this distinctive dependence on defect concentration when irradiated by neutrons at 4.6K (cf Fig. 6.1). Initially negative, R(LF) becomes positive at low doses as isolated point defects are formed. Abo e about 1000 ppm the clustered defects assert their interstitial scattering influence and R(LF) swings to increasingly negative values. (After Böning 1978.)

by a potential localized on a lattice site. Subsequently, the trend is reversed as the interstitial influence of the zones' scattering becomes dominant. This may be the two stages of precipitation that have been identified for such alloys: (1) initial nucleation of new zones and (2) a 'coarsening', when favoured zones grow at the expense of others. A similar behaviour, but which involves changes in the scatterer that are better understood, is sketched in Fig. 6.3. Post-irradiation annealing studies of Al have also been made using R(LF) (Kesternich et al. 1976). Whereas the electrical resistivity reflects essentially the changes in the concentration of defects, R(LF) reflects changes in their structure. The combination of these measurements becomes a promising method to identify the various stages of annealing.

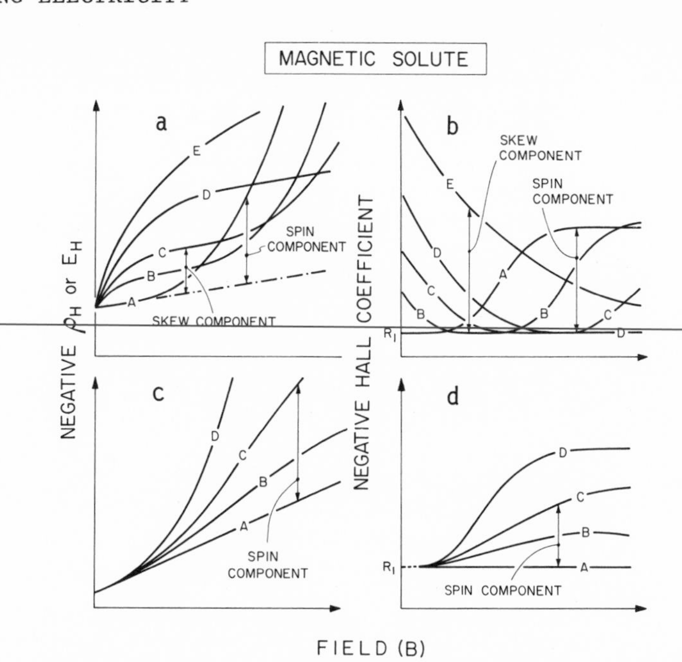

Fig. 6.4 A sketch of isothermal field dependences at 4K of the Hall resistivity (ρ_H), Hall field (E_H) or Hall coefficient for a nonferromagnetic alloy of a magnetic solute in a non-magnetic solvent metal. A→D or E represent increasing concentration. The skew component arises from spin-orbit coupling (Sec. 6.2.2) and the spin component from the separation by the applied field of the itinerant electron population into spin-up and spin-down groups with different dynamical properties. (McAlister and Hurd 1977.)

6.2 Magnetic Systems

6.2.1 Nonlinearity in the isothermal $\rho_{21}(\underline{B})$ of magnetic systems. It is convenient to distinguish systems with long-range ferromagnetic order from those where the concentration of magnetic ions is well below the ferromagnetic percolation limit. Qualitatively different behaviours of $\rho_{21}(\underline{B})$ are found in the two cases. The nonferromagnetic case shows a nonlinearity of $\rho_{21}(B)$ arising either from the so-called "skew component" due to spin-orbit coupling (Sec. 6.2.2) or from the "spin components" caused by the separation by the applied field of the itinerant electron population into two groups (Fert and Friederich 1976, McAlister and Hurd 1977). The spin-up and spin-down groups have different lifetimes between collisions because of the spin-dependent scattering by the magnetic ions. (The Pauli polarization of the itinerant electron population may also be a contributor but its effect is thought to be small by comparison.

The spin and skew components in $\rho_{21}(\underline{B})$ appear typically as sketched in Fig. 6.4. (a) shows a case like AuFe where these components coexist; (c) is like AgMn for which spin-orbit coupling is very weak or nonexistent so that only the spin component is observed. A→D represent alloys of increasing concentration. The skew component, when it exists, obliterates the spin component in alloys containing typically >~100 at.ppm. In such dilute alloys there is a further complication in the interpretation of the non-linear $\rho_{21}(\underline{B})$ because the high-field/low-field transition may be important (Sec. 5.1).

The isothermal $\rho_{21}(\underline{B})$ in a polycrystalline ferromagnet has typically an "elbow" between roughly linear sections. Such observations are among the oldest of modern measurements (see Jan 1957 for a review) and are treated empirically as two simply-additive components: the first is the normal Hall effect, linear in $\underline{B}$, and the second is the extraordinary (or anomalous) part proportional to the magnetization M_S. Thus (in mks):

$$\rho_{21} = R_o B + R_S M_S \tag{6.1}$$

where R_o and R_S are known as the ordinary and spontaneous (or extraordinary) Hall coefficients, respectively. A large literature has developed around this phenomenological treatment and has been covered in several reviews (Jan 1957, Hurd 1972, Coleman 1975). It seems fair to say that this empirical approach to understanding $\rho_{21}(\underline{B})$ in magnetic systems has been static for some years and not very informative. More interesting progress in the last couple of decades has come from attempts to understand the microscopic origins of R_S in ferromagnetic systems and the skew component in paramagnetic systems.

6.2.2 Spin-orbit coupling: skew and side jump-effects. Coupling between an itinerant electron's orbital moment and an ion's localized magnetic moment has two macroscopic manifestations: skew scattering and side-jump effects (Smit 1958, Berger 1970, Nozières and Lewiner 1973, Chazalviel 1975, Fert and Friederich 1976). Both produce voltage components in the same orientation as the normal Hall voltage from the Lorentz force. Skew scattering is the asymmetric scattering of an electron with respect to the plane containing its incident velocity and the ion's magnetic moment. Its contribution to the Hall effect (the "skew component" of Fig. 6.4) is well established in paramagnetic alloys at low temperatures (Fert 1973, 1974, 1977, Fert and Friederich 1976). Side-jump scattering is more controversial (Berger 1970). It has been explained either as symmetric scattering from a displaced scattering centre or in terms of the different delays suffered by electrons scattered

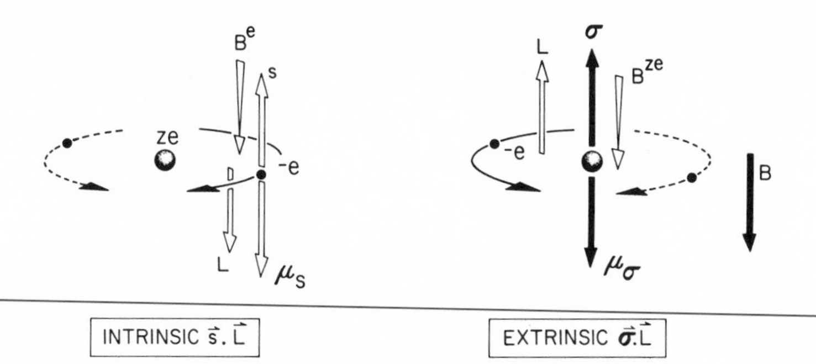

Fig. 6.5 Showing schematically the minimum energy configuration in two types of spin-orbit coupling for an electron of charge e (positive throughout) scattered by an ion of charge ze. The solid-line trajectory has in each case a lower energy than the dotted one.

to different sides of the scattering ion (Fert 1977, Chazalviel 1975).

The microscopic origins of these effects fall into two classes: band models, where the itinerant particles are responsible for both electrical and magnetic properties, and localized models, which distinguish between conduction electrons and 'magnetic' electrons. In a band model the coupling is the intrinsic interaction $\underline{s}\cdot\underline{L}$ between an itinerant electron's intrinsic spin $\underline{s}$ and its angular momentum $\underline{L}$; in a localized model it is the extrinsic interaction $\underline{\sigma}\cdot\underline{L}$ between the ion's total angular momentum $\underline{\sigma}$ and the scattered electron's orbital momentum $\underline{L}$. Figure 6.5 shows schematically how asymmetry arises in these cases. An electron with intrinsic magnetic moment $\underline{\mu}_s$ (spin s) is scattered by a charged ion ze. This carries a magnetic moment $\underline{\mu}_\sigma$ in the extrinsic case. Due to the angular momentum $\underline{L}$, an effective magnetic field $\underline{B}^e$ exists in the electron's frame of reference in the intrinsic case and couples with $\underline{\mu}_s$. The solid line is a lower energy trajectory than the dotted one. For a given incident spin, the scattering is thus asymmetric (or skew), as in the asymmetric Mott scattering of atomic theory (Joachain 1975). Similar asymmetry arises in the extrinsic case from the coupling between $\underline{\mu}_\sigma$ and $\underline{B}^{ze}$, where the latter is the effective field in the ion's frame due to the motion of the electron about it. Again, the solid line is the lower energy trajectory. This asymmetry is independent of s. Skew scattering could arise with nonmagnetic ions through the intrinsic mechanism because the itinerant electron population is spin polarized by the applied field, giving unequal scattering currents to either side of the

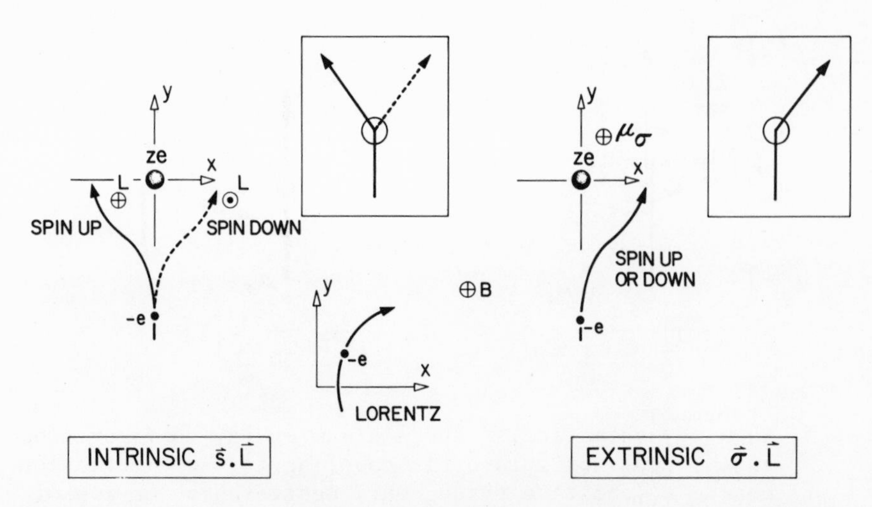

Fig. 6.6 The coupling depicted in Fig. 6.5 gives an electron a transverse velocity during scattering. Viewed macroscopically, as in the boxes, it emerges from the scattering centre's influence at an angle inclined to the incident direction. The sketch illustrates the cases for the orientation of vectors in Fig. 6.5.

ion, but no evidence of a skew contribution has been found in the Hall effect of diamagnetic systems. It seems that the intrinsic mechanism needs the unequal spin populations of the ferromagnet to give an appreciable effect.

Fert and Jaoul (1972) applied the simple mechanisms of Fig. 6.5 to resonant scattering through a virtual bound state about the ion, and show how the sign of the asymmetry can change as the position of the Fermi energy ε_f shifts with respect to the exchange-split levels of the state. Qualitatively (Bergmann 1979) the solid and dotted paths in Fig. 6.5 correspond to the m = $\pm$ levels of ℓ_z, for a given ℓ state of the spin-up half of the virtual level. The resonance coupling of these levels to the itinerant electrons depends on their energy separations from ε_f. If m = + is closer to ε_f than m = –, more electrons follow the dotted path in Fig. 6.5, and conversely if m = + is furthest from ε_f. Hence the sign of the skew component in $\rho_{21}(\underline{B})$ depends on the electronic structure of the virtual bound state.

Skew scattering gives the electron's centre of gravity a lateral velocity so the particle emerges on a trajectory inclined to the incident direction (Fig. 6.6), but side-jump scattering gives

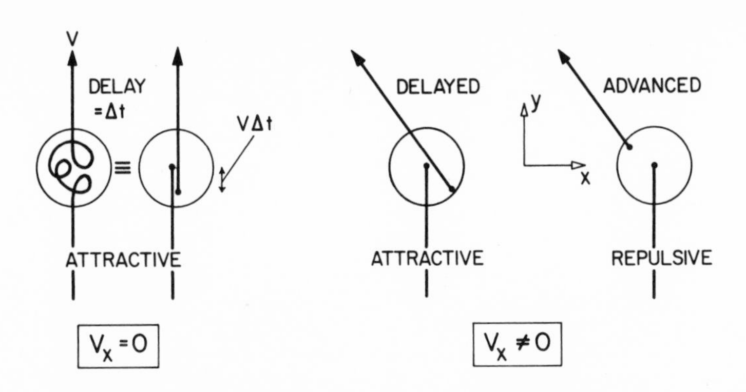

Fig. 6.7 A schematic explanation in terms of time delay of the origin of (left) "backwards jump" and "side-jump" during elastic electron scattering depicted in Fig. 6.6. Attractive or repulsive scattering potentials are assumed to operate within a sphere of influence defined by the circles. The straight-line trajectories are an external observer's view of the apparent effect of scattering. $\underline{B}$ is perpendicular to the page, as in Fig. 6.6.

a transverse displacement. Figure 6.7 illustrates this descriptive view (Berger 1970) of the nonclassical term first derived by Luttinger (1958) that is important when $\hbar/\tau\varepsilon_f$ becomes large. Left shows so-called transparent scattering by an attractive potential with no spin-orbit coupling and thus no transverse velocity v_x. The scatterer's range of influence is represented by the circle. The detailed behaviour within this range is not important; what matters to an external observer is that apparently the electron was delayed by a time Δt over that required to traverse the same region with no scattering potential. If we define that the scattering takes place at some instant, then we have an equivalent picture in which upon scattering the particle was jumped backwards through $v\Delta t$. The same arguments apply to a repulsive potential, where the particle is "forward jumped", or to one "side jumped" due to the transverse velocity v_x acquired from spin-orbit coupling (right of Fig. 6.7). Since the side-jump concept is a metaphorical explanation of a quantum mechanical effect, it was perhaps bound to lead to controversy (Smit 1973, 1978, Berger 1973, 1978, Lyo and Holstein 1974). Chazalviel (1975) has compared these conflicting views. This topic will be considered in more detail by Dr. Berger in a companion paper.

Skew scattering and repeated side jumps contribute to the Hall resistivity as shown in Fig. 6.8. The former gives a contribution to the Hall resistivity which varies as the solute concentration

Skew Repeated side jump

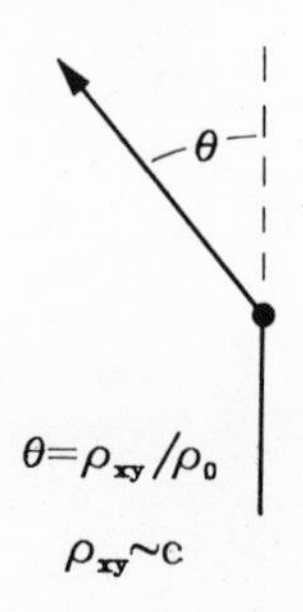

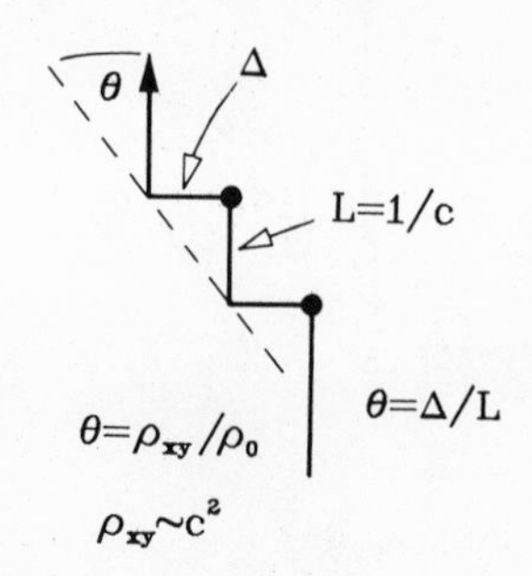

<u>SIDE JUMP</u>: C. of gravity of wave
packet gets lateral displacement

<u>SKEW</u>: C. of gravity of wave
packet gets lateral velocity

Fig. 6.8 Showing the contribution of skew scattering and repeated side-jump to the Hall resistivity ρ_{xy}. With $\underline{B}$ perpendicular to the page, θ is the Hall angle (Sec. 3.2.2) and ρ_0 is the electric resistivity which varies as c, the solute concentration. Since the mean free path L varies as 1/c, the Hall resistivity due to side-jump varies as c^2 and that due to skew as c. (Fert 1974.)

c, while the latter's contribution varies as c^2. [In the localized, free electron picture the side jump Δ amounts to only ~3 x 19^{-16} m at each scattering event—too small to account for the observed Hall angle—but is ~10^4 times greater when band effects are included (Berger 1970).] The temperature dependence of the skew and side-jump components is also different. The former varies like the magnetization, since to observe a net macroscopic skew scattering voltage the localized moments in the sample must be polarized, but the latter is also dependent on the mean free path as sketched in Fig. 6.8. Thus increasing the temperature reduces the skew component of the Hall voltage since it reduces the magnetization, but may increase that from side-jump because of the reduction in the mean free path.

These dependences on flux density and temperature make it generally possible to identify skew, side-jump and Lorentz contributions. A striking example is seen in Rhyne's (1968) results for $\rho_{21}(\underline{B},T)$ in monocrystalline Dy (Fig. 6.9). In the ferromagnetic range, increasing temperature <u>increases</u> the nonlinear (or

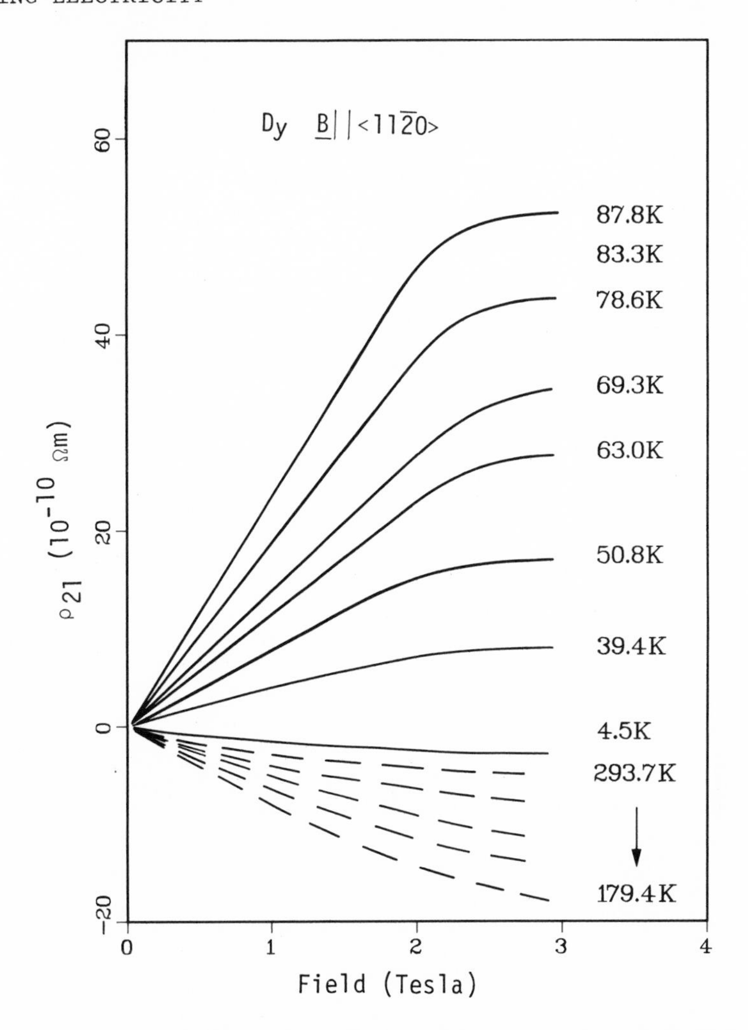

Fig. 6.9 Hall resistivity versus field for monocrystalline Dy. (Rhyne 1968.) The solid lines represent the ferromagnetic range below T_C, and the dashed ones are the paramagnetic range above T_N. The nonlinearity of $\rho_{21}(\underline{B})$ increases with increasing temperature in the ferromagnetic range because it is dominated by the side-jump component; the opposite temperature dependence is seen in the paramagnetic range because the skew component dominates. The skew and side-jump components have opposite signs in Dy, as in Fe and Co, but there is currently no explanation of this observation.

extraordinary) part of $\rho_{21}(\underline{B})$. Since the magnetization is reduced as temperature increases, this cannot be from skew scattering but must be due to the increasing side-jump component as the mean free path is reduced. When the temperature is reduced sufficiently to destroy ferromagnetism, $\rho_{21}(\underline{B})$ changes sign (as at 179.4K). In the paramagnetic state, increasing temperature reduces the nonlinearity in $\rho_{21}(\underline{B})$, as for the range 179K→294K. This is typical of the skew component in a paramagnetic system (Fert and Friederich 1976) which reduces as the magnetization is reduced. At 294K almost the same $\rho_{21}(\underline{B})$ is obtained as at 4.5K because in both cases there is no significant extraordinary effect: at 4.5K the mean free path is too long for any significant side-jump contribution to be evident, and at 294K the magnetization is reduced to the point that the skew component is obliterated. We conclude that the skew and side-jump components of $\rho_{21}(\underline{B})$ in Dy are negative and positive, respectively.

The signs of the skew and side-jump components of Fe, Co and Ni can be deduced (McAlister and Hurd 1979) from existing data by similar arguments, but to avoid encroaching even more on Dr. Berger's area we shall not go into details. Briefly, using the convention that the Lorentz component is negative, the skew component is found to be negative for scattering from Fe, Co and Ni ions while that from side jump is positive in Fe and Co but negative in Ni. It remains a theoretical problem to show why Ni is different from Co and Fe in this respect.

6.2.3 The extraordinary Hall effect in bulk Fe, Co and Ni. Knowing the signs of the skew and side-jump components, it is possible to interpret qualitatively the $\rho_{21}(\underline{B},T)$ behaviours found for bulk Fe, Co and Ni (McAlister and Hurd 1979). Again, we shall be brief for the reason given above, but Fe merits special consideration because it was one of the first metals studied by Hall. Figure 6.10 shows $\rho_{21}(\underline{B},T)$ for an Fe sample so impure that the low-field condition almost certainly exists over the whole experimental range. Increasing temperature increases the side-jump component which is positive in Fe, leading to the behaviour observed. (In a purer sample at low temperatures $\rho_{21}(\underline{B})$ becomes negative (Klaffky and Coleman 1974), probably because of the transition to the high-field condition in the experiment.) Hence we have some progress toward an explanation of the opposite signs of the transverse voltage in Au and Fe at room temperature, first noted by Hall. In Au the Lorentz force is the source of the effect but in Fe it is dominated at room temperature by spin-orbit coupling effects. Believers of side-jump may say the spin-orbit coupling has this particular manifestation because the mean free path is short, but they cannot yet account for the sign. We must give credit to Hall and Rowland for the durability not only of their techniques but also of some of the problems they left us.

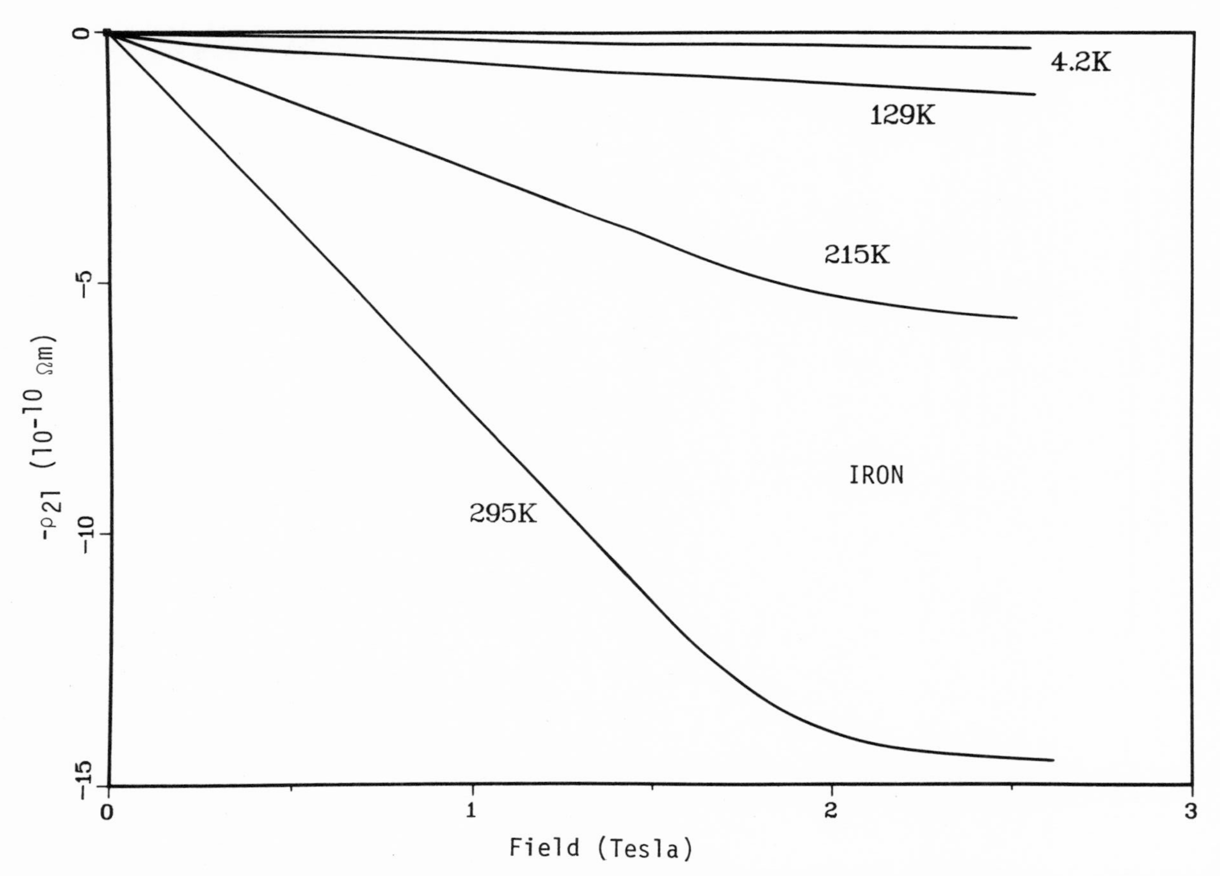

Fig. 6.10 Showing the isothermal field dependence of the Hall resistivity of bulk, polycrystalline Fe of low purity (RRR = 12). (After Volkenshteyn and Fedorov 1960.)

ACKNOWLEDGEMENTS

I have benefited over the years from helpful conversations and correspondence with many people, especially: J. Smit, A. Fert, C. Lewiner, R. A. Young, R. J. Douglas, and W. R. Datars. Many of the results and viewpoints given above come from collaboration with J. E. A. Alderson, S. P. McAlister and G. F. Turner.

REFERENCES

Alderson, J. E. A., Farrell, T., and Hurd, C. M., 1968, Phys. Rev., 174:729.
Alderson, J. E. A., Farrell, T., and Hurd, C. M., 1970, Phys. Rev. B, 1:3904.
Alderson, J. E. A., McAlister, S. P., and Hurd, C. M., 1977, Phys. Rev. B, 15:5484.
Alderson, J. E. A., and Hurd, C. M., 1973, Phys. Rev. B, 7:1226.
Alderson, J. E. A., Hurd, C. M., and McAlister, S. P., 1976, Can. J. Phys., 54:1866.
Ashcroft, N. W., 1969, Phys. Cond. Matter, 9:45.
Banik, N. C., and Overhauser, A. W., 1978, Phys. Rev. B, 18:1521.
Barnard, R. D., 1977, J. Phys. F: Metal Phys., 7:673.
Barnard, R. D., and Sumner, L., 1969, Philos. Mag., 20:399.
Berger, L., 1970, Phys. Rev. B, 2:4559.
Berger, L., 1973, Phys. Rev. B, 8:2351.
Berger, L., 1978, Phys. Rev. B, 17:1453.
Bergmann, G., 1975, Zeit. Phys., B21:347.
Blount, E. I., 1962, Phys. Rev., 126:1636.
Böning, K., 1970, Phys. Kondens. Mat., 11:177.
Böning, K., 1978, Crystal Lattice Defects, 7:215.
Böning, K., Pfänder, K., Rosner, P., and Schlütter, M., 1975, J. Phys. F: Metal Phys., 5:1176.
Böning, K., Pfänder, K., and Rosner, P., 1979, Zeit. Phys., B34:243.
Bradley, J., and Stringer, J., 1974, J. Phys. F: Metal Phys., 4:839.
Budd, H., 1962, Phys. Rev., 127:4.
Chambers, R. G., 1952, Proc. Phys. Soc. (Lond.), A65:458.
Chambers, R. G., 1956, Proc. Roy. Soc. (London), A238:344.
Chambers, R. G., 1960, Magnetoresistance in: "The Fermi Surface," W. A. Harrison and M. B. Webb, eds., John Wiley, New York, p. 100.
Chambers, R. G., 1969, Transport Properties: Surface and Size Effects, in: "The Physics of Metals," J. M. Ziman, ed., C.U.P., London, p. 175.
Chazalviel, J.-N., 1975, Phys. Rev. B, 11:3018.
Cohen, M. H., and Falicov, L. M., 1961, Phys. Rev. Lett., 7:231.
Coleman, R. V., 1975, AIP Conf. Proc., 29:520.
Coles, B. R., 1956, Phys. Rev., 101:1245.
Cooper, J. N., Cotti, P., and Rasmussen, F. B., 1965, Phys. Lett., 19:560.

Cracknell, A. P., 1971, "The Fermi Surfaces of Metals," Taylor and Francis, London.
Cross, R. C., 1978, Am. J. Phys., 46:771.
Douglas, R. J., and Datars, W. R., 1973, Can. J. Phys., 51:1770.
Douglas, R. J., and Datars, W. R., 1974, Can. J. Phys., 52:714.
Douglas, R. J., and Datars, W. R., 1975, Can. J. Phys., 53:1063.
Dugdale, J. S., and Firth, L. D., 1969, J. Phys. C: Solid State Phys., 2:1272.
Falicov, L. M., 1973, in: "Electrons in Crystalline Solids," Int. Atomic Energy Agency, Vienna, p. 207.
Falicov, L. M., Pippard, A. B., and Sievert, P. R., 1966, Phys. Rev., 151:498.
Falicov, L. M., and Sievert, P. R., 1965, Phys. Rev., 138:A88.
Fawcett, E., 1964, Adv. Phys., 13:139.
Fert, A., 1973, J. Phys. F: Metal Phys., 3:2126.
Fert, A., 1974, J. Phys. (Paris), 35:L107.
Fert, A., 1977, Physica, B86-88:491.
Fert, A., and Friederich, A., 1976, Phys. Rev. B, 13:397.
Fert, A., and Jaoul, O., 1972, Phys. Rev. Lett., 28:303.
Forsvoll, K., and Holwech, I., 1964, Philos. Mag., 10:921.
Garcia-Moliner, F., 1968, in: "Theory of Condensed Matter," Int. Atomic Energy Agency, Vienna, p. 229.
Garland, J. C., and Bowers, R., 1969, Phys. Rev., 188:1121.
Gerritsen, A. N., 1974, Phys. Rev. B, 10:5232.
Gerritsen, A. N., 1975, Phys. Rev. B, 12:4247.
Hall, E. H., 1879, Am. J. Math., 2:287.
Hurd, C. M., 1972, "Hall Effect in Metals and Alloys," Plenum, N.Y.
Hurd, C. M., 1975, "Electrons in Metals," John Wiley, New York.
Hurd, C. M., Alderson, J. E. A., and McAlister, S. P., 1976, Phys. Rev. B, 14:395.
Hurd, C. M., Alderson, J. E. A., and McAlister, S. P., 1977, Can. J. Phys., 55:620.
Hurd, C. M., and McAlister, S. P., 1977, J. Phys. F: Metal Phys., 7:969.
Jan, J.-P., 1957, Solid State Phys., 5:1.
Joachain, C. J., 1975, "Quantum Collision Theory," North-Holland, Amsterdam, fig. 18.3.
Katyal, O. P., and Gerritsen, A. N., 1969, Phys. Rev., 178:1037.
Kesternich, W., Ullmaier, H., and Schilling, W., 1976, J. Phys F: Metal Phys., 6:1867.
Kimura, H., and Honda, K., 1971, J. Phys. Soc. Jpn., 31:129.
Klaffky, R. W., and Coleman, R. V., 1974, Phys. Rev. B, 10:2915.
Lilly, D. A., and Gerritsen, A. N., 1974, Phys. Rev. B, 9:2497.
Luttinger, J. M., 1958, Phys. Rev., 112:739.
Lyo, S. K., and Holstein, T., 1974, Phys. Rev. B, 9:2412.
Mann, E., and Schmidt, H., 1975, Phys. Kond. Matter, 19:33.
Van der Mark, W., Ott, H. R., Sargent, D., and Rasmussen, F. B., 1969, Phys. Kond. Matter, 9:63.
McAlister, S. P., Alderson, J. E. A., and Hurd, C. M., 1977, Can. J. Phys., 55:1621.

McAlister, S. P., and Hurd, C. M., 1977, Phys. Rev. B, 15:561.
McAlister, S. P., and Hurd, C. M., 1979, J. Appl. Phys., 50:7526.
McAlister, S. P., Hurd, C. M., and Lupton, L. R., 1979, J. Phys. F: Metal Phys., 9:1849.
Meador, A. B., and Lawrence, W. E., 1977, Phys. Rev. B, 15:1850.
Miller, J. D., 1966, Physics Today, 29:39.
Nozières, P., and Lewiner, C., 1973, J. Phys. (Paris), 34:901.
Osamura, K., Hiraoka, Y., and Murakami, Y., 1973, Philos. Mag., 28:321.
Ozimek, E. J., Leisure, R. G., and Hsu, D. K., 1978, Phys. Lett., 66A:413.
Papastaikoudis, C., Rocoffylou, E., Tselfes, W., and Chountas, K., 1976, Zeit. Phys., B25:134.
Pfänder, K., Böning, K., and Brenig, W., 1979, Zeit. Phys., B32:287.
Pippard, A. B., 1965, "The Dynamics of Conduction Electrons," Blackie, London.
Pippard, A. B., 1966, Rept. Prog. Phys., 23:176.
Rhyne, J. J., 1968, Phys. Rev., 172:523.
Sarkissian, B. V. B., 1979, Philos. Mag., 39:413.
Sato, H., Babuchi, T., and Yonemitsu, K., 1978, Phys. Stat. Solidii, B89:571.
Schmidt, H., and Mann, E., 1979, Phys. Stat. Solidii, B94:95.
Shiozaki, I., 1975, J. Phys. F: Metal Phys., 5:451.
Smit, J., 1958, Physica, 24:39.
Smit, J., 1973, Phys. Rev. B, 8:2349.
Smit, J., 1978, Phys. Rev. B, 17:1450.
Smugeresky, J. E., Herman, H., and Pollack, S. R., 1969, Acta. Met., 17:883; erratum, 18:189.
Sorbello, R. S., 1974, J. Phys. F: Metal Phys., 4:1665.
Springford, M., 1971, Adv. Phys., 20:493.
Tsuji, M., 1958, J. Phys. Soc. Jpn., 13:979.
Visscher, P. B., and Falicov, L. M., 1972, Phys. Stat. Solidii, B54:9.
Volkenshteyn, N. V., Dyakina, V. P., Startsev, V. Ye., Azhazha, V. M., and Kovtun, K. P., 1974, Fiz. Met. Mettallov., 38:718 [Engl. trans.: Phys. Metals Metallogr., 38:37 (1974)].
Volkenshteyn, N. V., and Fedorov, G. V., 1960, Zh. Eksperim. i Teor. Fiz., 38:64 [Engl. trans.: Soviet Phys., JETP, 11:48 (1960)].
Volkenshteyn, N. V., Startsev, V. Ye., and Cherepanov, V. I., 1978, Phys. Stat. Solidii, b89:K53.
Yonemitsu, K., Takano, K., and Matsuda, T., 1978, Phys. Stat. Solidii, 88:273.
Young, R. A., 1968, Phys. Rev., 175:813.
Young, R. A., Ruvalds, J., and Falicov, L. M., 1969, Phys. Rev., 178:1043.
Ziman, J. M., 1961, Adv. Phys., 10:1.
Ziman, J. M., 1964, "Principles of the Theory of Solids," Cambridge University Press, London.

THE HALL EFFECT OF FERROMAGNETS[†]

L. Berger
Physics Department, Carnegie-Mellon University
Pittsburgh, Pennsylvania 15213

G. Bergmann
Institut für Festkörperforschung
Kernforschungsanlage, D-517 Jülich, West Germany

ABSTRACT

The history of the development of ideas concerning the Hall effect of magnetic metals is discussed. The Hall voltage depends on the magnetization rather than on the external field. Also, it arises from electron scattering rather than from the free motion of electrons. Two mechanisms are active for this "anomalous" Hall effect: skew scattering and side jump. While the first one is similar to Mott scattering, the second one represents a finite lateral displacement $\Delta y \simeq 10^{-10}$ m. of the electron on scattering. A corresponding displacement of the wavefronts of the scattered wave is discussed. Similar side jumps and forward jumps exist for electrons and photons in other parts of physics. The variation of the anomalous Hall effect with alloy composition in transition-metal series can be predicted on the basis of the simple "split-band" model. As an example, the case of amorphous Au-Fe, Au-Ni and Au-Co films is treated, and the predictions compared with experimental data.

FIRST MEASUREMENTS BY E. H. HALL ON FERROMAGNETS

In keeping with the nature of the present symposium, we are going to take the historical approach to the problem of the Hall effect of magnetic metals. Certain aspects will be emphasized at

† Work supported in part by the U.S. National Science Foundation, Grant DMR 78-24679.

the expense of others. As a result, important contributions by certain authors will sometimes be neglected, unfortunately.

Already in his second publication,[1] appeared in 1880, Hall describes measurements obtained at Johns Hopkins University on an iron foil, about 0.03 mm thick. He finds the Hall coefficient of iron to be about ten times larger than that of gold or silver (the two materials already studied), and of the opposite sign. Eighty years will elapse before the large magnitude of the Hall effect of ferromagnets finds an explanation.

In his third publication,[2] appeared in 1881, he studies nickel and cobalt. He states another important feature of the Hall effect of ferromagnets, namely the rapid increase with increasing temperature. He notes the similarity between this variation and that of the electrical resistance ρ itself.

In the same publication, he describes a third general characteristic of ferromagnets: the Hall voltage is not proportional to the external field B. Instead, it tends to become constant at field values comparable to those where the magnetization itself saturates. In Fig. 1a, we show Hall's experimental curve for the Hall resistivity $\rho_H = E_y/J_x$ of nickel.

In the present paper, the main current through the sample is assumed to be in the x direction.

ORDINARY AND ANOMALOUS EFFECT

Considerable work over many years was needed to confirm Hall's early findings. For example, Kundt[3] showed that the Hall resistivity is approximately proportional to the magnetization M.

In 1929, Smith and Sears, at Ohio State University, proposed[4] the relation

$$\rho_H = R_o \mu_o H + R_1 M \tag{1}$$

The "anomalous" Hall constant R_1 is usually at least one order of magnitude larger than the "ordinary" constant R_o. As a result, the term $R_o \mu_o H$ is difficult to measure. In order to solve this problem, Smith and Sears used Ni-Fe alloys of composition close to the one where R_1 vanishes and changes sign, so that R_o was actually comparable to R_1.

We can put Eq. (1) in an alternate form

$$\rho_H = R_o B + R_S M \tag{2}$$

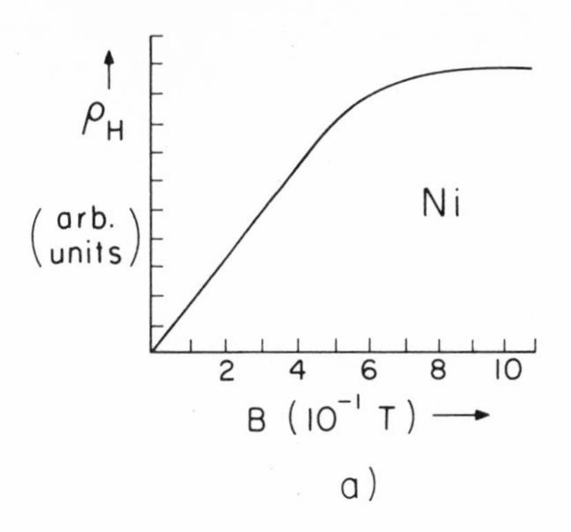

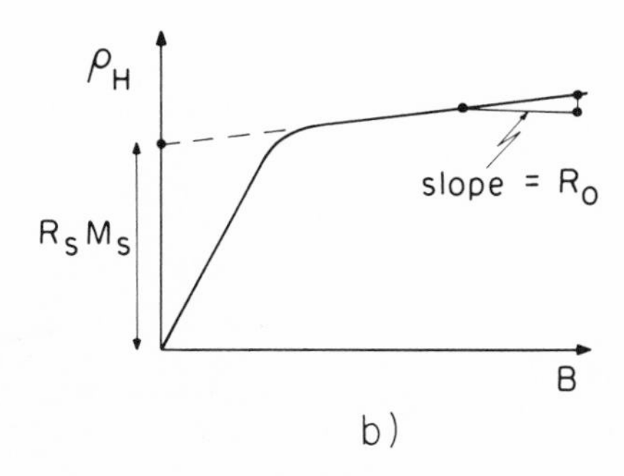

Fig. 1. a) Experimental curve for the Hall resistivity of nickel versus field B, as obtained by E. H. Hall in 1881. The curve shows a definite tendency towards saturation for $B > 0.6$ T.

b) Graphical determination of the two Hall coefficients R_o and R_s of a ferromagnet.

where $R_s = R_1 - R_o$. This form is preferable, since the external field outside a flat Hall sample is equal to B inside, and since the Lorentz force responsible for the ordinary coefficient R_o depends on B rather[5] than on H.

Between 1950 and 1965, E. M. Pugh and his collaborators[6] measured R_o in Ni-Cu, Ni-Fe, Ni-Co, Co-Fe, Ni-Pd, Fe-Cr, etc., at Carnegie Institute of Technology. This involves a determination of the (small) slope of ρ_H versus B above ferromagnetic saturation (Fig. 1b). As in non-magnetic metals, one can derive from R_o an effective number of electrons per atom $n^* = 1/R_o\, q\, N_A$, where q is the electron charge and N_A the number of atoms per unit volume. In pure nickel, n^* often exceeds 1 el./at. However, in Ni-Cu, Ni-Fe, Ni-Co, $n^* \simeq 0.3$ el./at. This low value is expected from the Mott picture of conduction, where 4s electrons with spin up (i.e.

majority spin) have the highest mobility, and play a dominant role. This conclusion has been confirmed by more recent work[7] by Campbell.

Important information on the Fermi surface of pure Ni, Fe, Co, can be derived[8] from ordinary Hall effect data in the high-field limit $\omega\tau >> 1$.

In the rest of this paper, we are going to leave aside the ordinary Hall effect, in order to concentrate on the truly ferromagnetic part of the effect.

RELATION BETWEEN ANOMALOUS EFFECT AND SCATTERING

When the first author was an undergraduate student at Lausanne University in the early nineteen fifties, Prof. Albert Perrier was head of the Physics Department. Many years before, in 1930, he analyzed data of anomalous Hall effect obtained by A. W. Smith[9] for Ni, Co, and Fe over a large range of temperature. He noted that ρ_H increases considerably with increasing temperature, in a way which is difficult to explain. Then he proposed[10] that the "Hall conductivity" $\gamma_H = J_y/E_x$ should be used to describe the anomalous effect, rather than $\rho_H \equiv E_y/J_x$. He showed that the thermal variation of γ_H is much smaller and simpler than that of ρ_H.

Correspondingly, Perrier[10] replaced the usual physical picture, where the longitudinal current density J_x creates a transverse Hall force or field E_y, by a different one: a longitudinal field E_x would directly create a special kind of transverse Hall current J_y, not associated with ohmic dissipation, which he called "spontaneous current" or "auto-current". In that respect, Perrier anticipated by thirty years the modern ideas described in the next two sections.

In the case $\rho_H << \rho$, Perrier derived the relation

$$\gamma_H \simeq \frac{\rho_H}{\rho^2} \simeq \frac{R_S M_S}{\rho^2} \tag{3}$$

where the small ordinary term of Eq. (2) is neglected. We assume here, as well as in the remainder of this paper, that saturation is is achieved ($M = M_S$).

J. P. Jan's doctoral thesis,[11] directed by Prof. Perrier and completed in 1952, added new data on temperature dependence for Ni and Fe. As Fig. 2 shows in the case of iron, γ_H varies much less than ρ_H. Also including later work,[12] the results can be summarized in the following manner, at $T > 100$ K:

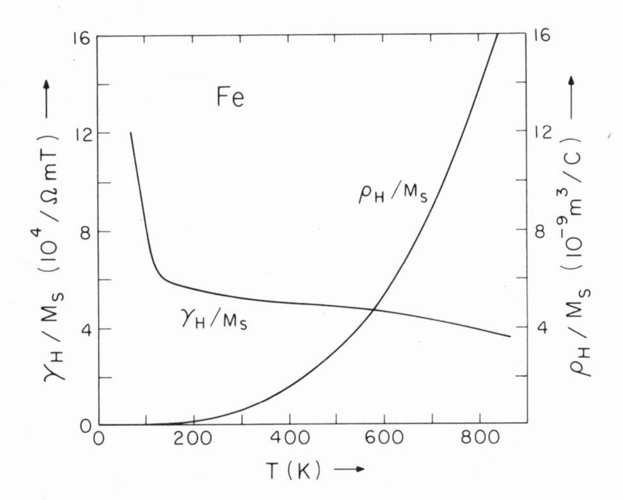

Fig. 2. Anomalous part of the Hall resistivity ρ_H, and of the Hall conductivity γ_H, of iron at saturation, according to data obtained by J. P. Jan in 1952. Both ρ_H and γ_H have been divided by the saturation magnetization M_s at the same temperature, in order to remove the effect of the thermal variation of M_s. Note that γ_H/M_s is almost independent of temperature above 100 K.

$$\rho_H \propto (\rho(T))^n$$

$$\text{Ni: } n \simeq 1.5 \qquad (4)$$

$$\text{Fe: } n \simeq 2.0$$

Note that a constant γ_H implies $n = 2$. In the case of gadolinium, later data[13] also indicate $n \simeq 2$.

What is even more significant is that a similar relation is obtained[14] if ρ is varied by dissolving various impurities in a pure metal, at constant temperature $\simeq$ 300 K. However, one should remember that the solutes may not only increase the rate of scattering but also change the band structure. In order for n to keep its simple meaning, the solute concentration should not exceed a few at.%.

The validity of Eq. (4) suggests strongly that the anomalous effect must be connected with electron scattering.

FORMAL THEORY OF THE ANOMALOUS EFFECT

From the experimental evidence presented above, we can guess the nature of the ingredients needed for a theory of the anomalous effect: since the effect depends on the magnetization, i.e. on electron spin, and since this spin must influence transport processes, the spin-orbit interaction should be involved. Also, the scattering potential created by impurities, phonons, or magnons will play a role, as explained in the last section.

Such a theory was developed by J. M. Luttinger[15] in 1958, assuming elastic scattering by impurities. If values $n > 1$ are to be obtained, the theory must go beyond a classical Boltzmann equation. In a quantum transport equation, the classical electron distribution f_k in k-space is replaced by a density matrix $f_{kk'}$. From $f_{kk'}$, the average electric current can be calculated. The final result is $\rho_H = a\rho + b\rho^2$, where a and b are constants, and where ρ is varied by changing the solute concentration. The first term is classical in nature (skew scattering), but the second one ($n = 2$) is not.

The dimensionless parameter which describes[16] the importance of quantum effects in transport is $\hbar/\tau\varepsilon_F$, where ε_F is the Fermi energy and τ the electron collision time. This can also be written as $\lambda_F/\pi\Lambda$, where λ_F is the Fermi wavelength and Λ the electron mean free path. While Λ was the only important characteristic length in classical transport, we conclude that a second length, roughly of atomic dimension, also plays a role in quantum transport. This is the "side jump" introduced in the next section.

Some of the many published theories[17] of the anomalous effect apply specifically to phonon scattering, or to magnon scattering. Unfortunately, there is little experimental evidence that the nature of the scatterers matters much. The formalism is usually frighteningly complicated.

INTUITIVE THEORY

Consider[18] an electron wavepacket approaching a central scattering potential (Fig. 3). Before scattering, the center of mass of the packet moves in a straight line through the periodic crystal. After scattering, the wavepacket is broken into a set of outgoing spherical waves. Since part of the wave is backscattered, the motion of the center of mass will be rather slow. Nevertheless, according to quantum mechanics, this motion again follows a straight line, at a constant speed (Fig. 3).

If the electron spin $\vec{S}$ is normal to the plane of the picture,

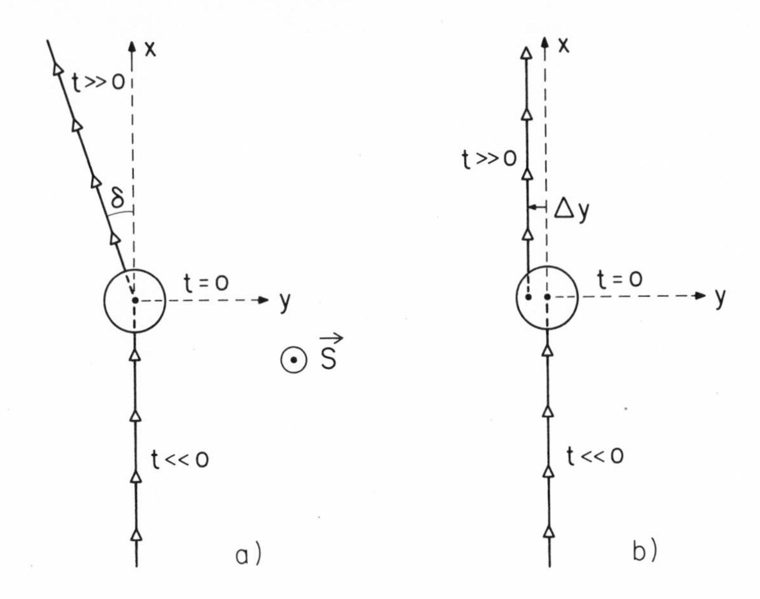

Fig. 3. a) The trajectory of a wavepacket is deflected by an angle δ after a collision (skew scattering).
b) Here, the trajectory after collision is translated sideways by a distance Δy, called side jump.

spin-orbit interaction removes any symmetry between right and left. As a result, the trajectory after scattering may well differ from the one before scattering. And past experience indicates that any effect not forbidden by symmetry usually exists.

For example, the new trajectory might be at an angle to the old one (Fig. 3a). This means that the electron acquires transverse momentum on scattering. The effect is called "skew scattering". It can be derived from a classical Boltzmann equation if the differential cross-section has a left-right asymmetry. However, it vanishes in the first Born approximation, i.e. for weak scatterers. In Eq. (4), it gives $n = 1$. Skew scattering was proposed[19] by J. Smit in 1958.

A second possibility[18,20,21] is that the new trajectory might be displaced by a finite distance from the old trajectory (Fig. 3b). This transverse displacement Δy is called a "side jump". If classical dynamics were used, Δy could not exceed the range of the scattering potential, of the order of an atomic distance $\simeq 10^{-10}$ m. in metals. In quantum mechanics, wave diffraction effects could possibly add a contribution to Δy of the order of a Fermi wavelength, also $\simeq 10^{-10}$ m. in metals. These simple considerations yield an upper bound for Δy.

Detailed calculations of the side jump have been performed[18,20.21] in the first Born approximation, for a static impurity potential. The side-jump Δy per collision (i.e. divided by the scattering probability) is found to be independent of the strength, range, or sign, of the scattering potential, and is predicted to be of order 10^{-11} - 10^{-10} m. This value is consistent with the upper limit quoted above, and is large enough to account for the experimental data of anomalous Hall effect in Fe or Ni around room temperature or above. Note that the side jump mechanism predicts n = 2 in Eq. (4), i.e. a constant γ_H:

$$\gamma_H = 2 \frac{Nq^2}{\hbar k_F} \Delta y \tag{5}$$

where N is the electron concentration per unit volume.

The transverse electron motion during the side jump can be described by a new kind of current localized inside the scattering centers, and running in the y direction. Thus is Perrier's 1930 idea of "auto-currents" vindicated.

The factor of two in Eq. (5) exists because, apart from the direct side-jump current, there is also a contribution of Δy to the transverse current through the potential energy of the electrons, as shown by Nozieres and Lewiner.[21]

The most important conclusion of this section may be that skew scattering alone, predicting n = 1, cannot explain the anomalous Hall effect of ferromagnets above 100 K. In iron and gadolinium $n \simeq 2$), the side-jump mechanism must be dominant. In nickel ($n \simeq 1.5$), both skew scattering and side-jump are probably present. The only materials where n=1 is observed are low-resistivity dilute alloys at 4 K, mentioned in the next section.

REMARKS ON SKEW SCATTERING

The original theory by Smit[19] considered free electrons and short-range potentials. To the scattering potential V(r) a spin-orbit interaction is added, of the form

$$H_{so} = -\lambda_{so} (\vec{k} \times \vec{S}) \cdot \vec{\nabla} V \tag{6}$$

For free electrons, the coupling parameter is $\lambda_{so} = \hbar^2/2m^2c^2$ as usual. However, the theory can be extended to 3d band electrons if the Fivaz effective spin-orbit hamiltonian is used. That interaction also has the form of Eq. (6), but with

$$\lambda_{so}(\varepsilon_F) = A_{so} \chi d^2 \sum_n \frac{|\text{matrix el.}|^2}{\varepsilon_n - \varepsilon_F} \tag{7}$$

where d is the distance between nearest-neighbor atoms, and where ε_n is the energy of a band state, $\chi \simeq 0.1$ is an overlap integral between nearest-neighbor atomic 3d states, and $A_{so} = 0.1$ eV is the atomic spin-orbit parameter for 3d electrons. This new λ_{so} is about 3×10^4 times as large as the free-electron value. Nozieres and Lewiner[21] derived a similar effective spin-orbit hamiltonian (their Eq.(21)), and their coupling constant contains the same energy differences in the denominator as our Eq.(7).

The Smit theory has the advantage of showing readily the analogy between skew scattering of free electrons and that of band electrons. In addition, it allows the study of the transition from weak scattering to strong scattering, including[23] the advent of resonances.

Since the Smit theory predicts skew scattering to change sign when the sign of the scattering potential is reversed, and since phonons can be represented by a potential of fluctuating sign, we conclude that skew scattering by phonons is probably very small.

In 1972, Fert and Jaoul[24] proposed a different skew-scattering theory, based on the Anderson model, where free electrons (or s electrons) are scattered into other free-electron states through an intermediate resonant 3d state bound to the scattering potential. This s-d-s scattering model is very appropriate for dilute transition-metal impurities dissolved in a noble metal. It may even apply to dilute impurities in nickel or iron, since conduction by high-mobility 4s electrons is probably dominant in skew scattering (though not in side jump). It cannot apply to weak scattering potentials, for which 3d resonances are absent.

In the case of localized 4f states in dilute rare-earth impurities dissolved in a noble metal, orbital angular momentum is not appreciably quenched. The scattering interaction between s electrons and 4f electrons can be written down in a systematic manner, including quadrupolar, exchange, and spin-orbit energy, and gives rise to skew scattering.[25] To this well-defined situation, accurate solutions can be found.

In semiconductors, the s-p wave functions of band electrons are known fairly accurately. Hence, skew scattering can be calculated[26] rather completely, using the second Born approximation. The scattering potential has a large range, here.

Data on skew scattering exist at 4 K for many different kinds of dilute impurities dissolved in nickel.[27] Data are much scarcer for impurities in iron.[28] Recent data concern paramagnetic rare-earth solutes in noble metals.[29] Skew scattering is characterized by a constant Hall angle, independent of impurity concentration, of order 10^{-2} radians.

Naively, one would expect skew scattering to play no role in the Hall effect of very pure metals in the high-field limit $\omega\tau >> 1$, since very few scatterers are present. This conclusion turns out to be incorrect for "compensated" metals (i.e., having equal numbers of holes and electrons) such as pure iron, or pure lead with magnetic residual impurities. There, skew scattering actually dominates[30,23] over the ordinary Hall effect at $\omega\tau >> 1$, so that the usual Lifshitz-Kaganov high-field theory fails. The Hall voltage is proportional to MB^2.

REMARKS ON THE SIDE JUMP

As mentioned earlier, the side jump has been calculated for a potential V(r) to which the spin orbit energy of Eq.(6) is added. Using the first Born approximation, one obtains [20,21] for the side jump per collision:

$$\Delta y = - \lambda_{so} S_z k_x \tag{8}$$

where $S_z = \pm \frac{1}{2}$ is the spin of the itinerant electron. For free electrons, a too small $\Delta y \simeq 10^{-15}$ m. is obtained. However, the effective spin-orbit interaction of Eq.(7) can be used in the case of 3d band electrons, giving $\Delta y \simeq 10^{-11}$-10^{-10} m., in agreement with experiments. Equation (8) predicts the side jump per collision to be independent of the strength, range, or sign of V(r). On the whole, this is in surprisingly good agreement with data at 300 K or above, which show[18] that γ_H is pretty much the same for phonon, impurity, or magnon scattering.

In the case of the ordinary Hall effect, as well as in the case of skew scattering, the contribution of each band is weighted according to the mobility of its electrons. As a result, high-mobility 4s electrons tend to give a dominant contribution. However, one can show[18] that there are no such weight factors in the case of the side-jump mechanism. Since scattering plays such an important role in that mechanism, low-mobility 3d electrons which undergo frequent collisions contribute as much or more as 4s electrons. The low degree of spin polarization of the 4s electrons further decreases their contribution. Thus, although a side jump could be calculated for the case of s-d-s scattering, we find very little motivation to do so. We prefer the simpler model with 3d itinerant electrons alone.

The coordinate of a band electron can be written as $\vec{r} = \vec{R} + \vec{q}$, where the operator $\vec{R} = (\hbar/i)\partial/\partial\vec{k}$ is the Wannier coordinate already present for a free electron, while $\vec{q}$ is a periodic operator arising from the existence of the periodic factor $u_{\vec{k}}(\vec{r})$ in the Bloch waves. The electron current becomes $d\vec{r}/dt = d\vec{R}/dt + d\vec{q}/dt$. While this decomposition of $\vec{r}$ is not invariant under a gauge transformation, the decomposition of $d\vec{r}/dt$ is free from that difficulty. Smit[31] was the first to show that $\langle d\vec{q}/dt \rangle = 0$ in the steady state. This result was confirmed by Lyo and Holstein.[32] This is a most important conclusion, as it implies that the d.c. side-jump current arises entire from $\langle d\vec{R}/dt \rangle$. Although $d\vec{q}/dt$ appears to contribute to the side-jump current in the Nozieres and Lewiner formulation,[21] their Table I shows clearly that $\langle d\vec{q}/dt \rangle$ ("polarization current") vanishes in the case of zero frequency.

FEATURES OF WAVE RESPONSIBLE FOR SIDE JUMP

The nature of the side jump can be clarified by a detailed analysis of the scattered wave. We assume a short-range central potential (range $R \ll \lambda_F$). Then the differential cross section is isotropic, of course. The effect of the spin-orbit interaction of Eq.(6) is to affect the phase of the wave.[18,21] The wavefronts are spherical, but centered on a point B displaced from the center A of the potential by a distance $\Delta y = -\lambda_{so} S_z k_x$ in the y direction (Fig. 4a). The most significant fact is that the flow lines of the probability current are straight lines diverging from point B. A wave packet is not needed to define this Δy, since the wavefronts already exist for a stationary state. Indeed, in that case the whole scattered wave is translated rigidly by that Δy. This is the side-jump value calculated by Lyo and Holstein[20] and by Nozieres and Lewiner,[21] using a stationary state to estimate the transverse current. This "asymmetric wavefront" value of Δy is the correct one to use in the side-jump theory, and is the one quoted by us in Eq.(8).

Superficially, another effect seems to exist: the "asymmetric time delay", which requires a wavepacket for its definition. The part of the scattered wavepacket moving towards left (Fig. 4b) emerges out of point B earlier than the part moving towards right. For parts moving from B in a direction at an angle θ to the y axis, the time of emergence is $t(\theta) = -k_F(d(\Delta y)/dk_F)\cos\theta/v_F = -\Delta y \cos\theta/v_F$, where v_F is the Fermi velocity, and Δy is given by Eq. (8), and where the energy variation of λ_{so} is neglected. As seen from point B, this effect causes an additional shift of the center of mass of the scattered packet by a distance $\Delta y/3$. Thus, the total side jump, including both wave-front and time-delay asymmetries would be $(4/3)\Delta y$. This is the value that any correct calculation with a wavepacket would give, immediately after scattering.

However, the part of the packet emerging earlier also collides earlier with other scattering centers, after travelling from B by a distance of one mean free path in any direction. When that happens, the wave packet is again centered on B (Fig. 4b). Thus, the effect of the asymmetric time delay is cancelled after that second collision. The correct Δy to be used in transport theory arises

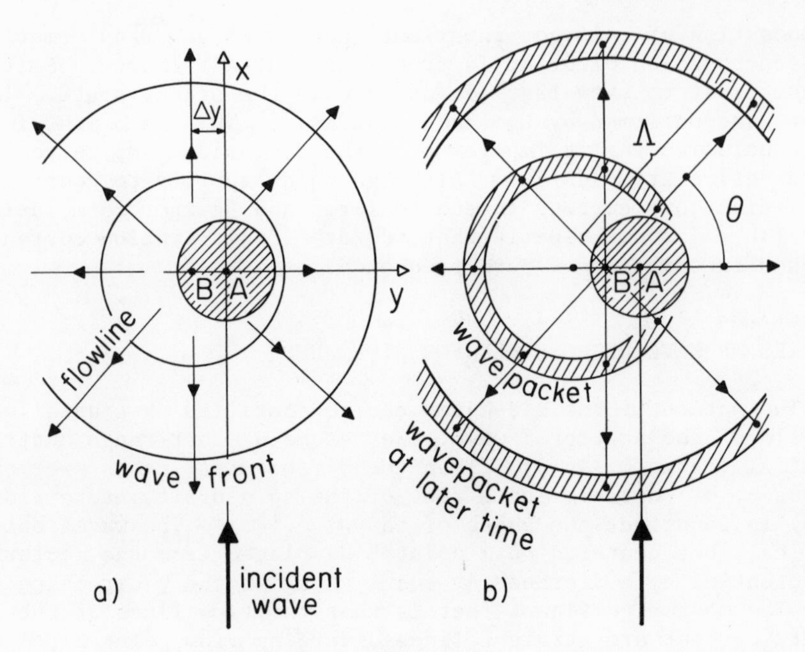

Fig. 4. a) The wavefronts of the scattered wave are centered on a point B, displaced from the center A of the scattering potential by a distance Δy in the transverse y direction.
b) The parts of a scattered wavepacket moving towards left emerge from B earlier than those moving towards right. This displaces laterally the center of mass, away from B. However, after the expanding wavepacket collides with other impurities, it is again centered on B, with uniform radius equal to a mean free path Λ.

from wavefront asymmetry alone, and is given by Eq.(8).

In the case of impurity scattering, all theories of the anomalous Hall effect have assumed the locations of impurities to be uncorrelated. This is not realistic, since other impurities are excluded from the volume occupied by a given impurity. One consequence of that exclusion is that the parts of the scattered wave emitted at B and moving towards right (Fig. 4) are protected from scattering again while they traverse the volume of the first impurity, centered at A. Effectively, their mean free path is increased by an amount $\Delta\Lambda = - \Delta y \cos\theta$, where Δy is given by Eq.(8).

This would cause an additional shift of the center of mass of the wavepacket at the time of the second scattering event, by $- \Delta y/3$, so that the total side jump would become $(2/3)\Delta y$ instead of Δy. It is not clear whether similar exclusion effects would exist for phonon or magnon scattering.

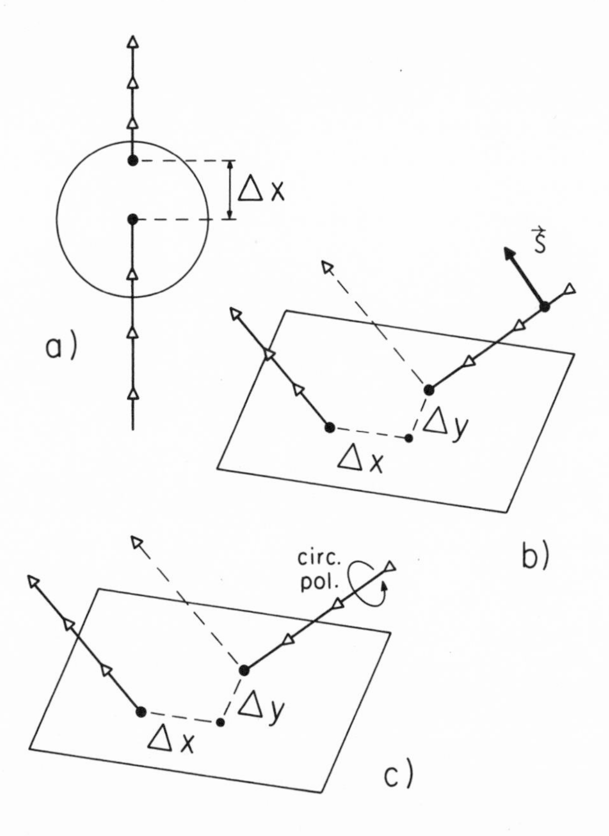

Fig. 5. a) Even in non-magnetic metals, the center of mass of a wavepacket undergoes a discontinuous forward jump Δx on scattering by an impurity potential.
b) A free electron reflected by a potential barrier undergoes a forward jump Δx and a side jump Δy, if the spin is in the incidence plane.
c) A beam of circularly polarized light undergoes a forward jump Δx and a side jump Δy, on total reflection.

SIDE JUMPS AND FORWARD JUMPS IN OTHER PARTS OF PHYSICS

Chazalviel and Solomon[33] have succeeded in detecting the anomalous Hall effect in semiconducting InSb, using an external field to polarize the spins in that Pauli paramagnet. Juretsches' spin resonance technique was used for that remarkable experiment. The side jump mechanism is dominant in that material.

Apart from the side jump, a forward jump also exists[18,34] in metals, even without spin-orbit interaction (Fig. 5a). After scattering by an impurity, the center of mass of a wavepacket differs from the extrapolated location before scattering, by a finite longitudinal displacement Δx. Thus a forward-jump current flows in parallel with the usual conduction current. This may explain[35] the "resistivity saturation" effects observed in high-resistivity alloys. The resistivity ρ is decreased below the usual value ρ_o according to the law

$$\frac{1}{\rho} = \frac{1}{\rho_o} + \frac{N q^2}{\hbar k_F} \Delta x \qquad (9)$$

where the last term is a constant independent of the electron mean free path. We can expect Δx to be of atomic dimension, in metals.

The existence of forward jumps and of side jumps has been proposed theoretically[36] for free electrons reflected by a potential barrier, if the electron spin $\vec{S}$ is parallel to the plane of incidence (Fig. 5b). The predicted displacements being only of the order of an electron Compton wavelength $\simeq 10^{-12}$ m., this effect has not been observed yet.

Even the photon enters this new dance of forward jumps and side jumps. A longitudinal displacement Δx of a ray of light totally reflected by a plane boundary was predicted by Picht[37] in 1929, and observed in 1947 by Goos and Hänchen[38] (Fig. 5c). If the light is linearly or, still better, circularly polarized, a displacement Δy transverse to the incidence plane is predicted (Fedorov, 1955).[39] This side jump was observed[40] by Imbert in 1968. Imbert's 1972 paper[40] contains many references to the literature, and many experimental details. Here, the displacements Δx and Δy are of the order of the wavelength of light.

VARIATION OF ANOMALOUS EFFECT WITH COMPOSITION OF ALLOYS

Although the side jump Δy is rather independent of the nature of the scatterers, it does depend on the band structure of the metal, and on the location of the Fermi level ε_F in these bands, through the effective spin-orbit parameter $\lambda_{so}(\varepsilon_F)$ appearing in Eqs.(8) and (7). Using these two equations, it is possible to explain the variation of the anomalous Hall effect as a function of composition in series of transition-metal alloys.

In the case of wide bands such as the s or p bands, a "rigid-band" model of alloys is reasonable. However, in the case of the 3d band, the opposite model is often better: this is the "split-band" model, valid[41-43] for narrow bands and for alloys made of components having sufficiently different nuclear charges. In that model, each component of the alloy has its own 3d band, distinct from others on the energy scale (Fig. 6). The bands of

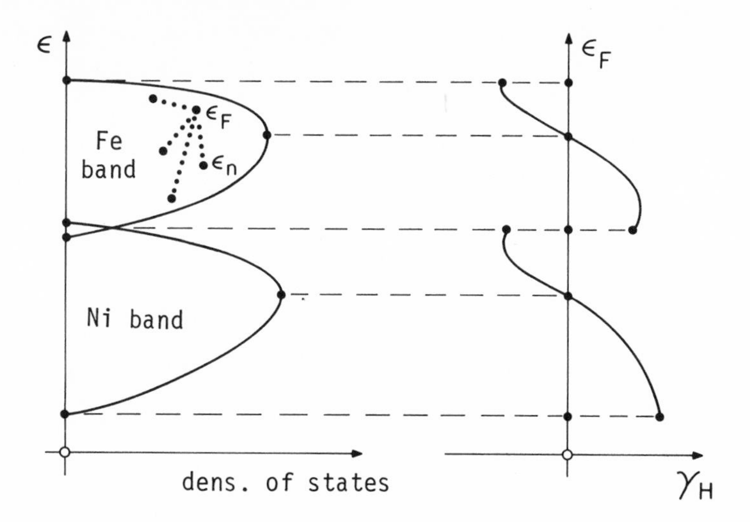

Fig. 6. At left: split-band model for Ni-Fe alloys, where each alloy component has its own 3d band. At right: the anomalous Hall conductivity γ_H is predicted to change sign when the Fermi level ε_F moves from the nickel band to the iron band, and also when ε_F crosses the middle of each band.

components having the nuclei most attractive to electrons are at the bottom. Localization of states on the energy scale does not imply complete localization in space on the corresponding type of atom.

There is good spectroscopic evidence for the approximate validity of this model, for both crystalline[44] and amorphous[45] alloys. The model is merely the extension to concentrated alloys of Friedel's "virtually-bound" state idea.

The sign of the anomalous Hall conductivity γ_H, and of the side jump Δy itself, reflects the sign of λ_{so} by Eqs. (5), (7) and (8). In turn, the sign of λ_{so} depends on the energy differences in the denominators of Eq.(7). When the Fermi level ε_F is in the upper half of a band, the states at ε_F are mostly mixed with states of lower energy ε_n in the band (Fig. 6). Thus $\varepsilon_n - \varepsilon_F$ is negative in most cases, and so is γ_H and Δy. On the other hand, when ε_F is in the lower half of a band, $\varepsilon_n - \varepsilon_F$ is positive, and so are γ_H and Δy (Fig. 6).

Thus, in a binary series such as Ni-Fe, γ_H is predicted to change sign up to three times (Fig. 6): once when ε_F crosses the boundary between the Ni band and the Fe band, and also when ε_F passes through the middle of each band. In Ni-Fe,[43] Ni-Fe-Cu,[42] Ni-Fe-V,[42] etc., and in amorphous Ni-Fe-B,[46] the sign change

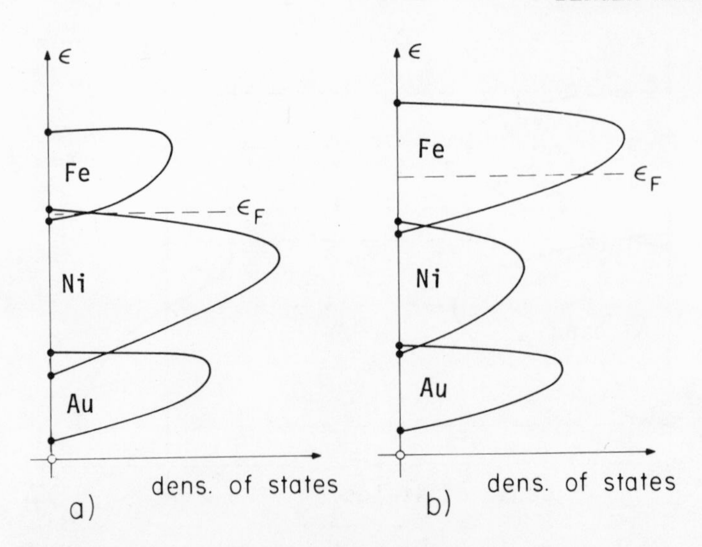

Fig. 7. a) 3d band of amorphous Au_{25} Fe_{25} Ni_{50}.
b) As we increase the iron content and decrease the nickel content, by going to Au_{25} Fe_{50} Ni_{25}, the number of states in the iron band increases while the nickel band shrinks.

between Ni and Fe bands has been observed experimentally. Its location agrees well with predictions. The sign changes at the middle of a band have not been observed satisfactorily yet.

The rules above, about the sign of γ_H, apply only[43] to alloys where the spin-up band is full, i.e. to those following the right-hand side of the Slater-Pauling curve. Otherwise, the contributions to γ_H from both spins would have to be combined.

The behavior of γ_H is very similar[43] to that of the orbital angular momentum of 3d states at the Fermi level, because the same energy denominators appear in both case (Eq.(7)).

APPLICATION TO AMORPHOUS Au-Fe, Au-Ni, AND Au-Co FILMS

As an example, we apply the split-band model to amorphous thin films. Our experience with crystalline materials indicates that the side-jump mechanism must be dominant in these materials, because of their very high resistivity. The very short mean free path must imply a considerable blurring and uncertainty in the value of k_F, as appearing in Eqs.(5) and (8). We are neglecting this

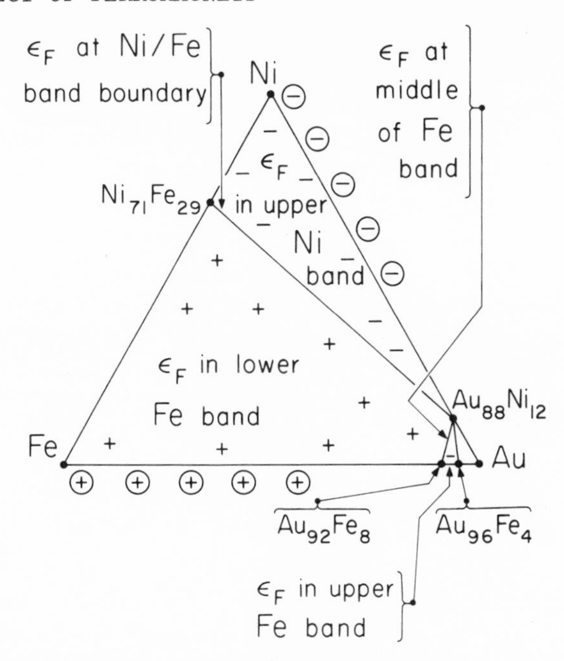

Fig. 8. Ternary phase diagram for amorphous Au-Fe-Ni. The predicted sign of the Hall conductivity is indicated by ordinary plus and minus signs. The measured sign is indicated by circled plus and minus signs in the case of binary Au-Ni and Au-Fe.

effect.

Consider the ternary Au-Fe-Ni amorphous system (Fig. 7a). We can think of the band of one constituent (e.g. iron) as a broadened version of the usual 3d atomic levels. Hence each iron atom present in the alloy brings 5 states of a given spin to the iron band. Thus the number of states in the iron band is 5 C_{Fe}, where C_{Fe} is the atomic concentration of iron.[42] We see (Fig. 7b) that each band grows or shrinks as the alloy composition varies, a picture very different from the one obtained with the rigid band model.

The total number of electrons, or rather number n_h of holes, available to fill these 3d bands is obtained[42] empirically from measurements of saturation magnetization, expressed as a number n_B of Bohr magnetons per atom. We have $n_h = n_B$ if the spin-up 3d band is full. In turn, this assumption is confirmed, at least for amorphous Au-Fe, by the fact that the slope of the magnetization data[47] versus composition has approximately the Slater-Pauling value, except close to pure iron. We summarize these data with the very rough formula $n_h = 2.88 - 3\,C_{Au} - 2C_{Ni}$, which has been generalized to Au-Fe-Ni. Here 2.88 is the number of 3d holes in pure amorphous iron, significantly larger than the value 2.55 for crystalline iron. The coefficients of other terms represent valence differences between iron and other components. Despite common misconceptions, these considerations do not require the validity of any specific 3d band model, "rigid" or otherwise.

We are now ready to calculate the alloy compositions for which ε_F lies at the boundary between Fe band and Ni band, i.e. compositions where the anomalous effect should change sign. This corresponds to the Fe band being just empty (Fig. 7a). In other words, the number of states in the Fe band is equal to the number of 3d holes in the alloy:

$$5\,C_{Fe} = 2.88 - 3\,C_{Au} - 2\,C_{Ni}$$

$$C_{Au} + C_{Ni} + C_{Fe} = 1 \qquad (10)$$

These equations correspond to a straight line in the ternary phase diagram (Fig. 8). The predicted sign of the Hall conductivity γ_H on each side of the line is indicated by plus and minus signs.

The absolute (rather than relative) sign of γ_H could be derived from theory, but it is simpler to obtain it, once for all, from data[43] on one metal, say crystalline nickel. This shows that $\gamma_H < 0$ when ε_F is in the upper half of a band.

Another sign change of γ_H should happen when ε_F is at the middle of the Fe band, and is given similarly by

$$1.5\,C_{Fe} = 2.88 - 3\,C_{Au} - 2\,C_{Ni}$$

$$C_{Au} + C_{Ni} + C_{Fe} = 1 \qquad (11)$$

These equations also give a straight line on Fig. 8. Note that the number of empty states in the Fe band is written as $1.5\,C_{Fe}$ instead of $2.5\,C_{Fe}$, in the left side. This reflects our belief that the sign change happens significantly above the actual middle of a band, due to a large density of states peak at the top of the 3d band of f.c.c. or amorphous alloys, as suggested on Fig. 6. This correction is not essential but is based on an unpublished analysis of data[46] on amorphous Ni-Fe-B: the fact that the "middle" of the Fe band is not observed in that system forces the ≃1.5 value to be correct, rather than 2.5.

The 3d band is completely full for alloys closer to pure gold than $Au_{96}Fe_4$, so that $\gamma_H = 0$ is predicted there (Fig. 8). This

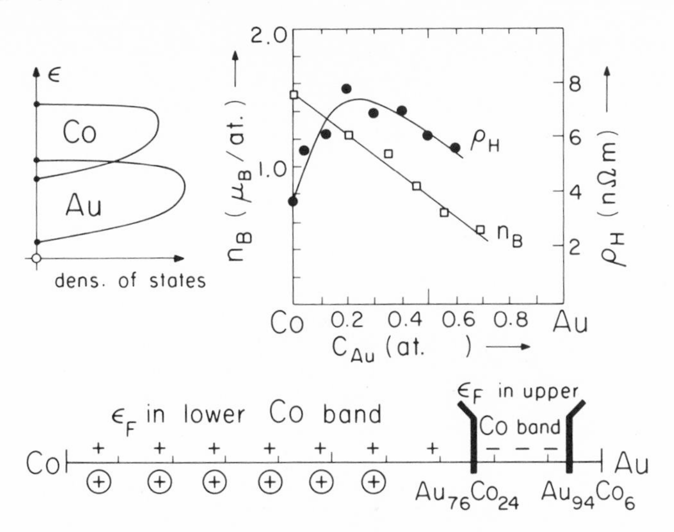

Fig. 9. a) 3d band of amorphous Au-Co.
b) Anomalous part of the Hall resistivity ρ_H of amorphous Au-Co at 4 K. plotted versus alloy composition. Also, saturation magnetization at 4 K, expressed as a number n_B of Bohr magnetons per alloy atom.
c) Phase diagram for amorphous Au-Co. The predicted sign of the Hall conductivity is indicated by ordinary plus and minus signs. The measured sign is indicated by circled plus signs.

line is obtained by putting zero in the left side of Eq.(10).

The second author has measured the anomalous Hall effect in amorphous Au-Ni and Au-Fe films.[48] The measured sign of γ_H is indicated by circled plus and minus signs in Fig. 8. Everywhere, it agrees with the predicted sign.

However, the present theory can only apply qualitatively to Au-Ni, since magnetization data show that their spin-up band is not quite full.

We end with a similar treatment of amorphous Au-Co films (Fig. 9a). The Fermi level is at the "middle" of the Co band when

$$1.5\ C_{Co} = 1.88 - 2\ C_{Au}$$

$$C_{Co} + C_{Au} = 1 \qquad (12)$$

As before, the right-hand side of the first Eq.(12) is a rough representation of magnetization data. Our data at 4.2 K are shown on Fig. 9b. As in the case of Au-Fe, the slope of the data is roughly consistent with the slope of the Slater-Pauling curve. These equations are similar to Eq.(11), and can be solved to give $C_{Co} = 0.24$, $C_{Au} = 0.76$ (Fig. 9c). The predicted sign of γ_H on each side of that point is indicated by plus and minus signs. Alloys closer than $Au_{94}Co_6$ to pure gold have a full 3d band, so that $\gamma_H = 0$ is predicted.

We have measured the anomalous part of the Hall resistivity of amorphous Au-Co films at 4 K (Fig. 9b). The measured sign is indicated by circled plus signs on Fig. 9c, and is in agreement with the predicted sign.

In most of the alloys considered in the present section, the nuclear charges (or rather the valences) of the various components differ by at least 2. This insures[43] that the 3d bands are reasonably split from each other. One glaring exception is Ni-Au where, however, the Fermi level never approaches the boundary between Ni and Au bands so that their imperfect degree of splitting is immaterial.

We acknowledge useful discussions with T. Holstein and R. C. O'Handley.

REFERENCES

1. E. H. Hall, Phil. Mag. 10:301 (1880).
2. E. H. Hall, Phil. Mag. 12:157 (1881).
3. A. Kundt, Wied. Ann. 49:257 (1893).
4. A. W. Smith and R. W. Sears, Phys. Rev. 34:1466 (1929).
5. J. Smit, Physica 17:612 (1951); 21:877 (1955).
6. A. I. Schindler and E. M. Pugh, Phys. Rev. 89:295 (1953); S. Foner and E. M. Pugh, Phys. Rev. 91:20 (1953); S. Foner, Phys. Rev. 99:1079 (1955); F. E. Allison and E. M. Pugh, Phys. Rev. 102:1281 (1956); S. Foner, Phys. Rev. 101:1648 (1956); S. Foner, F. E. Allison and E. M. Pugh, Phys. Rev. 109:1129 (1958); F. P. Beitel and E. M. Pugh, Phys. Rev. 112:1516 (1958); S. Soffer, J. A. Dreesen and E. M. Pugh, Phys. Rev. 140:A668 (1965); J. A. Dreesen and E. M. Pugh, Phys. Rev. 120:1218 (1960); E. R. Sanford, A. C. Ehrlich, and E. M. Pugh, Phys. Rev. 123:1947 (1961); A. C. Ehrlich, J. A. Dreeson and E. M. Pugh, Phys. Rev. 133:A407 (1964); G. C. Carter and E. M. Pugh, Phys. Rev. 152:498 (1966).
7. I. A. Campbell, Phys. Rev. Letters 24:269 (1970).
8. W. A. Reed and E. Fawcett, J. Appl. Phys. 35:754 (1964); R. V. Coleman, A.I.P. Conf. Proc. 29:520 (1975); plus references quoted there.

9. A. W. Smith, Phys. Rev. 30:1 (1910).
10. A. Perrier, Helv. Phys. Acta 3:317, 3:400 (1930).
11. J. P. Jan, Helv. Phys. Acta 22:581 (1949); 25:677 (1952). J. P. Jan and H. M. Gijsman, Physica 18:339 (1952).
12. C. Kooi, Phys. Rev. 95:843 (1954); A. I. Schindler and E. I. Salkovitz, Phys. Rev. 99:1251 (1955); J. M. Lavine, Phys. Rev. 123:1273 (1961).
13. N. V. Volkenshtein and G. V. Fedorov, Phys. Metals and Metallogr. 18:26 (1964).
14. W. Jellinghaus and M. P. DeAndres, Ann. Physik 7:189 (1961); W. Köster and W. Gmöhling, Z. Metallkunde 52:713 (1961); W. Köster and O. Römer, Z. Metallkunde 55:805 (1964).
15. J. M. Luttinger, Phys. Rev. 112:739 (1958).
16. R. E. Peierls, "Quantum Theory of Solids," Oxford U. Press, Oxford (1956), p. 141.
17. Y. Kagan and L. A. Maksimov, Sov. Phys. Solid State 7:422 (1965); L. E. Gurevich and I. N. Yassievich ibid. 5:1914 (1964); Yu. P. Irkhin and V. G. Postovalov, ibid. 8:346 (1966); Yu. P. Irkhin, A. N. Voloshinskii and Sh. Sh. Abelskii, Phys. Stat. Sol. 22:309 (1967); A. N. Voloshinskii and N. V. Rijanova, Fiz. Met. Metallov 34:21 (1972); 35:269 (1973); C. Lewiner, O. Betbeder-Matibet, and P. Nozieres, J. Phys. Chem. Solids 34:765 (1973); S. K. Lyo, Phys. Rev. B 8:1185 (1973); S. K. Lyo and T. Holstein, Phys. Rev. B 9:2412 (1974); S. K. Lyo, Phys. Rev. B 11:1260 (1975); S. K. Lyo, Phys. Rev. B 15:2791 (1977).
18. L. Berger, Phys. Rev. B 2:4559 (1970).
19. J. Smith, Physica 24:39 (1958).
20. S. K. Lyo and T. Holstein, Phys. Rev. Letters 29:423 (1972).
21. P. Nozieres and C. Lewiner, J. Phys. (Paris) 34:901 (1973).
22. R. C. Fivaz, Phys. Rev. 183:586 (1969).
23. A. K. Majumdar and L. Berger, Phys. Rev. B 7:4203 (1973).
24. A. Fert and O. Jaoul, Phys. Rev. Letters 28:303 (1972).
25. A. Fert, Physica 86-88B:491 (1977); A. Fert and P. M. Levy, Phys. Rev. B 16:5052 (1977).
26. P. Leroux-Hugon and A. Ghazali, J. Phys. C5:1072 (1972).
27. R. Huguenin and D. Rivier, Helv. Phys. Acta 38:900 (1966); J. W. F. Dorleijn and A. R. Miedema, Physica 86-88B:537 (1977); O. Jaoul, Doctoral Thesis, Univ. de Paris-Sud (1974).
28. A. K. Majumdar and L. Berger, Phys. Rev. B 7:4203 (1973).
29. A. Friederich and A. Fert, Phys. Rev. B 13:397 (1976); A. Fert and A. Friederich, A.I.P. Conf. Proc. 24:466 (1975).
30. L. Berger, Phys. Rev. 177:790 (1969); R. V. Coleman, A.I.P. Conf. Proc. 29:520 (1975).
31. J. Smit, Physica 21:877 (1955).
32. S. K. Lyo, Phys. Rev. B 8:1185 (1973).
33. J. N. Chazalviel, Phys. Rev. B 11:3918 (1975).
34. C. M. Hurd, Contemp. Phys. 16:517 (1975).
35. B. Chakraborty and P. B. Allen, Phys. Rev. Letters 42:736 (1979).
36. S. C. Miller and N. Ashby, Phys. Rev. Letters 29:740 (1972).

37. J. Picht, Ann. Physik 3:433 (1929); Physik Z. 30:905 (1929).
38. F. Goos and H. Hänchen, Ann. Physik 1:333 (1947); 5:521 (1949).
39. F. I. Federov, Dokl. Akad. Nauk SSSR 105:465 (1955).
40. C. Imbert, Phys. Rev. D 5:787 (1972); C. R. Acad. Sci. (Paris) B267:1401 (1968).
41. B. Velicky, S. Kirkpatrick and H. Ehrenreich, Phys. Rev. 175:747 (1968); S. Kirkpatrick, B. Velicky, N. D. Lang and H. Ehrenreich, J. Appl. Phys. 40:1283 (1969).
42. H. Ashworth, D. Sengupta, G. Schnakenberg, L. Shapiro, and L. Berger, Phys. Rev. 185:792 (1969).
43. L. Berger, Physica 91B:31 (1977).
44. H. D. Drew and R. E. Doezema, Phys. Rev. Letters 28:1581 (1972); K. Y. Yu, C. R. Helms, W. E. Spicer, and P. W. Chye, Phys. Rev. B 15:1629 (1977).
45. P. Oelhafen, E. Hauser, H. J. Güntherodt, and K. H. Bennemann, Phys. Rev. Letters 43:1134 (1979).
46. R. C. O'Handley and L. Berger, Inst. Phys. Conf. Ser. (London) 39:477 (1978); R. C. O'Handley, Phys. Rev. B 18:2577 (1978).
47. W. Felsch, Z. f. Angew. Physik 29:217 (1970).
48. G. Bergmann, Z. Physik B25:255 (1976); Solid State Comm. 18:897 (1976); G. Bergmann and P. Marquardt, Phys. Rev. B 18:326 (1978).

HALL EFFECT FROM SKEW SCATTERING BY MAGNETIC IMPURITIES

A. Fert
Laboratoire de Physique des Solides
Université Paris-Sud
91405 Orsay, France

A. Hamzić
Institute of Physics of the University
41001 Zagreb, Yugoslavia

The concept of skew scattering was introduced in 1958 by Smit[1] to explain the extraordinary Hall effect of the ferromagnetic metals. Studies of the skew scattering in dilute magnetic alloys were developped during the last ten years and have been very useful to understand its origins and mechanisms.

1. GENERAL

a. Skew Scattering

A skew scattering (Figure 1) is characterized by a difference between the probabilities of scattering to the right and to the left (right and left correspond to the two directions on the axis y perpendicular to the incident direction x and to the direction z of the moment of the scatterer). It can be also characterized by a difference of the probabilities of scattering between $\underline{k}$ and $\underline{k}'$ and between $\underline{k}'$ and $\underline{k}$. Skew scattering arises when the scattering potential includes an antisymmetric part H^A such that

$$\langle \underline{k}' | H^A | \underline{k} \rangle = - \langle \underline{k} | H^A | \underline{k}' \rangle = \text{imaginary}.$$

Let us assume a scattering potential

$$H = \sum_{\underline{kk}'} a^+_{\underline{k}'} a_k (V_{\underline{k}\underline{k}'} + i\, W_{k\, k'}). \tag{1}$$

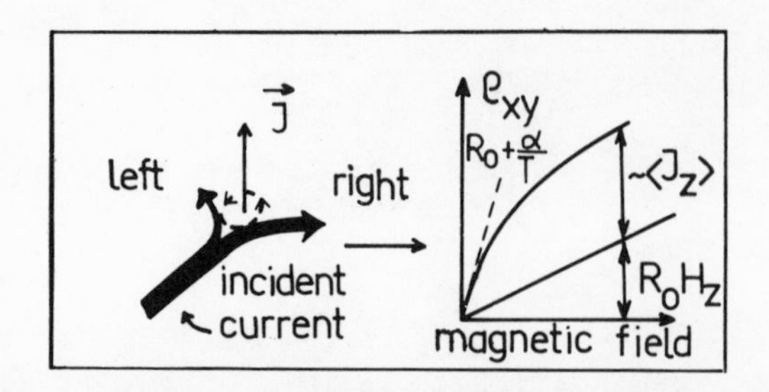

Fig. 1. Skew scattering by a local moment and resulting Hall resistivity in a magnetic alloy (standard case with $\rho^E_{xy} \sim \langle J_z\rangle$ and $\langle J_z\rangle \sim H/T$ for $\mu_B H \ll kT$)

Here $V_{\underline{kk}'}$ is real and symmetric, $W_{\underline{kk}'}$ is real and antisymmetric ($W_{kk'} = - W_{k'k}$).

In the Born approximation, the T matrix elements are:

$$T^{(1)}_{\underline{k}\,\underline{k}'} = V_{\underline{k}\,\underline{k}'} + i\,W_{k\,k'} \tag{2}$$

and the scattering probability is:

$$w^{(1)}\ (k\to k') = \frac{2\pi}{\hbar}\left|T^{(1)}_{\underline{k}\,\underline{k}'}\right|^2 \delta(\varepsilon_k-\varepsilon_{k'}) = \frac{2\pi}{h}(V^2_{\underline{kk}'}+W^2_{\underline{kk}'}) \times \delta(\varepsilon_k-\varepsilon_{k'}), \tag{3}$$

which is symmetric (non-skew).

Skew scattering is predicted only if the T matrix is calculated at higher order of the Born expansion. For example, let us consider the term in $T^{(2)}_{\underline{kk}'}$ of the form:

$$\sum_{\underline{k}''} \frac{V_{\underline{kk}''}V_{\underline{k}''\underline{k}'}}{\varepsilon_k-\varepsilon_{k''}+is} = -\,i\pi \sum_{\underline{k}''} V_{\underline{kk}''}V_{\underline{k}''\underline{k}'}\delta(\varepsilon_k-\varepsilon_{k''}) + \text{real part} \tag{4}$$

The imaginary and symmetric term of Eq.(4) will interfere in $|T^{(1)}+T^{(2)}|^2$ with the imaginary and asymmetric term $iW_{\underline{kk}'}$ of $T^{(1)}$ to give in the scattering probability a skew term of the form

$$-\frac{2\pi}{\hbar^2}\sum_{\underline{k}''} V_{\underline{kk}''}V_{\underline{k}''\underline{k}'}W_{\underline{kk}'}\ \delta(\varepsilon_{\underline{k}}-\varepsilon_{\underline{k}''})\,\delta(\varepsilon_{\underline{k}}-\varepsilon_{\underline{k}'}) \tag{5}$$

Thus there is no skew scattering in the Born approximation and skew terms of lowest order in the scattering probability are of third order with respect to the scattering potential; generally, the most important terms will be of first order with respect to the antisymmetric potential W and second order with respect to the symmetric potential V.

b. Origin of the Skew Scattering

We begin by considering the skew scattering by rare-earth ions. Their moment is associated with well-localized 4f electrons which are characterized by quantum numbers S (spin), L (orbital momentum) and J (total momentum). It is well known that, when L is different from zero, the Coulomb interaction between the conduction and 4f electrons includes both spin and orbital exchange terms.[2,3] The simplest of the orbital exchange terms is written as:

$$H_{orb.exch} \sim (2-g_J)\underline{\ell}\cdot\underline{J} \qquad (6)$$

where g_J is the gyromagnetic factor of the 4f electrons, $\underline{J}$ is their total angular momentum and $\underline{\ell}$ is the orbital angular momentum of the conduction electron. The other orbital exchange terms involve either $\underline{\ell}$ at higher odd order or both $\underline{\ell}$ and the conduction electron spin $\underline{s}$. As $\langle\underline{k}'|\underline{\ell}|\underline{k}\rangle = -\langle\underline{k}|\underline{\ell}|\underline{k}'\rangle$, the orbital exchange interaction has the antisymmetric character needed for skew scattering (see preceding paragraph). The contribution from orbital exchange to skew scattering has been clearly identified for noble metals containing rare-earth impurities.[4] However skew scattering has been also observed for noble metals containing gadolinium impurities ($L = 0$ or $g_J = 2$) for which orbital exchange does not exist. The skew scattering by gadolinium can be ascribed to a second mechanism, more precisely to combined spin scattering (due to the spin exchange $\underline{s}\cdot\underline{J}$) and spin-orbit scattering (due to intrinsic spin-orbit coupling $\underline{\ell}\cdot\underline{s}$ associated with the direct impurity potential).[4,5] Alternatively the combination of spin-orbit scattering can be also described as giving an effective $\underline{\ell}\cdot\underline{J}$ interaction similar to orbital exchange.[6] This double origin, orbital exchange and spin-orbit coupling associated with local or lattice potentials, seems to be very general in rare-earth systems (see section 2c). In some anomalous rare-earth alloys, the skew scattering has been ascribed to the Coqblin-Schrieffer interaction which actually is a form of exchange interaction including orbital terms.[8,9] Other origins of the skew scattering by rare-earths have been also proposed[7] but have never been identified in experimental results.

The skew scattering by magnetic impurities of transition metals appears—in most of the investigated systems—to be associated with the existence of local orbital moments and therefore related to some sort of orbital exchange between conduction and localized electrons (see section 3). Nevertheless, a contribution from the spin-orbit interaction in the conduction band should probably exist in some systems (metals with a conduction band of marked d character).

c. Contribution to the Hall Effect

The contribution from skew scattering to the Hall resistivity can be derived by introducing the skew part of the scattering probability in the transport equations. A general expression of the Hall resistivity can be obtained by using the Ziman variational method.[9] Let us first consider the simplest case with conduction electrons in plane wave states, with only non-spin-flip elastic skew scattering and with equal spin up and spin down currents. The Hall resistivity ρ_{xy} (i.e. $\rho_{xy} = E_y/i_x$) is expressed as

$$\rho_{xy} = R_o H + \frac{1}{4}(\rho_{xy}^{(+)} + \rho_{xy}^{(-)}) , \qquad (7)$$

where

$$\rho_{xy}^{(\sigma)} = - \left(\frac{\hbar}{8\pi^3 ne}\right)^2 \int \left(- \frac{\partial f^o}{\partial \varepsilon_k}\right) (\underline{k}.\underline{u}) (\underline{k}'.\underline{v}) w^a (\underline{k}\sigma \to \underline{k}'\sigma) d^3\underline{k}\ d^3\underline{k}' . \qquad (8)$$

$R_o H$ and $\rho_{xy}^{(\sigma)}$ are the contributions from ordinary Hall effect and skew scattering respectively. In Eq.(8), n is the number of conduction electrons per unit volume and per spin direction, $f^o(\varepsilon_{\underline{k}})$ is the Fermi-Dirac distribution function, $w^a(\underline{k}\sigma \to \underline{k}'\sigma)$ is the skew part of the scattering probability between $|\underline{k},\sigma\rangle$ and $|k',\sigma\rangle$ states, $\underline{u}$ and $\underline{v}$ are unit vectors along the x and y axes. The meaning of the factors $(\underline{k}.\underline{u})$ and $(\underline{k}'.\underline{v})$ is clear: $(\underline{k}.\underline{u})$ appears because an incident current along the x axis is associated with an extra-population on the Fermi surface which is proportional to $(\underline{k}.\underline{u})$; the factor $(\underline{k}'.\underline{v})$ appears because the Hall resistivity is related to the momentum which is transferred to the y axis. The Hall resistivity $\rho_{xy}^{(\sigma)}$ is proportional to the concentration of the magnetic impurities because w^a is proportional to it. The field dependence of $\rho_{xy}^{(\sigma)}$ arises from the dependence of w^a on the impurity magnetization. In many cases $\rho_{xy}^{(\sigma)}$ is simply proportional to this magnetization and the skew scattering contribution to the initial Hall coefficient is simply proportional to the initial magnetic susceptibility.

When, in an applied field, the spin up and the spin down electrons carry different currents (this may arise from spin scattering by localized moments) additional effects come in. The Hall resistivity is written[9]

$$\rho_{xy} = \left|1 + \left(\frac{i^+ - i^-}{i^+ + i^-}\right)^2\right| \times \left|R_o H + \frac{1}{4}(\rho_{xy}^{(+)} + \rho_{xy}^{(-)})\right| + \frac{i^+ - i^-}{i^+ + i^-} \times \frac{\rho_{xy}^{(+)} - \rho_{xy}^{(-)}}{2} \tag{9}$$

where i^+ and i^- are the currents for the two spin directions. Thus, the magnetization of local moments gives rise at the same time to skew scattering and to an enhancement of the ordinary Hall effect (this enhancement has been called the "spin effect"). The skew scattering is generally the predominant effect. Moreover, $(i^+ - i^-)$ tends to zero in the low field limit, so that the initial Hall coefficient $R = (\rho_{xy}/H)_{H \to 0}$ cannot be affected by the "spin effect" but only by the skew scattering.

Finally, if the skew scattering is associated with inelastic spin-flip scattering, the expression of the Hall resistivity becomes very involved (see Appendix of ref. 9).

Other ways of calculating the Hall resistivity have been described. When the antisymmetric part of the scattering probability has a very simple form the Hall resistivity can be derived by solving the Boltzmann equation;[2] the result is consistent with Eq. (8). Giovannini has also described a calculation of the Hall resistivity by the correlation function method.[10]

d. Selection Rules

Several rules simplify the selection of the scattering probability terms contributing to the Hall effect.

First, as this has been shown in 1a, the scattering probability cannot be skew if it is calculated in the Born approximation. Skew terms of the scattering probability are at least of third order with respect to the scattering potential and then they involve its antisymmetric part in first or third order.

Another selection rule applies when a partial wave expansion is used and the T matrix elements expressed as

$$T_{\underline{kk}'} \sim \sum_{\ell,m} e^{i\eta_{\ell m}} \sin(\eta_{\ell m}) Y^*_{\ell m}(\hat{\underline{k}}) Y_{\ell m}(\hat{\underline{k}}') \tag{10}$$

The resulting scattering probability

$$w(\underline{k}\rightarrow\underline{k}) \sim |T_{\underline{kk}'}|^2, \tag{11}$$

is composed of terms proportional to $Y^*_{\ell'm'}(\hat{k})Y_{\ell m}(\hat{k})Y^*_{\ell m}(\hat{k}')Y_{\ell'm'}(k')$ which, from Eq.(8), yield a contribution to the Hall resistivity proportional to

$$\int(\underline{k}.\underline{u})Y^*_{\ell'm'}(\hat{k})Y_{\ell m}(\hat{k})d\hat{k} \times \int(\underline{k}'\underline{v})Y^*_{\ell m}(\hat{k}')Y_{\ell'm'}(\hat{k}')d\hat{k}'. \tag{12}$$

These integrals cancel out unless $(\ell+\ell')$ is odd. In particular, the Hall resistivity cancels out if the scattering is restricted to one ℓ channel.

e. Skew Scattering and Side-jump in Dilute Magnetic Alloys

Orbital coupling of conduction electrons with magnetic impurities (such as Eq.(6)) can give rise, in principle, not only to skew scattering but also to scattering with side-jump (see ref. 11 and 12 and, in particular, Appendix B of the second reference). However, as the contributions from skew scattering and side-jump to the Hall resistivity are proportional to ρ and ρ^2 respectively, the skew scattering is largely predominant in dilute alloys at low temperatures (ρ small). The side-jump can be generally neglected in metals containing magnetic impurities expected for very high impurity concentrations.

f. Skew Scattering by Non-magnetic Impurities

All that we say in this paper implicitly concerns the skew scattering by magnetic impurities, that is skew scattering of non-polarized particles by polarized targets. We will not describe results on systems such as Cu:MnAu which correspond to skew scattering by Au impurities of spin up and spin down currents made unequal by spin scattering on Mn, that is to skew scattering of polarized particles by non-polarized targets.[13]

2. SKEW SCATTERING BY RARE-EARTH MAGNETIC IMPURITIES

a. Experimental Results

Skew scattering by rare-earth (RE) impurities in non-magnetic metals has been mainly studied for RE in silver and gold.[4] There are also some data for RE impurities in aluminum[4] and magnesium.[14]

The initial Hall coefficient R of several silver-based alloys

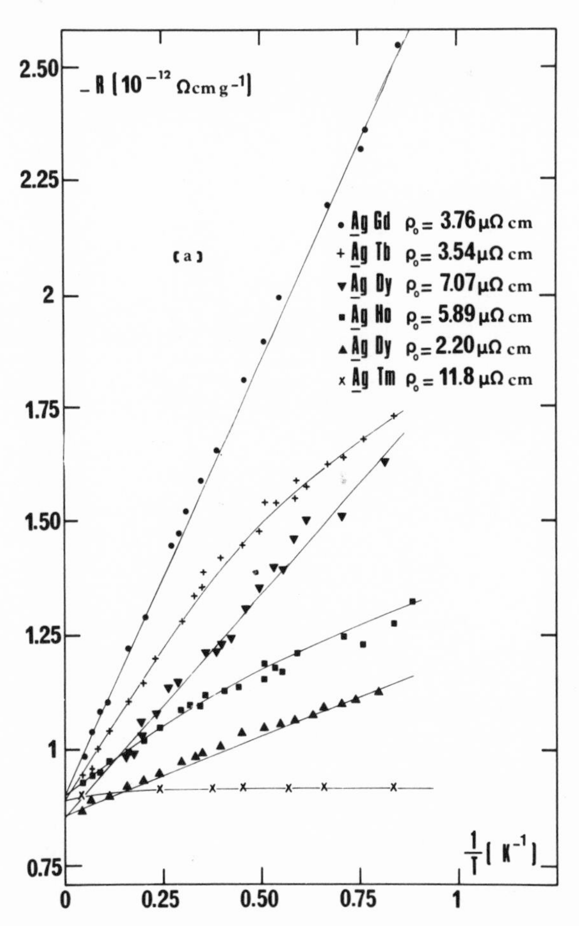

Fig. 2. Initial Hall coefficient R versus T^{-1} of Ag:Gd 0.71 at%, Ag:Tb 0.56 at%, Ag:Dy 1.06 at%, Ag:Ho 0.89 at%, Ag:Dy 0.33 at% and Ag:Tm 2 at% alloys (after Fert and Friederich[4]). R is the limit of ρ_{xy}/H when H tends to zero.

is shown in Fig. 2 as a function of T^{-1}. R is the sum of a constant R_o (the ordinary Hall coefficient which is independent of T in alloys at low temperatures[15]) and a skew scattering contribution proportional to the magnetic susceptibility of the RE impurities. This later contribution is proportional to T^{-1} at high enough temperatures when the susceptibility is close to the free ion susceptibility. At low temperatures, deviations from T^{-1} due to the crystal field are observed for non-S ions. Alloys containing various concentrations of the same RE element have been studied and the skew scattering turns out to be proportional to the residual resistivity ρ_o, i.e. to the concentration. In particular, at high

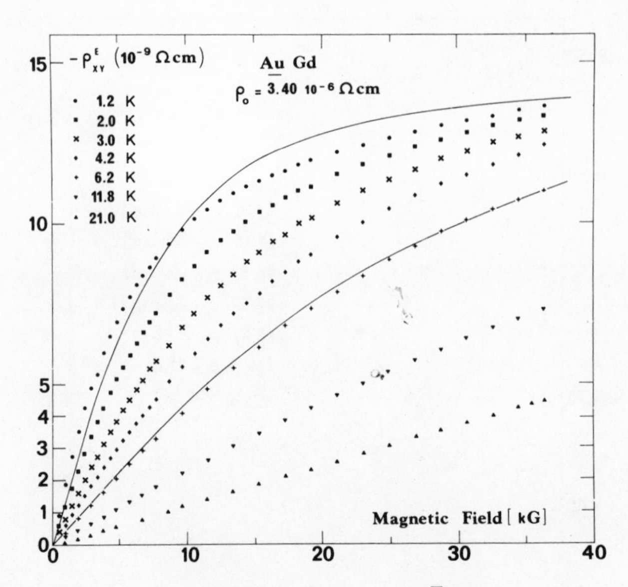

Fig. 3. Extraordinary Hall resistivity ρ^E_{xy} of a Au:Gd 0.5 at% alloy versus magnetic field at several temperatures. The ordinary Hall effect has been subtracted. The solid lines correspond to 14.3 $B_{7/2}$ (H/T) x 10^{-9} Ωcm for T = 2 K and T = 6.2 K respectively (after Fert and Friederich[4]).

enough temperatures R can be fitted to an expression of the form

$$R = R_O + \underline{a}\,\frac{\rho_o}{T}. \tag{13}$$

The coefficient a has been chosen to characterize the skew scattering by a RE element.

The Hall resistivity induced by skew scattering in a Au:Gd alloy is shown on Figure 3. Its field and temperature dependences are similar to those of the magnetization (compare with the Brillouin functions $B_{7/2}$(H/T)).

Figure 4 shows the variation of the skew scattering coefficient through the RE series in silver-based alloys. This variation can be accounted for by the competition between two mechanisms which are related to the k-f orbital exchange and to the spin-orbit coupling of the 5d electrons respectively, as explained in the next paragraph. The corresponding contributions to a are proportional to (2-g)gJ(J+1) and (g-1)gJ(J+1):

$$\underline{a} = \alpha_1(2-g)gJ(J+1) + \alpha_2(g-1)gJ(J+1), \tag{14}$$

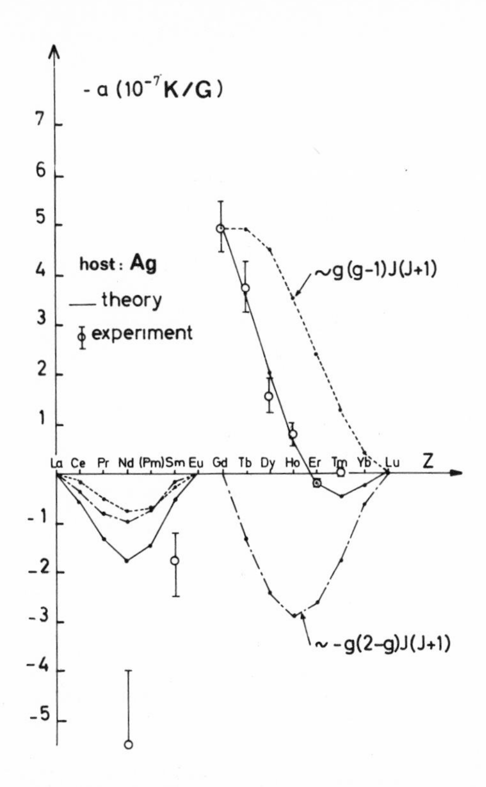

Fig. 4. Variation of the skew scattering coefficient a throughout the RE series in silver-based alloys.[4] The solid, dotted-dashed and dashed lines represent $\underline{a} = \alpha_1(2-g)gJ(J+1) + \alpha_2(g-1)gJ(J+1)$ (Eq.(14)) $\alpha_1(2-g)gJ(J+1)$ and $\alpha_2(g-2)gJ(J+1)$ are calculated with the values of α_1 and α_2 given in the text.

where J is the total angular momentum, and g the gyromagnetic factor of the 4f electrons; the coefficients α_1 and α_2 are expected to be approximately constant, at least through the heavy RE series. A good fit with the data of Figure 4 is obtained for

$$\alpha_1 = 0.43 \times 10^{-8} \text{ K/G}, \qquad \alpha_2 = - 1.57 \times 10^{-8} \text{ K/G}$$

With these values, Eq.(14) accounts for the rapid decrease of a at the right of Gd, for the change of sign between Ho and Er and also for the sign observed for light RE.

The experimental results in gold and aluminum-based alloys[4] are very similar to those in silver alloys and the variation of a can be interpreted in the same way (Eq.(14)). In magnesium alloys the interpretation of the experimental results is made difficult by the temperature dependence of the ordinary Hall coefficient.[14]

b. Theory of the Skew Scattering by RE Impurities in Noble Metals

The first attempt[4] to interpret the scattering by RE impurities in noble metals was based on an impurity scattering potential including:

i) An attractive potential characterized by the resulting phase shifts η_ℓ. The predominant phase shifts are η_o and η_2 which are associated with the attraction of electrons in 6s and 5d states (formation of a 5d virtual bound state by resonance with 5d levels).

ii) A k-f isotropic exchange interaction

$$H_{is.exch.} = -4\pi \sum_{\ell} \frac{(g-1)\Gamma^{(\ell)}}{2N} \sum_{m\,\underline{k}\,\underline{k}'} Y^*_{\ell m}(\hat{\underline{k}}) Y_{\ell m}(\hat{\underline{k}}') \times$$

$$\times \left| (a^*_{\underline{k}'+}a_{\underline{k}+} - a^*_{\underline{k}'-}a_{\underline{k}-})J_z + a^*_{\underline{k}'+}a_{\underline{k}-}J^- + a^*_{\underline{k}'-}a_{\underline{k}+}J^+ \right| . \quad (15)$$

iii) A k-f orbital exchange interaction of $(2-g)\vec{\ell}.\vec{J}$ type, but, for simplicity, restricted to the $\ell = 1$ partial waves

$$H_{orb.exch.} = - \sum_{\underline{k}\,\underline{k}'} \frac{i}{2N}(2-g)F_2(\underline{k}\times\underline{k}').\underline{J}\; a^*_{\underline{k}'\sigma}a_{\underline{k}\sigma} \quad (16)$$

(notation of Kondo, in Eq. 4.25 of ref. 2; F_2 is expected to be roughly constant in the RE series).

Interference between orbital exchange scattering and potential scattering gives rise to a first contribution to the skew scattering coefficient $\underline{a}$ which is proportional to $(2-g)F_2\langle J_z\rangle TH^{-1}$, i.e. to $(2-g)gJ(J+1)$. A second contribution arises if the spin-orbit splitting of the 5d levels is taken into account. This splitting results in orbit dependent resonant scattering (η_{2m} depending on m for a given spin direction). The contributions to $\underline{a}$ from interference with isotropic exchange scattering are proportional to $\Gamma^{(\ell)}(g-1)gJ(J+1)$. Eq.(14) is thus predicted.

The development of microscopic models of the k-f interaction in alloys of noble metals with RE has led to slightly modify the above interpretation. Estimates of the anisotropic terms of the k-f interaction (quadrupole and orbital terms) based on conduction electrons in plane wave states give values too small to explain the experimental data (in particular data on quadrupole scattering). Fert and Levy[5] have shown that the significant anisotropy found in these alloys can be explained by the formation of a 5d virtual bound state, i.e. by the admixture of 5d states into the conduction states. The conduction electrons can strongly feel the orbital anisotropy because the admixed 5d states lie closely to the 4f shell. It has been found that the k-f interaction arising from the 5d admixture is predominant. This k-f interaction affects only the $\ell = 2$ partial waves and can be expressed as a function of 4f-5d Slater integrals and of the parameters of the admixture : width of the virtual bound state and phase shifts. The Hall effect data have been interpreted in this model of the k-f interaction.[5,6] First, the 4f-5d orbital exchange gives rise to a k-f orbital exchange of the $(2-g)\vec{\ell}.\vec{J}$ type, similar to Eq.(16) but now restricted to $\ell = 2$ partial waves.[5] Secondly, when the spin-orbit splitting of the 5d virtual bound state is taken into account, the 4f-5d spin exchange gives rise to a pseudo k-f orbital exchange of the $(g-1)\vec{\ell}.\vec{J}$ type, again restricted to $\ell = 2$ partial waves.[6] Interference of both orbital scatterings ($\ell = 2$ channel) with potential scattering ($\ell=1$ channel to obey the selection rules of paragraph 1d) again leads to Eq.(4)

$$\underline{a} = \alpha_1(2-g)gJ(J+1) + \alpha_2(g-1)gJ(J+1)$$

A good agreement with the experimental value of α_2/α_1 in Ag-based alloys is obtained by assuming that the spin-orbit constant of the 5d states is slightly larger than that of atomic states. Otherwise the magnitude of α_1 and α_2, together with other experimental data on the residual resistivity, the thermoelectric power, the isotropic and anisotropic magnetoresistances, can be accounted for by a consistent set of parameters (4f-5d integrals, width of the virtual bound state, phase shifts).[6] We point out that it has been possible to analyze the skew scattering by RE impurities very quantitatively and that this analysis appears to be useful to probe fine details of the k-f interaction.

c. Discussion

The two causes of the extraordinary Hall effect (orbital exchange interactions and combination of spin-orbit and spin exchange interactions) seems to be very common in the RE systems. Not only in the dilute alloys but also in RE metals, compounds or amorphous alloys, the variation of the extraordinary Hall effect (EHE) roughly looks like its variation in silver based alloys (Figure 4). The EHE is generally positive for light RE (examples: Pr, Nd and Sm metals probably,[16] $PrAl_2$ and $NdAl_2$ compounds,[17] PrAu, PrAg, PrCu, NdAu, NdAg, NdCu and SmAu amorphous alloys).[18,23] It is positive for Gd and generally for the high spin elements just to the right of Gd in the series (examples: Gd metal,[19] Tb and Dy metals in their paramagnetic state,[20] $GdAl_2$ compounds,[17] GdAg, GdAu, GdCu, TbAg,

TbAu, TbCu, DyAu, HoAu and ErAu amorphous alloys).[18,23] It shows some tendency to become positive again when one goes toward the right of the series (examples: for the Ho and Er metals in their paramagnetic state the EHE is negative when the field is along <0001> but positive when the field is in the basal plane, the EHE is positive in the paramagnetic state of $TbAl_2$[17] and in DyAg, DyCu, HoCu, ErAg and ErCu amorphous alloys).[18] There are some exceptions in the pattern: the EHE seems to be negative in Tm metal and TmAu amorphous alloys.[22,23] Also, in Tb and Dy metals or $TbAl_2$, the data are more complex if one considers the ferromagnetic state in which the EHE changes its sign between low temperatures and T_c.[17,20,24] Nevertheless, the general features of the variation in the series suggest that, in dilute as in concentrated systems, there is some competition between the orbital exchange and the strong spin-orbit coupling of the 5d electrons. If, as it is generally admitted, the side-jump is important for the EHE of concentrated systems, this means that the competition also exists in the side-jump effects.

d. Skew Scattering in Anomalous Rare-earth Alloys

In some alloys containing Ce or Yb impurities the 4f level is close to the Fermi energy and the mixing between conduction and f states results in anomalous properties. The k-f interaction arising from the mixing has been expressed by Coqblin and Schrieffer.[22] It includes both spin and orbital couplings.

Skew scattering has been observed in an anomalous La:Ce alloy and has been interpreted in a model based on the Coqblin-Schrieffer interaction.[9] The same model could be appropriate to interpret the skew scattering also observed in $Ce_{1-x}Lu_xAl_2$ alloys.[23]

3. SKEW SCATTERING BY MAGNETIC IMPURITIES OF TRANSITION METALS

a. Transition Metal Impurities in Palladium and Platinum

There are numerous experimental data on the skew scattering in dilute alloys of transition metals with Pd[26,27] and Pt.[28] Most of these alloys are quasi-ferromagnetically ordered for relatively low concentrations of magnetic impurities so that the skew scattering can be studied either in paramagnetic or ordered systems.

In the paramagnetic alloys the skew scattering can be clearly identified in the temperature dependence of the initial Hall coefficient. For example, in Figure 5, the inital Hall constant of Pt:Co alloys approximately fits, for $T \lesssim 10$ K, with the expression

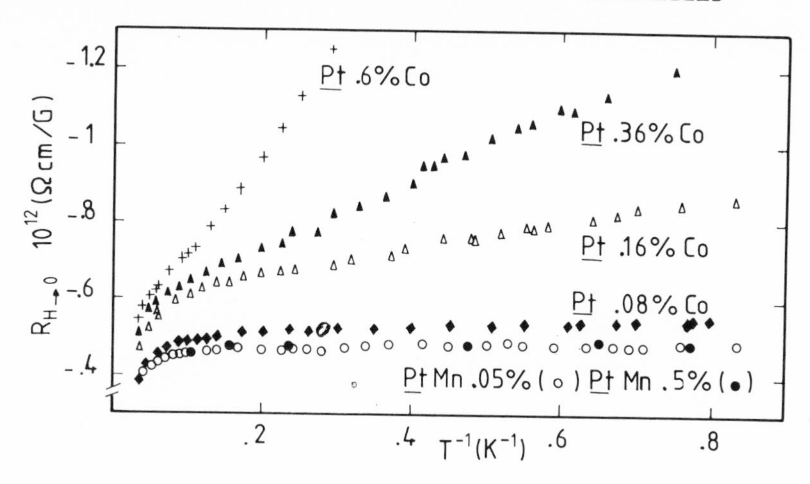

Fig. 5. Initial Hall coefficient of Pt:Co and Pt:Mn alloys versus 1/T (after Hamzić et al.).[26]

$$R = R_o + \frac{A}{T} \tag{15}$$

The A/T term has to be ascribed to skew scattering. The coefficient A increases somewhat faster than the Co concentration, in agreement with the concentration dependence of the magnetic susceptibility. In contrast, for the Pt:Mn alloys of Figure 5, R is practically temperature and concentration independent below 10 K, which shows the absence of skew scattering by the Mn impurities. (We note that the skew scattering contribution can be clearly identified only at low temperatures. Above 10 K, in both Pt:Co and Pt:Mn, R shows a behaviour which is not due to skew scattering but to the variation of the ordinary Hall coefficient with the temperature. This effect, cause by significant phonon scattering is well known in non-magnetic alloys.)[15]

The Hall resistivity of alloys of higher concentrations is shown in Figure 6. In Pt:Co and Pt:Fe the skew scattering results in a Hall resistivity term which saturates when the spontaneous magnetization saturates in a small applied field. Above this saturation field, there remains a linear increase mainly due to the ordinary Hall effect. For comparison, we also show in Figure 6 the Hall resistivity of Pt:Mn alloys which exhibits only ordinary Hall effect. Other examples of skew scattering contributions to the Hall resistivity in magnetically ordered alloys are shown in Figure 7. Note the approximate proportionality of the spontaneous Hall resistivity ρ^s_{xy} to the concentration, or alternatively to the residual resistivity ρ_o. In such alloys the skew scattering by a given

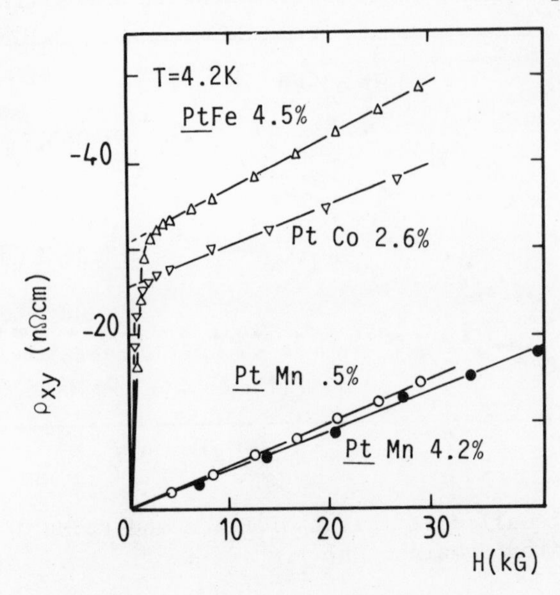

Fig. 6. Hall resistivity of Pt:Co, Pt:Fe and Pt:Mn alloys versus magnetic field at 4.2 K (after Hamzić et al.).[28]

impurity type can be conveniently characterized by the spontaneous Hall angle

$$\phi_H = \rho_{xy}^s/\rho_o \tag{16}$$

Skew scattering has been observed for Co and Ni impurities in Pd[26,27] and Co and Fe impurities in Pt[28] (with ϕ_H approximately constant in the dilute limit). In contrast there is no skew scattering by dilute Fe and Mn impurities in Pd[26] and Mn impurities in Pt[28] (ϕ_H vanishingly small in the dilute limit). The spontaneous Hall angle of the dilute Pd and Pt based alloys are given in Table I together with data on other physical properties which are related to orbital magnetism: resistivity anisotropy, magnetocrystalline anisotropy, hyperfine field at the impurity nucleus. There is a clear correlation between skew scattering and the existence of a local orbital moment. In Pd:Co and Pd:Ni, the existence of a local orbital moment results in skew scattering, strong resistivity anisotropy and magnetocrystalline anisotropy and positive hyperfine field. In Pt:Co and Pt:Fe skew scattering is associated with strong resistivity anisotropy and magnetocrystalline anisotropy but the hyperfine field is negative. This could be due either to the local orbital moment being too small to make the hyperfine positive (an orbital moment up to $0.2\mu_B$ could be reconciled with the observed hyperfine fields) or to the orbital character being located on Pt atom surrounding the impurities. Finally, Pd:Fe, Pd:Mn, Pt:Mn do

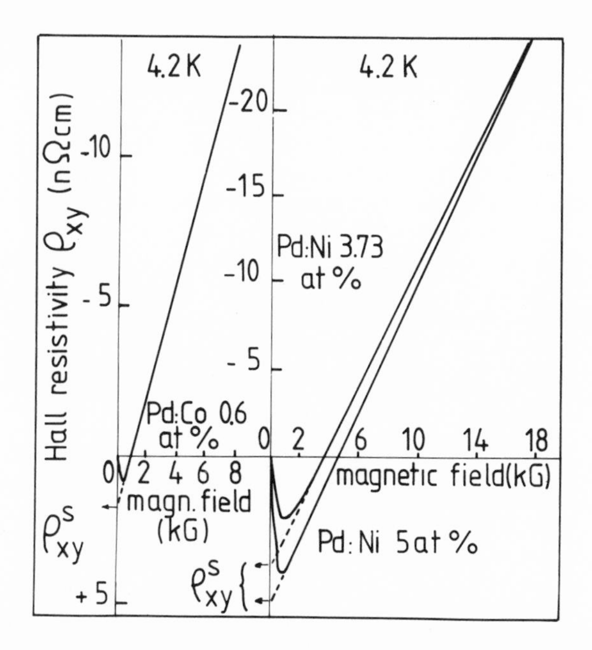

Fig. 7. Hall resistivity versus magnetic field at 4.2 K for Pd:Co and Pd:Ni alloys at 4.2 K (after Hamzić).[29] The spontaneous Hall resistivity ρ^s_{xy} is the Hall resistivity associated with the saturation of the spontaneous magnetization (in about 1 kG) and is determined by the extrapolation represented by a dashed line (subtraction of the ordinary Hall effect). An extrapolation from much higher fields, as discussed in ref. 27, can give different results but is an irrelevant way to determine the extraordinary Hall effect when 1 kG is sufficient to magnetize the sample.

not exhibit any orbital effects, which shows that only spin moments exist.

The correlation between skew scattering and orbital effects indicates that it is caused by orbital exchange interactions. Unfortunately, in alloys where the impurity d shell is far from its ionic structure and affected by some mixing with conduction states, there is not yet an appropriate model of the orbital magnetism and, what is more, of the orbital exchange interaction between localized and conduction electrons. In an ionic approach, it is not possible to find crystal field parameters consistent with the whole set of data on the orbital magnetism of the Pd (or Pt) based alloys. Senoussi[35] obtains a better agreement in an extended Wolff-Glodston model.

Table I. Spontaneous Hall angle, resistivity anisotropy, magnetocrystalline anisotropy coefficient K_1/M and hyperfine field at the impurity nuclei in Pd and Pt based alloys. The data are from refs. 26, 28-34. For Pd:(Mn,Fe,Co,Ni) and Pt:(Fe, Co), we give the limit values of the spontaneous Hall angle and the spontaneous resistivity anisotropy in the less concentrated alloys (but concentrated enough to present a quasi-ferromagnetic order at temperatures of few Kelvin; the values for Pd:Ni are obtained by an extrapolation over the range c < 2.3 at.% in which Pd:Ni is non-magnetic). In Pt:Mn the absence of extraordinary Hall effect and resistivity anisotropy has been observed in paramagnetic and spin glass alloys. K_1/M is given for Pd:(Fe,Co,Ni), Pt:(Fe, Co) with 1.7, 1.5, 5, 4.5, and 3.9 at.% respectively at 4.2 K or 10 K.

Alloy	Pd:Mn	Pd:Fe	Pd:Co	Pt:Ni	Pt:Mn	Pt:Co	Pt:Fe
$\phi_H \times 10^3$	~ 0	~ 0	+1.2	+3.5	~ 0	- 5	- 1.5
$\frac{\rho_{\parallel} - \rho_{\perp}}{\rho_o} \times 10^2$	~ 0	~ 0	+1.4	-3	~ 0	+ 1	+ 0.7
K_1/M (G)		-65	-900	+1100		+185	+675
H_{hf}(kOe)	-360 -400 -370	-300	+210 +200 +230	+175	-365	-200 -210	-315

We have focussed up to now on dilute alloys where the extraordinary Hall effect can be thought to be due to skew scattering by localized moments in a non-magnetic metal. At high concentrations—not very high actually in nearly ferromagnetic Pd—the magnetization of the conduction band becomes significant and the host itself is a ferromagnetic metal. Then the Hall effect due to skew scattering by local moments tends to be covered up by other contributions associated with the ferromagnetism of the host. The concentration range where local moment effects are predominant is characterized by a concentration independent ϕ_H. As it can be seen in Figure 8 this range is limited to x < 0.8 at% for Pd:Co and to x < 8% for Pd:Ni; at higher concentrations, contributions from the ferromagnetism of the Pd band change the sign of ϕ_H; in Pd:Fe and Pd:Mn there is no skew scattering in the dilute limit but only band effects which progressively increase with x. In the Pt-based alloys, the contributions from the band ferromagnetism turn out at much higher concentrations.

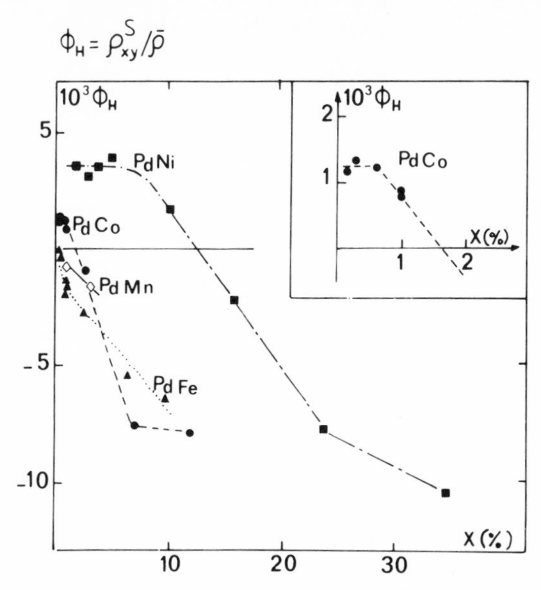

Fig. 8. Spontaneous Hall angle of Pd:Ni (■), Pd:Co (o), Pd:Mn (◊), Pd:Fe (Δ) alloys as a function of their concentration x (after Hamzić et al.).[26]

b. Transition Metal Impurities in Copper, Silver and Gold

The skew scattering by magnetic impurities in normal metals has been first identified in Au:Fe alloys[36] and since the Hall effect of many Cu-, Ag- and Au-based alloys has been investigated.[13,36-40] Nevertheless several questions remain open, in particular about the consequences of the Kondo effect on the skew scattering. We will focus on data on dilute alloys rather than on spin-glass concentrated systems.

The strongest skew scattering effects have been observed in Au:Fe and Au:Cr alloys. In Au:Cr the skew scattering effects are associated with quadrupole scattering effects which make the magnetoresistance anisotropic.[13] Examples of experimental results on the Hall effect are given in Figure 9 (Hall resistivity of Au:Fe alloys) and Figure 10 (initial Hall coefficient of Au:Cr alloys). As the skew scattering of Au:Fe and Au:Cr alloys is much stronger than that observed in any alloy containing S-ion impurities, it has been ascribed to local orbital magnetism. Its sign and magnitude is fairly well accounted for by a virtual bound state model in which a small orbital moment is induced by the spin-orbit splitting of the d energies.[13,36] This model predicts for the Hall angle at saturation

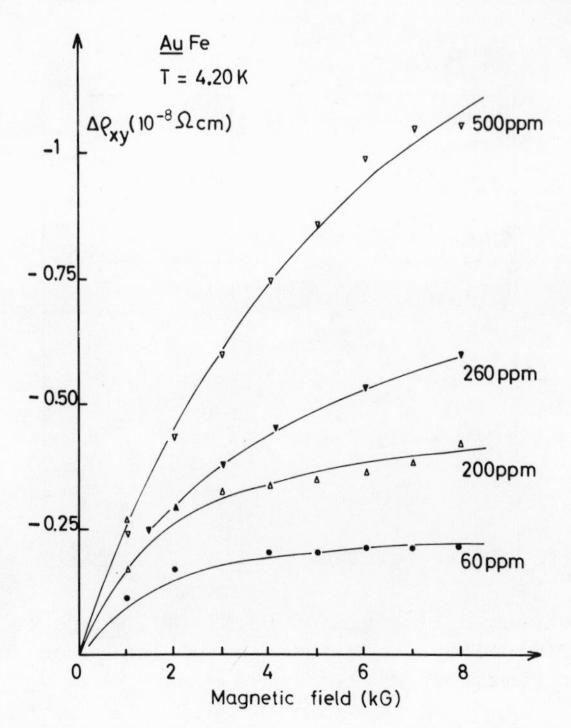

Fig. 9. Hall resistivity of Au:Fe dilute alloys versus magnetic field at 4.2 K. The concentrations in ppm are indicated on the curves (after Fert et al.).[13]

$$\phi_H = \frac{i_+}{i}\phi_+ + \frac{i_-}{i}\phi_- \,, \tag{17}$$

with

$$\phi_\pm = \pm\frac{3}{5}\frac{\lambda_\pm}{\Delta}\sin(2\eta^\circ_{2\pm}-\eta_1)\sin\eta_1 . \tag{18}$$

Here i_+ and i_- are the spin up and spin down currents, $\lambda_\pm$ are effective spin-orbit coefficients, Δ is the width of the virtual bound states and the η are phase shifts.

However, several points are difficult to explain in the experimental results. The Hall resistivity curves of Figure 9 saturate faster than the magnetization of the Fe impurities. In Au:Cr a fit with a Brillouin function is better but is far from being perfect. Secondly, it can be seen on Figure 10 that a dependence $R-R_0 \sim cT^{-1}$ is fairly well obeyed in most of the alloys between 30 K and 10 K whereas the variation R(T) definitely departs from

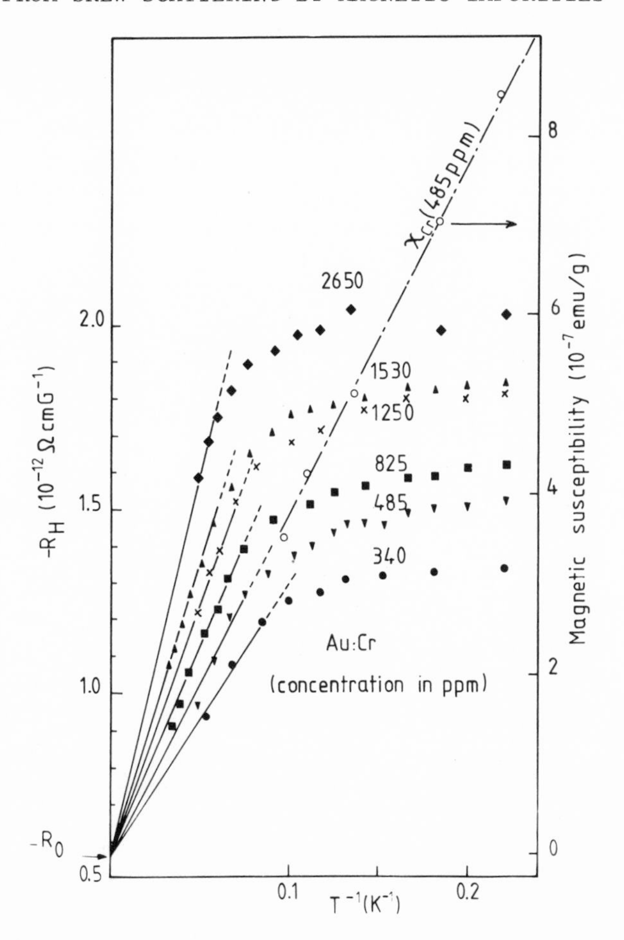

Fig. 10. Initial Hall coefficient of Au:Cr dilute alloys versus T^{-1}. Straight lines have been drawn to make apparent the range where $R-R_o \sim AT^{-1}$ with A increasing with the concentration (indicated in ppm on the curves). For comparison, we have plotted (open symbols) the variation as a function of T^{-1} of the magnetic susceptibility[42] of the sample containing 485 ppm (after Fert et al.).[13]

the variation $\chi(T)$ below 10 K, R(T) tending to be constant at low temperatures. A similar flattening of R(T) is also observed below 4 K in dilute Au:Fe alloys. This could suggest a compensation of the orbital moment by the Kondo effect at temperatures well above the compensation temperature T_K of the spin moment ($T_K \lesssim 0.1$ K in Au:Cr, $T_K \sim 0.5$ K in Au:Fe). But, to our knowledge, there are neither other experimental results also suggesting this behaviour nor theoretical models predicting it.

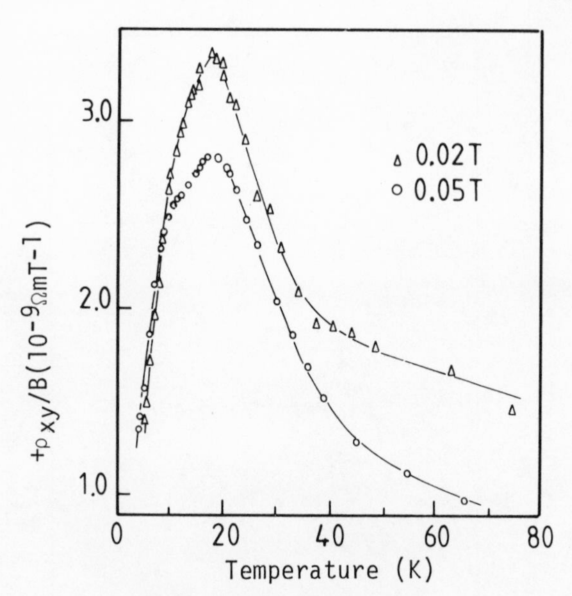

Fig. 11. Temperature dependence of ρ_{xy}/B in the Au:Co 12 at% in the fields indicated (after McAlister).[40]

Skew scattering effects have been also observed in Au:Mn alloys. They could arise either from the existence of a small orbital moment on Mn or from the spin-orbit coupling in the conduction band. In contrast, there is very little skew scattering in Ag:Mn and Cu:Mn (it is too small to be clearly identified). In Au:Co the existence of skew scattering depends on the impurity concentration, as localized magnetic moments do not develop in the dilute limit but are induced by interaction effects.[40] The change of sign of the skew scattering in Au:Co with respect to Au:Cr and Au:Fe is consistent with Eq.(18).

Finally, several studies have been devoted to the relation between the skew scattering and the spin-glass properties in concentrated alloys.[38-40] In spin-glass systems, the variation of the Hall coefficient as a function of the temperature exhibits a maximum similar to the maximum observed in the magnetic susceptibility (Figure 11). The information obtained from the susceptibility and the Hall resistivity respectively have been compared and discussed.[38-40] Hall effect data obtained on Au:Fe alloys have also suggested the existence of side-jump effects in the concentrated alloys.[41]

4. CONCLUSION

The first interest of the skew scattering by magnetic

impurities in metals is to allow us a better understanding of the mechanisms of the EHE. In rare-earth systems, we have seen that the results on the dilute alloys have shown the double origin of the skew scattering in orbital exchange and spin-orbit interactions. In dilute alloys of transition metals the skew scattering seems to be related to the orbital character of local moments; when the concentration increases, a crossover to band effects is observed. From a more general point of view, the studies of skew scattering are of interest to reveal the existence of orbital moments (transition metal impurities), to determine the orbital dependence of the exchange interactions and to investigate the spin-orbit effects in the conduction band. Information on orbital magnetism, orbital interactions and spin-orbit effects seems to be today of great interest to understand the magnetic properties of the rare-earth compounds and probably the spin-glass systems.[44]

ACKNOWLEDGEMENTS

R. Asomoza, I. A. Campbell, A. Friederich, P. M. Levy, S. Senoussi have been closely involved in many aspects of these studies. Discussions with M. Christen, B. Giovannini and C. M. Hurd have also been very useful to us.

REFERENCES

1. J. Smit, Physica 24:39 (1958).
2. J. Kondo, Progr. Theor. Phys. 27:772 (1962).
3. L. L. Hirst, Adv. Phys. 27:231 (1978).
4. A. Fert and A. Friederich, Phys. Rev. B 13:397 (1976).
5. A. Fert and P. M. Levy, Phys. Rev. B 16:5052 (1977).
6. G. Lacueva, P. M. Levy and A. Fert, to be published.
7. J. Kondo, Progr. Theor. Phys. 28:846 (1962); Y. Kagan and L. A. Maksimov, Sov. Phys. Sol. St. 7:427 (1965); F. E. Maranzana, Phys. Rev. 160:421 (1967).
8. B. Coqblin and J. R. Schrieffer, Phys. Rev. 185:847 (1969).
9. A. Fert, J. Phys. F: Metal Phys. 3:2126 (1973).
10. B. Giovannini, Low Temp. Phys. 11:849 (1973).
11. L. Berger, Phys. Rev. B 2:4559 (1970).
12. P. Nozières and C. Lewiner, J. de Phys. (Paris) 34:901 (1973).
13. A. Fert, A. Friederich and A. Hamzić, J. of Mag. Mag. Mat., to be published.
14. J. Bījvoet, G. Merlijn and P. Frings, J. de Phys. (Paris) 40 (Supp au n°5):C5-38 (1979).
15. J. E. A. Alderson and C. M. Hurd, Can. J. Phys. 48:2162 (1970).
16. C. J. Kevane, S. Legvold and F. H. Spedding, Phys. Rev. 91:1372 (1953); N. V. Volkenshtein and G. V. Fedorov, Fiz. Tverd. Tela 7:3213 (1965), Soviet Phys. Sol. St. 7:2599 (1966).

17. M. Christen, Thèse (Université de Genève, 1978); M. Christen, B. Giovannini and J. Sierro, Phys. Rev. B 20:4624(1950).
18. R. Asomoza, J. B. Bieri, A. Fert, J. C. Ousset, 1980 Conference on Liquids and Amorphous Metals, to be published.
19. R. S. Lee and S. Legvold, Phys. Rev. 162:431 (1967).
20. J. J. Rhyne, Phys. Rev. 172:523 (1968); J. Appl. Phys. 40:1001 (1969).
21. J. R. Cullen and J. J. Rhyne, J. Appl. Phys. 41:1178 (1970).
22. N. V. Volkenshtein and G. V. Federov, Fiz. Met. Metallov. 20:508 (1965); Phys. Met. Metallogr. 20:28 (1965).
23. T. R. McGuire, R. J. Gambino and R. C. Taylor, J. Appl. Phys. 50:7653 (1979).
24. R. Giovannini, Phys. Lett. 36A:381 (1971); A. Fert, J. de Phys. Lettres (Paris) 35:L107 (1974).
25. T. Hiraoka, J. Phys. Soc. Jap. 45:476 (1978).
26. A. Hamzić, S. Senoussi, I. A. Campbell, A. Fert, Sol. St. Comm. 26:617 (1978).
27. C. M. Hurd, S. P. McAlister and C. Couture, J. Appl. Phys. 50:7531 (1979).
28. A. Hamzić, S. Senoussi, I. A. Campbell and A. Fert, 1979 International Conference on Magnetism (Munich), to be published.
29. A. Hamzić, Thèse (Orsay, in preparation).
30. S. Senoussi, I. A. Campbell and A. Fert, Sol. St. Comm. 21:269 (1977).
31. D. M. S. Bagguley and J. A. Robertson, J. Phys. F 4:2282 (1974); H. Fujiwara, T. Tokunaga and A. Nakagima, J. Phys. Soc. Jap. 34:1104 (1973).
32. J.Flouguet, Progress in Low Temp. Phys. 711:649 (1978).
33. Le Dang Khoi, P. Veillet and I. A. Campbell, J. Phys. F 6:L197 (1976).
34. M. Sheng, E. R. Seidel, F. J. Litterst, W. Gierisch and G. M. Kalvius, J. de Phys. 35 (supp. n°12):C6-527 (1974).
35. S. Senoussi, Thèse (Orsay 1978).
36. A. Fert and O. Jaoul, Phys. Rev. Lett. 28:303 (1972).
37. C. M. Hurd and J. E. A. Alderson, Phys. Rev. B 7:1233 (1973).
38. S. P. McAlister and C. M. Hurd, Phys. Rev. Lett. 37:1017 (1976).
39. C. M. Hurd and S. P. McAlister, Phys. Rev. 15:514 (1977).
40. S. P. McAlister, J. Appl. Phys. 49:1616 (1978).
41. C. M. Hurd and S. P. McAlister, Phys. Rev. Lett. 41:1513 (1978).
42. J. J. Prejean, private communication.
43. P. M. Levy, J. de Phys. 40 (supp n°5):C5-8 (1979).
44. J. J. Prejean and P. Monod, to be published; P. M. Levy and A. Fert, to be published.

HALL EFFECT IN SINGLE CRYSTALS OF IRON

R. V. Coleman, R. W. Klaffky* and W. H. Lowrey†

Physics Department
University of Virginia
Charlottesville, Virginia 22901

I. INTRODUCTION

In this paper we review the results of measurements of the Hall effect in pure iron in the temperature range 1 - 247 K. For most temperatures measurements were made in the field range 0 - 50 kOe while in the range 1.2 - 4.2 K measurements have been made in fields up to 220 kOe. Previous experiments[1-3] have reported measurements at selected temperatures in the above range and measurements for a continuous set of temperatures in this range and in fields up to 150 kOe were reported by Klaffky and Coleman.[4] The field range has been extended to 220 kOe by Coleman, et al.[5]

At temperatures above ~100 K both the ordinary Hall coefficient R_o and the ferromagnetic Hall coefficient R_s are smooth functions of temperature and are in essential agreement with previous experiments and theories. Below 100 K the Hall resistivity develops a non-linear dependence on magnetic field above saturation and this complicates the analysis and separation of the ordinary and ferromagnetic Hall coefficients. The Hall resistivity changes sign from positive to negative as the temperature is reduced below 80 K and this corresponds to a reversal in the sign of the ordinary Hall coefficient R_o. Separation of the ordinary coefficient R_o and the ferromagnetic Hall coefficient R_s have been affected by using relations of the form $\rho_H(B) = R_oB + 4\pi M_sR_s$ above 70 K and $\rho_H(B) = R_oB = 4\pi M_sR_s + CB^2$ below 70 K. The coefficients obtained below 70 K using this procedure exhibit several anomalous features

* Brookhaven National Laboratory, Upton, New York 11973
† Kirtland Air Force Base, Albuquerque, New Mexico 87117

and we will examine these in detail.

Examination of Kohler plots of the Hall resistivity over the temperature range below 80 K lead to the conclusion that some of these anomalous effects may come from a mobility transition corresponding to carriers making the transition from the low-field limit, $\omega_c\tau << 1$, to the high-field limit, $\omega_c\tau >> 1$, as recently suggested by Majumdar and Berger.[2] Extensive analysis will be given of these effects and the Kohler plots are used to take into account the $\omega_c\tau$ dependence of the Hall resistivity. Data taken at 4.2 K in magnetic fields up to 150 kOe show a linear behavior above 100 kOe and extrapolation of the data for applied fields above 100 kOe to B = 0 shows a zero intercept indicating that the ferromagnetic Hall effect approaches zero at 4.2 K.

For fields applied in the <100> directions extensive open orbit networks exist on the Fermi surface of iron and at low temperatures and high fields the ordinary Hall effect again reverses sign from negative to positive for fields above 150 kOe. This behavior is highly sensitive to rotation of the field and will be analyzed in connection with other Fermi surface features which appear in both the dc magnetoresistance and in the de Hass Shubnikov oscillations.

The magnetoresistance data have also been used to deduce extensive information on the topological features of the Fermi surface. In particular the existence of quantum interference orbits has been established and used to define the Fermi surface more precisely. A review of these results will be included along with comparison to band structure calculations.

II. EXPERIMENTAL METHODS

Single crystal iron whiskers grown by the hydrogen reduction[6,7] of $FeCl_2$ were used for the experiments. For the high temperature ranges two <111> axial whiskers with diameters on the order of 0.5 mm and residual resistivity ratios of 3166 and 4000 were used to measure the Hall voltage and are designated by sample numbers RWK-4 and RWK-5. For temperatures of 4.2 K and below a variety of crystals with orientations along <100>, <110> and <111> have been used with residual resistance ratios in the range 1000 to 10,000. The Hall leads were spot welded to the faces of the crystal and the magnetic field was applied perpendicular to the direction defined by the Hall leads. The Hall voltage probes were short lengths of nichrome wire which were in turn soldered to copper leads. All leads were thermally grounded to the copper heater block which maintained the sample temperature. Two pairs of Hall leads were attached to each specimen as a check for reproducibility, one pair near the center of the specimen and one pair nearer to the end. The magnetoresistance leads were spot welded to the same face with a gauge length of 5-10 mm. All thermal voltages were less than one microvolt and were essentially independent of temperature indicating that thermal gradients were not a serious problem. The heated copper block provided long term temperature stability of .01 K and was controlled by an Artronix temperature controller using a platinum resistor as a sensor above 40 K while below 40 K a 100 Ω 1/8 watt Allen-Bradley carbon resistor was used as the sensor.

The voltages were recorded using a Keithley 148 nanovoltmeter and source driving a recorder. With the use of careful shielding and grounding techniques noise levels were reduced to a few nanovolts. For magnetic fields up to 50 kOe a standard helium cryostat and superconducting solenoid system were used with the sample holder mounted in a vacuum can inserted into the solenoid. For fields up to 230 kOe the Bitter solenoids at the Francis Bitter National Magnet Laboratory were used.

The voltage from the transverse Hall pair of leads was measured for positive and negative directions of the magnetic field and for positive and negative current directions corresponding to each field direction. These four values of the voltage were then summed with appropriate signs in order to extract the transverse odd or Hall voltage, the thermal voltage, and the magnetoresistance plus transverse even voltage. The magnetoresistance term arises from the small misalignment of the transverse contacts which is unavoidable in a transverse two probe measurement. The transverse even voltage appears to become appreciable only at low temperatures and above 50 kOe so that the magnetoresistance extracted in the above procedure for measurements up to 50 kOe could be compared with the direct magnetoresistance measured in the standard four lead arrangement. The agreement was good over the whole temperature range and provided a good check on the accuracy of the extraction procedure. Consistency was also good between the two pairs of Hall leads ex-except at a few low temperature points where the end pair appears to have been influenced by domain effects connected with demagnetization structure near the end of the specimen. The data presented in this paper is always taken from the center pair of leads where variations due to end effects should be minimal.

The low field points are influenced by multidomain effects and at low temperatures these can produce very large effects in the magnetoresistance as discussed by Shumate, et al.[8] and of course contribute to an averaged Hall voltage rather than that characteristic of a single domain with internal field equal to the saturation magnetization field $4\pi M_s$. Therefore all data points reported here begin above saturation which in these specimens was assured for applied fields above 15 kOe. The magnetic induction was calculated from $B = H + 4\pi M_s - \alpha 4\pi M_s$ where the demagnetizing coefficient α was equal to .498 and .497 for specimens RWK-4 and RWK-5 respectively. Demagnetization factors for the other crystals used in these experiments were comparable.

The de Haas Shubnikov oscillations were detected by superimposing an ac modulation field on the dc field and detecting the ac voltage signal generated from the magnetoresistance variation. A PAR model 190 transformer coupled to a PAR model HR-8 lock-in amplifier was used to detect the signal. The output was fed to a DEC 11/34 computer and analyzed using a fast Fourier transform program. For fields up to 80 kOe a modulation amplitude up to 300 G was used while for fields up to 150 kOe a modulation field up to 2500 G was used.

III. EXPERIMENTAL RESULTS

A. Hall Effect as a Function of Temperature

The transverse-odd or Hall voltage is generally given by the following phenomenological equation which expresses the Hall resistance as

$$V_{Hall}/I = 4\pi M_s R'_s + R'_o B \ . \tag{1}$$

The Hall resistivity is given by

$$\rho_H = \frac{E_y}{j_x} = 4\pi M_s R_s + R_o B \tag{2}$$

where R_o and R_s are the Hall coefficients. The term $4\pi M_s R_s$ is associated with the ferromagnetic Hall effect (also referred to as the anomalous, extraordinary, or spontaneous Hall effect) and is considered to depend only on M_s the saturation magnetization. The term $R_o B$ is associated with the ordinary Hall effect and is generally taken to be linearly dependent on B in most experiments. The ordinary Hall effect reflects the balance between electron and hole cyclotron orbits in either the high-field or low-field condition and this balance may be sensitive to both temperature and field through a dependence on the exact values of $\omega_c\tau$ for the hole and electron orbits respectively as well as possible anisotropies in the scattering mechanisms and Fermi surface. These effects can introduce a field dependence greater than linear as well as possible sign changes. In addition, the ferromagnetic Hall effect can possibly exhibit a high-field limit term[9] proportional to $M_s B^2$. The experimental data definitely show strong non-linear terms for certain field and temperature ranges.

The values of the Hall voltage for magnetic field values well above those required to produce saturation (H > 15 kOe, B > 25 kG) have been recorded in the temperature range 1 to 247 K. Representative curves drawn from a least squares computer fit of V/I to a function of the form $a + bB + cB^2$ are shown in Fig. 1. A non-linear dependence on B is clearly evident at high fields for temperatures below 78 K and detailed plots of the data at 51 and 45 K are shown in Fig. 2.

The Hall resistivity, ρ_H, measured at 56 kG as deduced from the data of Fig. 1 is shown as a function of temperature in Fig. 3. At high temperatures ρ_H is positive and is increasing with increase of temperature due to the ferromagnetic Hall constant which is proportional to the square of the resistivity, ρ. Below 85 K, ρ_H begins to decrease rapidly and changes sign from positive to negative between 77 and 72 K for both specimens measured. The decrease in ρ_H continues down to 40 K where a plateau region is observed between 40 and 20 K. Below 20 K ρ_H continues to decrease attaining a minimum value at the lowest temperature measured, ~1 K. Both specimens show minimum values of ρ_H at 1 K considerably below the value measured by Dheer[1] on his highest purity specimen and this is consistent with the fact that his specimen had an RRR equal to about 1/10 that of the present specimens.

Below 80 K the Hall resistivity develops substantial curvature as a function of magnetic field making separation of the exact coefficients imprecise. However, the curvature can be approximated for purposes of fitting the data by adding a term proportional to B^2 so that the Hall resistivity is given by

$$\rho_H(B) = R_o B + 4\pi M_s R_s + CB^2 \tag{3}$$

The derivative with respect to the magnetic induction will therefore be given by

$$\frac{d\rho_H}{dB} = 2\ CB + R_o = R_o(B) \tag{4}$$

When curvature is not present $d\rho_H/dB = R_o$ as obtained from a conventional linear analysis. In general $d\rho_H/dB$ will be field dependent and we have designated this by a field dependent quantity $R_o(B)$. Plots of this quantity as a function of temperature for three different magnetic field values are shown in Fig. 4.

At temperatures above 90 K, $d\rho_H/dB$ is independent of field and equal to R_o. The value of R_o increases linearly with temperature above 90 K as observed in previous experiments.[1,3]

At temperatures below 80 K a field dependent term is clearly present in $d\rho_H/dB$ and reflects the strong field dependent curvature present in the Hall resistance. Near 30 K the field dependence

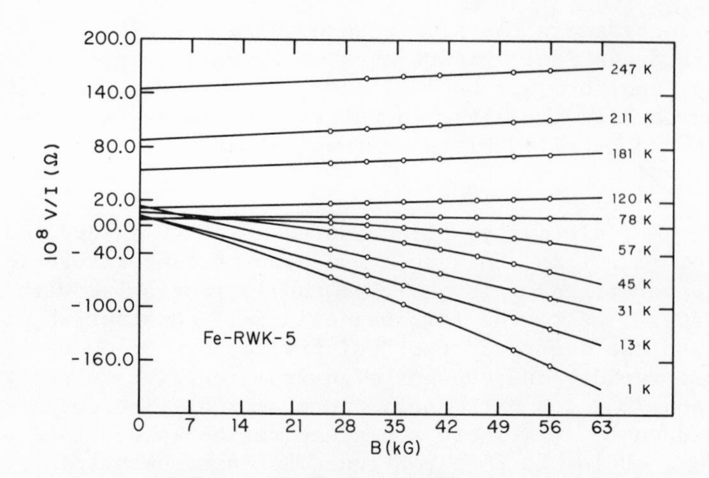

Fig. 1. Plots of the Hall resistance versus magnetic induction for representative temperatures measured in the present experiment. Solid curves drawn through the data points are obtained by a least squares computer fit to a function of the form $V/I = a + bB + cB^2$. Sample RWK-5 with RRR = 4000. (Ref. 4)

nearly vanishes indicating a return to a linear dependence of the Hall resistance on magnetic field at this temperature. The curvature developing in the Hall resistance as a function of magnetic field below 80 K suggests that a term proportional to $\omega_c\tau$ may be present. The carriers in this range of temperature and field are probably making a transition from the low-field, $\omega_c\tau < 1$, to the high-field, $\omega_c\tau > 1$, condition and this transition in turn effects the Hall resistivity. The intermediate maximum in $d\rho_H/dB$ just below 30 K followed by a further strong decrease at lower temperatures indicates that an additional effect possibly associated with scattering mechanisms or with further complexity in the low-field to high-field transition is playing a role.

The traditional computation of the ferromagnetic Hall coefficient would proceed by extrapolation of the Hall resistance curve to B = 0 and the value of the intercept obtained would be assigned to a term $4\pi M_s R_s$. For the present data we have carried out such a procedure using both a linear extrapolation, $\rho_H(B) = R_o B + 4\pi M_s R_s$, and an extrapolation including a term proportional to B^2, $\rho_H(B) = R_o B + 4\pi M_s R_s + CB^2$, in a field range up to B = 56 kG. The resulting values of R_s as a function of temperature are shown in Figure 5. R_s is positive for all temperatures measured and above 80 K shows a nearly quadratic increase with temperature. The high temperature values of R_s are in good agreement with the results of Volkenstein and Fedorov[3] for polycrystalline iron and somewhat

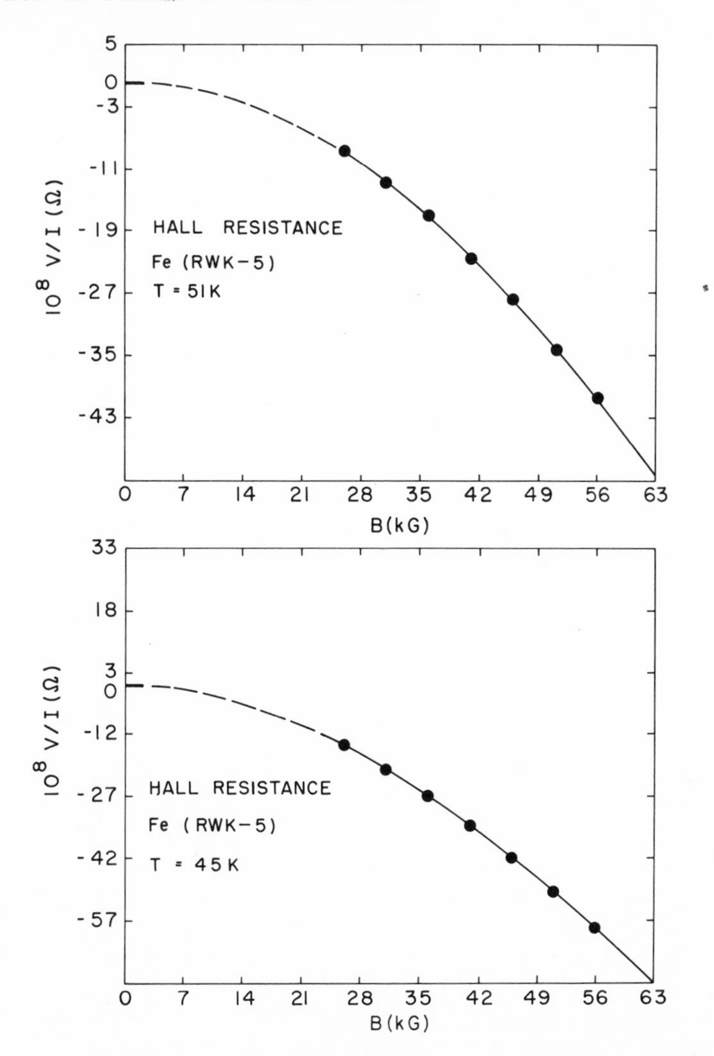

Fig. 2. Hall resistance V/I measured at temperatures of 51 and 45 K where a large nonlinear field dependence is present. Solid curves are drawn from a computer fit of the function $V/I = a + bB + cB^2$. Dashed curves are extrapolations to B=0 using values calculated from Kohler plots. (Ref. 4)

higher than Dheer's results[1] for <111> axial iron whiskers.

Below 80 K both the linear and quadratic analyses give an increase in R_s to a relative maximum at 40 K. This anomalous maximum in R_s is a reflection of the variation in the field and temperature dependent curvature of the Hall resistivity just below 40 K which also produced the intermediate maximum in $R_o(B)$. In fact at 31 K

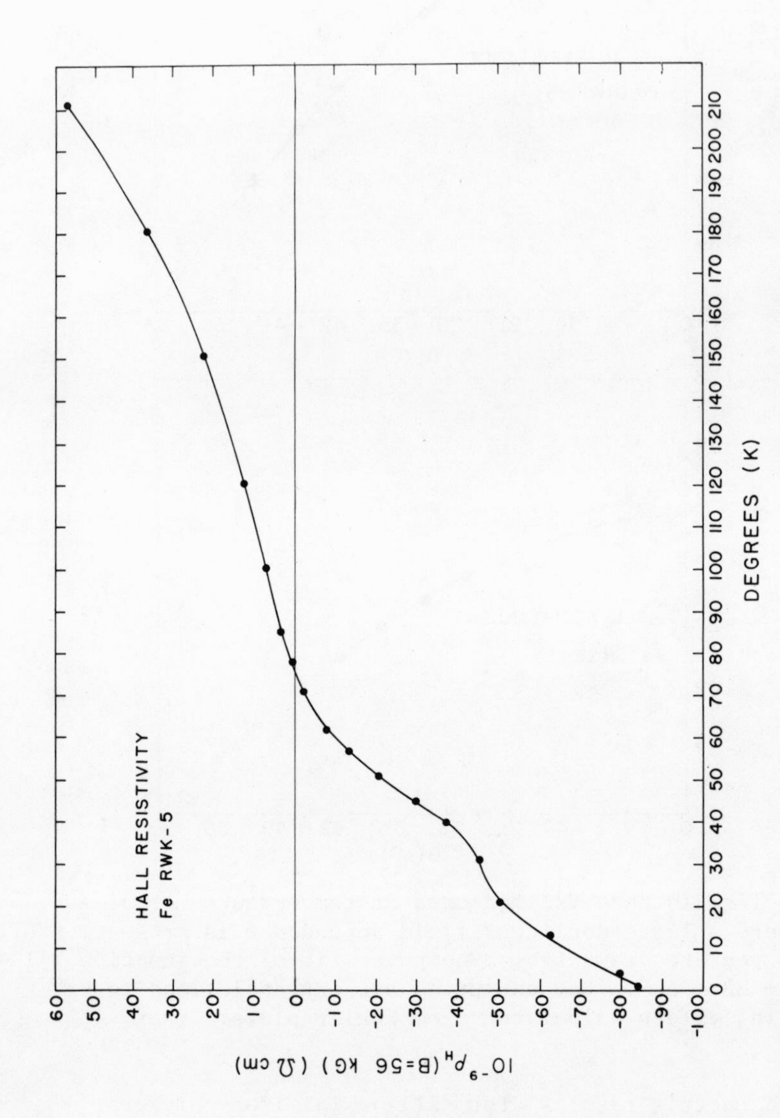

Fig. 3. The Hall resistivity ρ_H measured at B - 56 kG plotted as a function of temperature in the range 1 - 210 K. Sample RWK-5 with RRR = 4000. (Ref. 4)

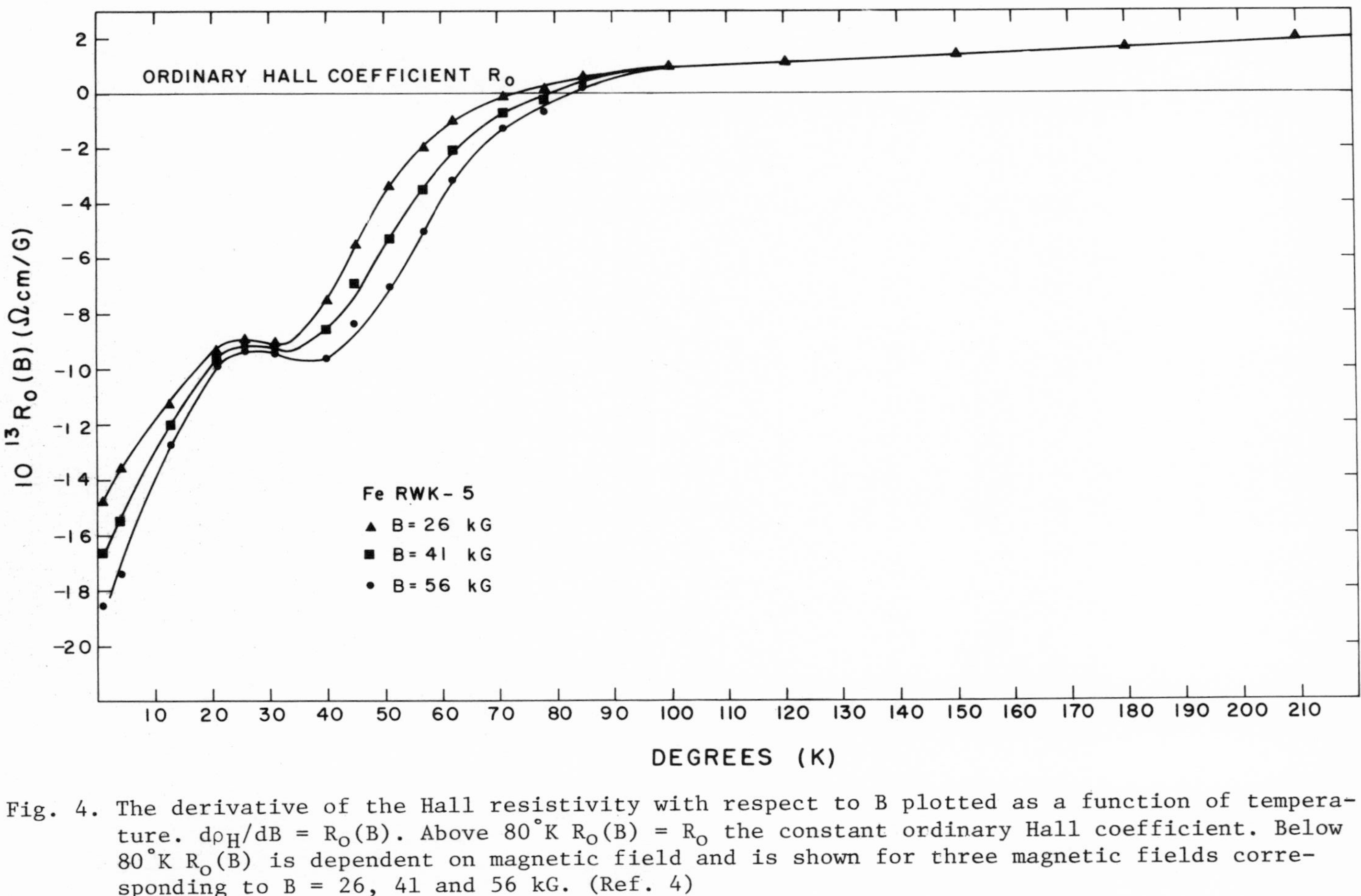

Fig. 4. The derivative of the Hall resistivity with respect to B plotted as a function of temperature. $d\rho_H/dB = R_o(B)$. Above 80°K $R_o(B) = R_o$ the constant ordinary Hall coefficient. Below 80°K $R_o(B)$ is dependent on magnetic field and is shown for three magnetic fields corresponding to B = 26, 41 and 56 kG. (Ref. 4)

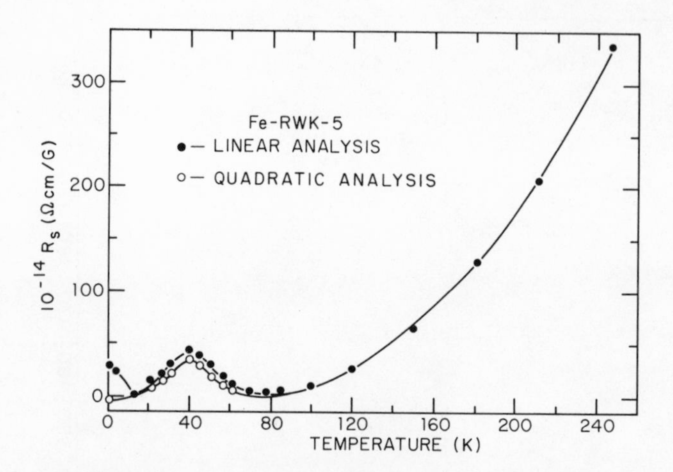

Fig. 5. The ferromagnetic Hall coefficient R_S plotted as a function of temperature in the range 1 - 247 K. This coefficient is calculated from the intercept of the Hall resistivity obtained by extrapolating to $\bar{B}$ = 0. Two approximate fits have been used for extrapolation. ● $\rho_H = R_oB + 4\pi M_sR_s$ (linear analysis), O $\rho_H = R_oB + 4\pi M_sR_s + CB^2$ (quadratic analysis). (Ref. 4)

the curvature has vanished and the Hall resistivity is a linear function of magnetic field up to 45 kOe (56 kG). Linear extrapolation of the curve at 31 K would give a large non-zero intercept as seen in Fig. 1. Further discussion of the analysis and extrapolation procedure will be given in Section IV.

Preliminary data on the Hall effect up to 150 kOe has been obtained for sample RWK-5 at 4.2 K and 63.5 K. AT 4.2 K in the applied field range 20 - 100 kOe the Hall voltage continues to show a strong nonlinear term, but for fields above ~100 kOe the Hall voltage shows a linear dependence on magnetic field as shown in Figure 6. Within the accuracy of the data a straight line projection of the high field data back to the origin also gives a zero intercept.

The field dependence of the Hall voltage at 63.5 K has also been measured in applied fields to 150 kOe and results are shown in Figure 7. In this case the data can be fit by a single B^2 term over the entire field range and this is represented by the solid line. Essentially no linear term is present and this is consistent with the observation that R_o is very small in this temperature range. The total Hall voltage is also small in this temperature range and noise generated when running in the high field solenoid makes it difficult to measure the small transverse signals below about 50 kOe. In Figure 7 we have therefore used values for $H < 50$ kOe measured in the superconducting solenoid and have shifted the ordinate of the high field curve to match the low field values. This simply

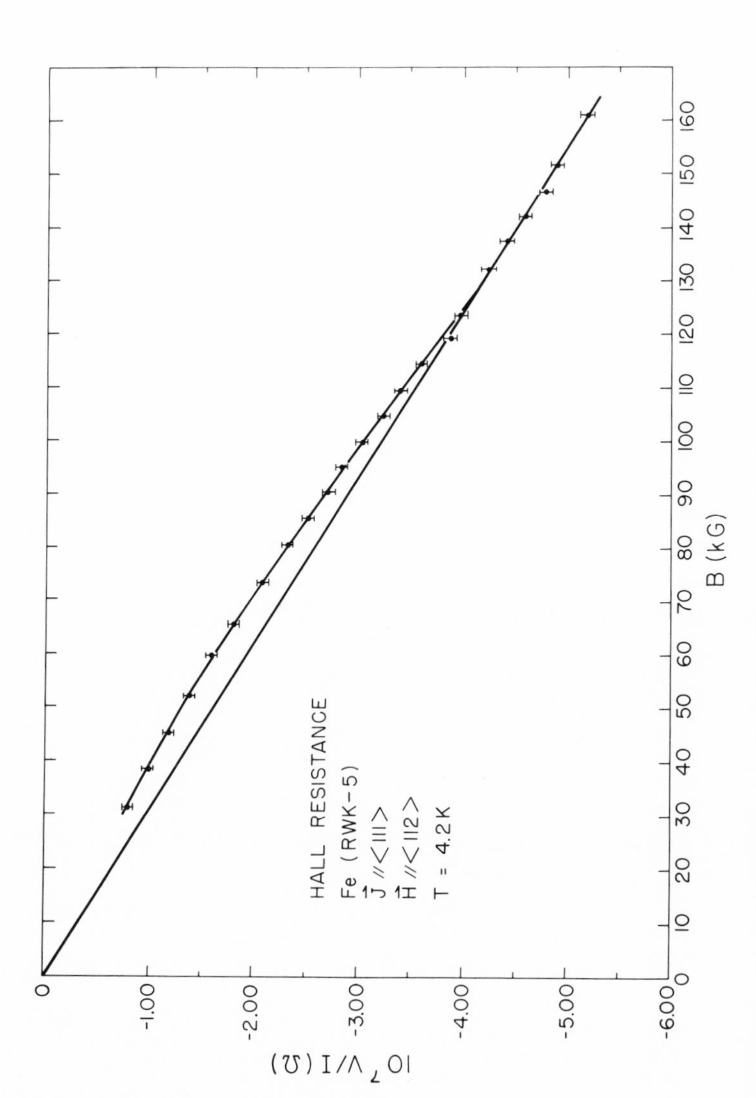

Fig. 6. Hall resistance at 4.2 K plotted as a function of magnetic induction in the range 30-160 kG. Straight solid line passing through the origin represents extrapolation to B = 0 of the data points above B - 120 kG. (Ref. 4)

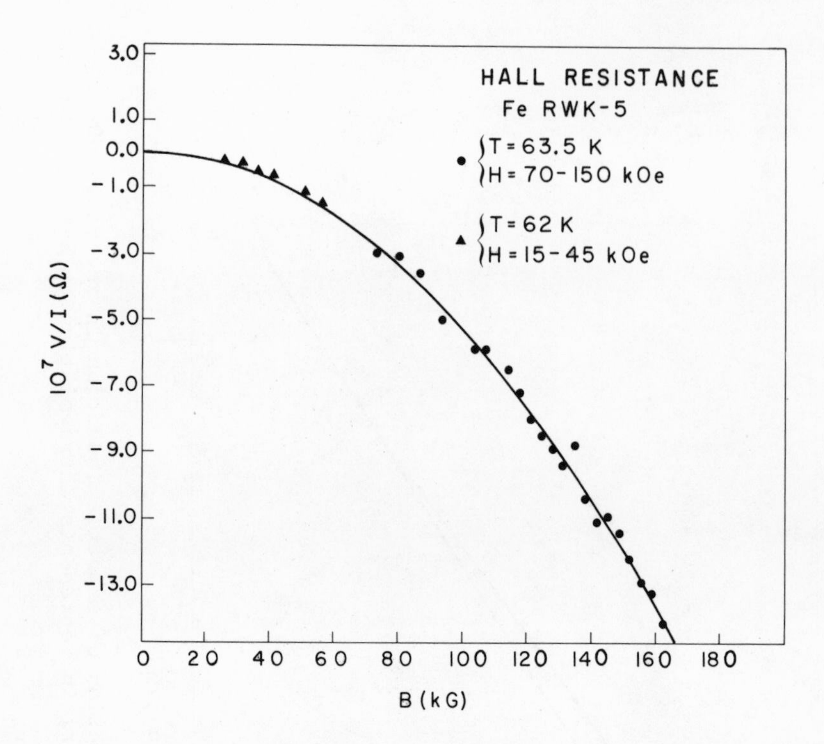

Fig. 7. Hall resistance at 63.5 K plotted as a function of magnetic induction in the range 70 - 160 kG. Solid curve represents a least squares computer fit to the data points. Points from 20 - 60 kG were taken at 62 K. The fit is well represented by a single B^2 term over the entire range. (Ref. 4)

means that the values obtained for H = 50 - 150 kOe are not accurate enough in this temperature range to give a meaningful extrapolated value of the zero field intercept although a reasonably good estimate of the field dependence at high fields is obtained and this can probably be assigned to the transition in values of $\omega_c\tau$.

B. High-Field Hall Effect

In high magnetic fields at low temperatures a major new observation[5,10] is that for current along [100] and field along [010] the sign of the Hall voltage at 4.2 K reverses as a function of field for values of field corresponding to $\omega_c\tau$ values greater than 100. This behavior is demonstrated in Figure 8 for four different purities of <100> axial crystal. The data of Figure 8 are plotted on a Kohler plot where ρ_H/ρ_o is plotted as a function of B/ρ_o rather than B. ρ_H is the Hall resistivity and ρ_o is the zero field residual

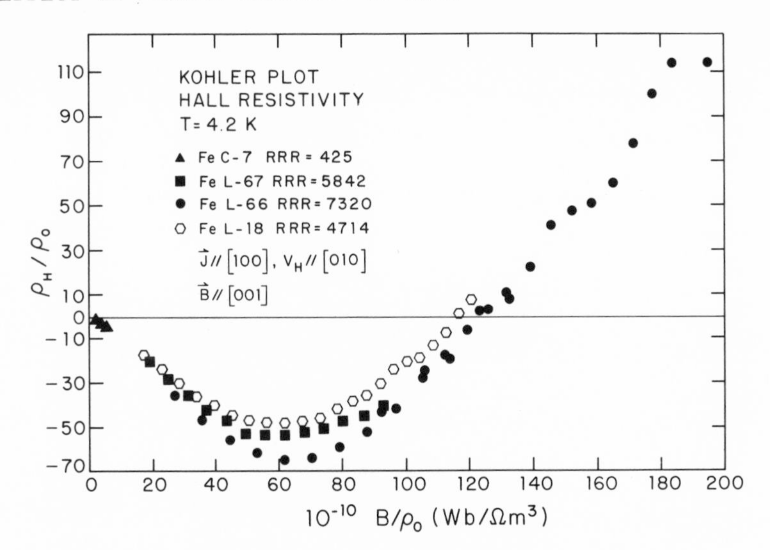

Fig. 8. Kohler plot of Hall resistivity for a field range from 20 to 200 kG. ρ_H and B have been divided by the residual resistivity ρ_o of each specimen measured at 4.2 K. The minimum and sign change occurs at the same value of B/ρ_o for each specimen and therefore occurs at the same value of $\omega_c\tau$. (Ref. 5)

resistivity. The occurrence of the minimum and sign reversal at the same values of B/ρ_o for the specimens with different values of ρ_o means that the onset of new contributions to the Hall voltage occurs at the same $\omega_c\tau$ values rather than at the same B value for the different specimens. The ordinary Hall coefficient R_o therefore reverses sign as a function of field. The previous work[4] has shown that the contribution of the anomalous Hall effect at temperature of 4.2 K and below is negligible compared to the ordinary Hall voltage.

The sign reversal discussed above is only observed for field directions very close to [100] and current along [010]. If the field is rotated a few degrees away from [100] the sign reversal is no longer observed as shown in Figure 9. This would indicate that the Hall effect at the precise [100] field direction is dominated by the unique orbit topology that develops for the deep [100] minimum observed in the magnetoresistance. Comparison with the Hall resistivities measured for other field and current directions is shown in Figure 10. For current along [110] and field again in the [100] direction, the Hall resistivity follows essentially the same curve as observed for the [100] current direction although a smaller range of $\omega_c\tau$ is covered due to the lower purity of the crystal. Whether a reversal of the Hall voltage would occur at sufficiently

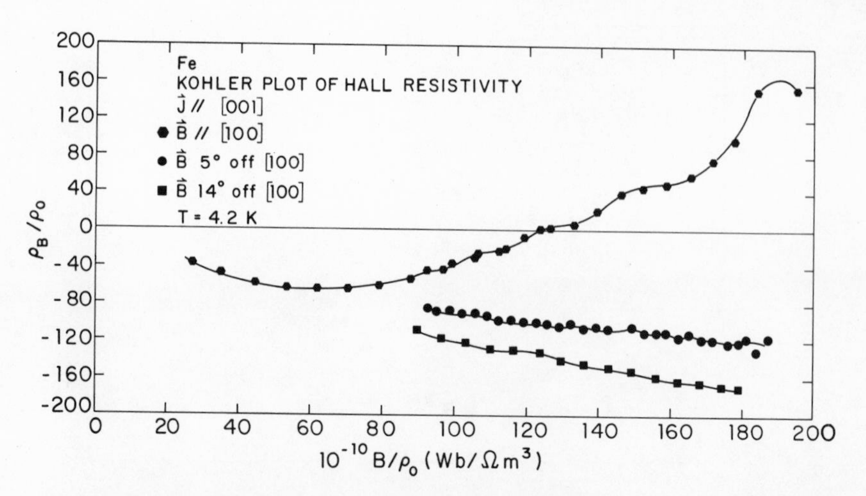

Fig. 9. Kohler plot of Hall resistivity showing sensitivity to field orientation near the [100] direction. Minimum and sign change in ρ_H are only observed within a few degrees of [100] for fields in the range 20 to 200 kG. (Ref. 5)

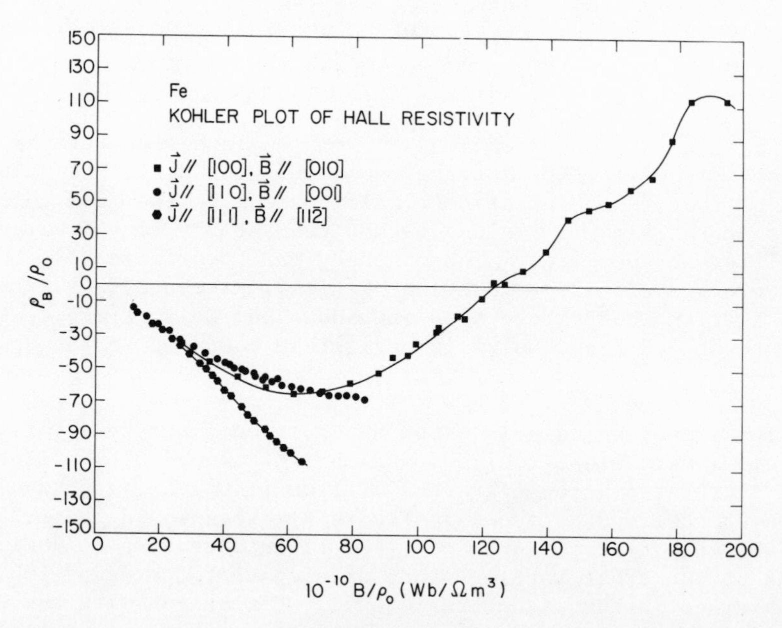

Fig. 10. Kohler plot of Hall resistivity observed for different current and field directions in the field range 20 to 200 kG. At low fields ρ_H is negative and shows the same magnitude for all three cases. Different behaviors are observed in the high-field limit. (Ref. 5)

high $\omega_c\tau$ cannot be checked at present, but is probable. The lowest curve of Figure 10 corresponds to field in the [112] direction and current in the [111] direction. In this case a more negative Hall resistivity is observed and no trend toward sign reversal is evident although it is possible that at much higher $\omega_c\tau$ values a reversal could also occur for this combination of field and current directions. The magnetoresistance shows a lesser minimum for this current and field orientation. At the highest range of $\omega_c\tau$ the Hall voltage for the [100] field direction shows an increasing oscillatory component at a frequency in the range 1.2 to 1.4 MG corresponding to the dominant frequency observed in the dc magnetoresistance.

C. dc Magnetoresistance

The dc magnetoresistance has been measured in fields up to 230 kG and the general features have been previously reported.[11] The transverse magnetoresistance rotation diagrams show a large anisotropy with deep minima occurring for field directions lying in {100} and {110} planes and lesser minima occurring for field directions lying in other low index planes as shown in the stero diagram of Figure 11. In fields of 200 kG the transverse magnetoresistance rotation diagrams of the highest purity crystals show anisotropies corresponding to a factor of ~6 in the magnetoresistance. The deepest minima are observed for field direction along <100> and examples are shown in the rotation diagram of Figure 12 for a <100> axial crystal. The field dependence of the transverse magnetoresistance at these minima follows a power law $\Delta\rho/\rho_o \sim B^n$ where $n > 1$ at low fields and $n < 1$ at high fields. For the highest values of $\omega_c\tau$ and for field in the <100> direction the lowest observed value of n has been 0.26. For fields in general directions not corresponding to minima the transverse magnetoresistance does not saturate and the values of n remain in the range $1 < n < 2$. Examples of the field dependence of transverse magnetoresistance for the two types of field direction are shown in Figure 13 corresponding to a range of $\omega_c\tau$ from less than 1 to 250.

As pointed out in previous papers[11,12] the values of n measured for a given field direction over a range of purity and field show changes that suggest changes as a function of field in the character of the orbits contributing to the magnetoresistance. For most field directions the value of n at low fields (> the magnetic saturation field of ~15 kG) has been observed to approach a value of 1.7 to 1.9 and then to decrease at much higher field. Such behavior can result from magnetic breakdown[13] at high fields with a consequent change in orbit topology, from intersheet scattering[14] at selected points on the Fermi surface, or from direct effects of the change in $\omega_c\tau$ particularly if a complex Fermi surface and orbits with heavy cyclotron masses are involved. The field dependence at the minima is further complicated by the presence of large amplitude

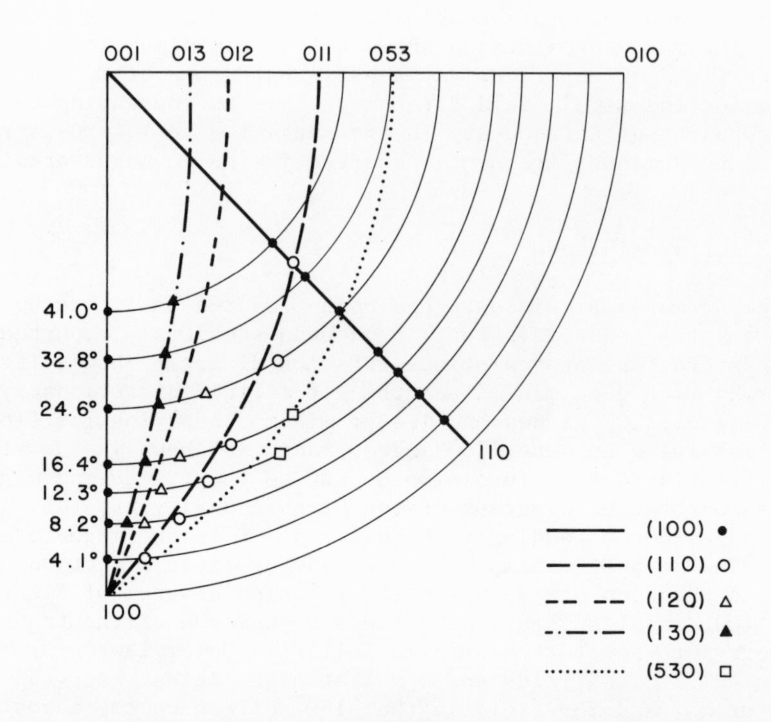

Fig. 11. Steroplot of field orientations at which deep minima are observed in the magnetoresistance rotation diagram for current along the [001] direction. Field sweeps are at 90° to [001] and at selected angles to [001] as indicated in figure. Minima occur when field crosses (100), (110), (130) and (530) planes. (Ref. 5)

low frequency dHS oscillations which contribute oscillatory components with periods of 20 kG or more at 200 kG. An example of a field sweep is shown in Figure 14 where above 120 kG rather abrupt changes in slope are observed. Some of this structure suggests a nonoscillatory change in slope possibly characteristic of magnetic breakdown[13] although a beat pattern of several low frequency oscillations might also contribute.

The slope of the magnetoresistance is also extremely sensitive to orientation and some of the observed variation can arise from small changes in orientation although for one setting of the crystal the slope changes as shown in Figure 14 are reproducible. The field

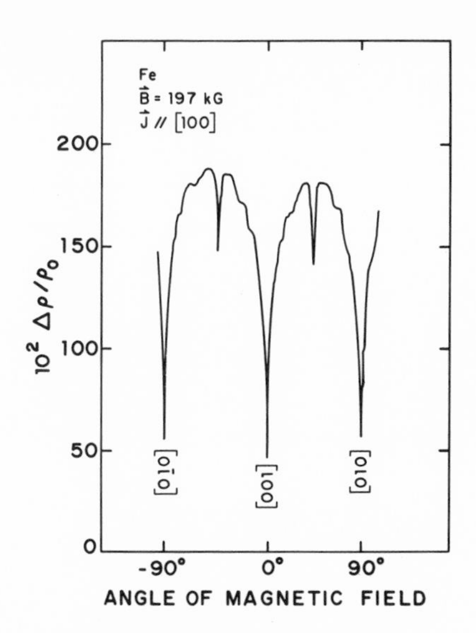

Fig. 12. Field rotation diagram recorded at 197 kG and 4.2 K for current along [100]. Deep minima in <100> field directions represent a reduction of ~80% in the magnetoresistance while the <110> minima show less than a 25% reduction. At this field value the higher order minima are practically quenched. (Ref. 5)

dependence of magnetoresistance measured at the <100> for two of the highest purity <100> axial crystals are shown in Figure 15. For fields below 90 kG the exponents n observed over the same field range are approximately the same for all specimens while above 90 kG variations are observed which do not scale directly with purity unless special care is taken to orient the specimen. The highest purity specimen shows the closest approach to saturation with n=.23. This sensitivity to orientation near the <100> pole is also observed in results from the Hall measurements and from the oscillatory magnetoresistance and this will be discussed in later sections.

The longitudinal magnetoresistance shows a linear magnetic field dependence for most of the $\omega_c\tau$ range studied in the present experiments. Only the highest ratio <100> specimen shows a slight departure toward saturation in the range above $\omega_c\tau = 250$.

For most field directions the frequencies of the largest amplitude oscillations observed in the dc magnetoresistance are in the range 1.2 to 1.4 MG. Higher frequencies with varying amplitudes are also observed and for the highest purity specimen with field in the <100> direction frequencies in a range up to 48 MG have

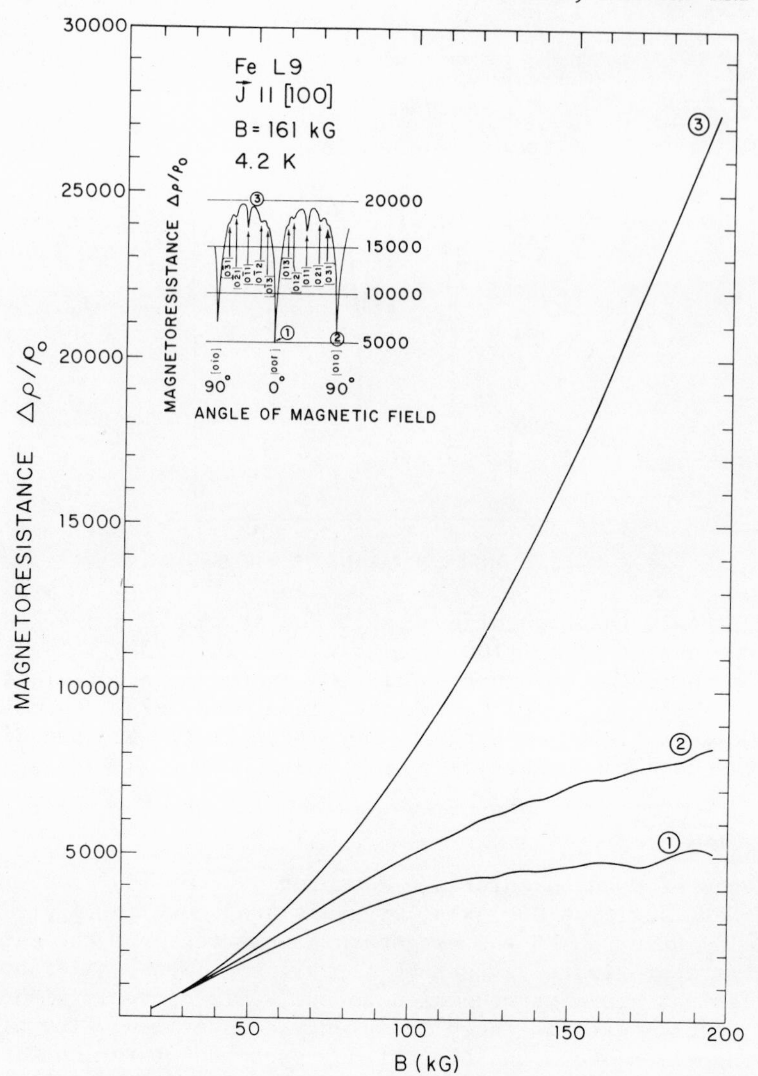

Fig. 13. Field dependence of magnetoresistance observed at 4.2 K for current along [100]. Field orientations correspond to (1) H//[001], (2) H//[010] and (3) H parallel to maximum in rotation diagram of Fig. 12.

been resolved. Figure 16 shows a dc magnetoresistance sweep in the range 100 to 200 kG for field in the <100> direction. The frequencies have been analyzed using a fast Fourier transform computer program and also by a standard graphical construction. The graphical constructions give frequencies of 1.3±.2, 4.9±.2 MG and 11.2±.2 MG. The first two frequencies are essentially in agreement with the previous analysis of magnetoresistance oscillations

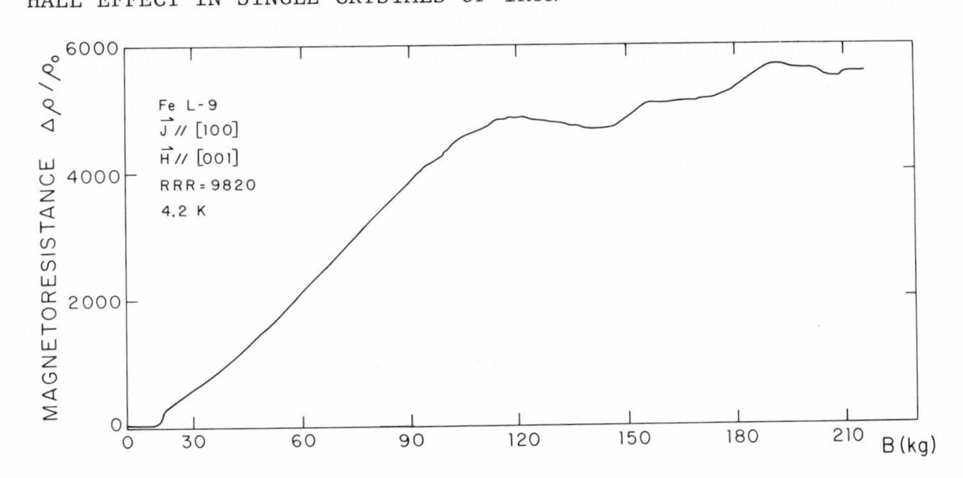

Fig. 14. Field dependence of magnetoresistance observed at 4.2 K for current along [100], irregular oscillations are observed above 120 kG.

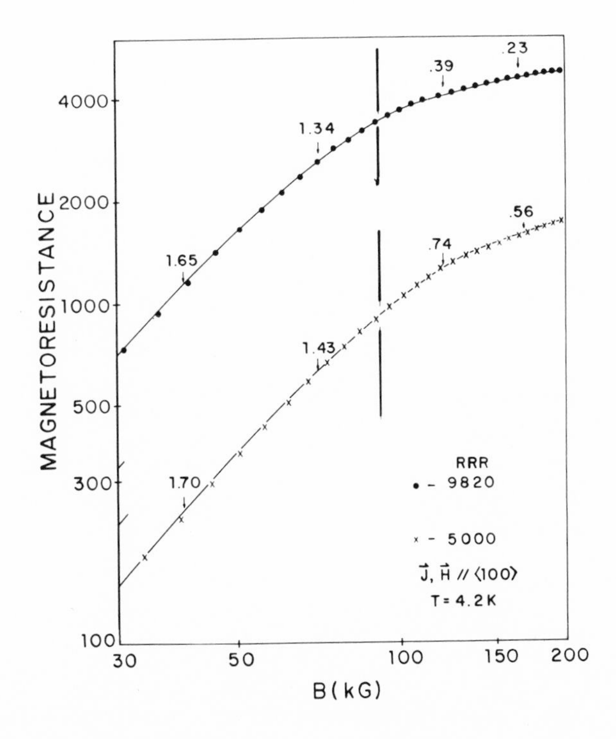

Fig. 15. $\ell n\ \Delta\rho/\rho_o$ versus ℓn B for <100> axial crystals of two different purities. Numbers along curve are values of the exponent n in the relation $\Delta\rho/\rho_o$=const. B^n. (Ref. 5)

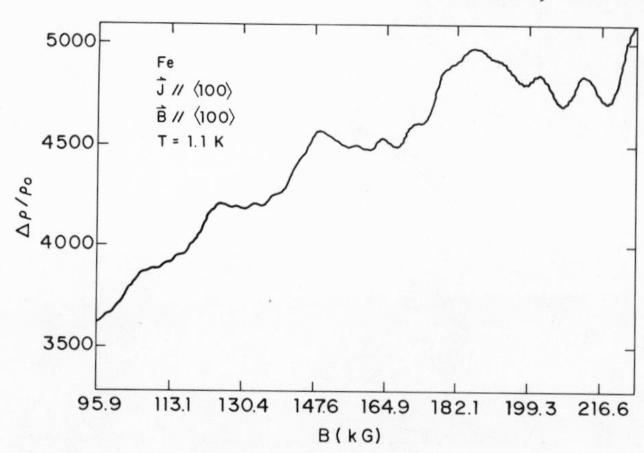

Fig. 16. Quantum oscillations observed in the dc magnetoresistance for the <100> field direction. Field range 95-230 kG, frequencies are ~1.1, 4.7 and 11.4 MG. (Ref. 5)

given by Coleman, et al.[11] and by Angadi, et al.[15] The new frequenc at 11.2 MG is observed only if the crystal is carefully oriented wit field along <100>.

D. ac Magnetoresistance

Use of an ac modulation field and second harmonic detection is a much more sensitive method for recording the oscillatory component of the magnetoresistance than extracting it from the dc magnetoresistance sweep. The absence of the rising dc background magnetoresistance also improves the quality of the fast Fourier transform analysis and allows a more detailed separation of frequencies. The ac modulation technique has been applied over magnetic field ranges between 30 and 160 kG using a large number of field and current orientations. Crystals with residual resistance ratios in the range 5,000 - 10,000 were used in the ac experiments. A large number of frequencies have been resolved and Figure 17 shows examples of the direct oscillatory data as a function of magnetic field for several low index field directions. Examples are shown for the two major dc field ranges used for the ac modulation experiments. The range 30 - 80 kG was covered using superconducting solenoids which allowed high sensitivity detection and lower field modulation. The ranges 30 - 160 kG and 100 - 160 kG were covered by modulating the Bitter solenoids and required a higher modulation field and less sensitive detection. The data was recorded on paper tape and was analyzed using the same fast Fourier transform computer program as used for the dc data and the frequency analysis resulting from this program will be discussed below.

(a)

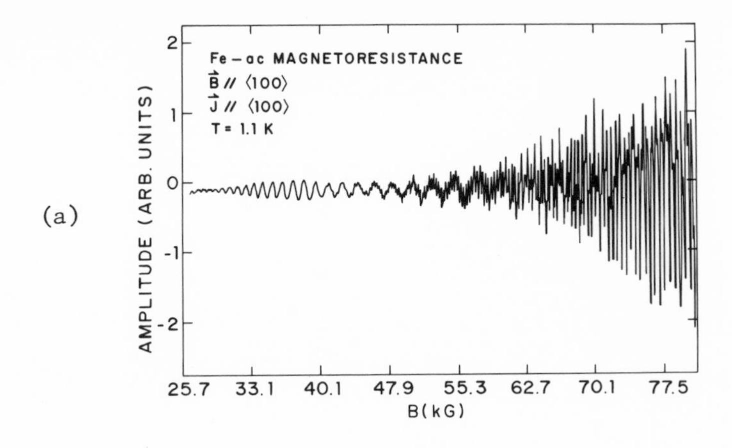

(b)

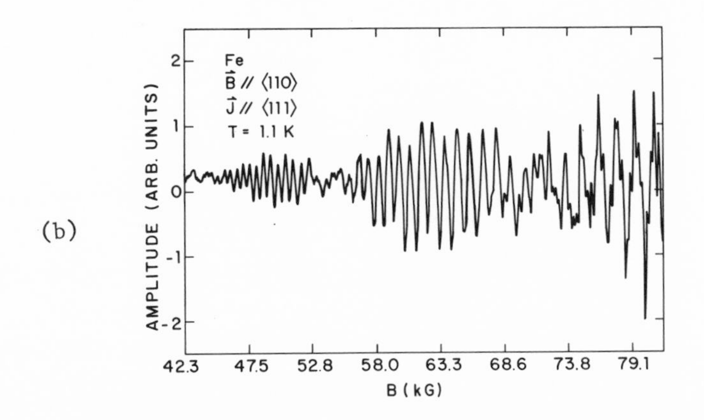

Fig. 17. Representative examples of magnetoquantum oscillations observed in Fe single crystals. Figures are recorder tracings of the lock-in amplifier output as a function of magnetic field.
a) B// <100> for field range 25-80 kG, modulation was ~300 G.
b) B// <110> for field range 42-80 kG, modulation was ~300 G.

The most comprehensive data has been obtained for the low index field directions <100> and <110>. The consistency is quite good on the major frequencies and most variations result from differences in the residual resistance ratio of the crystal or changes in modulation amplitude, detection sensitivity and field range of the sweep. In addition some frequencies are extremely sensitive to

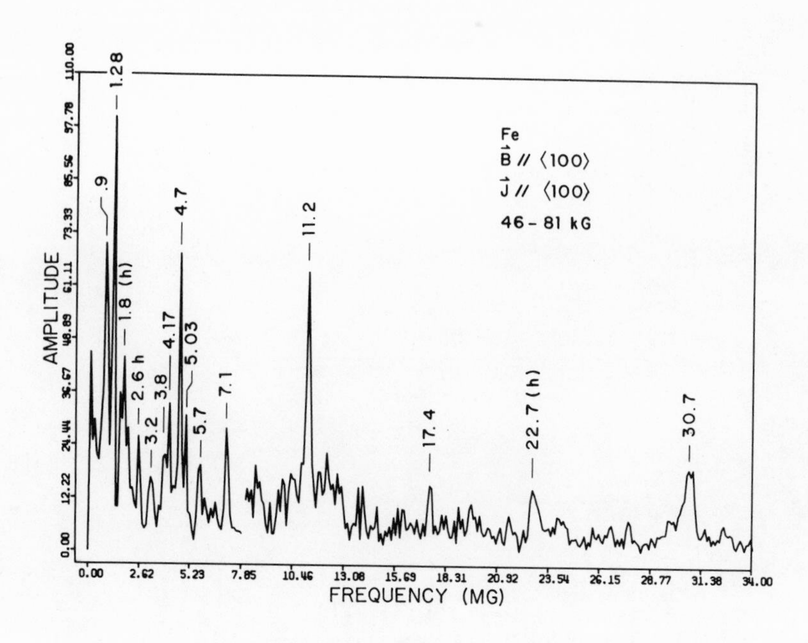

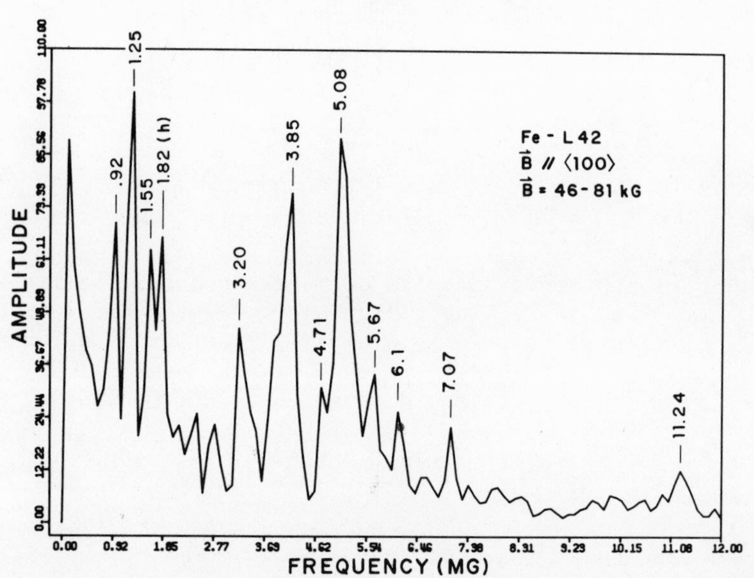

Fig. 18. Fast Fourier transforms obtained for different field ranges and modulation amplitudes with field parallel to <100>. a) 46-81 kG, modulation ~300 G, strong interference frequency at 11.2 MG is observed. b) 49-155 kG, modulation ~2500 G, low frequencies are emphasized. (Ref. 5)

orientation and the relative amplitude can change from run to run due to slight differences in orientation due to resetting of the crystal in the magnetic field. As pointed out in the section on dc magnetoresistance the 11.2 MG frequency observed for the <100> field direction is such a case and both it and the second harmonic at 22.4 MG can completely dominate the ac magnetoresistance sweep when the crystal is precisely oriented in the <100> field direction.

The lowest frequencies observed in the ac results are in the range .9 to 1.5 MG. For field orientation along <100> the ac data and analysis generally resolve several separate frequencies below 1.5 MG rather than the single dominant oscillation apparent in the dc data at low frequency. Figs. 18(a) and 18(b) show fast Fourier transforms obtained for field in the <100> direction on the highest ratio specimens measured in the present experiments. Frequencies of approximately .9, 1.2 and 1.28 MG are resolved. These frequencies are sufficiently reproducible from run to run to establish the presence of three separate frequencies although the precise values are probably not determined to better than $\pm$ .1 MG. The pair of frequencies at 1.28 and 1.55 MG can be identified with the ellipsoids at N formed from the minority spin bands and a corresponding pair of frequencies are observed in de Haas van Alphen data.[16] The .9 frequency has not been observed in de Haas van Alphen data and cannot be immediately identified with a calculated feature of the band structure although small interference orbits can be identified on the hybridized topology.

The ac results show a number of closely spaced frequencies in the range 3.8 to 5.6 MG. For field in the <100> direction a pair of frequencies at approximately 3.8 MG and 4.1 MG is resolved in many of the runs where resolution is maximum although this close frequency spacing is nearly at the limit of our resolution. A single fairly strong frequency is observed at 5.1 MG. The three frequencies cited above are also observed in de Haas van Alphen measurements[17] and there is good agreement with the above values. The magnetoresistance runs show additional frequencies at 4.7 and 5.6 MG. These show weaker amplitudes in the Fourier transforms, but are detected in most of the runs for the <100> field direction. The 4.7 MG frequency is variable in amplitude from run to run, but is observed with sufficient resolution to show that it is separate from the 5.1 MG frequency. In the magnetoresistance data the 4.7 MG frequency becomes more dominant in the crystals with RRR's less than 5000 while the 5.1 MG frequency appears to dominate in the transforms from the higher ratio crystals. Previous dc magnetoresistance results by Coleman, et al.[11] and Angadi, et al.[15] both reported frequencies in the range 4.9 MG while the de Haas van Alphen results measured the 5.1 MG frequency. The present results clarify this point by showing that the two frequencies can be simultaneously present as shown in the high resolution transform in Fig. 18(a).

A frequency is observed at 7.1 MG for <100> field directions. A frequency near this value was also observed in de Haas van Alphen measurements[16] and assigned to a harmonic, but we can interpret it as a difference frequency. The ac magnetoresistance results show a weak Fourier component at 10.2 MG in a number of the transforms which we identify as the second harmonic of the 5.1 MG frequency.

For field in the <100> direction the de Haas van Alphen results identify two frequencies at 20.6 and 23.8 MG. These are assigned to the intermediate H centered hole pocket arising from the majority spin bands.[16] These are detected with reproducible but weak amplitudes in the transforms of the ac magnetoresistance data.

The second harmonic of the 11.2 MG frequency appears at 22.3 MG and is clearly distinct from the frequencies associated with the hole pocket. The fundamental at 11.2 MG is a new frequency and must arise from the special orbit topology affecting the magnetoresistance for field precisely oriented along <100>.

For fields in the <100> direction the ac magnetoresistance results also show higher frequencies at 17.4 and 30.7 MG as indicated in the Fourier transform of Fig. 18(a). These show weaker amplitudes, but are reproducible and have also been confirmed in graphical plots construed directly from the recorded field sweep data. The 17.4 MG frequency has not been observed in the de Haas Van Alphen experiments to date. A de Haas van Alphen frequency at 15.5 MG is identified with the small H centered majority spin hole pocket and the 2nd harmonic would occur at 31 MG. The dHS frequency at 30.7 MG might be so identified, but the fundamental is not resolved for some reason.

The present ac magnetoresistance experiments do not resolve higher frequencies primarily because of experimental limitations. Lower temperatures in the range below 1 K will be required to substantially decrease thermal broadening of the Landau levels and improvement of the electronic field control particularly in the high field range will be required. The crystals themselves are clearly of a quality that offers no limitations to the observation of the much higher frequency Fermi surface sections.

This summary of magentoresistance data will be used in reference to later discussion of the Fermi surface topology. Further deatils of the ac magnetoresistance experiments can be found in reference 5.

IV. DISCUSSION

The temperature and magnetic field dependence of the Hall effect show a number of features which will require a fairly detailed analysis of all the transport mechanisms which can contribute in a ferromagnetic metal. We will discuss the major mechanisms playing a role in both the ordinary and ferromagnetic Hall effect and attempt to interpret the main features observed in the experiments.

A. Ordinary Hall Effect

The major feature to be explained in the ordinary Hall effect is the change in sign from positive to negative observed as the temperature is decreased in the range near 70 K.

The mobilities of holes and electrons can be quite different when carriers in different bands exhibit widely different $\omega_c\tau$ values for a given magnetic field and temperature due to different effective masses or relaxation times. These differences can be expected to play a role when the experimental conditions are such that the specimen makes a transition from the low-field, ($\omega_c\tau \ll 1$), to the high-field, ($\omega_c\tau \gg 1$), limit over the range of temperature in question or when the basic scattering mechanism changes as a function of temperature. Particular sections of the Fermi surface may also behave locally as electron-like or hole-like depending on the value of $\omega_c\tau$. For example Bragg reflections can come into play as $\omega_c\tau$ increases and these can also change the relative character of the orbits. Further discussion of this general problem can be found in the book by Hurd.[18] The estimates of $(\omega_c\tau)_{Av}$ obtained for the specimen RWK-5 give $(\omega_c\tau)_{Av} \approx 0.4$ at 50 K and $(\omega_c\tau)_{Av} \approx 0.1$ at 80 K when B = 56 kG. This estimate would tend to indicate that the sign change in R_o occurs at temperatures where most of the carriers are still in the low-field limit. However, the estimates of $(\omega_c\tau)_{Av}$ are only approximate and the fact that only about 5% of the d electrons are judged to be itinerant[19] makes it difficult to evaluate the role of mobility precisely. A previous study[3] on a much less pure specimen (RRR = 11.5) does not show a sign change in R_o as a function of temperature and suggests that a strong temperature dependence of mean free path must play some role.

Fivaz[20] has explored the idea that the temperature dependent motion of the Fermi level through a critical crossing point in the band structure could account for the sign change by changing the number of high mobility electrons. Such motion would result from a temperature variation of the ferromagnetic splitting. The accidental degeneracy of the levels Δ_2 and Δ_5 in the minority (spin-down) bands of iron would be a good case for the occurrence of such a mechanism. Band structure calculations[21] and recent experimental analysis[5,22] place the Δ_2 - Δ_5 crossing just below the Fermi level. However, measurements by Lonzarich and Gold[23] have shown that the temperature dependent exchange splitting in iron is very small and would not be expected to cause a significant shift of Fermi level.

An alternative explanation for the sign change in R_o has been developed by Cottam and Stinchcombe[24] based on anisotropy of the Fermi surface and relaxation times. They use a two band model in the low-field limit and introduce parameters to describe the effects of anisotropies in the Fermi surface and relaxation times. In relatively pure specimens they argue that at high temperatures phonon-electron scattering will dominate, at intermediate temperatures Baber or electron-electron scattering[25] will dominate, while at the lowest temperatures impurity scattering will dominate. When appropriate values of the anisotropy parameters are used to describe a d band and an s band the model leads to positive values of R_o at high and low temperatures with negative values at intermediate temperatures in the range below 60 K.

The data up to 150 kOe at 4.2 K extend the range of $\omega_c\tau$ considerably and this data is plotted on a Kohler plot as shown in Figure 19. For values of B/ρ_o at 4.2 K corresponding to 100 kOe and above the Hall resistivity becomes linear and the resulting Kohler plot is of course linear as well. Extrapolation of this linear portion to B = 0 gives a zero intercept as shown by the solid line in Figure 19 indicating that $R_s = 0$ at 4.2 K. Data from previous experiments on pure Fe and on an Fe + 0.25% Co alloy are also included in Figure 19. Discussion of deviations from Kohler's rule at higher temperatures can be found in reference 4.

B. Ferromagnetic Hall Effect

The ferromagnetic Hall coefficient R_s is associated with that part of the transverse voltage which depends on the magnetization of a magnetic metal. Many theoretical papers have been written on the subject and all agree that some aspect of the spin-orbit coupling interaction is responsible for the effect. In the presence of the spin-orbit interaction electrons with spin polarizations parallel and antiparallel to the magnetization are deflected in opposite directions at right angles to the electric current. If the two spin populations are unequal then a net transverse current appears which is cancelled by the resulting Hall voltage.

In the original work of Karplus and Luttinger[26] the transverse current was associated with the force due to the spin-orbit part of the periodic potential and gave rise to a nonvanishing effect only if the carriers in a given band were coupled to other bands through the spin-orbit interaction. Strachan and Murray[27] made a somewhat more complete calculation of these effects using conventional quantum-mechanical transport theory to consider the motion of the electrons under the influence of electric field components

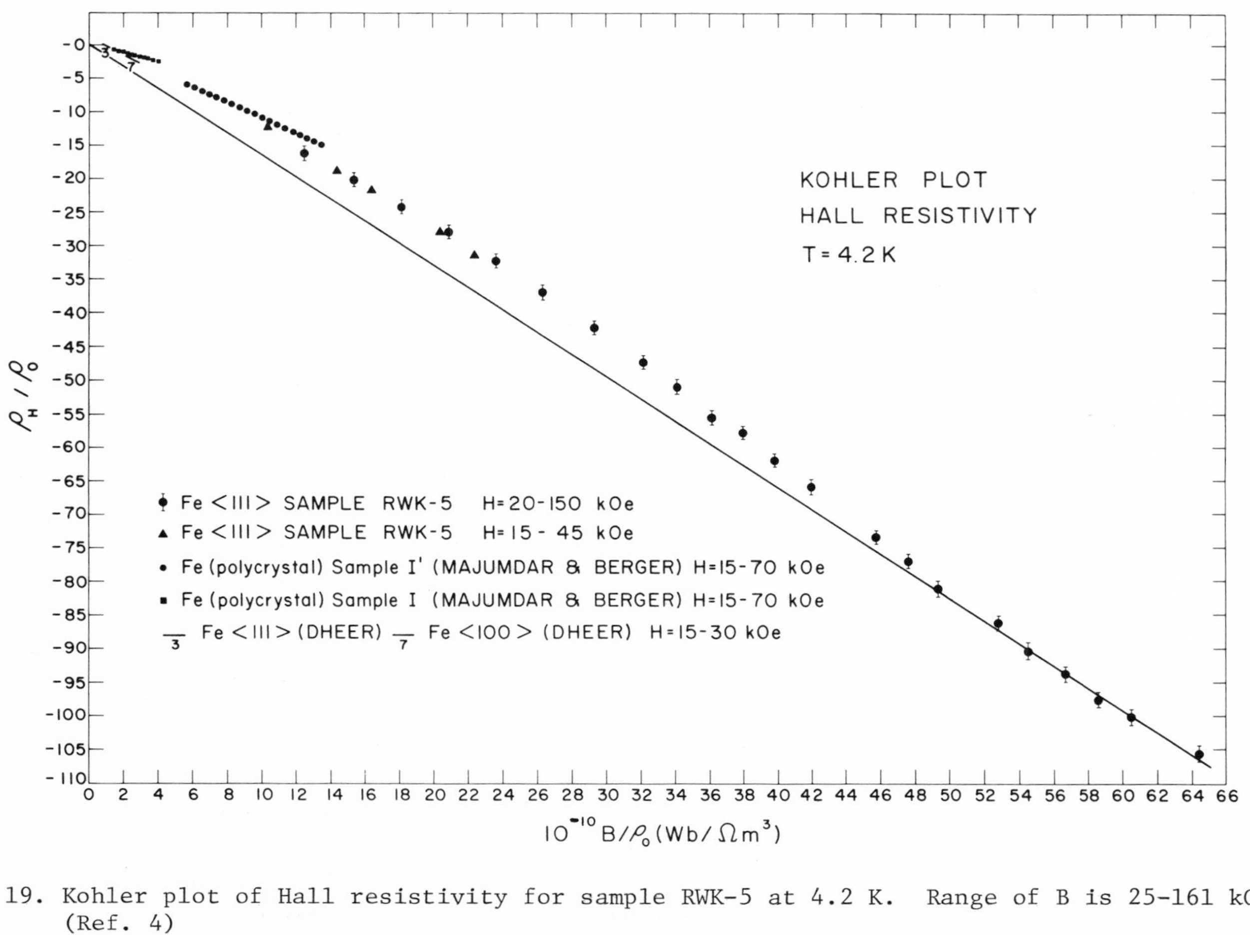

Fig. 19. Kohler plot of Hall resistivity for sample RWK-5 at 4.2 K. Range of B is 25-161 kG. (Ref. 4)

and spin-orbit coupling allowing for all possible intra- and inter-band effects. Subsequent treatments of the problem have been carried out by Smit,[28] Kondo,[29] and Maranzana[30] in which the spin-orbit part of the perturbing potential of the scattering center as well as the central periodic part has been considered. This analysis results in a skew or asymmetric scattering of the electrons and gives a nonvanishing result with the required symmetry only when the scattering probability is calculated in the second Born approximation. In general, calculations of skew scattering in the second Born approximation using plane waves as unperturbed wave functions give results which are much too small to explain the magnitude of the experimental results.

Leroux-Hugon and Ghanzali[31] have reconsidered the skew scattering by using wave functions which properly include the periodic part of the spin-orbit interaction and when calculated in the second Born approximation the transition probability leads to a transverse resistivity in the Hall geometry which they estimate to be of the right order of magnitude to agree with experiment. The influence of the periodic part of the spin-orbit interaction on the skew scattering has also been considered in a theory by Fivaz.[32]

In addition to the contributions considered in the above theories Berger[33] has recently considered nonclassical terms in the Botlzmann equation which become important for very short mean free paths corresponding to conditions where the parameter $h/\varepsilon_F\tau$ is appreciable. ε_F is the Fermi energy and τ is the relaxation time of the electrons. These terms probably become important at high temperatures or in concentrated alloys. When calculated in the presence of spin-orbit coupling these terms result in a side-jump displacement at every scattering of the electron on impurities or phonons. This displacement is on the order of 10^{-11} m and results because the impurity or phonon distorts the wave function locally and creates a local current density. For short mean free paths and a random distribution of impurities or phonons interference phenomena can be neglected and the first order Born approximation is sufficient to obtain a contribution to the ferromagnetic Hall resistivity of the right order of magnitude.

In high purity metals at reasonably high temperatures, the electron scattering is dominated by the electron-phonon interaction and Leribaux[34] was the first to calculate the ferromagnetic Hall coefficient specifically in the electron-phonon interaction regime. Leribaux used Kubo's formal solution of the transport problem[35] to calculate the antisymmetric component of the ferromagnetic Hall conductivity tensor and derived an expression for the transverse conductivity that resulted from a term of zero order in the electron-phonon interaction. The results have been specifically applied to iron by evaluating matrix elements of the spin-orbit operator between wave functions constructed to properly represent Wood's[36] dispersion curves for iron. The main contributions came from bands close to the Fermi surface and using symmetry arguments Leribaux suggested that the major contribution results from the spin-up band that is nearly spherical about the Γ point. Leribaux's expression for the ferromagnetic Hall coefficient is

$$R_s = \frac{20.9}{4\pi M_s(T)} \rho_{xx}^2 [1 + 1.12 \times 10^{-8} T^2] \; \Omega \text{ cm/G} \tag{5}$$

This essentially assigns all of the temperature dependence to ρ_{xx}^2 as was also a consequence of the original Karplus and Luttinger theory where the inversion of the conductivity tensor automatically results in a Hall resistivity proportional to the resistivity squared.

Berger's theory of the side-jump mechanism also leads to a dependence of R_s on ρ^2 and his expression based on an estimated value of the effective spin-orbit enhancement factor is

$$R_s = \left(\frac{\sigma_{xy}}{4\pi M_s}\right) \rho^2 = [10^2/4\pi M_s] \; \rho^2 \, \Omega \text{ cm/G}. \tag{6}$$

The proportionality factor is therefore approximately a factor 5 greater than that derived in Leribaux's theory.

In the present experiment the high temperature data in the range 100 - 247 K obeys an expression of the form

$$R_s = (1.44 \times 10^3/4\pi M_s) \; \rho^{1.94} \tag{7}$$

where R_s is in (Ω cm/G), M_s in (Gauss) and ρ in (Ω cm) as indicated in the plot of Figure 20.

The expected dependence on ρ^2 is clearly confirmed, but the constant of proportionality is larger than estimated in either theory. However, the estimates made in the theories are probably only valid to within an order of magnitude and the present results are not unreasonable with respect to such estimates. Our result is essentially in agreement with that of Dheer[1] who found a somewhat lower coefficient of proportionality, his average value being equal to $9.3 \times 10^2/4\pi M_s$.

Below 100 K the values of R_s are difficult to extract precisely due to the field dependent curvature developing in the Hall resistivity. By using either a linear or quadratic extrapolation as previously shown in Figure 5 an anomalous maximum near 40 K is observed for R_s. In the linear extrapolation an anomalously high value of R_s is also observed at 4 K and below as was previously reported by Dheer.[1] The 4 K anomaly is clearly a result of the

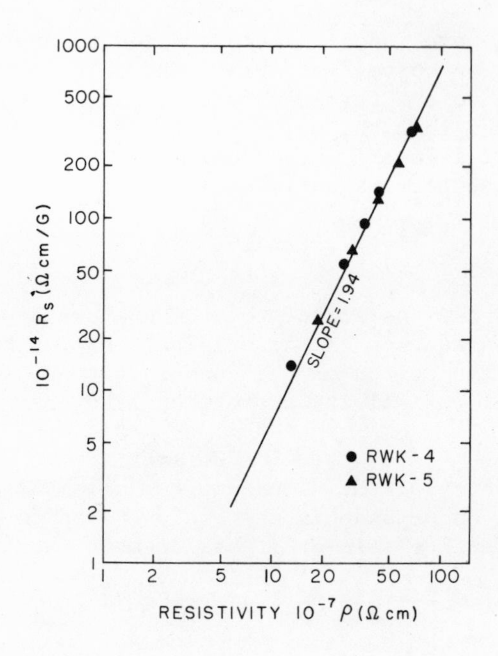

Fig. 20. Log-log plot of the ferromagnetic Hall coefficient R_s as a function of resistivity. Temperature range for data points is 80 - 247 K. (Ref. 4)

curvature and is removed in a quadratic analysis or in an extrapolation using only data for B greater than 100 kG at 4.2°K.

The anomalous maxima at 40 K is a reflection of the deviation from a smooth Kohler plot in this region and cannot be easily removed unless one ignores the vanishing of the curvature at 31 K and approximates the behavior with a smooth Kohler plot with continuous downward curvature over this region.

C. Fermi Surface and Band Structure

A substantial number of band structure calculations[21,37-41] have been made for iron and the general Fermi surface topology is represented in Figure 21. The unhybridized majority (spin up ↑) and minority (spin down ↓) sheets are shown and consist of four sheets belonging to the majority spin Fermi surface and four sheets belonging to the minority Fermi surface. Analysis of the de Haas van Alphen data and comparison to band structure calculations lead Gold, et al.[17] to conclude that the minority Fermi surface gave rise to a "jack" configuration similar to the Fermi surfaces found in molybdenum and tungsten. This consists of a central electron sheet, electron balls located along Δ and a hole octahedron

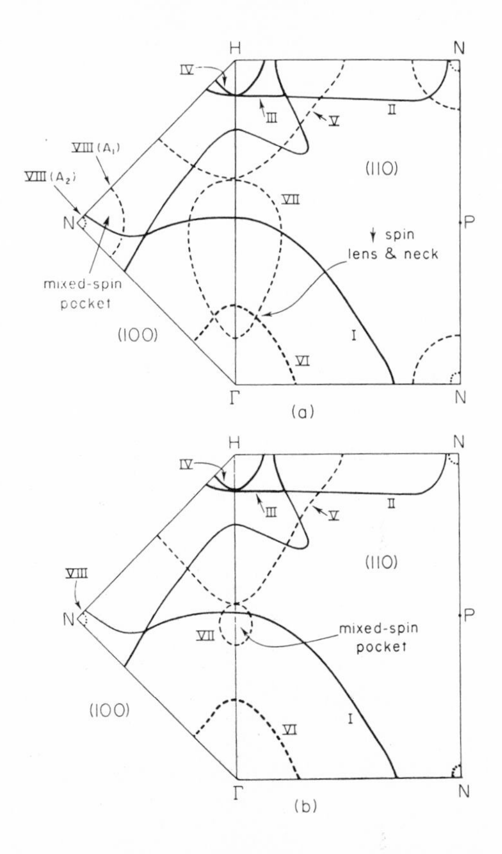

Fig. 21. Fermi surface cross sections in <100> and <110> planes. Extremal dimensions are consistent with existing de Haas van Alphen data where precise assignments are known. Dashed lines represent minority spin down Γ Fermi surface and solid lines represents majority spin up ↑ Fermi surface. a) Model with large minority electron ball (VII) overlapping minority electron surface centered on (VI). b) Model with small minority electron ball (VII); no overlap with surface VI exists. This model is consistent with the band structure calculations of Callaway and Wang.[21] Both models have been constructed by Lonzarich.[22]

centered on H. In this model the electron balls intersect the hole octahedron and spin-orbit coupling results in the formation of a small electron "lens" located inside a neck. In addition to the "jack" the minority Fermi surface contains small ellipsoids at N first detected in magnetoresistance[11,15] and later confirmed by dHvA data.[42]

In contrast to the above model some band structure calculations predict a relatively small electron ball so that no intersection with the hole octahedron occurs. In this case no "lens" and "neck" are formed. Reasonably consistent identifications of the observed dHvA and dHS frequencies can be made using the model with the relatively small electron ball and certain experimental details are more favorable to this alternative model which would require a revision of the detailed assignments for a number of the frequencies in the range 1 to 20 MG. These have previously been interpreted in terms of the "neck" and "lens".

The majority spin Fermi surface consists of hole arms running along <110> a large electron surface centered on Γ and two concentric hole pockets centered on H. The detailed topology near the point N is still subject to uncertainty. Gold, et al.[17] proposed that the large majority electron surface has substantial intersection with the arms and hybridization pinches them off so that the arms are no longer continuous near N. In some calculations and models the minority hole ellipsoids at N are quite large so that they intersect the arms. Hybridization would then be expected to produce small mixed spin pockets as well as necks. At present the experiments generally support small ellipsoids at N.

D. Hall Effect at the High-Field Limit

As pointed out in section IIIB the field dependence of the ordinary Hall resistivity shows a minimum which occurs at the same value of $\omega_c\tau$ when τ is varied by approximately a factor of two. Sign reversals of this type have been observed in cadmium[43] and cannot be attributed to magnetic breakdown since the onset of magnetic breakdown should be observed at a critical value of magnetic field rather than at a critical value of $\omega_c\tau$. In the case of cadmium Young, et al.[14] examined the influence of small-angle intersheet scattering between different sheets of the Fermi surface and were able to obtain excellent agreement between the theory and observed experimental results on both the magnetoresistance and Hall effect. In the intersheet scattering mechanism a local scattering time is introduced which describes a tunneling process between the two sheets at a local "hot spot". In contrast to magnetic breakdown the probability of scattering decreases with increase of magnetic field since the time an electron spends in the region of the "hot spot" is decreased.

Whether this mechanism explains the results in iron for the <100> field direction is uncertain. In cadmium the second band hole sheet exhibits six pinched off necks which are ideal regions for the occurrence of intersheet scattering. The {100} Fermi surface of iron is much more complex and the many gaps appear to be magnetic breakdown gaps which contribute stronger open orbit effects at high fields. Although complex the topology does not offer a unique set of scattering points where the dominant orbit topology can be changed by cutting off the scattering process.

The general experimental observations on iron suggest that the high-field Hall effect for the <100> field direction is dominated by the open orbit network and that the dominant holelike contribution is quenched as soon as the field is rotated so as to appreciably reduce the thickness of the open orbit region. The sign reversal could be associated directly with a straightforward critical coherence length of the electron on the open orbit network. The slow approach to saturation of the magnetoresistance is consistent with fields in excess of 100 kG being required for the open orbits to dominate even in the purest specimens. The Hall effect minima occur in the range of 75 to 100 kG and are therefore consistent with the onset of saturation in the magnetoresistance. The transition from electronlike to holelike behavior would therefore occur at the same average electron coherence length on the open orbit network. This would occur at the same $\omega_c\tau$ value for the specimens of different purity as observed.

E. de Haas Shubnikov Oscillations, Interference Orbits and Open Orbits

The Shubnikov de Haas experiments detect frequencies which agree with all of the de Haas van Alphen frequencies reported for the smaller Fermi surface sheets. In addition, new frequencies which can be associated with interference orbits are clearly resolved and for some field orientations dominate the oscillatory behavior.

The strongest new frequency observed is at 11.2 MG for field oriented along <100>. The second harmonic is observed at 22.4 MG and can exhibit an amplitude comparable to that of the fundamental. The absence of this frequency is dHvA and the high sensitivity to field orientation suggest interpretation as an interference orbit.

Interference orbits[44] result when there is a junction of Fermi surface sheets where the propagating electron state is split into a transmitted and reflected state each of which propagate along well defined trajectories, first diverging and then intersecting at a subsequent junction where coherent recombination can occur. Junctions of this type are characterized by a fractional magnetic breakdown probability which determines the transmittance and mixing properties of the junction. The total phase difference of the interfering state is magnetic field dependent and introduces an oscillatory interference term in the transverse magnetoresistance. These can be particularly strong when electrons are moving on open trajectories which traverse with interference region.

The hybridized Fermi surface of iron contains numerous gaps induced by spin-orbit coupling and provides many open electron trajectories where interference contributions to the magnetoresistance might be observed. In the presence of the ferromagnetic exchange splitting of the bands the spin-orbit splitting is also a function of the magnetic field direction[45] and this will complicate the calculation of the precise magnetic breakdown gaps. An approximate rule states that the spin-orbit effects are strongest when $\bar{M}$ is either parallel or antiparallel to the vector connecting the zone center Γ to the particular point of degeneracy being considered. For the <100> field direction this implies that most of the spin-orbit splittings will be greatly reduced or quenched in first order. This makes the size of the magnetic breakdown gaps uncertain, but introduces a sensitive angular dependence of the gap due to the rapid increase of the spin-orbit splitting as the field is rotated from <100>.

Assuming small but finite magnitudes for most of the gaps resulting from hybridization of the Fermi surface sheets shown in Figure 21 when the field is parallel to <100> gives rise to a Fermi surface topology which generates interference orbits with areas close to 11.2 MG as observed in the experimental data. These orbits are shown in Figure 22 where the complete {100} zone cross-sections centered on H are shown.

The double interferometer generated by this topology is very similar to one observed in magnesium and extensively analyzed by Stark and Friedberg.[44] In the case of magnesium two identical magnetic breakdown (MB) gaps H_1 form the ends of the interferometer with with a second different MB gap H_2 at the exact center. A feedback loop through a third MB gap H_3 was also present. The iron interferometer shown in Figure 22 has a similar geometery with respect to MB gaps H_1 but has two gaps H_2 at the center of the interferometer. It also mixes with a second interferometer thorugh the MB gaps H_3 rather than forming a closed feedback loop through H_3. The small electron ball connected through gaps H_2 does however form a small feedback loop to the interferometer. The transmission probabilities for the magnesium interferometer have been modeled in detail by Stark and Friedberg[44] and a similar analysis could be attempted for the present case although modification of detail will be required. Although exact correspondence with the magnesium interferometer cannot be established unless the required models are analyzed several results of the magnesium analysis are worth noting. Depending on the values of H_1, H_2 and H_3 it is possible for certain field ranges to generate a ratio fundamental/second harmonic which is on the order of one. It is also crucial that two open trajectories couple at the interference region in order to observe interference oscillations and this is certainly satisfied in the iron case. The open orbit toplogy generated for field in the <100> direction is shown in Figure 23. The double interferometer is

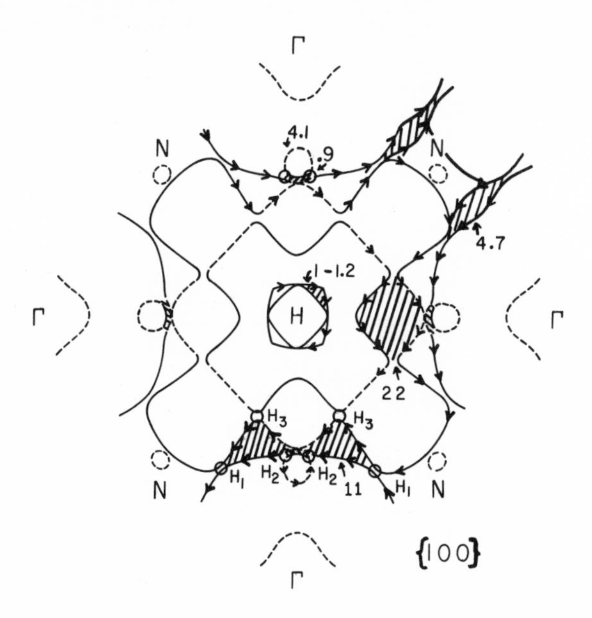

Fig. 22. Cross sections of the Fermi surface in the {100} plane showing complete section of zone around H drawn according to the model of Fig. 21(b). Shaded areas show possible interference orbits at frequencies close to 0.9, 1.2, 4.7, 11 and 22 MG. The minority electron ball can hybridize with the large majority electron surface to form a small interference orbit of 0.9 MG. A double interferometer is formed as shown in the lower part of the figure involving spin-orbit gaps H_1, H_2 and H_3. The majority hole arms are shown hybridized with the majority electron surface to form an interference orbit at 4.7 MG. (Ref. 5)

coupled into open orbits in both the <100> and <110> directions. This is also the network probably responsible for the sign change observed in the high-field Hall effect.

V. CONCLUSIONS

The Hall effect in iron is characterized by many contributions which make it one of the most complex properties in the magneto-transport studies of pure ferromagnetic metals. Most of the unusual features develop at lower temperatures where long mean free paths and high magnetic fields make it necessary to consider the detailed effects of Fermi surface topology and band structure. In this paper we have demonstrated the existence of strong non-linear effects and sign changes which result from a combination of scattering

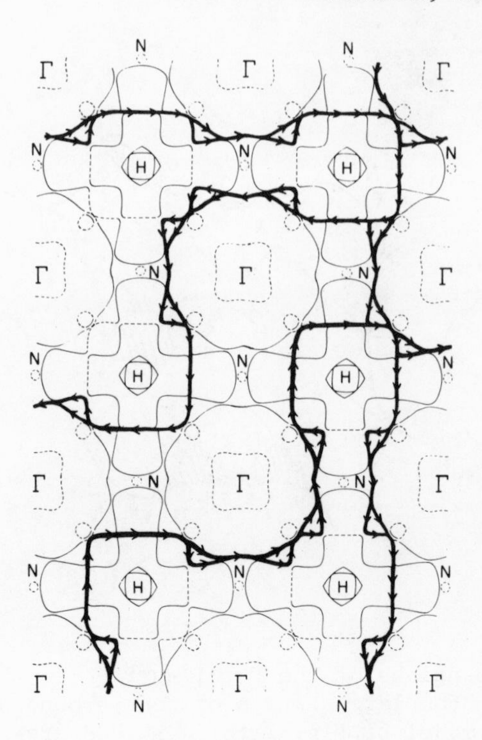

Fig. 23. <100> and <110> open orbits on the revised Fermi surface topology of the model in Fig. 21(b). In this case all orbits are hybridized orbits utilizing sections of both majority and minority Fermi surface. The open orbits in this model are also well coupled to the double interferometer. (Ref. 5)

mechanisms, magnetic breakdown and open orbit topology. In the high-field limit the Hall effect is extremely sensitive to field orientation and for the <100> field direction a field induced sign change has been demonstrated which appears to be connected with a scattering mechanism involving coherence on an open orbit network. A review of recent de Haas Shubnikov results have also been included showing evidence for interference orbits arising from the hybridized Fermi surface. A complete resolution of the details will require further experimental work but a reasonably complete picture of the specialized orbit topology and the connection to the Hall effect and magnetoresistance has been established.

REFERENCES

1. P. N. Dheer, Phys. Rev. 165:637 (1967).

2. A. K. Majumdar and L. Berger, Phys. Rev. B 7:4203 (1973).
3. N. V. Volkenshtein and G. V. Fedorov, Zh. Eksperim. i Teor. Fiz. 38:64 (1960) [English transl: Soviet Phys.-JETP 11:48 (1960)].
4. R. W. Klaffky and R. V. Coleman, Phys. Rev. 10:2915 (1974).
5. R. V. Coleman, W. H. Lowrey and J. A. Polo, Jr., Phys. Rev. B (to be published).
6. R. V. Coleman, Metall. Rev. 9:261 (1964).
7. S. S. Brenner, "The Art and Science of Growing Crystals," J. J. Gilman, ed., Wiley, New York (1963), Chap. 2, p. 30.
8. P. W. Shumate, Jr., R. V. Coleman and R. C. Fivaz, Phys. Rev. B 1:394 (1970).
9. L. Berger, Phys. Rev. 177:790 (1969).
10. R. V. Coleman, AIP Conference Proceedings No. 29, J. J. Becker, G. H. Lander and J. J. Rhyne, eds., AIP (1975).
11. R. V. Coleman, R. C. Morris and D. J. Sellmyer, Phys. Rev. B 8:317 (1973).
12. R. V. Coleman, W. H. Lowrey, R. C. Morris and D. J. Sellmyer, International Colloquium Proceedings C.N.R.S. No. 242, Grenoble (1974), p. 381.
13. L. M. Falicov and Paul R. Sievert, Phys. Rev. 138:A88 (1965).
14. Richard A. Young, J. Ruvalds and L. M. Falicov, Phys. Rev. 178:1043 (1969).
15. M. A. Angadi, E. Fawcett and Mark Rsaolt, Can. J. Phys. 53:284 (1975).
16. A. V. Gold, J. Low. Temp. Physics 16:3 (1974).
17. A. V. Gold, L. Hodges, P. T. Panousis and D. R. Stone, Int. J. Magn. 2:357 (1971).
18. Colin M. Hurd, "The Hall Effect in Metals and Alloys," Plenum Press, New York (1972).
19. M. B. Stearns, Phys. Rev. B 8:4383 (1973).
20. R. C. Fivaz, J. Appl. Phys. 39:1278 (1968).
21. J. Callaway and C. S. Wang, Phys. Rev. B 16:2095 (1977).
22. G. G. Lonzarich, in: "Electrons in Metals," M. Springford, ed.
23. G. Lonzarich and A. V. Gold, Can. J. Phys. 52:694 (1974).
24. M. G. Cottam and R. B. Stinchcombe, J. Phys. C: Solid State Physics 1:1052 (1968).
25. N. F. Mott, Adv. Phys. 13:325 (1964).
26. R. Karplus and J. M. Luttinger, Phys. Rev. 95:1154 (1954).
27. C. Strachan and A. M. Murray, Proc. Phys. Soc. 73:433 (1959).
28. J. Smit, Physica 24:39 (1958).
29. J. Kondo, Prog. Theor. Phys. 27:772 (1962).
30. F. E. Maranzana, Phys. Rev. 160:421 (1967).
31. P. Leroux-Hugon and A. Chazili, J. Phys. C: Solid State Physics 5:1072 (1972).
32. R. C. Fivaz, Phys. Rev. 183:586 (1969).
33. L. Berger, Phys. Rev. B 2:4559 (1970).
34. H. R. Leribaux, Phys. Rev. 150:384 (1966).
35. R. Kubo, J. Phys. Soc. Japan 12:510 (1967).
36. J. H. Wood, Phys. Rev. 126:517 (1962).

37. S. Wakoh and J. Yamashita, J. Phys. Soc. Jap. 21:1712 (1962).
38. K. J. Duff and T. P. Das, Phys. Rev. B 1:192 (1968).
39. R. Maglic and F. M. Mueller, Int. J. Magn. 1:289 (1971).
40. R. A. Tawil and J. Callaway, Phys. Rev. B 7:4242 (1973).
41. M. Singh, C. S. Wang and J. Callaway, Phys. Rev. B 11:287 (1975).
42. G. Lonzarich, Ph.D. Dissertation, University of British Columbia (1973).
43. C. G. Grenier, K. R. Efferson and J. M. Reynolds, Phys. Rev. 143:406 (1966).
44. R. W. Stark and C. B. Friedberg, Phys. Rev. Letters 26:556 (1971); R. W. Stark and C. B. Friedberg, J. Low Temp. Physics 14:111 (1974).
45. L. Hodges, D. R. Stone and A. V. Gold, Phys. Rev. Letters 19:655 (1967).

HALL EFFECT IN AMORPHOUS METALS

T. R. McGuire, R. J. Gambino and R. C. O'Handley

IBM, Thomas J. Watson Research Center
Yorktown Heights, New York 10598

I. INTRODUCTION

This article reviews the Hall effect in amorphous metals and amorphous metallic alloys from an experimental point of view. The theory of the Hall effect in amorphous materials is similar to that in liquid metals which is covered in the article by Ballentine (1980) in this volume. Some theoretical papers directed specifically at disordered metals are Samoilovitch and Kon'kov (1950) and Kondorskii et al. (1975 a,b).

An attempt has been made to be complete in our citations of experimental work published in this field although we have chosen for discussion only those results which allow some physical insight. Formulae have been presented in SI (Systeme Internationale) units and, where possible, in the more popular Gaussian cgs units. Conversion between the two systems is described. Data are presented and discussed in the units used in the references from which they are cited.

A. Brief Historical Survey

Investigations of amorphous metals can be traced back at least to the work of Brill (1930) on electrodeposited Ni-P films and possibly as far back as the work of Wurtz in 1845 (Brenner and Riddell 1946). More extensive investigations on electrodeposited NiP films were conducted at the National Bureau of Standards (Brenner et al. 1950) and later by Goldenstein et al. (1957). Amorphous superconducting films of Ga, Bi and Sn were studied by Buckel (1952,1954). The first splat-quenched samples were reported by

Duwez' group in the Au-Si system (Klement et al. 1960). Isolated reports of magnetism in amorphous samples can be found (eg Brenner et al. 1950) but the earliest documented deliberate efforts to observe ferromagnetism in amorphous metals are found in three sources (Mader and Nowick 1965; Bagley and Turnbull 1965; and Tsuei and Duwez 1966). A strongly ferromagnetic liquid-quenched amorphous metallic alloy $Fe_{80}P_{12.5}C_{7.5}$ was reported by Duwez and Lin (1967) and shortly thereafter Lin (1969) published what are believed to be the first Hall effect measurements in an amorphous metal. The extensive literature on the Hall effect in thin films of amorphous Gd-Co, the prototypical amorphous ferrimagnet, can be traced back to the work of Ogawa (1974). More recent references on the developments in amorphous metals can be found in numerous papers in the proceedings of the Third International Conference on Rapidly Quenched Metals (ed. Cantor 1978), and in the proceedings of the 1979 INTERMAG-MMM Conference (IEEE Trans. MAG 15 Nov 1979 and Jour. Appl. Phys. 50 Nov. Part II 1979).

B. Nature of Amorphous Metals

1. Structure. The most important tool for studying the structure of amorphous materials is X-ray scattering. Electron and neutron scattering have particular advantages for very thin films and for magnetic materials respectively, but X-rays have the most general utility. If an amorphous material is studied in the usual Bragg diffraction geometry using monochromatic radiation, the scattered intensity as a function of Bragg angle generally shows a few broad maxima. In this respect the diffraction pattern of an amorphous material strongly resembles that of a liquid. The pattern is very different from the multiple sharp intensity maxima observed in the diffraction pattern of a polycrystalline solid.

In crystal structure analysis, the symmetry elements of a periodic lattice can be used to full advantage. The symmetry operations of the unit cell are readily obtained from the symmetry of the single crystal diffraction patterns. A structural model of the unit cell is constructed incorporating these symmetry elements and the diffraction pattern of the model is calculated and compared with the observed pattern. The structure is then refined by making small adjustments in the atomic positions in the model and again comparing the calculated and observed intensities. If the initial model structure is a reasonably good approximation of the actual structure it is possible to obtain convergence of the calculated and observed intensities after a number of iterations.

Many of the techniques used in crystal structure determination are not available in analyzing the structure of amorphous solids or liquids. The general principles used are the same, however.

A model is constructed and the scattered intensity calculated for the model is compared with the observed scattering pattern. The amount of information in the scattering pattern of an amorphous material is very limited, however, so it is usually not possible to be sure the model structure is unique. Furthermore, in the crystalline case, translational symmetry insures that the position of every atom in the perfect crystal is defined. In the amorphous alloy, we can only determine an average structure, e.g. the average nearest neighbor coordination of an average atom.

As a consequence of this ambiguity in structure determination, other analytical tools must usually be used to distinguish between various possible models. High resolution electron microscopy has played an important role in this respect. Other techniques which have provided structural information are extended X-ray absorption fine structure (EXAFS), NMR, Rutherford back scattering and photoemission.

Most of the structural models for amorphous metals were originally used for liquid metals. The radial distribution function (RDF) of amorphous cobalt is compared with that of liquid cobalt in Figure 1. The RDF is obtained from the scattered intensity by a Fourier transformation of the intensity as a function of K ($K \equiv 4\pi \sin \theta/\lambda$). It represents the distribution of scattering electron density as a function of distance from an average atom. The first maximum occurs at a radius which corresponds to the first nearest neighbor distance. Note that this peak is broad, suggesting a distribution of first nearest neighbor (nn) distances. The second and higher nn peaks are increasingly weak. This is a consequence of the absence of long range order, i.e. no translational symmetry.

The first model of a disordered metal was a mechanical model constructed by Finney (1970). The Finney model consisted of hard spheres compressed into contact with each other. The Finney model was intended to simulate a liquid metal but Cargill (1970a) suggested using it for amorphous metals. Cargill (1970b) applied the Finney model to amorphous Ni-P. Knowing the coordinates, atomic scattering factors and atomic radius of the metal, the RDF of the model was calculated and compared to the experimental RDF. The agreement was generally good, however the intensity distribution in the second peak was not well reproduced by the model calculation.

Computer modeling was first used by Bennett (1972). Starting with a seed cluster, atoms are added one at a time, each atom is brought into hard sphere contact with one of the atoms already in the cluster. An amorphous structure could then be constructed by this procedure, but the splitting of the second peak in the RDF was still not well reproduced.

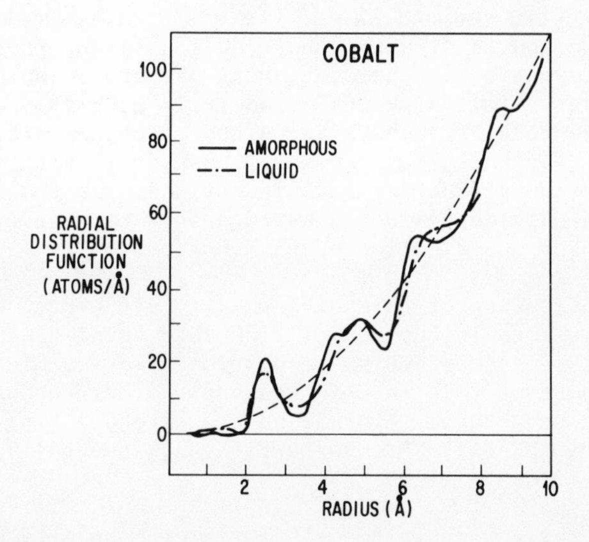

Fig. 1. Comparison of radial distribution functions for amorphous and liquid cobalt (from Leung and Wright 1974).

The most successful structural model of amorphous metals is the relaxed dense random packing (DRP) model of Cargill and Kirkpatrick (1976). The algorithm used to generate this model is similar to the Bennett model except that the atom being added to the cluster is allowed to move to the nearest position where it can make contact with three atoms in contact with each other to form a nearly perfect tetrahedron. The model is extremely successful in that it explains the split second peak in the RDF.

If the hard spheres in these models are replaced by spheres with more realistic pair-wise interaction potentials it is possible to relax the structure to obtain agreement with the observed density. Relaxation also improves the local tetrahedral order (Barker et al. 1975). We will use the relaxed DRP model in subsequent discussions of amorphous alloy structure.

The relaxed DRP model can also be obtained by another algorithm which may offer some advantages in terms of computational simplicity and physical insight. The model is shown schematically in two dimensions in Fig. 2. The starting point in this case is a perfect single crystal. Each atom is moved away from its equilibrium position in a direction, and by a distance, selected at random. All the atoms are then relaxed to new positions determined by their repulsive interaction with their neighbors. The resulting amorphous structure appears to be independent of the initial crystal structure (Pickart 1979). The Pickart model is very much like the

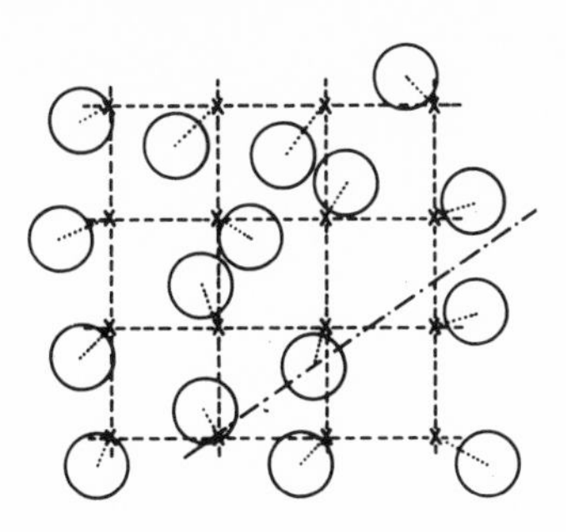

Fig. 2. Schematic representation of an amorphous solid. A two dimensional lattice is shown. Random displacements of atoms from their crystalline sites are shown by dotted lines. (from Ohkawa and Yosida 1977).

process envisioned in the formation of an amorphous alloy by ion implantation whereas the Bennett-Cargill-Kirkpatrick model is more like the formation of an amorphous alloy by vapor quenching.

The DRP model gives the following picture of an amorphous metal. The average nearest neighbor coordination is close to 12, which is the value for close-packed crystals (fcc or hcp). Nearest neighbor distances are usually in good agreement with the interatomic distances calculated using Goldschmidt radii corrected for 12-fold coordination. In DRP structures there are a number of types of interstitial spaces that are not possible in close-packed crystals. These spaces, known as Bernal holes, are in general larger than the tetrahedral and octahedral interstices found in crystalline close-packed structures. The Bernal holes may play a role in amorphous structures in accommodating the metalloid atoms in transition metal-metalloid (TM-M) alloys (Polk 1972) and in the incorporation of inert gases in sputtered amorphous rare earth-transition metal (RE-TM) alloys (Cuomo and Gambino 1977). It has also been suggested that the surroundings of the metalloid atoms are probably similar to those in crystalline compounds of the same composition. (Polk 1972, Chen 1973, Raj 1978).

In addition to atomic-scale structure, some amorphous alloy samples have been shown to have microstructural features on a scale of a few tens to a few hundreds of angstroms (Cargill 1975). In this size range, small-angle scattering and electron microscopy are important analytical tools (Herd 1978, Dirks and Leamy 1977). The development of anisotropic microstructures during vapor deposition has been modeled by Henderson et al. (1974) and by Dirks and Leamy (1977).

Perpendicular magnetic anisotropy has been observed in vapor deposited thin films of amorphous RE-TM alloys (Chaudhari et al.

1973). The magnitude of the anisotropy field in these films is too large to be explained by simple microstructural shape effects (Gambino et al. 1974). Models based on short range atomic ordering (e.g. pair ordering) have been successful in explaining the magnitude of the anisotropy (Cargill and Mizoguchi 1978) and the relationships between pair ordering anisotropy and process conditions (Gambino and Cuomo 1978, Nishihara et al. 1978, Muller and Perthel 1978). Anisotropy in the atomic distribution sufficient to account for the observed anisotropy is too small to be detected in the RDF. Direct structural confirmation of these models has not been possible but a number of experiments using ion implantation and magnetic susceptibility support the short range atomic ordering model (Mizoguchi et al. 1977, Ali et al. 1977, Mizoguchi et al. 1978). Other models for the anisotropy are based on compositionally inhomogeneous microstructures. (Cargill and Mizoguchi 1978). These models have been successfully applied to evaporated thin films contaminated with reactive gases (Brunsch 1977, Herd 1977). Perpendicular easy axis anisotropy can significantly reduce the field needed to obtain the saturation value of the spontaneous Hall effect. Thus the sensitivity of a Hall device to small fields can be increased by taking advantage of the perpendicular anisotropy (McGuire and Gambino 1977).

To summarize the present understanding of the structure of amorphous metals: the major features are dense random packing of spheres giving a local coordination number close to 12 but only very weak correlations beyond first nearest neighbors. In this sense, a fcc metal which has been randomized by small displacements of all the atoms from their equilibrium positions is a good conceptual model (shown schematically in two dimensions in Fig. 2). The metalloid atoms have a lower coordination number than the metals and probably retain some remnant of the local coordination observed in compounds of the same composition. Finally, the absence of long range crystalline order does not preclude the occurrence of structural anisotropy either in terms of small deviations from a totally random distribution of alloy constituents or columnar microstructural features. The nature and magnitude of the structural anisotropy is strongly dependent on the method of fabrication and the process parameters.

2. Stability. It is quite probable that given a sufficiently high quench rate and a sufficiently low substrate temperature, any metal can be prepared in the amorphous state. Amorphous phases that are stable over a reasonably large temperature range and that can be prepared using experimentally accessible quench rates are of the most interest for studies of Hall effect. It is convenient to discuss amorphous alloy formation in terms of two parameters: a stability criterion defined as the crystallization temperature normalized to the equilibrium melting point (T_x/T_m) and the glass

forming tendency (GFT) which we will define as the inverse of the minimum quench rate that will produce an amorphous phase (GFT = $(dT/dt)^{-1}$ minimum). Giessen (1976) has recently reviewed the empirical relationships between alloy composition and GFT. A comprehensive theory for amorphous alloy stability is not yet available.

Pure elemental amorphous phases have been prepared by vapor quenching. For example, Fe Co and Ni have been prepared by evaporation on to liquid helium cooled substrates. The phases crystallize at 60K or below (Whyman and Aldridge 1975). The low stability of amorphous metallic elements is often observed, i.e., T_x/T_m is generally less than 0.05. Alloying appears to be essential to produce an amorphous metal with a crystallization temperature above room temperature. Not all alloys form amorphous phases with high crystallization temperature. For the most part discussions of GFT and stability are concerned with predicting which alloy systems and what composition ranges within these systems will yield amorphous alloys.

Stability depends on a barrier to atomic displacements. The atomic displacements of concern are those which can transform a DRP structure into one with long range order. Since the DRP structure can be produced from close-packed crystalline structures by small random displacements, it follows that small displacements back toward the equilibrium positions will lead to crystallization. From this model it is not surprising that the pure elemental metals do not form stable amorphous phases. If the Bernal holes in a DRP structure of metal atoms are filled by other atoms, stable amorphous alloys can be formed. The concept of stabilizing a DRP structure by using small atoms in a matrix of larger atoms is known as the principle of jamming. The transition metal-metalloid systems are a particular example of jamming but the same mechanism is probably important in all amorphous alloys as evidenced by the strong correlation between GFT and stability and the radius ratio of the alloy constituents. Nowick and Mader (1965) observed that in vapor deposition of binary alloys, amorphous phases were only produced if the radius ratio of the constituent elements differed from unity by more than 10%. Giessen (1976) reports that all of the several dozen systems that readily form amorphous alloys have radius ratios that satisfy the condition $r_1/r_2 < .88$ or >1.12. He states that the radius ratio criterion may be the only stabilizing factor in some systems.

It has also been observed that ternary or higher order multicomponent systems are often more stable than binary systems. This has been referred to as the principle of confusion. It probably stems from the fact that the more complex the system, the larger the critical displacement required for crystallization. Alternatively, if the equilibrium phase is a complex structure with a large unit cell, the alloy constituents have to diffuse large distances to find equilibrium atomic sites. This situation appears to confer stability even in a binary system, e.g. Mo-Co (Wang 1978). An electronic criterion for amorphous phase stability has been suggested by Nagel and Tauc (1975). They suggest that amorphous phases are most stable for certain ranges of electron/atom ratio. It is not clear if the correlations they observe are intrinsic to amorphous state stability or simply reflect the requirements for the formation of equilibrium phases often with complex unit cells, e.g. the Hume-Rothery phases.

It is clear that covalent bonding has a stabilizing influence in elemental amorphous semiconductors. The random network model of amorphous Si and Ge consists of tetrahedra connected at their apices (Polk 1971, Steinhardt et al. 1974). The local coordination and bond angles satisfy sp^3 hydridization so the energy of the random network is not very different from that of the ordered diamond cubic lattice. In order to crystallize the material, however, these strong covalent bonds must be broken so the amorphous network structure is relatively stable.

It has been suggested that covalent bonding contributes to the stability of some metallic glasses. In particular, the fact that many metallic glasses contain transition metal elements which are expected to have considerable covalent bond character via hybridized d orbitals has been cited as evidence of the importance of covalency. Geissen has shown recently, however, that stable metallic glasses can be formed in alloy systems containing only simple metals (e.g., Ca-Mg) provided that the size criterion is satisfied. This observation indicates that covalent bonding is not a dominant factor (St. Armand and Geissen 1978).

Turnbull and coworkers have discussed metallic glass formation in terms of nucleation theory. Much of this work has been reviewed by Spaepen and Turnbull (1976). They have emphasized the importance of the melt-crystal interfacial tension in the occurrence of a nucleation barrier (see section II A). From their theory, they calculate that if T_x/T_m is 0.5, a minimum quench rate of 10^6 C/sec is required. They also point out that even if homogeneous nucleation is completely suppressed, crystallization can still occur by hetrogeneous nucleation. For example, vapor quenching of elemental amorphous metals onto liquid helium cooled substrates is apparently fast enough to suppress homogeneous nucleation and produce an amorphous film. The barrier to subsequent crystallization, however, is very low as evidence by the fact that these films crystallize below 100K.

The composition range in alloy systems which will yield an amorphous alloy must be defined in terms of the quench rate used. For example, in the rare earth Co and Fe systems, amorphous alloys can only be obtained by liquid quenching in a very narrow range near a deep eutectic at about 75% RE. The quench rate in most rapid melt quenching systems is 10^6 to 10^7K/sec. With very small samples and specialized quenching methods it is possible to achieve liquid quench rates of 10^{10} K/sec. During vapor deposition, however, it is estimated that the thermal accommodation time is of order 10^{-12} sec giving an effective quench rate of 10^{14} K/sec. It has been shown that amorphous films can be obtained by vapor quenching from about 10 to 90% RE in these RE-TM systems (Orehotsky 1972). As previously mentioned even the pure transition metals can be prepared in the amorphous state by vapor quenching on liquid helium cooled substrates.

To obtain an amorphous alloy that is stable at least up to room temperature, it is generally necessary to work in the 10 to 90% composition range in an alloy system in which the constituent elements differ in size by at least 10%. If liquid quenching is used, it is usually necessary to select compositions close to a deep eutectic. Some examples of types of amorphous systems which have been studied are listed below:

TM-M Ni-P, Pd-Si
TM-TM Zr-Cu, Nb-Ni, Mo-Co
RE-TM Gd-Co
Act-TM U-V, U-Cr
AE-B Mg-Zn, Ca-Al
AE-AE Ca-Mg, Sr-Mg

As can be seen from the above list, the occurrence of amorphous alloys is not a rare phenomenon nor is it isolated to any particular part of the periodic table.

C. Transport in Amorphous Metals

The literature on the electrical conductivity in amorphous metals, including liquid metals, is extensive and our purpose here is to present a simplified view of this subject. The resistivity can be one or two orders of magnitude greater in the amorphous state than in the crystalline state. From simple transport theory where there is a single type carrier, an electron with charge e, with mass m and number density n the resistivity can be expressed by the relation $\rho = mv/ne^2\lambda$. Here v is the electron velocity and ρ is proportional to the inverse of the mean free path λ. The high resistivity of the amorphous metal is related to the decrease in the electron mean free path. If a typical value for a v of 2×10^8 cm/sec is used then mean free paths of 2 to 3 Å are calculated for resistivities of 150 to 250 $\mu\,\Omega$ cm.

Mean free paths have also been estimated for crystalline materials from the study of ρ as a function of thickness in thin films (Mitchell et al. 1964 also Williams et al. 1968). For

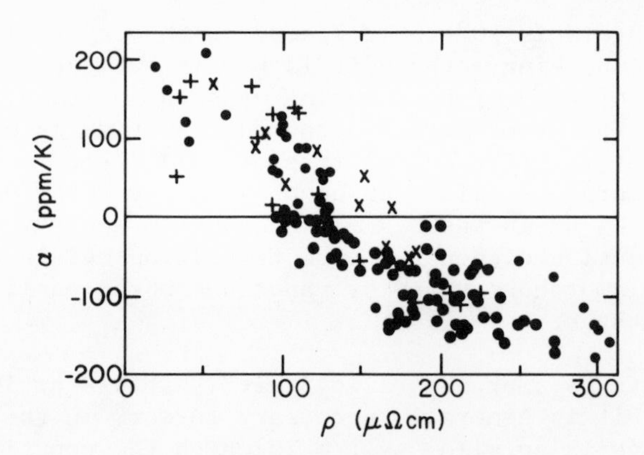

Fig. 3. The temperature coefficient of resistance versus resistivity for bulk alloys (+), thin films (●), and amorphous (x) alloys. (Materials from Mooij, 1973)

permalloy at room temperature $\lambda \approx 150$ to 300 Å with $\rho \approx 18$ to 29 $\mu\ \Omega$ cm. Scaling this measured mean free path to amorphous alloys with $\rho \approx 200\ \mu\ \Omega$ cm suggests mean free paths of 22 to 27 Å, i.e., several times the distance between atoms in a disordered material. An upper limit to metallic resistivity in amorphous materials may be when $\lambda \approx 3$ Å, the distance between adjacent atoms. This would correspond to a $\rho \approx 1500\ \mu\ \Omega$ cm.

Mott (1978) gives an upper limit for metallic resistivity of $\rho = 20\alpha/e^2$ where α is the lattice constant in a strong scattering material. For a mean free path comparable to the lattice constant $\lambda \sim 3 \times 10^{-8}$cm, then ρ has a limited value between 1000 and 3000 $\mu\Omega$ cm which is consistent with estimate given in the above paragraph. Higher values of ρ are associated with Anderson localization where a hopping mechanism for conduction becomes dominant. The alloys listed in this paper all fit the category of metallic conduction in terms of the value of ρ.

The temperature dependence of ρ for amorphous materials has several features which need comment. Mooij (1973) has measured a series of disordered transition metal alloys and plots $\alpha = 1/\rho\ d\rho/dT$ the temperature coefficient of resistance, between 25 and 75°C vs ρ as shown in Fig. 3. As a general trend, resistivities below 100 $\mu\ \Omega$ cm have negative α's. Similar features for α have been noted for Gd-TM amorphous alloys (McGuire et al. 1976). For alloys with ρ between 100 and 200 $\mu\ \Omega$ cm there is variation in the sign of α. In many cases alloys in this resistance range have a minimum associated with ρ (Hasagawa 1972). These minimums are small amounting to less than 1% reduction in ρ from the value at 4.2 K.

One way to view a positive α is that the value of ρ is low enough so that scattering due to phonons becomes an important contribution and ρ increases with temperature. On the other hand the negative α according to Ohkawa and Yosida (1977) means that the lattice disorder can decrease under certain conditions. Consider a configuration of metallic atoms highly distorted from the regular lattice as shown in Fig. 2. Such a distortion brings about a large anharmonicity to the lattice vibrations. This acts to diminish the distortion when the amplitude of the thermal vibrations is increased because the movement causes the average position of the atom to be closer to the crystalline position.

Application of the theory of liquid metals to the resistivity and its temperature dependence in amorphous metals gives another explanation for the sign of α (Nagel 1978). In simple liquid metals the resistivity is determined largely by the value of the structure factor $S(q)$ at $q = 2k_F$. With increasing temperature the first peak in $S(q)$ (which occurs at a value corresponding to 1.7 e/atom) is depressed and broadened. If $2\ k_F$ lies near the peak in $S(q)$ then $\alpha < 0$; if $2\ k_F$ lies on the wings of the peak in $S(q)$ then $\alpha > 0$. Extension of this model to amorphous transition metal alloys requires the addition of a term for the more important scattering due to the 3d-virtual states near E_F. Again the temperature dependence of this term is dominated by $S(2k_F)$ and the conclusions for $\rho(T)$ are the same (Guntherodt and Kunzi 1978a). An interesting experimental support for this model is provided by the effects of structural relaxation on the resistivity (Lin et al. 1979). At the present time the temperature dependence of the scattering and its relationship to the spontaneous Hall effect remains a problem for future investigations.

D. Simple Models of the Hall Effect

Professor Hall's name is given to two effects, the ordinary Hall effect (OHE) and the extraordinary or spontaneous Hall effect (SHE). In each case a transverse field E^H_y is observed when a magnetic field H_z is applied normal to a thin sample carrying a current density j_x (Fig. 4). The OHE is a consequence of Lorentz force acting on the current carriers whereas the SHE arises from spin-orbit interactions (SOI) and other coupling (Fert and Friederich (1976). If one examines the relative magnitudes, the signs, and the temperature dependences of the two Hall effects, one is quickly reminded of their different origins.

Nevertheless, a degree of unity is regained by considering the two Hall effects to follow from the covariant electrodynamic forms of the fields seen by a charge moving in the presence of transverse electric and magnetic fields, $E_\perp$ and $B_\perp$.

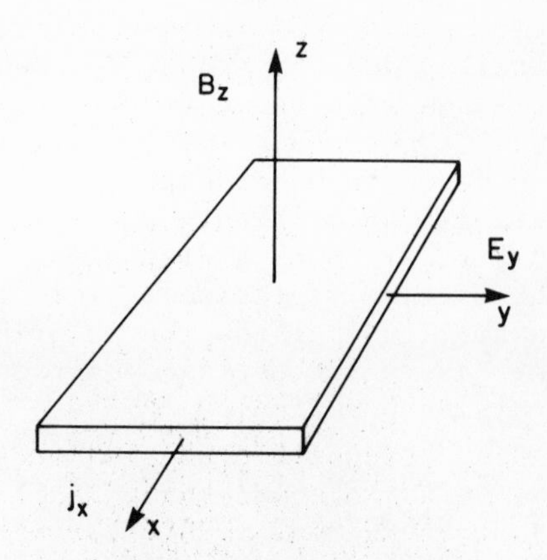

Fig. 4. Coordinates used herein for Hall voltage E_y in a thin sample in the xy plane. Applied current is j_x and applied magnetic field H_z gives rise to induction B_z.

$$E'_\perp = \gamma(E_\perp + v \times B_\perp) \qquad 1(SI)$$
$$B'_\perp = \gamma(B_\perp - v \times E_\perp/c^2) \qquad 2(SI)$$
$$E'_\perp = \gamma(E_\perp + v \times B_\perp/c) \qquad 1(cgs)$$
$$B'_\perp = \gamma(B_\perp - v \times E_\perp/c) \qquad 2(cgs)$$

The prime indicates a field seen in the rest frame of the moving particle and the relativistic correction $\gamma = \sqrt{1-v^2/c^2}$ which is about equal to one for typical Fermi velocities.

1. Ordinary Hall effect. Equation 1 is the Lorentz force per unit charge responsible for the OHE. Current carriers are deflected by the Lorentz force due to the component of their drift velocity normal to the magnetic induction B (not H, see Panofsky and Phillips 1962). Once sufficient charge accumulates at the edges of the sample, the field due to that charge $E_\perp = E^{OH}$ establishes a steady state by cancelling the transverse drift due to the v x B term. Then $E'_\perp = 0$ and $E^{OH} = -v \times B$. Using $v_x = j_x/ne$ for the drift velocity and $B_\perp = B_z$ we have in the coordinates of Fig. 4:

$$E_y^{OH} = 1/(ne)j_xB_z \qquad 3(SI)$$

$$E_y^{OH} = 1/(nec)j_xB_z \qquad 3(cgs)$$

The constant of proportionality 1/(ne) is the ordinary Hall coefficient R_o which, in simple metals, takes on the sign of the charge carriers.

The Fermi surface of an amorphous metal is expected to be more free-electron like than is that of the same metal in the crystalline state because of the rotational symmetry imposed by disorder. We thus expect Eq. 3 to obtain more often in amorphous metals than in crystalline metals (see Ziman 1979 for more detailed considerations of transport in disordered materials).

2. Spontaneous Hall effect. Just as a charge moving in a magnetic field sees an electric field (Eq. 1) and hence is subject to a force, so a magnetic moment moving in an electric field sees a magnetic field (Eq. 2) whose gradient gives rise to a force on the moment. The interaction of the electronic spin with the electric field of the screened nucleus defines its spin-orbit interaction through Eq. 2. The SOI gives rise to a left-right asymmetry in the scattering of spin-polarized carriers. This asymmetry is observed as the SHE:

$$E_y^{SH} = R_s j_x \mu_o M \qquad 4(SI)$$

$$E_y^{SH} = R_s j_x 4\pi M \qquad 4(cgs)$$

where the constant of proportionality R_s is called the spontaneous Hall coefficient and is related to the strength of the SOI.

The spontaneous Hall effect is most easily observed in ferromagnetic materials and its coefficient is generally much larger than that of the OHE.

a) Skew scattering and side jump. Two types of scattering events are distinguished in the SHE literature (see Berger and Bergmann 1980 and references therein). One is referred to as skew scattering and is characterized by a constant spontaneous Hall angle

$$\theta_S = \rho_H/\rho$$

at which the scattered carriers are deflected from their original trajectories (Fig. 5a). Thus for skew scattering

$$\rho_H \propto \rho.$$

The other scattering mechanism is quantum mechanical in nature and results in a constant laterial displacement Δy of the charge carrier's trajectory at the point of scattering (Fig. 5b). The side jump Δy of the carrier wave packet is approximately related to the skew scattering parameter θ_S by the mean free path λ:

$$\mathrm{Tan}\theta_S \approx \theta_S \simeq \Delta y/\lambda \propto \rho\Delta y$$

Thus for the side jump mechanism:

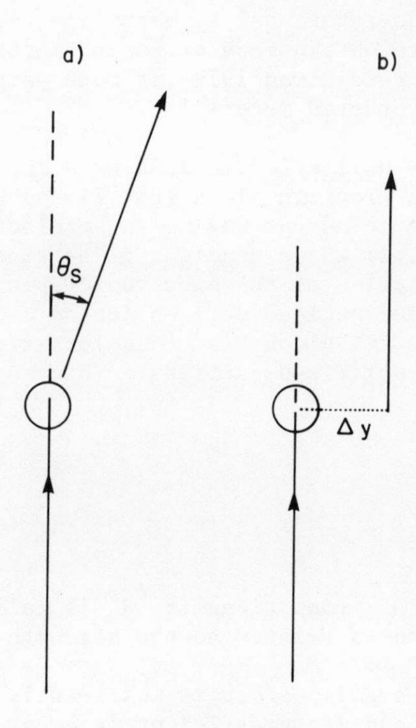

Fig. 5. Schematic comparison of a) skew scattering characterized by a constant spontaneous Hall angle θ_s and b) side-jump, characterized by a constant lateral displacement Δy of carrier trajectory.

$$\rho_H \propto \rho^2$$

Dilute impurities and spin disorder are predicted to give rise to $R_s \propto \rho$, whereas concentrated spin defects and phonons are expected to give rise to $R_s \propto \rho^2$ (Kagan and Maksimov 1965, Kondorskii 1966). Because of the different dependence on resistivity of these mechanisms the SHE is usually attributed to the skew scattering when ρ is small (low temperatures and/or pure metals) and to the side jump when ρ is large (high temperatures, concentrated alloys and disordered materials). Intermediate cases often exist (Dorleijn 1976).

b) Spin-orbit interactions. Two types of spin-orbit interaction can be distinguished (see for example Bethe and Salpeter 1957 or Condon and Shortley 1963). The most familiar one is the intrinsic spin-orbit interaction ISOI involving the coupling between the spin and orbital angular momenta of a particle moving in an electric field, e.g. $L^{3d} \cdot S^{3d}$ for a 3d electron. The other is the spin-other-orbit interaction SOOI coupling the spin and orbital angular momenta of two charged particles in relative motion e.g. $L^k \cdot S^{4f}$ for a conduction electron and a 4f electron. These interactions can differ from each other in both sign and magnitude depending not only on the respective spin and orbital quantum numbers but also in the case of SOOI on the signs of the charges and on the degree of overlap of the wave functions of the two particles.

Karplus and Luttinger (1954) investigated the ISOI for itinerant, spin-polarized electrons. This model appears to be most applicable to 3d transition metals. Maranzana (1967) and several Russian authors (Kondorskii et al. 1964 and Kagan and Maksimov 1965 for example) have pointed out the importance of the SOOI in Rare-earth metals and alloys where the magnetic electrons carry no current and, therefore, the ISOI would be zero. Kondo (1962) has proposed another mechanism for skew-scattering based on the s-d (or s-f) exchange between localized and itinerant electrons. Estimates of the magnitude of this effect suggest it is much too small to account for the observed values of R_s and it has, therefore, received little attention in the literature.

3. Temperature dependence. Because the carrier type and density are practically temperature independent in metals, so is R_o. An exception may occur if the band structure (and hence possibly the nature and density of carriers) is altered as may happen at a structural phase transition.

The spontaneous Hall coefficient generally follows the temperature dependence of the resistivity (ρ or ρ^2 as in section 2a above). In crystalline metals the resistivity is a strongly increasing function of temperature. If the metal is magnetically ordered at low temperatures, the resistivity may show an additional component due to magnetic-disorder scattering on approaching T_c (Fig. 6a). For this case R_s is expected to increase sharply below T_c then remain approximately constant or increase above T_c (Allison and Pugh 1956). See Figure 6b. Of course ρ_H will decrease sharply at T_c because of the temperature dependence of the magnetization. In amorphous metals $d(\ln \rho)/dT < 0.05$ over most of the temperature range below crystallization T_x (Fig. 6c). In this case R_s (Fig. 6d) is also expected to be independent of temperature up to T_x (Samoilovitch and Kon'kov 1950).

4. Sign of R_o and R_s. For simple metals a free electron model suffices to indicate the nature of the carriers and hence the sign of R_o. If the Fermi surface is non-spherical, the band structure must be examined in k space to predict R_o.

R_s is most easily observed in ferromagnetic metals. In these materials the Fermi surface is clearly non-spherical. Also the charge carriers may be of either spin direction thus giving

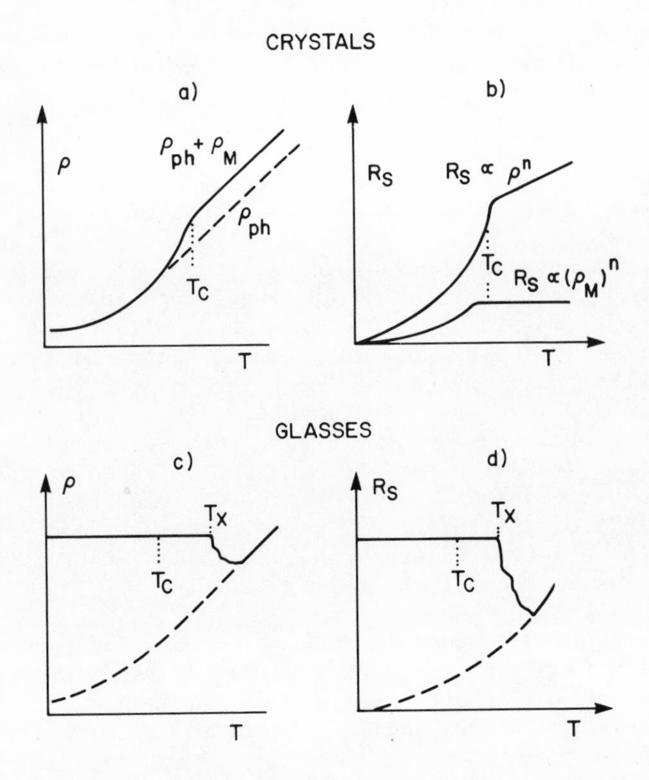

Fig. 6. Schematic representation of the temperature dependence of resistivity ρ and spontaneous Hall constant R_s in crystals (a and b) and in glasses (c and d). Curie and crystallization temperatures are designated T_c and T_x respectively. Phonon and magnetic contributions to ρ are indicated by ρ_{ph} and ρ_M respectively.

possibly competing spin-orbit forces for the two sub-bands. Despite these complications reasonable success has been had in interpreting the sign of R_s in certain binary transition metal alloy systems where the density of states splits each spin sub-band into two sub-bands (separated on an energy scale by a gap) associated with the two component metals. In this model, the expectation value of the z-component of orbital angular momentum at E_f can change sign with alloy concentration as the Fermi energy passes through the gap between the sub-bands or through the middle of one of the sub-bands. Therefore at these singular points the SOI and R_s are predicted, and often observed, to change sign (Berger 1976, 1977; Kondorskii et al. 1975 a,b). See Sec. IV B 2.

E. Data Analysis

1. Units. The two most widely used units for the Hall coefficients are SI units, $m^3/C = m^3/As$, and the hybrid unit Ohm-cm/G (which combines the practical quantities volt and amp with the cgs units centimeter and Gauss). Two numbers for the same quantity expressed in these two units are simply related by R(SI) = 100 x R(hybrid) because the unit m^3/C = Ohm-m/Tesla is one hundreth of the unit Ohm-cm/G. (For further discussion of units, see O'Handley, this volume.)

In SI units care must be taken to clarify the meaning of the magnetization. One recommended convention is (Bennett et al. 1976)

$$B = \mu_o(H + M)$$

and

$$B = \mu_o H + J$$

where B and J are in Vs/m^2 (Tesla) and H and M are in A/m. Then combining Eqns. 3 and 4 we have

$$\rho_H = R_o B + R_s M \mu_o \qquad 5(SI)$$

or

$$\rho_H = R_o B + R_s J \qquad 6(SI)$$

And in Gaussian units we always have:

$$\rho_H = R_o B + R_o 4\pi M. \qquad 7(cgs)$$

We follow the units used in the references which we cite.

2. Separating R_o and R_s. A common technique in extracting R_s (or $R_1 = R_o + R_s$) from experimental ρ_H vs. H_a data is to use the approximation $(d\rho_H/dH)_{H=0} = \mu_o R_s$. The justification for this equation, which must be understood in interpreting results, particularly when the data spans T_c, is outlined elsewhere in this volume (O'Handley 1980), where it is shown that $1/\mu_o(d\rho_H/dH)_o \equiv R_H$ gives $R_o + R_s$ ($=R_s$ if $R_o \ll R_s$) for $T < T_c$, but it gives $R_o + \chi R_s$ when $T > T_c$.

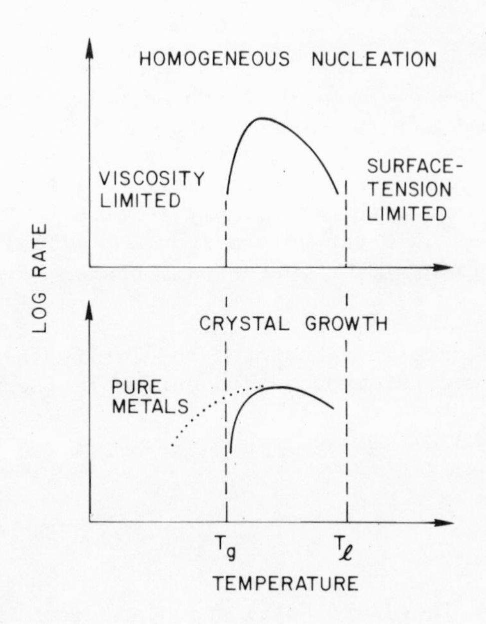

Fig. 7. Simple representation of homogenous nucleation and crystal growth in undercooled metallic melt. T_g and T_ℓ indicate glass transition and liquidus temperatures respectively.

II. SPECIAL EXPERIMENTAL CONSIDERATIONS

A. Preparation

Fabrication conditions for amorphous materials must be such as to preclude crystallization. Crystallization requires the presence of crystal nuclei and their growth. The rate of crystallization therefore depends on the product of the rate of nucleation and the rate of growth of crystallites. Figure 7 shows schematically how the crystallization rate is appreciable in a temperature range bounded by the liquidus temperature T_1 above and by the glass transition temperature T_g below (Spaepen and Turnbull 1976). Nucleation is limited at high temperatures by the surface tension of the melt-nucleus boundary (the nuclei collapse or dissolve spontaneously at high temperatures) while crystal growth (except in pure metals) and nucleation are sharply curtailed below T_g by increased viscosity. In order to form an amorphous material one must either quench a liquid sample through this region at a rate greater than that which would allow observable crystallization to occur or alternatively condense a vapor onto a substrate cooled sufficiently to preclude crystallization. The quench rate required varies widely with the materials used and also depends strongly on the presence of glass forming elements (B,C,P,Si etc.) as well as on impurities.

(See section I B 2).

1. Liquid quenching. Liquid quenching techniques are reviewed in numerous articles, most recently by Davies (1978). Liquid quenching owes its success to the invention of the gun technique (Duwez et al. 1960, 1963) with its high quench rate. However, two other techniques are more widely used today: the piston and anvil technique (Pietrokowski 1963) and continuous roll casting (Pond and Maddin 1969; Chen and Miller 1970). In the piston and anvil method a droplet of the melt to be quenched is rapidly compressed between two flat metal surfaces. The resulting samples are also typically 40 μm thick and roughly round with a diameter of approximately 2 cm. Continuous roll casting techniques employ a crucible with a small orifice at its bottom which directs a jet of molten alloy onto the surface (inside or out) of a rapidly rotating cylinder. The resulting samples are also typically 40 μm thick and, when continuous operation can be achieved, they are produced as ribbons at several hundred meters per second (Kavesh 1977).

2. Sputtering and vapor deposition. In addition to the melt quenching methods widely used in the preparation of metal-metalloid alloys, amorphous thin films have been prepared by a number of vapor quenching techniques. The most common vapor deposition methods are evaporation and diode sputtering. Vacuum evaporation for the preparation of amorphous alloys was first studied systematically by Nowick and Mader (1965). Substrate temperature and deposition rate are the most critical process parameters. Sloope and Tiller (1965) showed that in the deposition of amorphous germanium, amorphous films are only obtained above a critical deposition rate which depends linearly on T_{sub}^{-1} the inverse substrate temperature. Chaudhari et al. (1975) showed that a similar relationship holds in the deposition of amorphous Gd-Co alloys. Another difficulty with vacuum evaporation is composition control. Amorphous alloys typically contain several elements with very different vapor pressures so that evaporation of the alloy causes fractionation (Taylor 1974). To obtain homogenous films, therefore, it is usually necessary to have a separate evaporation source for each constituent and to control each via a feedback loop. Multiple electron beam heated sources have been used successfully in the preparation of ternary RE-TM amorphous alloys (Hasegawa and Taylor 1975).

Sputtering has the advantage that the target composition can be reproduced in the film so complex mixtures can be more easily prepared. The film composition can also be adjusted away from the target composition in a controllable manner by bombarding the surface of the growing film during deposition, a process known as bias sputtering (Cuomo et al. 1974 and 1975, Dove et al. 1976). Many variations of the basic sputtering process have been used for amorphous metals. Getter sputtering has been used by Hauser (1975) to avoid contamination of films being deposited onto low temperature substrates. Troide sputtering has been used to achieve very high deposition rates making it practical to prepare thick films (c.a. 1 mm) which are very useful for neutron diffraction studies (Allen 1975). Magnetron and ion beam sputtering have been used for some special studies (Albert et al. 1977, Harper et al. 1980).

One disadvantage of sputtering as compared to evaporation is that the sputtered films always contain some of the inert gas used to sustain the glow discharge. By properly selecting the process conditions, this contamination can be kept down to about 1 atom %. (Cuomo 1977). A second difficulty with sputtering occurs when the desired alloy composition is difficult to fabricate into a homogeneous target. Very often, the amorphous alloy of interest does not coincide with a single phase equilibrium composition. In this case it may be necessary to use hot pressed targets to obtain a random distribution of the various phases present. The most comnonly used solution to this problem is to fabricate a composite target. For example, to prepare a Gd-Fe alloy, an iron plate is used which is partially covered with pieces of gadolinium. In this way the alloy composition can be adjusted at will by varying the fractional surface coverage of Fe and Gd. (Imamura et al. 1976). Good homogeneity has been demonstrated by this technique.

Experience with metal-metalloid systems suggests that any alloy composition that can be prepared in the amorphous state by melt quenching can also be prepared by sputtering. In general it is necessary to heat sink the substrates so that the film temperature does not rise too near the crystallization temperature during deposition. In the high rate triode method, for example, the heat load on the substrate is very high so that special heat sinking methods must be used.

As mentioned in the section on history, some of the first amorphous magnetic alloys were prepared by electro- and electroless deposition. Both of these methods have been studied more extensively since the more widely used liquid- and vapor-quenching methods were developed (Cargill 1970b, Clements 1975). More recently, a number of amorphous alloys have been prepared by ion implantation (Poate 1978, Grant 1978).

B. Sample Forms

The amorphous samples made by the techniques described above are well suited to transport measurements. Splat-quenched samples are typically 40 μm in thickness and a few cm in diameter. Continuous liquid quenching yields ribbon samples also of approximately 40 μm thickness with widths from a few mm up to several cm. These materials can be punched, cut, etched or milled by electric discharge to form suitable test strips for resistivity and Hall measurements. Electrical contact can be made by soldering or spot welding but the effects of these techniques on the amorphous structure have never been clearly determined. Pressure contacts with copper or other deformable conductor have been found to give reproducible results without affecting the sample.

In addition to the importance of accurate resistivity data in itself, further significance is attached to the resistivity in interpreting the magnitude of R_s. A difficulty comes from accurately measuring the thickness of thin samples particularly when surface irregularity is a significant fraction of the thickness. O'Handley (1978b) and Marohnic (1978) seem to have solved the problem for ribbons by taking the thickness from the independently determined density and from measurements of the weight and the two large dimensions of the ribbons. Only when resistivity is determined in this way is it possible to reduce the scatter in a compositional study to the extent that a plausible trend in the data can be seen. Resistivities determined in this way are generally approximately 20% smaller than those obtained by other means.

III. EXPERIMENTAL RESULTS

A. Non-rare-earth Amorphous Metals

Tables 1a, b, c, d, and e summarize the published Hall data for non-rare-earth amorphous metals. We believe it is complete up to the time of this symposium. Limited space prevents us from discussing each contribution in detail. We will try to distill from the approximately forty published papers in this field what we feel are the important physical results.

1. Ordinary Hall effect.

a) Paramagnetic glasses. Because of their disordered structure, amorphous metals are expected to have a spherical Fermi surface and therefore to be nearly free-electron like. The ordinary Hall coefficients of amorphous metals should, then, be interpreted more easily than those of comparable crystalline metals.

An early example of this is the work of Bergmann (1972) in which amorphous films of $Sn_{86}Cu_{14}$, $Ga_{95}Ag_{05}$, $Bi_{98}Tl_{02}$ and $Pb_{75}Bi_{25}$ were studied. The ordinary Hall coefficients were found to be unlike those of the metals in the crystalline state and resembled more those of the metals in the liquid state. The Sn and Ga based samples showed R_o's which gave carrier concentrations (3.4 and 2.7 electrons per atom respectively) close to their number of free electrons (3.6 and 2.9 respectively), while the Bi and Pb rich samples showed R_o's corresponding to carrier densities (10.7 and

TABLE I. Published Hall data on amorphous metals and alloys (designated with prefix a-). Unless otherwise specified, data for pure amorphous alloys are at 4.2 K and the mixed systems (Table Ie) are at room temperature. Data on related liquid (prefix l-) and crystalline materials are included for comparison. Data for materials followed by parentheses are extrapolations of data for the materials in those parentheses. Numbers in parentheses after carrier concentration per unit volume n indicate number of carriers per atom or per formula unit derived from R_o unless otherwise indicated. This number cannot be calculated from R_o or n without the density.

Table 1a PARAMAGNETIC METALS AND ALLOYS

Material	$\left(\frac{R_o}{10^{-12}}\right)$ ohm-cm/G	$\left(\frac{n}{10^{+22}}\right)$ cm^{-3}	$\left(\frac{R_s}{10^{-10}}\right)$ ohm-cm/G	$\left(\frac{\rho}{10^{-6}}\right)$ ohm-cm	n_B (μ_B)	Ref.
a-Sn	-.426	14.7(3.4)		47		a
a-Ga	-.394	15.9(2.6)		28		a
a-Bi	-.2	31.2(10.75)		135		a
a-Pb	-.22	28.4(8.9)		86		a
a-PdSi	-.93	6.7				b
a-$Pd_{.80}Si_{.20}$(RT)	-.78	8(1.2)				c
a-$Pd_{.81}Si_{.19}$	-.86	7.27(4/Si)		80		d,e
a-$Pd_{.78}Cu_{.06}Si_{.16}$	-7.5	.83		76		f
l-$Pd_{.81}Si_{.19}$	-.8	7.8	100			d

a) Bergmann 1976, b) Willens 1973, c) Butvin and Duhaj 1976a, d) Guntherodt et al. 1977b, e) Guntherodt et al. 1973a, f) Rao et al. 1978.

Table 1b IRON

Material	R_o 10^{-12} ohm-cm/G	n 10^{+22} cm^{-3}	R_s 10^{-10} ohm-cm/G	ρ 10^{-6} ohm-cm	n_B (μ_B)	Ref.
a-Fe	+6		.7	100	1.4	a
a-Fe(4.2K)	+19	.33	2.7	90	1.2	b
a-Fe(AuFe)			3	130		c
a-Fe(FeSi)			0.7			d
a-$Fe_{.5}Au_{.5}$	-1.5	4.17	.65			e,c
a-$Fe_{.80}P_{.13}C_{.07}$	>0		6			f
a-$Fe_{.8}B_{.2}$			5	120	2.0	g,h
a-FeB			5.4	122		i
a-$Fe_{.77}B_{.23}$	3	2.1	4.5	243		j
a-$Fe_{.78}B_{.22}$	1	6.25	4.6	199		j
a-$Fe_{.79}B_{.21}$	1	6.25	3.7	191		j
a-FeB			4.3-4.7	130-155	12.5kG	k
a-$Fe_{.78}B_{.12}Si_{.1}$	33.7	.19	6.14	175		l
a-$Fe_{.7}Si_{.3}$			8	760		d
2615			6.3	145		m,n
2605			5.5	118		m,n
2605A			5.8	130		m,n
l-Fe	3.6d			130		o,p
l-Fe(l-FeAu)	3.5d					q
l-$Fe_{.80}Ge_{.20}$	2.7	2.3		180		o
Cryst Fe	+.23	27	+.072		2.2	r

a) Aldridge et al. 1976, b) Raeburn et al. 1978, c) Bergmann et al. 1978, d) Shimada et al. 1978, e) Bergmann 1979b, f) Lin 1969, g) O'Handley 1977, h) O'Handley 1978b, i) Marohnic et al. 1978, j) Stobiecki et al. 1978, k) McGuire et al. 1979b, l) Obi et al. 1977, m) Malmhall et al. 1978c, n) Malmhall et al. 1979c, o) Fischer et al. 1976b, p) Busch et al. 1974, q) Guntherodt et al. 1973, r) Volkenshtein 1960.

Table Ic COBALT and Id NICKEL

Material	$\left(\begin{matrix} R_o \\ 10^{-12} \\ \text{ohm-cm/G} \end{matrix}\right)$	$\left(\begin{matrix} n \\ 10^{+22} \\ \text{cm}^{-3} \end{matrix}\right)$	$\left(\begin{matrix} R_s \\ 10^{-10} \\ \text{ohm-cm/G} \end{matrix}\right)$	$\left(\begin{matrix} \rho \\ 10^{-16} \\ \text{ohm-cm} \end{matrix}\right)$	n_B (μ_B)	Ref.
a-Co	-1.59	3.9(.48)	1.45	41		a
a-Co			.44			b
a-$Co_{.8}B_{.2}$			1.0	90	1.2	c,d
a-CoB			1.2-2.2	93-155	10kG	e
a-$Co_{.78}B_{.12}Si_{.10}$	+3.05	2.05	2.62	150	11kG	f
Sol para Co	+.42	1.5(2)	1.67(3)	95	2.2	g,h
l-Co	-1.62	3.86(.5)	5.2	115	2.0	g,h
Cryst Co	-.84	7.4	+.006		1.7	i
a-Ni				115		b
a-Ni(NiAu)	-1.3		-.64			j
a-$Ni_{.83}Au_{.17}$	-1.3	4.8	-.65			j
a-$Ni_{.65}Au_{.35}$	-1.2	5.2(.6)	-.48			k
a-$Ni_{.78}B_{.12}Si_{.10}$	-1.17	5.3	0	117	0	l
l-Ni(NiGe)	-1.15d	5.4		80		m,n,o
l-$Ni_{.65}Au_{.35}$	-1.15d	5.4		88		m
Cryst Ni	-0.46	13.6	-.06		0.6	i

a) Bergmann 1977, b) Whyman et al. 1975, c) O'Handley 1977, d) O'Handley 1978b, e) McGuire et al. 1979b, f) Obi et al. 1977, g) Guntherodt et al. 1977b, h) Guntherodt et al. 1978b, i) Volkenshtein 1960, j) Bergmann 1976b, k) Bergmann 1976a, l) Stobiecki et al. 1979, m) Guntherodt et al. 1973, n) Guntherodt et al. 1978a, o) Busch et al. 1974.

Table Ie MIXED TRANSITION METAL-BASED AMORPHOUS ALLOYS

Material	R_o (10^{-12} ohm-cm/G)	n (10^{+22} cm^{-3})	R_s (10^{-10} ohm-cm/G)	ρ (10^{-16} ohm-cm)	n_B (μ_B)	Ref.
2826A	+2.7		(3.2)	180		a,b
2826A	3.0		4.24	172		c
2826A	+2		5.5(4.2K)			d
2826A			5.5	175(100K)		e,f,g
2826	+6.6		3.35	154		h
2826			5.4	165		e,g
2826			5.4	160		f
$Fe_{40}Ni_{40}B_{20}$	>0		3.5			i
$Fe_{40}Ni_{40}B_{20}$			2.9	129		g
2826MB			4.0	155		e,g
2826B	+7.4		2.33	138		h
2826B			4.8			j
2826B	+2		5.1	155		e,g,j
$Ni_{53}Fe_{27}P_{14}B_6$			4.2	164		k
Fe-Ni-B			4.9 - 0	130-116		m,n
Fe-Ni-B-Si	34- -1.17		6.1 - 0	175-117		o
Fe-Co-B			4.9-.9	120- 91		m,n
Fe-Co-B-Si	34- +3		6.1-2.6	175-150		o
Fe-Ni-$B_{.20}$			5.4-2.7	122-116		p
Fe-Ni-$P_{.14}B_{.06}$			5.9-4.7	143-136		p

a) Fischer et al. 1976a, b) Fischer et al. 1976b, c) Malmhall et al. 1977, d) Malmhall et al. 1978a, e) Malmhall et al. 1978c, f) Malmhall et al. 1979a, g) Malmhall et al. 1979c, h) Rao et al. 1976, i) Marohnic et al. 1977, j) Malmhall et al. 1978b, k) Malmhall et al. 1979b, m) O'Handley 1977, n) O'Handley 1978b, o) Obi et al. 1977, p) Marohnic et al. 1978.

Allied Chemical Trade names and compositions: Metglas Alloys 2615 ($Fe_{80}P_{16}C_3B_1$), 2605 ($Fe_{80}B_{20}$), 2605A ($Fe_{78}B_{20}Mo_2$), 2828 ($Fe_{40}Ni_{40}P_{14}B_6$), 2826A ($Fe_{32}Ni_{36}Cr_{14}P_{12}B_6$), 2826B ($Ni_{49}Fe_{29}P_{14}B_6Si_2$), 2826MB ($Fe_{40}Ni_{38}B_{18}Mo_4$).

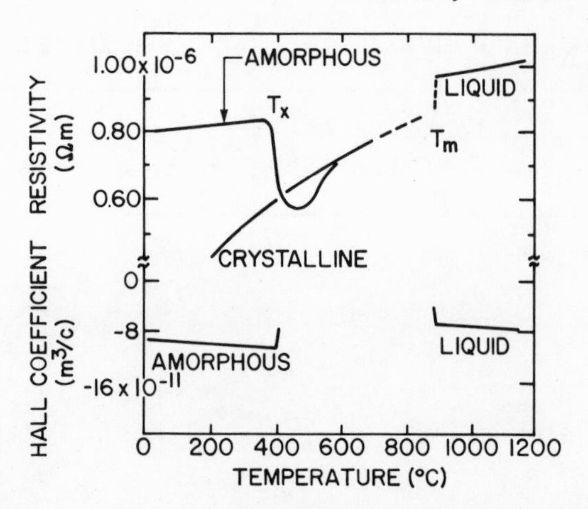

Fig. 8. Comparison of resistivity and Hall coefficient for $Pd_{81}Si_{19}$ in amorphous and liquid states. T_m and T_x indicate melting and crystallization temperatures respectively. (From H-J. Guntherodt et al. 1977b).

and 8.9 electrons per atom respectively) approximately twice their number of free-electrons (5 and 4.5 respectively). See Table Ia.

The structural similarity between the amorphous and liquid states of metals is also revealed by the temperature dependences of ρ and R_o in the two states. Guntherodt's group (1977b) showed

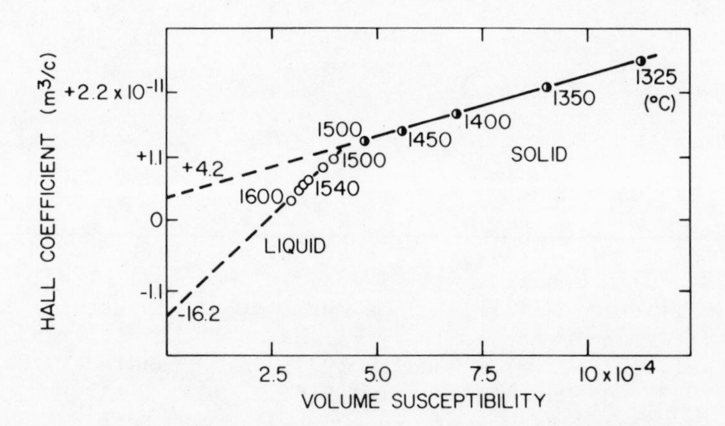

Fig. 9. Hall coefficient as a function of magnetic susceptibility for cobalt in paramagnetic solid and liquid states. Numbers by datum points give temperatures in degrees C: two numbers of right of ordinate give Hall coefficients for $\chi = 0$. (After H - J. Guntherodt et al. 1977b).

this with measurements of ρ and R_o on amorphous $Pd_{81}Si_{19}$. Their results are shown in Fig. 8. The observed R_o implies approximately four conduction electrons per silicon atom in both the amorphous (density=10.25 g/cm^3) and liquid PdSi. Butvin and Duhaj (1976a) have argued from their measurements of R_o in Pd and PdSi that Si tends to fill the holes in the Pd d-band and that this filling effect increases strongly with increasing silicon content. Amorphous PdCuSi also shows $R_o < 0$ (Rao 1976).

The substitution of small amounts of Co for Pd in $Pd_{80}Si_{20}$ causes ρ_H to increase, going positive for x_{Co} = 7 at% and showing a spontaneous-magnetization-like curvature with applied field at x_{Co} = 11 at% at room temperature (Butvin and Duhaj 1976b). The positive contribution of Co to ρ_H below the onset of long-range order is attributed by Butvin and Duhaj to clustering of the cobalt moments. Their ρ_H vs. H data in samples showing a spontaneous-magnetization-like behaviour show a positive slope above the knee in the curve up to their highest field, 15 kG. Because for pure a-Co $R_o < 0$ (Table Ic) it might be assumed that this positive slope comes from the SHE in Co and from the large high-field susceptibility at low Co concentrations. This example would then point out the difficulty of trying to extract R_o from insufficiently-high high-field data.

b) Ferromagnetic glasses. Many attempts to obtain R_o in ferromagnetic materials from data above T_c are also plagued by the strong lingering effects of χR_s (Lin 1969, Obi et al. 1977, Aldridge et al. 1976, Malmhall et al. 1978b, and M. Fischer et al. 1976a). Proper results are best illustrated by the data of Guntherodt et al. (1977b) for solid and liquid paramagnetic cobalt. They plot R_H (=$R_o + \chi R_s$) vs χ (see section I E 2) as shown in Fig. 9. Note that R_s contributes even at these elevated temperatures. R_s will be discussed below. Ni-Au R_o changes from +0.4 x 10^{-12} Ohm-cm/G to -1.6 x 10 Ohm-cm/G on heating through the melting point. It is possible that the positive paramagnetic value is a consequence of the band structure of γ-cobalt. We take the liquid value to be closer to the free-electron value. Solid amorphous cobalt is expected to behave much like its liquid phase shown in Fig. 9. Indeed, Bergmann (1977) has found R_o = -1.59 x 10^{-12} Ohm-cm/G from high field data up to 80 kOe (Fig. 10).

The studies on amorphous Ni and Ni-rich alloys have consistently found $R_o < 0$ (Table Id). Bergmann (1976a,b) has studied the Ni-Au system from $Ni_{65}Au_{35}$ where R_o = -1.2 x 10^{-12} Ohm-cm/G to $Ni_{83}Au_{17}$ where R_o = -1.3 x 10^{-12} Ohm-cm/G, and by extrapolation to pure Ni where R_o = -1.3 x 10^{-12} Ohm-cm/G. He uses plots of ρ_H vs $1/(T-T_c)$ which is proportional to χ and extrapolates to χ=0 in one case (1976a) and in the other he employs fields up to 100 kOe (1976b). See Fig. 10. These values closely resemble those of Guntherodt and coworkers (1973) for liquid $Ni_{65}Au_{35}$ (-1.15 x 10^{-12})

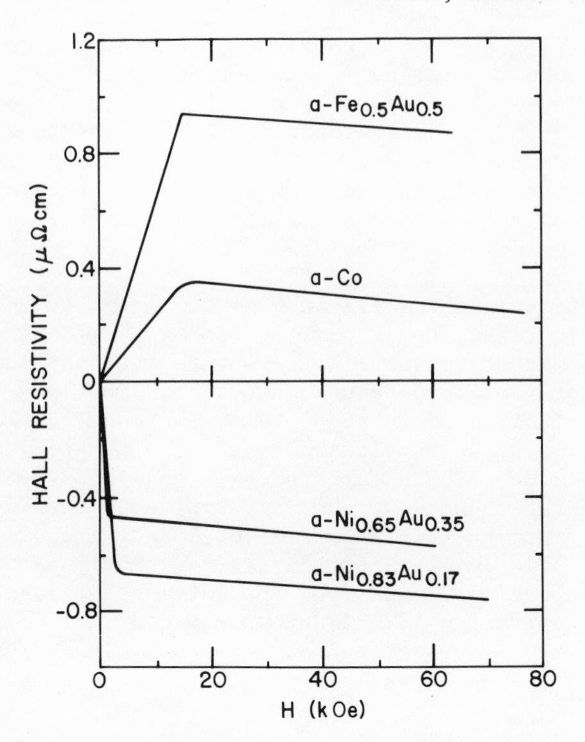

Fig. 10. Comparison of Hall resistivity vs applied field for amorphous thin films based on Fe, Co and Ni. (Bergmann 1976a, 1976b, 1977, 1979b, and Bergmann and Marquard 1978).

and liquid Ni (-1.15×10^{-12}) by extrapolation from data on Ni-Ge.

The situation is less clear with amorphous Fe and Fe-rich alloys. Positive R_o's are found where only moderately high fields (~2T) were used (Obi et al. 1977, Raeburn and Aldrige 1978) or above T_c where $4\pi\chi R_s$ was not adequately removed (Lin 1969). The ρ_H vs $1/(T-T_c)$ data of Stobieki and Hoffmann (1979) for Fe-B above T_c also show $R_o > 0$ when χ goes to 0. However, Bergmann's work (Bergmann and Marquard 1978, Bergmann 1979b) at very high fields (>5T) in $Fe_{1-x}Au_x$ films shows $R_o = -1.5 \times 10^{-12}$ (Fig. 10) and R_o remains negative as x approaches zero.

Numerous studies exist on amorphous alloys containing two or more transition metals (Table Ie). The OHE data generally show $R_o > 0$ but the problems associated with removal of the SHE are not always properly addressed.

One interesting point is made by Fischer et al. (1976a) in drawing a comparison between liquid and amorphous metals after studying the alloy $Fe_{32}Ni_{36}Cr_{14}P_{12}B_6$ (Metglas 2826A). They assume

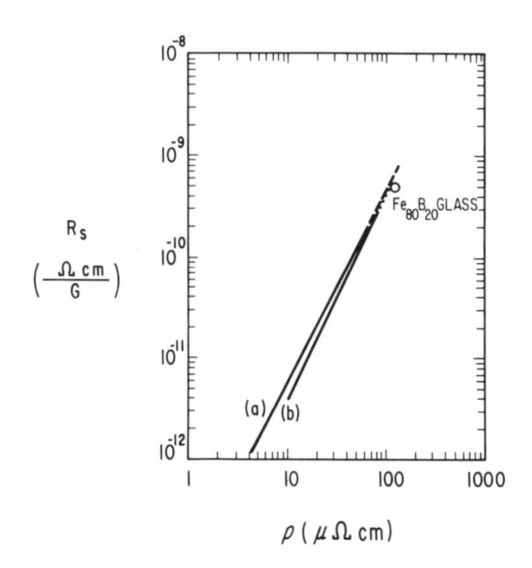

Fig. 11. Spontaneous Hall coefficient vs resistivity for crystalline Fe-Si alloys (Kooi 1954) (a) and for Fe-TM alloys (Jellinghaus and Andres 1961) (b) and for amorphous $Fe_{80}B_{20}$ glass (from O'Handley 1978b).

the mean electronic configuration of Cr-Fe-Ni to be close to that of iron and take trivalent B and pentavalent P to have an effect similar to quadrivalent Ge. Certainly this is an oversimplified approach that incorrectly assumes the rigid-band model to apply to the transition metal combination; moreover it neglects the strong effects of metalloid size on metallic glass properties (Kazama et al. 1978). Nevertheless, they make the interesting observation that liquid $Fe_{80}Ge_{20}$ shows nearly the same values of ρ, $d\rho/dT$, R_o, and χ as the metallic glass $Fe_{32}Ni_{36}Cr_{14}P_{12}B_6$. The first two of these properties are well predicted by the theory of liquid metals. The fact that R_o changes sign across the liquid Fe-Ge series (Fischer et al. 1976b) may help explain the apparent inconsistencies in the sign of R_o for amorphous Fe-base alloys.

In summary, the data suggest conduction by electrons in amorphous Co and Ni. In a-Fe and Fe-rich alloys a more complicated situation exists that may depend on other components present.

2. Spontaneous Hall effect. The SHE is generally much larger for metals and alloys in their amorphous state than in their crystalline state. This appears to be a direct consequence of the high resistivity of the amorphous state. This is illustrated in Fig. 11 by data for crystalline Fe, Fe-Si and for amorphous $Fe_{80}B_{20}$. The spontaneous Hall coefficient of the amorphous metal

$Fe_{80}B_{20}$ is consistent with the series of data for the crystalline alloys considering the high resistivity of the former.

The assumption in Fig. 11 is that any band structural differences due to the additives are small and hence R_s is simply proportional to ρ^n (with n = 2 in this case). But the band structure is expected to be severely altered by the addition of 20% boron or in alloys between transition metals differing in atomic number by more than 2. Compositional differences in the SHE are reviewed next and are discussed in Sec. IV B in terms of the split-band model.

a) Compositional dependence. Amorphous Fe and Fe-rich alloys show the highest spontaneous Hall coefficients of the three transition metal species studied. The data of Table Ib are generally in accord for the Fe-rich glasses with two exceptions. Aldridge and Raeburn (1976) find $R_s = +0.7 \times 10^{-10}$ Ohm/cm/G for a-Fe and Shimada and Kojima's data (1978) for $Fe_{1-x}Si_x$ extrapolate for x = 0 to a number close to this. Bergmann's low value for R_s in a-$Fe_{.5}Au_{.5}$ is a consequence of the large Au concentration (see Fig. 10). As gold concentration approaches zero R_s approaches 3×10^{-10} Ohm-cm/G (Bergmann and Marquard 1978).

It is interesting to note that whereas Shimada finds that R_s decreases in a-$Fe_{1-x}Si_x$ thin films as x goes to zero, Bergmann and Marquard (1978) find R_s increases in $Fe_{1-x}Au_x$ thin films as x goes to zero and McGuire et al. (1980) find R_s increases in a-$Fe_{1-x}B_x$ thin films as x goes to zero. More significantly, Shimada's data extrapolate to $R_s < 1 \times 10^{-10}$ Ohm/cm/G whereas Bergmann's and McGuire's data both extrapolate to 5×10^{-10} Ohm-cm/G for x = 0 (Fig. 12). It should not matter if these data were taken at room temperature; R_s should be essentially the same there as at 4.2 K in amorphous metals.) One might gloss over these differences except for the fact that a similar duality exists for the saturation moment (4.2 K) of a-Fe. Some data for a-Fe-B and a-Fe-P (Hasegawa and Ray 1978, Kazama et al. 1978) extrapolate to $n_B = 2.2\ \mu_B$/atom for a-Fe whereas Felsch (1969,1970) and Fukamichi et al. (1978) find n_B of several Fe-rich glasses containing variously B, P, Ge, Si, and O extrapolate to a lower value near 1.4 μ_B. The Curie temperatures of a-Fe obtained by extrapolation are invariably low. Taken together these data suggest that more than one structural short-range order can exist in a-Fe: one leading to high moment, to $R_s=5\times10^{-10}$ Ohm-cm/G, and low T_c and the other giving lower moment, $R_s<1\times10^{-10}$ Ohm-cm/G, and low T_c. One or the other of these local orders may be more stable depending on metalloids, impurities and/or techniques used to fabricate the sample (Durand 1976).

R_s is universally found to be > 0 in a-Co and Co-rich amorphous alloys although its magnitude is considerably smaller than that found in a-Fe (Fig. 10). McGuire et al. (1979b) find R_s increases with increasing x in a-$Co_{1-x}B_x$ films, contrary to the trend in their

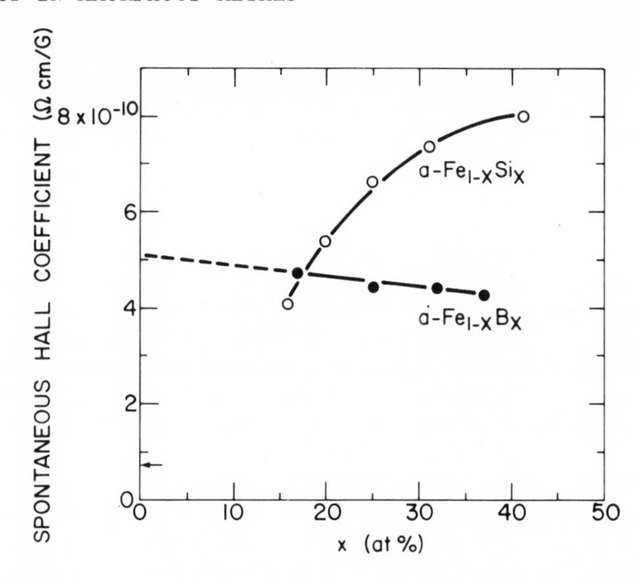

Fig. 12. Spontaneous Hall coefficient vs composition for a-$Fe_{1-x}Si_x$ (Shimada and Kojima 1978) and for a-$Fe_{1-x}B_x$ (McGuire et al. 1979b) alloy films. Arrow to right of ordinate indicates R_s value measured by Aldridge and Raeburn (1967) for a-Fe films.

a-FeB films (Fig. 12).

Amorphous Ni and Ni alloys show $R_s < 0$ (Fig. 10), consistent with the values observed in crystalline materials. Apparently the presence of metalloids B (O'Handley 1978b) and B-Si (Obi et al. 1977) shift R_s to more positive values. It has been observed by these authors that this shift in the $R_s = 0$ composition is in the opposite sense expected from d-band filling in a rigid-band model. See Sec. IV B below for an explanation.

b) Temperature dependence. As mentioned in the introduction, R_s is expected to be either increasing or roughly constant with T above T_c (see Fig. 6). Unfortunately, most of the data displayed through T_c show $R_H = R_o + 4\pi\chi R_s$ so that a misleading impression can be given of the temperature dependence of R_s above T_c. An excellent example of the absence of any sharp decrease in R_s above T_c is found in the complete set of data of Guntherodt et al. (1977b) shown in Fig. 9 above. This figure allows $R_H = R_o + 4\pi\chi R_s$ to be partitioned into its ordinary and spontaneous parts for both the solid paramagnetic and liquid phases of Co. The values so obtained are $R_s = 1.67 \times 10^{-10}$ and 5.2×10^{-10} Ohm-cm/G respectively. Note that except for the abrupt change at the melting point, both R_o and R_s are approximately independent of temperature over the ranges studied in each phase. Moreover, the solid paramagnetic value of R_s is very close to that observed in ferromagnetic a-Co.

Table Ic compares the R_s values of the high temperature phases with those in low temperature a-Co and a-Co-rich alloys. The values for R_s in the table have not been normalized to account for the lower Co content in the alloys. The values are remarkably close considering the different temperatures and sources of the data.

Two cases exist in the literature where $\chi(T)$ and $R_H(T)$ data above T_c are presented for the same composition (2826A in both cases) (Malmhall et al. 1977, and Fischer et al. 1976b). In the former case the susceptibility data are derived from the Hall data assuming R_o and R_s constant through T_c. Their derived susceptibility agrees well with their independent SQUID measurements of $\chi(T)$ reported elsewhere (Figueroa et al. 1976). In the latter case, a plot of R_H vs χ gives $R_o = +3\times10^{-12}$ and $R_s = 3.2\times10^{-10}$ Ohm-cm/G (when χ is put in the proper units) in reasonable agreement with Malmhall's data. This is further evidence of the absence of an abrupt drop in R_s at T_c.

B. Amorphous Alloys Containing Rare Earths

1) Gd-Co ferrimagnet. Probably the first rare earth amorphous alloy in which the anomalous Hall effect was studied was the ferrimagnetic Gd-Co. Measurements on Gd-Co were evidently stimulated by the proposed use of this material for bubble domain devices (Gambino et al. 1974). Gd-Co, because of the complexity caused by two interacting magnetic ions, gave rise to several models to explain the Hall behavior.

An understanding of Gd-Co begins with the observation that the spontaneous magnetization is the difference between the two antiparallel systems Gd and Co. This arrangement is shown schematically in Fig. 13a for an alloy with approximate composition $Gd_{0.2}Co_{0.8}$. Illustrated in this figure is the compensation temperature (T_{comp}) where the magnetizations of the two magnetic systems just balance each other. Below T_{comp} the Gd moment is dominant while above T_{comp} the Co moment is greatest. The Hall effect must be explained in this context.

The first papers on the Hall effect in Gd-Co were those of Ogawa et al. (1974), Okamoto et al. (1974) and Shirakawa et al. (1976). Figure 14 shows a typical result in this case for $Gd_{17}Co_{83}$ with T_{comp} at 100°K illustrating the sharp change in sign for the

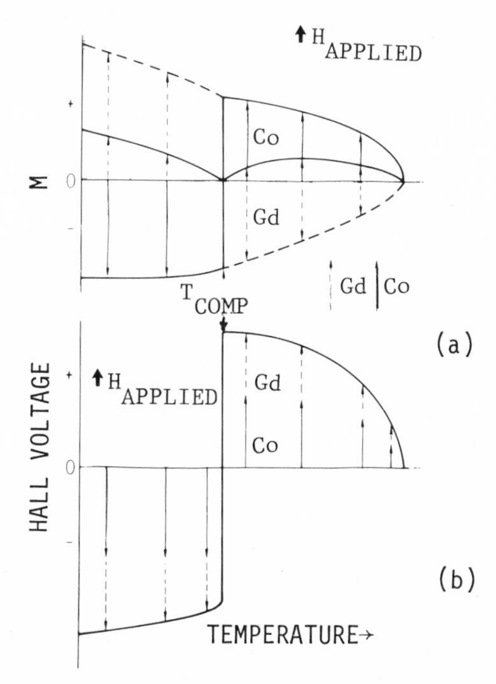

Fig. 13. (a) Schematic diagram of magnetization versus temperature for a two-subnetwork system as represented by Gd and Co. (b) Schematic diagram of Hall voltage versus temperature for a two-subnetwork system.

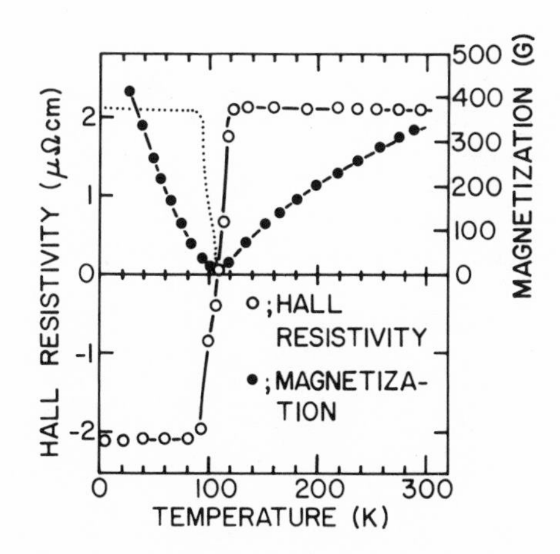

Fig. 14. Temperature dependences of magnetization and ρ_H of a $Gd_{17}Co_{83}$ sputtered film. (From Shirakawa et al. 1976).

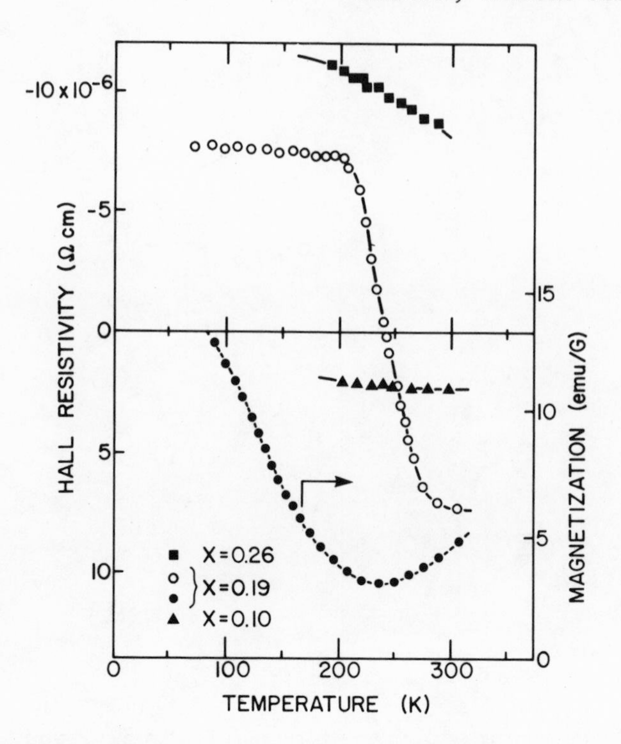

Fig. 15. Hall resistivity of Gd_xCo_{1-x} sputtered films and magnetization of a $Gd_{0.19}Co_{0.81}$ film as functions of temperature. (From Ogawa et al. 1976).

Hall voltage at T_{comp}. Figure 15 can be understood in terms of the change in compensation temperature which occurs with composition as measured by the Hall effect. McGuire et al. (1977) have shown that the sharpness of the transition depends on the homogeneity of the sample. The Hall coefficient R_s as a function of composition for $Gd_{1-x}Co_x$ at room temperature is plotted in Fig. 16 (Stobiecki 1978) showing the dispersive character of R_s as a compensation composition is approached. Asomoza et al. (1977) studied Gd-Co and other amorphous ferrimagnetic alloys as listed in Table II. Their work illustrates that other RE-TM alloys have a behavior for the Hall resistivity similar to that of Gd-Co.

An understanding of the change in sign of ρ_H at T_{comp} can be gained from Fig. 13b. Consider first the Gd magnetic moment which has associated with it a negative Hall scattering for spin up below T_{comp}. However, above T_{comp} the Gd moment is now spin down and its Hall voltage changes sign correspondingly. The same analysis can be made for the cobalt except spin up Co has a positive ρ_H opposite in sign to Gd. These two Hall effects should add together.

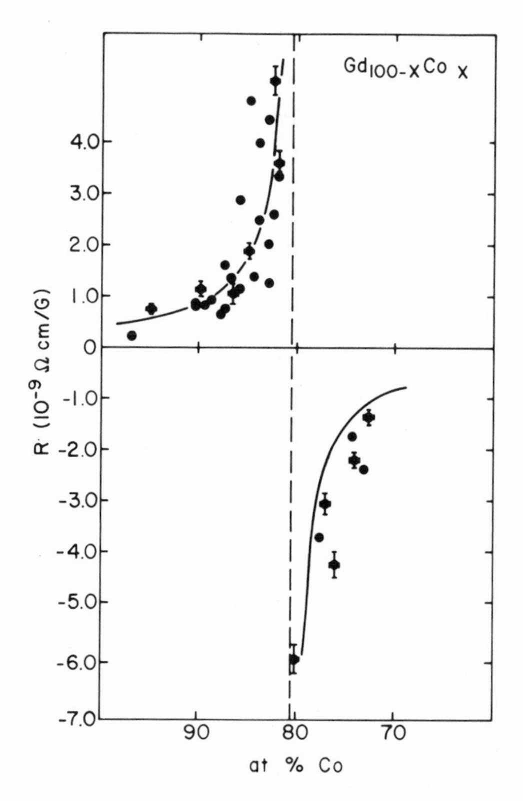

Fig. 16. The spontaneous Hall coefficient as a function of composition for $Gd_{100-x}Co_x$ measured at room temperature. Compensation point at room temperature at approximately x = 80. (From Stobiecki and Hoffmann 1980).

It is of interest that Ogawa et al. (1976) and also Ratajczak et al. (1979) assigned the change of sign of ρ_H to the Gd system alone. This was based on the observation that the magnitude of the Hall resistivity does not depend on the magnetization of the sample but varies in proportion to the amount of Gd present in Gd_xCo_{1-x}. As x increases the absolute value of the Hall resistivity increases. This is illustrated for three compositions shown in Fig. 15.

In contrast Okamoto et al. (1974) and Skirakawa et al. (1979) assign a dominant role to the cobalt. This interpretation came about because the temperature dependence of ρ_H and the Kerr magneto-optic rotation are similar. It is known from previous work that the main contribution to the Kerr effect is the Co and not Gd.

Asomoza et al. also arrive at the same conclusion namely that it is only the Co which contributes to ρ_H in Gd_1Co_3 and other RE_1Co_3

TABLE II

Alloy	$4\pi M$ (G)	Temp (K)	T_c (K)	$\rho(\mu\Omega cm)$	$\rho_H(\mu\Omega cm)$	Ref.
$Gd_{0.15}Ni_{0.85}$	500	77	300		negative	a
$Gd_{0.25}Ni_{0.75}$	12400		40		positive	b
$Gd_{0.21}Co_{0.79}$		77	>300	250	-5	c
$Gd_{0.23}Co_{0.77}Ar_{0.10}$	1650	295		460	+1.15	d
$Ho_{0.3}Co_{0.7}$	10000	77	-900	-	3.8	e
$Dy_{0.33}Co_{0.66}$	-	-	-	-	-5.5	f

a) Mumura (1976); b) Asomoza (1977); c) Shirakawa (1976); d) Gron (1979); e) Ratajczak (1977); f) Asomoza (1979).

alloys. Evidence cited for this view includes the sign of the Hall voltage, the temperature dependence which is constant between 4.2°K and room temperature, and analysis of data from Gd(Co-Ni) alloys. Studies by Mimura et al. (1976) also cite the strong effect of Ni in Gd-Ni to suggest the transition metal plays the more important role.

It is not the purpose here to try to place the above comments on Gd-Co in critical assessment. One obvious possibility is that Gd and Co both contribute to ρ_H. The schematic diagram shown in Fig. 13b shows that this is a straightforward approach since Gd and Co have opposite signs for ρ_H and point in opposite directions so that the Hall scattering is additive. McGuire et al. (1977) have taken this approach. Their model uses the tangent of the Hall angle ρ_H/ρ as the quantitative factor to be considered. As a first approximation the Hall angle is simply the algebraic sum of each component of the alloy based on the atomic fraction of each component. As is noted in section IV D 1, the 3d transition metals Hall angle is not independent of concentration because the moments of these metals decrease due to charge transfer. The values of ρ_H/ρ for the components must be based on single magnetic ion alloys such as Gd-Au, Y-Co or Fe-P. From these one can calculate values for Gd-Co and also Gd-Fe. Agreement is best for Gd-Fe and poor for Gd-Co. The problems that arise for interpreting the Hall effect in $Gd_{1-x}(TM)_x$ amorphous alloys will be further discussed in sections III B-3, and IV D.

2. Alloys of rare earths with non-magnetic metals. Rare earths with non-magnetic components such as Au, Cu, and Al are considered

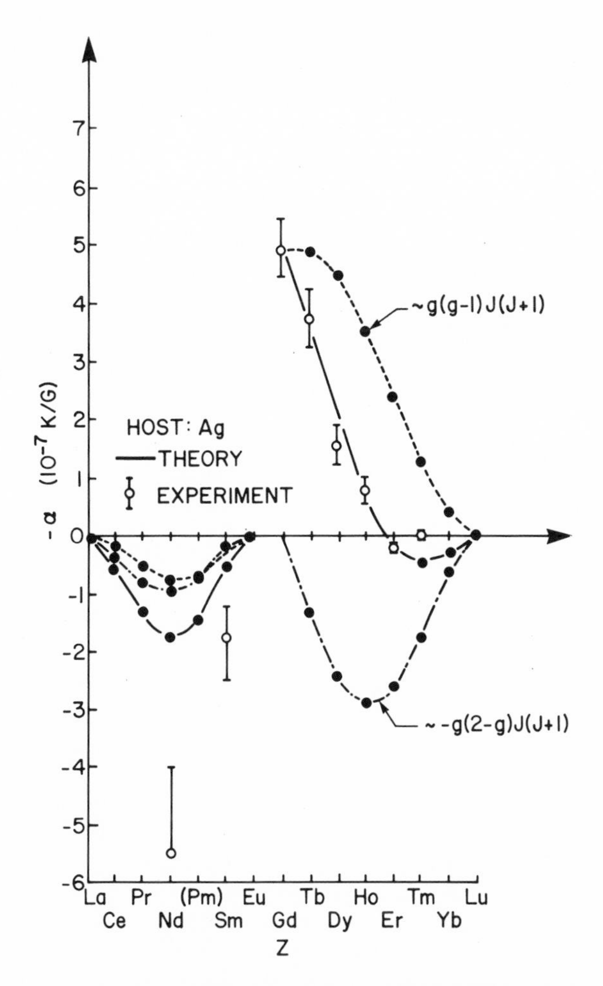

Fig. 17. Coefficient α which characterized the skew scattering of each rare earth impurity (0.1 to 2.0%) is plotted for the RE series. The solid, dashed and dash-dotted lines are theoretical plots as described in the text. (From Fert and Friederich 1976).

in this section. As far as we are aware all such amorphous alloys exhibit predominately positive exchange interactions. (Heiman et al. 1975, Poon and Durand 1977, Gruzalski and Sellmyer 1979, McGuire and Gambino 1979). The alloys do not always show ferromagnetic order. For example Gd-Al (Mitzoguchi et al. 1977), Gd-Cu (McGuire et al. 1978) and Dy-Cu (von Molner et al. 1980) all show compositions where spin glass features are found. Even when ferromagnetism occurs all rare earth ions other than Gd^{3+} and Eu^{2+} are influenced by

TABLE III

Alloy	T_c	θ	C_M	$4\pi M_s$		ρ		ρ_H		ρ_H/ρ	$R_s \times 10^{10}$
				Meas.	Theory	297K	4.2K	meas.	pure		
(at. %)	(K)	(K)	(Kcm^3)	(kGauss)		($\mu\Omega cm$)		($\mu\Omega cm$)		(%)	($\Omega cm/G$)
$Pr_{.47}Au_{.53}$	< 4.2	7	1.6	1.6	6.9	167	168	+0.96	+6.99	+4.16	+ 6.48
$Nd_{.385}Au_{.615}$	10	15	1.5	2.00	6.2	230	234	+2.24	+18.50	+7.90	+16.58
$Sm_{.43}Au_{.57}$		4		0.6	1.5	173	177	+ .83	+ 4.45	+2.50	+17.80
$Eu_{.2}Au_{.8}$	< 4.2	10	4.3	3.8	6.9	213	198	+ .30	+ 2.72	+1.37	+ 1.76
$Gd_{.537}Au_{.463}$	99	109	9.8	17.6	19.9	205	210	−5.66	−11.88	−5.65	− 4.81
$Tb_{.47}Au_{.53}$	> 4.2	31	11.4	9.4	20.4	190	197	−1.27	− 5.87	−2.98	− 1.79
$Dy_{.46}Au_{.54}$	15	18	14.1	10.2	22.5	132	138	−0.40	− 1.91	−1.39	− .52
$Ho_{.44}Au_{.56}$	11	24	10.1	13.5	22.0	93	95	−0.30	− 1.10	−1.16	− .29
$Er_{.58}Au_{.42}$	9.5	10	12.5	9.9	17.4	191	199	−0.03	− .07	− .04	− .03
$Tm_{.62}Au_{.38}$		3	8.3	8.0	20.1	194	207	−0.09	− .34	− .17	− .12

TABLE IV

Alloy	T_c	θ	C_M	$4\pi M_s$		ρ		ρ_H		ρ_H/ρ	$R_s \times 10^{10}$
				Meas.	Theory	297K	4.2K	meas.	pure		
(at. %)	(K)	(K)	(Kcm^3)	(KGauss)		($\mu\Omega cm$)		($\mu\Omega cm$)		(%)	($\Omega cm/G$)
$Pr_{.40}Cu_{.60}$	< 4.2	8	2.0	2.3	7.1	86	82	+0.43	+ 3.31	+ 4.03	+ 3.31
$Nd_{.44}Cu_{.56}$	< 4.2			3.0	7.8	114	116	+2.07	+12.12	+10.4	+10.86
$Gd_{.42}Cu_{.58}$	77	85	8.9	14.8	16.5	176	175	−1.96	− 5.22	− 2.48	− 2.11
$Tb_{.503}Cu_{.497}$	23	39	11.8	7.8	24.0	147	151	−0.26	− 1.59	− 1.05	− 0.48
$Dy_{.45}Cu_{.55}$	16	20	19.5	12.8	25.3	141	145	+0.15	+ .66	+ 0.45	+ 0.18
$Ho_{.44}Cu_{.56}$	8	8	17.7	11.8	22.0		220	+0.45	+ 1.91	+ 0.86	+ 0.51
$Er_{.51}Cu_{.49}$	< 4.2	11	8.8	8.4	17.9	176	178	+0.52	+ 2.17	+ 1.22	+ 0.89
$Tm_{.61}Cu_{.39}$		−3	9.4	10.2	21.7	149	159	+0.31	+ 1.08	+ 0.67	+ 0.40

random crystalline fields which cause a fanning of the spins over a hemispherical range called by Coey (1977) "asperomagnetic". Formally the spontaneous Hall effect is only influenced by the degree of polarization as given by the measured $4\pi M$ or χH. Thus the type of order should not be of importance but as we will discuss this may not be the case. The alloys that have been studied are listed in Tables III, IV and V.

The most complete investigations have been done on RE-Au, -Ag and -Cu systems. Fert and Friederich (1976) have reported on dilute rare earths in Au and Ag. Although these are crystalline alloys they fit the description of amorphous because the rare earths are randomly distributed in the Au or Ag matrix. The resistivity increases by $7\mu\Omega$cm for each one percent of RE ion in these alloys. We do not know at this time what the effect would be on the Hall resistivity if the Au or Ag were also amorphous in these dilute alloys.

The Hall effect for the dilute RE alloys are shown in Figs. 17 and 18 where the coefficient α is defined by $R = R_o + \alpha\rho_o T^{-1}$ and can be obtained graphically from a plot of R vs. 1/T. It is not possible from the data given in the figure to go directly to the value for R_s since the susceptibility is needed. We calculated R_s for $Gd_{.005}Au_{.995}$ assuming a Curie law with the free ion moment for the Gd. From the data in the paper $R_o \approx 0.7 \times 10^{-12}$ Ohm-cm/G, and $R \approx 3.1 \times 10^{-12}$ at 1°K. Under these conditions $R_s \approx 4.7 \times 10^{-11}$ Ohm-cm/G for $Gd_{.005}Au_{.995}$. Tables III and IV list another Gd-Au composition to compare with this value. In the ferromagnetic amorphous Gd-Au, R_s is 10 times greater than for the dilute composition.

We emphasize two features of the data of Fert and Friederich in Figs. 17 and 18. First is the cross-over going from positive Hall angle for Sm to negative Hall angle for Gd. The second feature is the shift in the RE-Ag alloys toward more positive values of Hall angle. For example Er-Au is negative and Er-Ag is positive in α. Fert and Friederich analyze their results in terms of orbital skew scattering proportional to $g(2-g)J(J+1)$ and spin skew scattering proportional to $g(g-1)J(J+1)$. Values for these two expressions are of opposite sign for the heavy rare earths and give the theoretical fit shown in the figures. It is important to note that the skew scattering mechanism in these alloys is based on the linear relationship found for α with ρ_o for different compositions of Dy-Au, and Dy-Ag (ρ_o is the residual resistivity).

We now compare the dilute alloys with the concentrated amorphous alloys of RE-Au and RE-Cu (McGuire and Gambino 1979, 1980). These data, as listed in Tables III and IV and shown in Fig. 19 are normalized to the conditions that represent the pure rare earth with all spins parallel. Comparing with the dilute systems (Figs.

TABLE V

Alloy	x at. frac. Gd	y at. frac. Gd	$4\pi M$ (kG) 4.2°K	T_c (K)	θ (K)	ρ μΩcm room temp	ρ μΩcm 4.2°K	ρ_H μΩcm 4.2°K	$\frac{\rho_H}{\rho}$
GdMg	.73	.79	15.8	130	133	191	196	- 4.3	-.022
Gd-Ti	.56	.70	17.2	88	116	137	142	- 2.6	-.018
GdV	.54	.73	14.7	105	144	174	181	- 4.5	-.025
GdCr	.50	.74	15.0	75	102	202	216	- 3.8	-.017
GdNb	.56	.70	10.5	48	90	170	174	- 2.2	-.012
GdGe	.55	.64	17.8	71	101	173	286	- 4.7	-.018
GdSi	.58	.70	12.9	87	111	964	1276	-21.4	-.017
GdAu	.54	.69	17.6	99	109	205	210	- 5.6	-.026

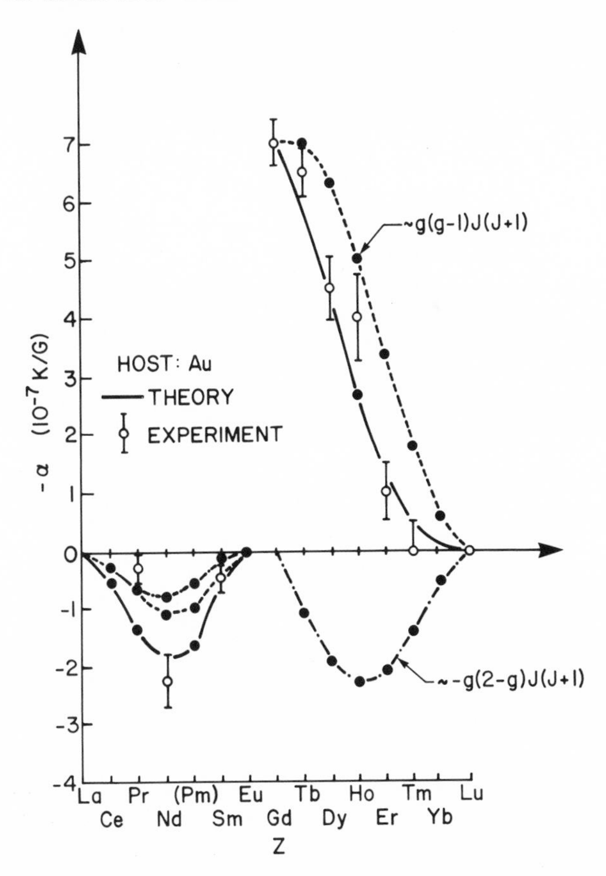

Fig. 18. Coefficient α which characterizes the skew scattering by each rare earth impurity (0.5 to 2.0%) in Au is plotted for the RE series. The dashed and dash-dotted lines are theoretical plots as explained in text. (From Fert and Friederich 1976).

(17 and 18) we note that there are significant similarities. One is the shift in sign going from positive Hall angle (ρ_H/ρ) for Eu^{2+} to negative ρ_H/ρ for Gd^{3+}. A second similarity is the positive enhancement of ρ_H/ρ for the a-RE-Cu alloys compared to the a-RE-Au alloys which is similar to the positive shift in the crystalline dilute RE-Ag alloys compared to the dilute RE-Au. For the RE-Au systems McGuire and Gambino (1980) interpret the ρ_H/ρ data in terms of the exchange energy. Shown on the figure normalized to Gd are the θ values for each alloy as determined from the Cure-Weiss law $[\chi_m = C_m/T-\theta)]$ where θ in molecular field theory is proportional to the sum of the exchange interactions. The fit is best for the heavy rare earths where the values of θ follow the measured Hall angle ρ_H/ρ.

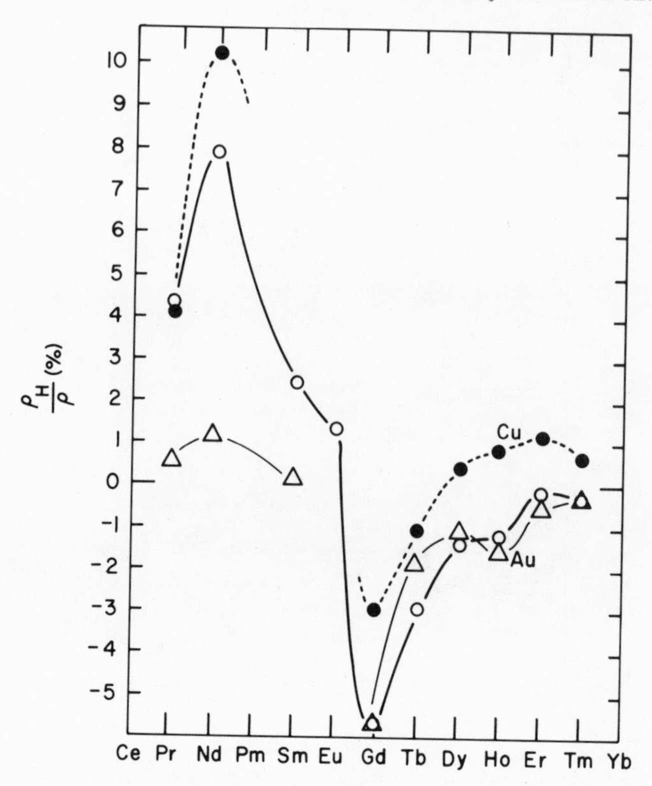

Fig. 19. The tangent of the Hall angle in percent for the rare earth elements as determined from RE-Au and RE-Cu amorphous alloys by extrapolating data at 4.2 K and 20 kG to complete saturation for the pure element. The Δ points are obtained by using the Curie-Weiss θ values normalized to Gd (see text). (From McGuire and Gambino 1980).

The change in sign of ρ_H/ρ from positive for the light rare earths to negative for the heavy rare earths is attributed to the spin direction of the RE atom which changes sign in the rare earth metals when going from light to heavy rare earths. In the rare earth metals and alloys the RKKY exchange mechanism is believed to be dominant. This exchange is between conduction electron spins and the localized spins of the rare earth. The conduction electrons are polarized by this exchange and the strength of the exchange is related to the Curie temperature or Curie Weiss θ. The increase in spin polarization of the conduction electrons enhances the intrinsic spin-orbit coupling (see section I D, 2a and b) and consequently those alloys with the highest θ also have the largest ρ_H/ρ. Going from light rare earths to the heavy rare earths is then simply related to the reversal of the rare earth spin and its associated conduction electron spin polarization coupled through the RKKY

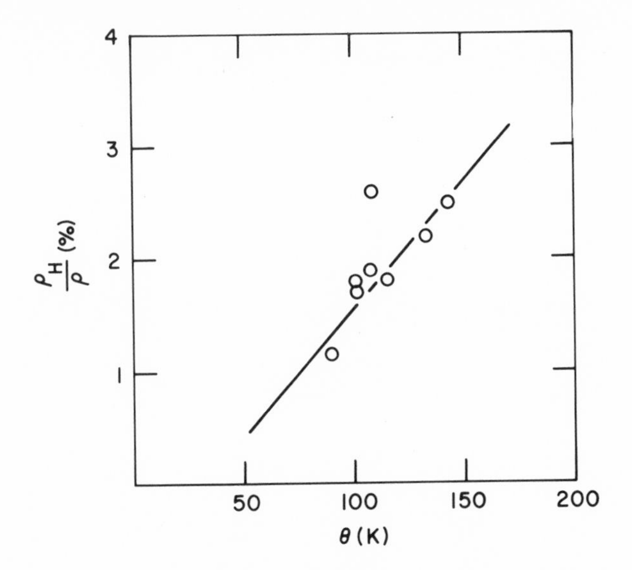

Fig. 20. The tangent of the Hall angle in % at 4.2 K vs. Curie-Weiss θ for Gd alloys listed in Table V. (From McGuire and Gambino 1980).

exchange.

Shown in Fig. 20 are several alloys of Gd(M) where M is a non magnetic metallic atom as listed in Table V. Again we observe that ρ_H/ρ is proportional to the Curie-Weiss θ giving further evidence of the importance of the exchange energy in its effect on the Hall scattering.

3. RE-TM ferrimagnets. In this section we examine the Hall effect in ferrimagnetic systems and also examine the comments made in section III B - 1 to see if they can be qualified in any way. Table II lists several alloys mainly Gd-TM that have been measured. Figure 21 plots Gd-Fe and Gd-Co and Figure 22 shows Gd-Ni and Gd-Mn (McGuire et al. 1976, 1977, 1978, 1979).

A decrease in ρ_H/ρ is observed in both the Fe and Co systems as Gd is added to the transition metal. If we assume a linear relationship should exist and each species scatters in proportion to its concentration then for a-$Gd_{1-x}Co_x$ we expect the relation $\rho_H/\rho = (\rho_H/\rho)_{Gd}(1-x) + (\rho_H/\rho)_{Co}x$ with suitable values chosen for $(\rho_H/\rho)_{Gd}$ and $(\rho_H/\rho)_{Co}$. For the a-Gd-Fe alloys this relationship works reasonably well, however for Gd-Co the measured values lie well below those calculated. Part of this difference may arise from the moment of the Co decreasing because of charge transfer and in fact the Co may have zero moment for $x < 0.4$ (see section IV D 1).

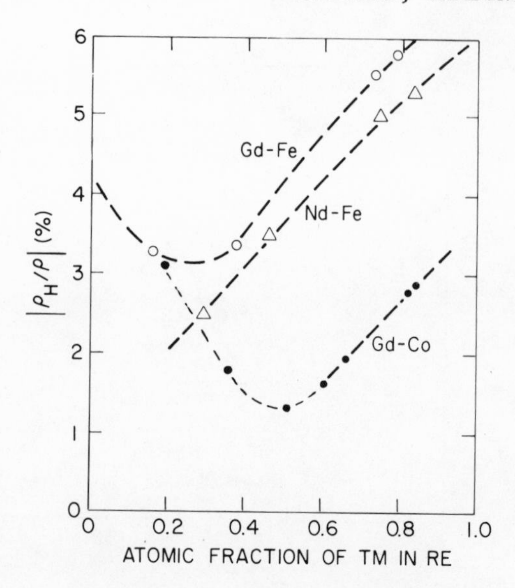

Fig. 21. Tangent of the Hall angle $|\rho_H/\rho|$ in percent shown vs. atomic fraction of the transition metal in the rare earth ($RE_{1-x}TM_x$). Temperature is 4.2 K. (From McGuire et al. 1977 and Taylor et al. 1978).

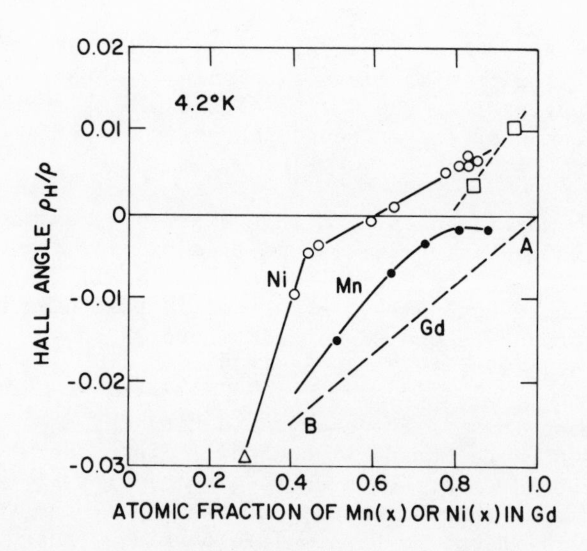

Fig. 22. Tangent of the Hall angle ρ_H/ρ vs. atomic fraction of Mn or Ni in Gd. Data points marked □ are Y-Ni and point Δ is for Gd-Au. Temperature is 4.2 K (from McGuire et al. 1978, 1979).

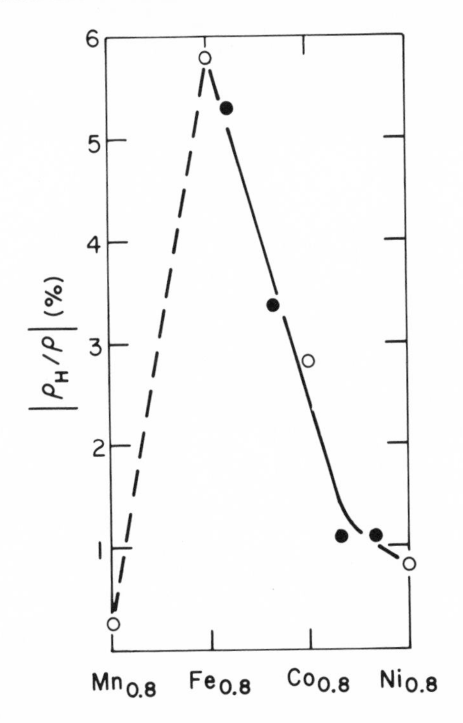

Fig. 23. Hall resistivity ratio $|\rho_H/\rho|$ in percent is plotted for nominal $Gd_{.2}(TM)_{.8}$ alloys as open circles (listed in Table I). The dark circles are for mixed 3d transition metal alloys; from left to right $Gd_{.25}Fe_{.60}Co_{.15}$, $Gd_{.22}Fe_{.30}Co_{.48}$, $Gd_{.19}Co_{.67}Ni_{.14}$ and $Gd_{.16}Co_{.33}Ni_{.57}$.

For the Co rich samples the decrease of ρ_H/ρ may also be influenced by the decrease in Curie temperature which takes place. At this time it does not seem possible to definitely assign values of ρ_H/ρ to both Co and Gd.

The data for Gd-Fe also follows the same pattern as Gd-Co but there is less of a drop in the values of ρ_H/ρ on the Fe rich end. It is therefore possible to assign ρ_H/ρ for Gd and Fe in reasonable agreement with the atomic fraction of each component showing that both Gd and Fe contribute to ρ_H/ρ and not one or the other alone. The measurements for Gd-Ni and Gd-Mn shown in Fig. 22 indicate that both Mn and Ni have an effect opposite to that of Fe and Co and therefore have a negative sign of ρ_H/ρ. Because Gd-Ni and Gd-Mn are ferrimagnetic the Ni and Mn moments are antiparallel to that of the Gd and give a positive contribution to ρ_H/ρ. This can be seen by comparing the measured values to the line AB in Fig. 22 which now represents ρ_H/ρ for Gd using the linear model we previously discussed. In addition the moment of the Ni and Mn and also the exchange interaction must be changing significantly with composition. Again we are in the same position as with Gd-Co and cannot make a definite assignment for ρ_H/ρ. By the way of summing up the behavior of $Gd_{1-x}(TM)_x$ amorphous alloys we show the plot of ρ_H/ρ at $x \approx 0.8$ in Fig. 23 for the various transition metals, the largest value being for the Fe alloy.

We again call attention to the feature that in many amorphous alloys there is a fanning or dispersion of the spins over a spherical or hemispherical region of space. Many of the RE-Au and RE-Cu compositions fulfilled this condition. In the RE-TM case, we show one example, that of Nd-Fe. This is plotted in Fig. 21. The values for ρ_H/ρ fall slightly below the Gd-Fe and reflects the conditions that both the Nd and Fe spins are fanned out (Taylor et al. 1978). Detailed analysis based on magnetization and Mössbauer spectroscopy indicate that there is some fanning of the Fe spins but that the major dispersion is in the Nd system i.e., 70 to 80% of the Nd is in a disordered state. The Hall effect supports the model that only a partial magnetic contribution comes from the Nd.

IV. FURTHER DISCUSSION

A. Side Jump vs Skew Scattering

Figure 5 is a schematic representation of the skew scattering and side jump mechanisms in the absence of a Hall field. These processes, outlined in Section ID-2, are treated in more detail elsewhere in this book where further references can be found (Berger and Bergmann 1980).

Because of the short mean free path of the carriers in amorphous metals we expect the side jump mechanism (R_s proportional to ρ^2) to dominate the SHE. This ρ^2 relation holds provided the band structure (i.e. the composition) is not significantly altered. It is difficult to verify this relation for amorphous metals in the usual way by varying temperature because $d\rho/dT$ is generally small in these materials. Furthermore, small changes in composition produce a very small effect on resistivity because it is already so large due to the frozen-in disorder. One exception to these rules is the series of $Fe_{1-x}Au_x$ films of Bergmann and Marquard (1978) which for $x = 0.5$ show $\rho = 55$ μΩcm at 5K. $d\rho/dT$ is appreciable and as x approaches zero ρ approaches 130 μΩcm. They find ρ_H proportional to ρ when plotting data for $Au_{.5}Fe_{.5}$ with T as a parameter over the range 5 to 231 K. However a plot of their ρ_H and ρ data with composition as a parameter shows ρ_H varies more like ρ^2.

It is tempting to want to compare $R_s/\rho^2 = A$ for different compositions but this overlooks the changes in band structure which may also alter R_s. This approach has nevertheless been taken by Rao et al. (1976) and by Marohnic et al. (1978) with marginal success. The first group finds A = .014 $(Ohm\text{-}cm\text{-}G)^{-1}$ in 2826 and .013 $(Ohm\text{-}cm\text{-}G)^{-1}$ in 2826B compared to .04 $(Ohm\text{-}cm\text{-}G)^{-1}$ in permalloy. The second group finds A varies from .036 to .02 in Fe-Ni-B and from .029 to .023 in Fe-Ni-P-B glasses. The parameter A can be calculated from the data in the various tables. Caution must be exercised in interpreting these numbers because of the wide variations reported for the resistivity. When the materials are of similar composition and ρ is measured by the same technique in each case, reasonable agreement is expected in the values for A. Shimada and Kojima (1978) find $R_s/\rho^{1.3}$ is nearly constant for a series of Fe-Si glasses.

Some authors have followed R_s and ρ through various annealing and/or crystallization stages and found relations between R_s and ρ to hold throughout. For example Malmhall finds it is possible to fit R_s either with $\rho^{2.6}$ or with a ρ + bρ^2 when Metglas 2826A is crystallized (Malmhall et al. 1978a). However, due to the phase separation and the dramatic changes in band structure which may occur during these processes, it is difficult to attach significance to these results now.

Perhaps a more reliable approach is to take the value for the ratio R_s/ρ^2 (assuming accurate resistivity values as described in Sec. IIB-1) and calculate therefrom the magnitude of the side-jump using Berger's formula (Majumdar and Berger 1973). This has been done for $Fe_{80}B_{20}$ glass and the result, Δy - 1.0 Angstrom, is comparable to the value for crystalline Fe and Fe base alloys (as implied by Fig. 11), despite the differences in magnetic moment and resistivity (O'Handley 1978b). Malmhall et al. (1979a) have also calculated the side jump for amorphous and crystalline Metglas 2826 and found them both to be close to 1 Angstrom.

It appears that the side-jump mechanism dominates the SHE in concentrated amorphous metals, as expected although the low-temperature data for Fe-Au show ρ_H proportional to ρ.

B. Compositional Dependence of R_s

1. Band filling and the rigid-band model. Data from a variety of sources and for different physical properties of metal-metalloid systems can be interpreted consistently by assigned a number to each metal-metalloid combination and interpreting it, within the rigid-band model, to be the amount of charge transferred from the vicinity of the metalloid atom to the transition metal d band. However, there is appreciable controversy over the concept of charge transfer in these materials and, moreover, the rigid-band model does not apply when the band structure is appreciably altered as

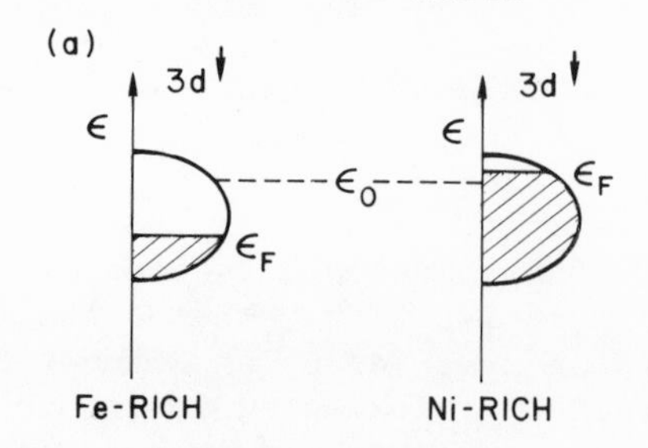

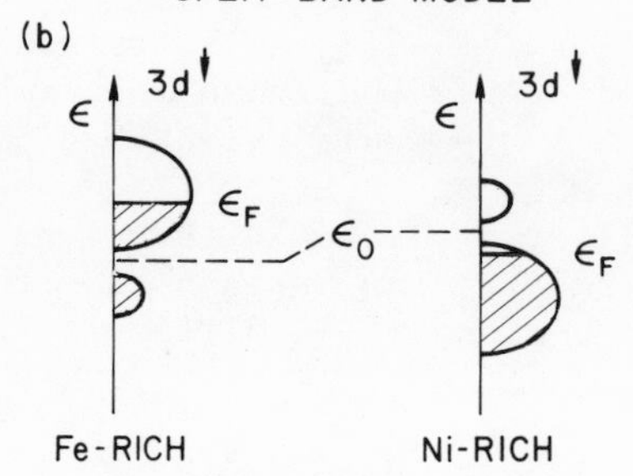

Fig. 24. Schematic representation of minority-spin state densities in Fe-rich and Ni-rich alloys in the simple rigid-band model (a) and in the split-band model (b). Fermi energy is indicated by E_f and a singular point (see text) is represented by E_o. (From O'Handley 1978b).

it must be by the presence of approximately 20 at% metalloid. Perhaps a more realistic interpretation of such a number then is that it represents the degree of change caused in the band structure by the metalloid, the change being such that the magnetic moment is reduced (O'Handley et al. 1977, O'Handley and Boudreaux 1978), the Knight shift is reduced (Hines et al. 1978) and the isomer shift is enhanced (Chien 1979) etc. relative to the TM-only cases.

If we try to apply these charge transfer or band alteration numbers to the interpretation of R_s within the rigid-band model, we meet with unambiguous failure (O'Handley 1978b, Obi et al. 1977). This would be done by assuming that $R_s = 0$ when E_f coincides with some point in the band structure at which the z-component of orbital angular momentum vanishes (Berger 1976, 1977). Then the presence of additional charge from the metalloid atoms would cause E_f to coincide with that singular feature for lower average atomic number of the transition metal components (Fig. 24a). For example, whereas $R_s = 0$ at the crystalline permalloy composition, $Ni_{82}Fe_{18}$, the rigid-band model suggests that $R_s = 0$ at a more iron-rich composition when glass formers are present. Again see Fig. 24a. The shift in $R_s = 0$ with the addition of metalloids is, in fact, observed to be in the opposite direction. We do not take this as evidence that the charge transfer (or band alteration) is in the opposite sense because the preponderance of the data support the original sense (O'Handley and Boudreaux 1978). Instead we use a more accurate model of the band structure.

2. Split-band model. When a binary alloy $A_{1-x}B_x$ is formed such that the centers of gravity of the A and B energy bands E_A and E_B differ in energy by more than their average width ΔE, its band structure splits into two parts distinct on the energy scale, each part identified with one of the two species. The condition $|E_A - E_B| > \Delta E$ is generally met for concentrated alloys for which $|Z_A - Z_B| \geq 2$. Figure 24b schematically depicts this situation. The singular point at which $\langle L_z(E_f)\rangle = 0$ is now the gap or boundary in the band structure. It is a simple matter to see that the coincidence of E_f and E_o occurs for higher average TM atomic number when metalloids are added to the TM matrix. In fact for charge transfer equal to that assigned to boron in Ni (O'Handley and Boudreaux 1978) we can predict $E_f = E_o$ near $(Ni_{.95}Fe_{.05})_{80}B_{20}$ (O'Handley 1978b). This is consistent with $R_s = 0$ observed by Obi et al. (1977) in $Ni_{78}B_{12}Si_{10}$ and by O'Handley (1978b) extrapolating to $Ni_{80}B_{20}$, whereas amorphous Ni without metalloids present shows $R_s < 0$ (Bergmann 1976b).

The split-band model suggests that $R_s = 0$ because the expectation value of the z-component of orbital angular momentum of the 3-d electrons vanishes. That is, R_s arises from an intrinsic spin-orbit interaction (ISOI) involving L_z^{3d} viz $L_z^{3d}\cdot S^{3d}$ and not the orbital angular momentum of conduction electrons e.g. $L^{4s}\cdot S^{3d}$ as in the SOOI. This is in contrast to the interaction assumed to obtain in the rare earth metals, $L^{5d}\cdot S^{4f}$ (Maranzana 1967).

The split-band model has also been confirmed to apply to the occurrence of zeros in magnetostriction (for which $\langle L_z(E_f)\rangle = 0$ is sufficient) in amorphous Fe-Ni-B alloys (O'Handley 1978b, O'Handley and Berger 1978).

3. Phase-shift model. Bergmann (1976b) has used the phase-shift model of Fert and Jaoul (1972) to explain the negative SHE in amorphous Ni. According to this model, the scattering contribution to R_s is proportional to $\sin(2\delta^{+-})$ where δ^+ and δ^- are the scattering phase shifts for the spin-up and spin-down bands respectively. The full 3d+ band has $\delta^+ = \pi$ and, therefore, R_s is proportional to $\sin(2\delta^-)$ with δ^- given by $n_d^-\pi/5$ where n_d^- is the number of spin-down electrons. Assuming for Fe $n_d^- = 2$ and for Ni $n_d^- = 4$ one expects a sign change between Fe and Ni as observed. If one wishes to include the effects of metalloids in this model care must be taken to avoid

a simple rigid-band picture which (as shown in Sec. IV B-1) gives the wrong sign for the shift in $R_s = 0$. A split-band model must be used.

C. Temperature Dependence

1. Which resistivity? The various contributions to the total resistivity are more or less prominent in different temperature regimes. Similarly, the scattering for each of these mechanisms contributes to R_s in a characteristic way (see Sec. I D-2a) whose prominence will vary with temperature. Theoretical attention is often focussed on the magnetic disorder resistivity ρ_M (Maranzana 1967, Kondo 1962 (Fig. 6a) to explain the sharp increase in R_s just below T_c in crystalline materials (Fig. 6b). However, the question may arise as to whether the temperature dependence of R_s is a consequence only of magnetic disorder scattering $R_s \propto (\rho_M)^n$ or rather results from any scattering of the carriers $R_s \propto \rho^n$.

We know that ρ_M in amorphous metals is comparable in absolute magnitude to that in crystalline metals even though the resistivity of the former is much larger: the magnetic contribution to ρ at T_c is only weakly visible in some of the amorphous alloys for which $T_c < T_x$ (Malmhall 1979c). Nevertheless R_s is much larger in amorphous metals than in comparable crystalline metals and it is larger in proportion to the total resistivity (see Fig. 11). Then at least in this case the temperature dependence of R_s derives from the scattering due to structural disorder, analogous to scattering by phonons, as well as to spin-disorder. Also because $R_s \propto \rho^n$ below T_c we expect that above T_c, R_s should follow the solid curve in Fig. 6b: $R_s \propto (\rho_{ph} + \rho_M)^n$.

It is interesting to note that below T_c dRs/dT does not always have the same sign as $d\rho/dT$. Also a cusp is sometimes observed in $R_s(T)$ at T_c whereas a kink is observed in $\rho(T)$ near T_c (Malmhall et al. 1979b, 1979c).

2. Critical exponents. Malmhall, Rao and coworkers have used the temperature dependence of the Hall resistivity above T_c to study the critical behavior of $\chi \propto [(T/T_c)-1]^\gamma$ in several metallic glasses. They established that this method gives correct critical exponents for crystalline Ni (Malmhall 1979c). The results of their studies are summaried in Table VI. In general they find the critical exponents to be constant over a broader temperature range in amorphous metals than in crystalline metals and they often find the magnitude of γ to be larger in the former than in the latter materials. The magnitude of γ is taken as an indication of enhanced magnetization fluctuations above T_c.

Despite their check on the validity of the Hall-effect-derived

TABLE VI. Critical exponent γ for various metallic glasses (derived from Hall effect data) and Curie temperatures above which γ was determined. Data for Ni were also obtained by the Hall effect. Metglas compositions are listed at the end of Table Ie.

Material	γ	Tc(K)	References
2826A	1.75	225	c
2826A	1.7(a)	252(a)	d
2826A	1.4(b)	295(b)	d
2615	1.1	561	e
2605A	1.5	506	f
2826	1.6	506	f
2826MB	1.7	610	f
2826B	1.7	372	e,g
$Ni_{53}Fe_{27}P_{14}B_6$	1.54	368	h
crys. Ni	1.3	629	f

(a) as received
(b) annealed 3 hrs at 600K
(c) Malmhall et al. 1977
(d) Rao et al. 1978
(e) Malmhall et al. 1979c
(f) Malmhall et al. 1979a
(g) Malmhall et al. 1978b
(h) Malmhall et al. 1979b

critical exponents (provided by their measurements on crystalline Ni as in Table VI) these measurements disagree with conventional magnetization data which generally give critical exponents for metallic glasses in close agreement with values for crystalline metals and with the results of the three-dimensional Heisenberg model. Poon and Durand (1977) summarize and discuss these results. There is evidence that the higher values for γ apply for $T/T_c - 1 > 0.03$ whereas the values near 1.3 obtain closer to T_c; such transitions in critical behaviour are not uncommon (Rao personal communication 1979). Exact calculations and approximations for γ in several models are found in Table 3.4 on p. 47 of Stanley's book (1971) and range from 1 to 2 with $\gamma = 1.4$ being the value approximated for the 3-dimensional Heisenberg model which is followed by most crystalline magnetic metals.

D. Subnetwork Scattering Model

1) Effects of moment reduction in Gd-TM alloys. In some cases we have treated the Hall effect with certain simplifying assumptions regarding the properties of binary alloys. These include: 1) The

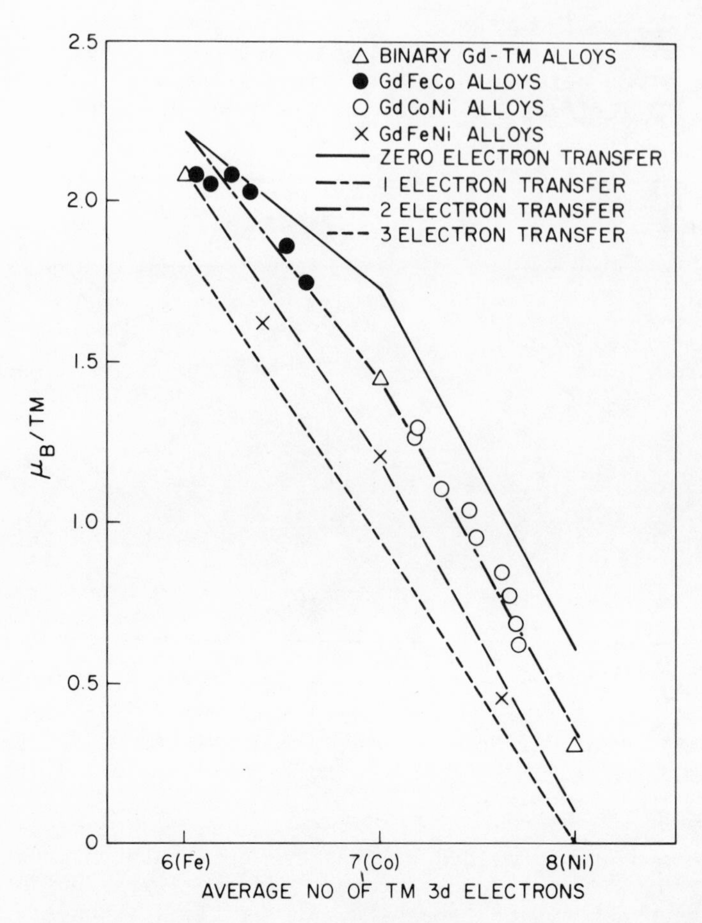

Fig. 25. Magnetic moment of the 3d transition metal subnetwork at 4.2 K as a function of composition for GdFeCo, GdCoNi and GdFeNi alloys. The lines are calculated assuming electron transfer to the transition metals from Gd. (From Taylor and Gangulee 1980)

Hall angle ρ_H/ρ for the rare earths does not change with concentration because the moment remains fixed while for the 3d transition metals ρ_H/ρ does change because the 3d magnetic moment is reduced by charge transfer or band structure alterations (see section IV B1). 2) For the binary Gd-TM alloys there is evidence that ρ_H/ρ depends on the strength of the exchange interaction.

For a more quantitative discussion of the moment reduction we show in Fig. 25 the magnetic moments for binary and ternary $Gd_{.2}(TM)_{.8}$ amorphous alloys (Taylor and Gangulee 1980). Except

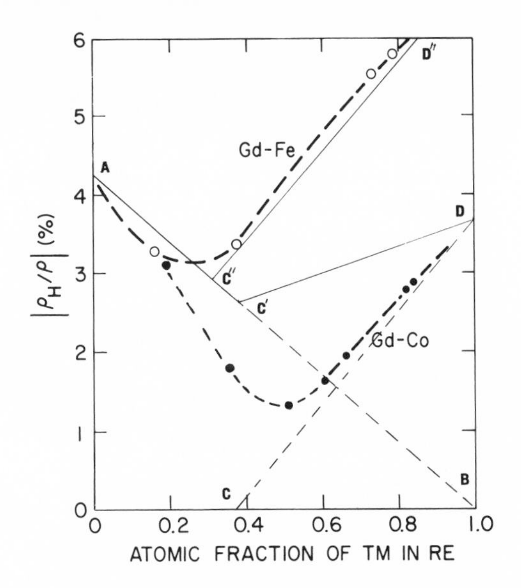

Fig. 26. Data of Figure 21 of Gd-Fe and Gd-Co with theoretical lines representing decrease of ρ_H/ρ for each component with dilution. Line A B is for Gd. Line C D is for Co. Line A C' D is the sum of the Gd and Co values for $|\rho_H/\rho|$. A similar analysis gives the line AC"D" for Gd-Fe. (see text)

for ternary alloys with Fe-Ni the best fit suggests approximately one electron per Gd transferred to the 3d metal. Even $Gd_{.2}Fe_{.8}$ appears to be close to this one electron transfer but this may be due in part to spin fanning. On the basis of one electron transfer we would expect zero moment for the 3d metal at $Gd_{.69}Fe_{.31}$, $Gd_{.63}Co_{.37}$ and $Gd_{.38}Ni_{.62}$.

These values of composition for zero moment can be used to make a linear fit for the Gd-Co and Gd-Fe data as shown in Fig. 26. The Co subnetwork contribution is assumed to follow the dashed line CD which goes to zero Hall angle at $Co_{.37}$. The Gd follows the line AB. The sum of these two lines AC'D is then the total Hall angle of Gd-Co. It is seen that the experimental values lie considerably below this line. In comparison a similar analysis of Gd-Fe (with $\rho_H/\rho = 0$ at $Fe_{.31}$) leads to the summation given by AC"D". The fit of the Gd-Fe data is surprisingly good. This difference between Gd-Co and Gd-Fe can possibly be attributed to the strong decrease in the exchange interaction found in Gd-Co as compared to Gd-Fe.

In Fig. 27 we plot the Curie temperatures of Gd-Fe and Gd-Co. Over a large range of compositions for Gd-Fe the Curie temperature which is a measure of the exchange interaction, only varies between

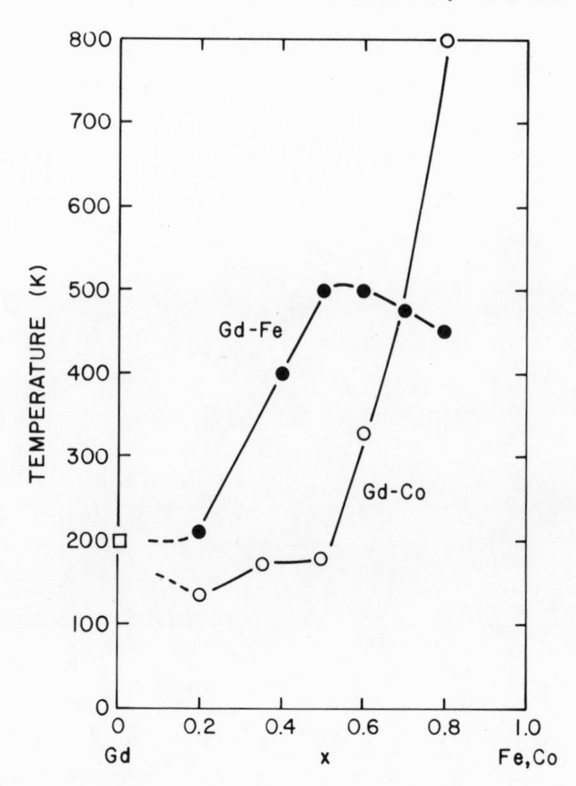

Fig. 27. Curie temperatures for $Gd_{1-x}Fe_x$ and $Gd_{1-x}Co_x$ as a function of x. The value shown for $Gd_{.2}Co_{.8}$ is estimated.

400 to 500°K. In comparison, that of Gd-Co drops sharply from 800 to 200°K over this same range. This decrease in the strength of the exchange interaction may be an important consideration in understanding the drop in ρ_H/ρ for Gd-Co, but the effect must be analyzed using three exchange parameters so more must be known about single magnetic atom behaviour first.

2) Effects of fanning. The decrease in ρ_H/ρ attributed to the decrease in magnetic moment due to charge transfer cannot be distinguished from disordered magnetic structures using the Hall effect measurement alone. For example it is known that YFe_2 (Coey 1978) has a random magnetic structure called asperomagnetic where the Fe spins fan out over a hemispherical surface. Taylor and Gangulee also analyze Gd-Fe data in the same context with either Fe antiferromagnetic or fanning out to reduce the magnetization of Fe system. In some alloys such as Nd-Fe both the Nd and Fe systems are believed to be asperomagnetic (1978). Consequently any source of moment change can cause a comparable change in the spontaneous Hall effect and it is necessary to have supporting measurements

such as neutron diffraction or Mössbauer to propose a unique model for the Hall scattering.

E. Applications of the Hall Effect

1. Magnetic bubble and recording technologies. The SHE of amorphous films has a number of applications in magnetic bubble technology. A number of authors have used SHE loops to characterize amorphous bubble films (Okamoto et al. 1974, Kobliska and Gangulee 1977). Layering effects in amorphous films were studied by Lutes et al. (1977) using both SHE and magneto-optical Kerr loops.

Several authors have suggested that it should be possible to propagate magnetic domains by an electric current in a magnetic material. (Berger 1974, Carr 1974). This effect, known as the domain drag effect, is caused by the perturbed current distribution in the vicinity of the wall as a consequence of the Hall effect. Domain drag propagation of bubble and stripe domains in amorphous Gd-Co-Mo (DeLuca et al. 1978) provided the first experimental demonstration of the effect. Subsequent work has shown that propagation velocities as high as 100 m/sec can be achieved in amorphous Gd-Co-Au films using the domain drag effect (DeLuca et al. 1980).

The use of SHE in amorphous films for magnetic-recording readheads has also been suggested (Gambino and McGuire 1976). By using films with a perpendicular easy axis anisotropy, low field Hall detectors with a sensitivity comparable to or better than InSb can be achieved.

2. Mean free path of carriers. Bergmann (1978a, 1979a) has made clever use of the asymmetrically scattered electrons in an amorphous ferromagnetic film to measure directly the mean free path in overlayed non-magnetic crystalline metal films. A fraction of the electrons asymmetrically scattered in a magnetic film (taken to be in the xy plane in Fig. 1) have a component of their velocity in the z direction. Some of these will enter the thin non-magnetic film which is deposited above the ferromagnetic film; they carry with them a component of velocity in the y direction due to their last scattering event which was asymmetric. The magnitude of this transverse y drift in the non-magnetic film depends on the mean free paths in each of the two layers. Bergmann has measured the spontaneous Hall field carried in this way from amorphous Fe films ($\lambda \approx 6$ Å) into paladium or indium films of varying thicknesses. The thicknesses dependence of v_y in the non-magnetic overlays is found to agree with his calculations (1979a) for this effect if λ=47 Å(Pd) and 240 Å(In), which are reasonable.

3. Magnetic dead layers. Aldridge et al. (1976) have pointed out that the spontaneous Hall effect is a convenient way to measure

the spontaneous magnetization of soft magnetic materials in thin film form. Bergmann (1978b, 1979b) has carried this technique further with a novel study of the magnetization of sub-monolayer-and-up magnetic films on non-magnetic substrates. His aim was to examine the onset of magnetic ordering as a function of magnetic-film thickness in order to shed new light on the old problem of magnetic dead layers (see Gopel 1979 for a recent list of references on dead layers).

Bergmann found that spontaneous magnetization appears in Fe films less than 1 monolayer thick on a-$Pb_{.75}Bi_{.25}$ substrates whereas Ni overlays on the same substrates showed spontaneous magnetization only above three atomic layers.

4. Annealing and crystallization of amorphous metallic alloys. Whyman and Aldridge (1974) first studied the crystallization of amorphous cobalt films using the Hall effect. Their group later found films of Cr, Co (Whyman and Aldridge 1975), and Fe (Raeburn and Aldrige 1978) to crystallize at approximately 60K while Ni films could not be made amorphous under clean vacuum conditions.

Malmhall, Rao and co-workers have found the Hall effect to be a useful probe of structural relaxation as well as of crystallization in amorphous metallic alloys which are stable to much higher temperatures than the pure amorphous metals mentioned above. The Hall resistivity is more sensitive to these changes than is the Ohmic resistivity and also afford a precise measure of T_c at the same time. Structural relaxation of the amorphous state can cause a significant Curie temperature increase (first reported by Chen et al. in 1976 and by Egami in 1978). These increases in T_c have been confirmed using the Hall effect (Malmhall et al. 1978a). The increase in T_c with annealing is explained on the basis of stronger attractive interactions between like TM atoms than between unlike TM atoms, and also depends on the like pairs having stronger exchange interactions than the unlike pairs (Malmhall et al. 1978b and 1979b).

As mentioned in Sec. IV A above, R_s vs ρ is well behaved through crystallization (Malmhall et al. 1978a).

V. CONCLUSIONS

Studies of the Hall effect in amorphous metals provide new insights into the effect itself or the materials. We summarize some of these below.

A. New Insights into the Hall Effect

1. R_o appears to be independent of temperature in both the solid paramagnetic and the liquid states although its magnitude is different in each case.

2. The spontaneous Hall constant does not drop sharply above T_c but is only weakly dependent on temperature even into the liquid state.

3. The temperature dependence of R_s below T_c arises not from magnetic disorder scattering only but rather from the total resistivity.

4. The broad range of alloys available in the amorphous state allows new insights into the interplay of the intrinsic spin-orbit interaction and spin-other-orbit interactions in the Hall effect of rare-earth-containing materials.

5. Both orbital- and spin-asymmetric scattering contribute to the SHE in disordered metals containing RE atoms. The compositional dependence of the SHE in concentrated RE alloys is very similar to that in dilute systems.

6. The spontaneous Hall coefficient R_s scales with the Curie temperature in concentrated amorphous RE alloys.

7. R_s changes sign at T_{comp} in ferrimagnetic systems and asymmetric scattering from both sub-networks is important.

B. New Insights into Amorphous Materials

1. The resistivity and ordinary Hall coefficient in amorphous materials closely follow the low temperature extrapolations of those parameters measured in the liquid state.

2. The ordinary Hall coefficients of amorphous metals are free-electron-like more often than are those of the comparable crystalline metals.

3. The large magnitude of the spontaneous Hall coefficient in amorphous metals is due to their large resistivities.

4. The spontaneous Hall effect in amorphous metals appears to be the result of side jump rather than skew scattering.

5. There appear to exist two different short range orders in amorphous iron, one giving rise to a moment of 2.2 μ_B/Fe atom, $R_s \approx 5 \times 10^{-10}$ Ohm-cm/G and a low T_c and the other giving rise to

a much smaller moment, $R_s < 1 \times 10^{-10}$ Ohm-cm/G and a low T_c.

ACKNOWLEDGEMENTS

The authors wish to thank Professor Peter de Chatel for his critical reading of the manuscript and Professor Luc Berger for helpful discussions regarding certain sections.

REFERENCES

Albert, P. A., and Guarnieri, C. R., 1977, J Vac Sci Technol, 14:138.
Aldridge, R. V., and Raeburn, S. J., 1976, Phys Lett, 56A:211.
Ali, A., Grundy, P. J., and Stephin, G. A., 1976, J Phys D Appl Phys, 9:L69.
Allen, R. P., Dahlgren, S. D., and Merz, M. D., 1976, in: "Rapidly Quenched Metals," N. J. Grant and B. C. Giessen, eds., MIT, Cambridge, p. 37.
Allison, F. E., and Pugh, E. M., 1956, Phys Rev, 102:1281.
Asomoza, R., Campbell, I. A., Jouve, H., and Meyer, R., 1977, J Appl Phys, 48:3829.
Asomoza, R., Campbell, I. A., Fert, A. Liénard, A., and Rebouillat, J. P., 1979, J Phys F, 9:349.
Bagley, B. G., and Turnbull, D., 1965, Bull Am Phys Soc, 10:1101.
Baker, J. A., Hoarc, M. R., and Finney, J. L., 1975, Nature, 257:120.
Ballentine, L. E., 1980, this volume.
Bennett, C. H., 1972, J Appl Phys, 43:2727.
Bennett, L. H., Page, C. H., and Schwartzendruber, L. J., 1976, AIP Conf Proc, 29:xix.
Berger, L., 1970, Phys Rev, B2:4459.
Berger, L., 1972, Phys Rev, B5:1862.
Berger, L., 1974, J Phys Chem Solids, 35:947.
Berger, L., 1976, AIP Conf Proc 34:355.
Berger, L., 1977, Physica, B91:31.
Berger, L., and Bergmann, G., 1980, this volume.
Bergmann, G., 1972, Z Phys 255:1976.
Bergmann, G., 1976a, Sol St Comm 18:897.
Bergmann, G., 1976b, Zeit fur Phys B25:255.
Bergmann, G., 1977, Phys Lett, 60A:245.
Bergmann, G., 1978a, Phys Rev Lett, 41:264.
Bergmann, G., 1978b, Phys Rev Lett, 41:1619.
Bergmann, G., 1979a, Phys Rev, B19:3933.
Bergmann, G., 1979b, Phys Today, Aug:25.
Bergmann, G., and Marquard, P., 1978, Phys Rev, B18:326.
Bethe, H. A., and Salpeter, E. E., 1957, "Quantum Mechanics of One- and Two-Electron Atoms," Springer Verlag, Berlin, p. 181ff.
Brunsch, A., and Schneider, J., 1977, J Appl Phys, 48:2641.
Brenner, A., and Riddell, G., 1946, J Res Nat Bur Stds, 37:31.

Brenner, A., Couch, D. E., and Williams, C. K., 1950, J Res Nat Bur Stds, 44:109.
Brill, R., 1930, Z Kristallog, 75:217.
Buckel, W., and Hilson, R., 1952, Z Phys, 131:420.
Buckel, W., 1954, Z Phys, 138:136.
Busch, G., Guntherodt, H-J., Kunzi, H. U., Meier, H. A., and Schlapbach, L., 1974, J de Phys, C4, suppl to 35:22.
Butvin, P., and Duhaj, P., 1976a, Czech J Phys, B26:208.
Butvin, P., and Duhaj, P., 1976b, Czech J Phys, B26:469.
Cantor, B. (ed), 1978, "Rapidly Quenched Metals III," Vols. 1 and 2, The Metals Soc., Chameleon Press, London.
Cargill, III, G. S., 1970a, J Appl Phys, 41:12.
Cargill, III, G. S., 1970b, J Appl Phys, 41:2249.
Cargill, III, G. S., 1975, in: "Sol St Phys," F. Seitz, D. Turnbull and H. Ehrenreich, eds., Academic Press, New York, Vol. 30, p. 227.
Cargill, III, G. S., 1975, in: "Rapidly Quenched Metals," N. J. Grant and B. C. Giessen, eds., MIT, Cambridge, p. 293.
Cargill, III, G. S., and Kirkpatrick, S., 1976, AIP Conf Proc, 31: 339.
Cargill, III, G. S., and Mizoguchi, T., 1978, J Appl Phys, 49:1753.
Carr, Jr., W. J., 1974, J Appl Phys, 45:394.
Chaudhari, P., Cuomo, J. J., and Gambino, R. J., 1973, IBM J Res and Dev, 17:66.
Chaudhari, P., Cuomo, J. J., and Gambino, R. J., 1975, U.S. Patent #3, 965, 463.
Chen, H. S., and Miller, R. C., 1970, Rev Sci Instrum, 41:1237.
Chen, H. S., and Park, B. K., 1973, Acta Met, 21:395.
Chen, H. S., Sherwood, R. C., Leamy, H. J., and Gyorgy, E. M., 1976, IEEE Trans MAG 12:933.
Chien, C-L., Musser, D., Gyorgy, E. M., Sherwood, R. C., Chen, H. S., Luborsky, F. E., and Walter, J. L., 1979, Phys Rev, B20:283.
Clements, W. G., and Cantor, B., 1976, in: "Rapidly Quenched Metals," N. J. Grant and B. C. Giessen, eds., MIT, Cambridge, p. 267.
Coey, J. M. D., 1978, J Appl Phys, 49:1646.
Condon, E. U., and Shortley, G. H., 1963, "Theory of Atomic Spectra," Cambridge University Press, Cambridge, G.B., p. 211.
Cuomo, J. J., Gambino, R. J., and Rosenberg, R. J., 1974, J Vac Sci Technol, 11:34.
Cuomo, J. J., and Gambino, R. J., 1975, J Vac Sci Technol, 12:79.
Cuomo, J. J., and Gambino, R. J., 1977, J Vac Sci Technol, 14:152.
Davies, H., 1978, in: "Rapidly Quenched Metals, III," B. Cantor, ed., The Metals Soc., Chameleon Press, London, Vol. 1, p. 1.
DeLuca, J. C., Gambino, R. J., and Malozemoff, A. P., 1978, IEEE Trans MAG 14:500.
DeLuca, J. C., Gambino, R. J., Malozemoff, A. P., and Berger, L., 1980, to be published.
Dorleijn, J. W. F., 1976, Phillips Res Rep, 31:287.
Durand, J., 1976, IEEE Trans MAG 12:945.
Dove, D. B., Gambino, R. J., Cuomo, J. J., and Kobliska, R. J.,

1976, J Vac Sci Technol, 13:965.
Duwez, P., Willens, R. H., and Klement, W., 1960, J Appl Phys, 31:1136.
Duwez, P., and Willens, R., 1963, Trans Met Soc AIME, 227:362.
Duwez, P., and Lin, S. C. H., 1967, J Appl Phys, 38:4096.
Duwez, P., 1976, Annual Rev of Mat Sc, 6:83.
Egami, T., 1978, Mat Res Bull, 13:557.
Felsch, W., 1969, Z Phys, 219:280.
Felsch, W., 1970, Z Angew Phys, 29:217.
Fert, A., and Jaoul, O., 1972, Phys Rev Lett, 28:303.
Fert, A., and Friederich, A., 1976, Phys Rev, B13:397.
Figueroa, E., Lundgren, L., Beckman, O., and Bhagat, S. M., 1976, Sol St Comm, 20:961.
Finney, J. L., 1970, Proc Roy Soc (London), 479.
Fischer, M., Guntherodt, H-J., Hauser, E., Kunzi, H. U., Liard, M., and Muller, M., 1976a, in: "Rapidly Quenched Metals," N. J. Grant and B. C. Giessen, eds., MIT, Cambridge, p. 329.
Fischer, M., Guntherodt, H-J., Hauser, E., Kunzi, H. U., Liard, M., and Muller, R., 1976b, Phys Lett, 55A:423.
Fukamichi, K., Kikuchi, M., Hiroyoshi, H., and Masumoto, T., 1978, in: "Rapidly Quenched Metals III," B. Cantor, ed., The Metals Soc., Chameleon Press, London, Vol. 2, p. 117.
Gambino, R. J., Chaudhari, P., and Cuomo, J. J., 1974, AIP Conf Proc, 18:578.
Gambino, R. J., and McGuire, T. R., 1976, IBM Tech Disclosure Bulletin, 18:4214.
Gambino, R. J., and Cuomo, J. J., 1978, J Vac Sci Technol, 15:296.
Giessen, B. C., "Metallic Glasses," 1976, Am Soc for Metals, Metals Park, Ohio, p. 22.
Goldenstein, A. W., Rostoker, W., and Schlossberger, F., 1957, J Electrochem Soc, 104:104.
Gopel, W., 1979, Surf Sci, 85:400.
Grant, W. A., 1978, in: "Proc AVS Sym on Ion Implantation," W. L. Brown, ed., AIP, New York, p. 1644.
Gruzalski, G. R., and Sellmyer, D. J., 1979, Phys Rev, B20:194.
Guntherodt, H-J., and Kunzi, H. U., 1973, Phys Kond Mat, 16:117.
Guntherodt, H-J., Kunzi, H. U., Liard, M., Muller, M., Muller, R., and Tsuei, C. C., 1977a, in: "Amorphous Magnetism II," R. Levy and R. Hasegawa, eds., Plenum Press, New York, p. 257.
Guntherodt, H-J., Kunzi, H. U., Liard, M., Muller, R., Oberle, R., and Rudin, H., 1977b, "Liquid Metals," The Inst of Physics, Bristol, p. 342.
Guntherodt, H-J., and Kunzi, H. U., 1978a, in: "Metallic Glasses," J. J. Gilman and H. J. Leamy, eds., Am Inst Met, Metals Park, Ohio, p. 247.
Guntherodt, H-J., Muller, M., Oberle, R., Hauser, E., Kunzi, H. U., Liard, M. and Muller, R., 1978b, in: "Transition Metals," M. G. J. Lee, J. M. Perz, and E. Fawcett, The Inst of Phys, Bristol, p. 436.
Harper, J. M. E., and Gambino, R. J., 1980, J Vac Sci and Technol,

to be published.
Hasegawa, R., 1971, Phys Letters, 36A:425.
Hasegawa, R., and Taylor, R. C., 1975, J Appl Phys, 46:3606.
Hasegawa, R., and Ray, R., 1978, J Appl Phys, 49:4174.
Hauser, J. J., 1975, Phys Rev, B12:5160.
Heiman, N., Lee, K., and Potter, R. I., 1975, AIP Conf Proc, 29:130.
Henderson, D., Brodsky, M., and Chaudhari, P., 1974, Appl Phys Letters, 25:641.
Herd, S. R., 1978, J Appl Phys, 49:1744.
Hines, W. A., Kabacoff, L. T., Hasegawa, R., and Duwez, P., 1978, J Appl Phys, 49.
Imamura, N., Mimura, Y., and Kobayashi, T., 1976, IEEE Trans MAG 12:55, 12:1724.
Jellinghaus, W., and De Andres, M. P., 1961, Ann Phys, 7:189.
Kagan, Yu, and Maksimov, L. A., 1965, Sov Phys Sol St, 7:422.
Karplus, R., and Luttinger, J. M., 1954, Phys Rev, 95:1154.
Kaul, S. N., 1977, Phys Rev, B15:755.
Kavesh, S., 1977, in: "Metallic Glasses," J. J. Gilman and H. J. Leamy, eds. Am Inst Met, Metals Park, Ohio, p. 36.
Kazama, N. S., Mitera, M., and Masumoto, T., 1978, in: "Rapidly Quenched Metals, III," B. Cantor, ed., The Metals Soc, Chameleon Press, London, Vol. 2, p. 164.
Klement, W., Willens, R. H., and Duwez, P., 1960, Nature, 187:869.
Kobliska, R., and Gangulee, A., 1977, in: "Amorphous Magnetism II," R. Levy and R. Hasegawa, eds., Plenum Press, New York, p. 447.
Kondo, J., 1962, Prog Theo Phys, 27:772.
Kondorskii, E. I., 1966, Fiz Met Metallog, 22:168.
Kondorskii, E. I., Cheremushkina, E. A., and Kirbaniyazov, N., 1964, Sov Phys Sol St, 6:422.
Kondorskii, E. I., Vedyayev, A. V., and Granovskiy, A. B., 1975a,b, Fiz Metallog, 40:455, 40:903.
Kooi, C., 1954, Phys Rev, 95:843.
Leung, P. K., and Wright, J. G., 1974, Phil Mag, 30:185.
Lin, S. C. H., 1969, J Appl Phys, 40:2175.
Lin, C. H., Bevk, J., and Turnbull, D., 1979, Sol St Comm, 29:641.
Lutes, O. S., Holmen, J. O., Kooger, R. L., and Aadland, O. S., 1977, IEEE Trans MAG 13:1615.
Mader, S., and Nowick, A. S., 1965, Appl Phys Lett, 7:57.
Majumdar, A. K., and Berger, L., 1973, Phys Rev, B7:4203.
Malmhall, R., Rao, K. V., Backstrom, G., and Bhagat, S. M., 1977, Physica, 86-88B:796.
Malmhall, R., Backstrom, G., Bhagat, S. M., and Rao, K. V., 1978a, J Non-Cryst Sol, 28:159.
Malmhall, R., Backstrom, G., Rao, K. V., Bhagat, S. M., Meichle, M., and Salamon, M. B., 1978b, J Appl Phys, 49:1727.
Malmhall, R., Backstrom, G., Rao, K. V., and Bhagat, S. M., 1978c, in: "Rapidly Quenched Metals III," B. Cantor, ed., The Metals Soc, Chameleon Press, London, Vol. 2, p. 145.
Malmhall, R., Backstrom, G., Bhagat, S. M., and Rao, K. V., 1979a, J Phys F, 9:317.

Malmhall, R., Backstrom, G., Rao, K. V., and Egami, T., 1979b, J Appl Phys, 50:7656.
Malmhall, R., Bhagat, S. M., Rao, K. V., and Backstrom, G., 1979c, phys stat sol (a), 53:641.
Maranzana, F. E., 1967, Phys Rev, 160:421.
Marohnic, Z., Babic, E., and Pavuna, D., 1977, Phys Lett, 63A:348.
Marohnic, Z., Babic, E., Ivkov, J., and Hamzic, A., 1978, in: "Rapidly Quenched Metals III," B. Cantor, ed., The Metals Soc, Chameleon Press, London, Vol. 2, p. 1949.
McGuire, T. R., Taylor, R. C., and Gambino, R. J., 1976, AIP Conf Proc, 34:346.
McGuire, T. R., Gambino, R. J., and Taylor, R. C., 1977, J Appl Phys 48:2965.
McGuire, T. R., Gambino, R. J., and Taylor, R. C., 1977, IEEE Trans MAG 13:1598.
McGuire, T. R., and Gambino, R. J., 1978, IEEE Trans MAG 14:838.
McGuire, T. R., Mizoguchi, T., Gambino, R. J., and Kirkpatrick, S., 1978, J Appl Phys, 49:1689.
McGuire, T. R., and Taylor, R. C., 1979, J Appl Phys, 50:1605.
McGuire, T. R., and Gambino, R. J., 1979, J Appl Phys, 50:763.
McGuire, T. R., and Gambino, R. J., 1980, J of Mag and Mag Mat, ICM Munich, 1979, to be published.
McGuire, T. R., Aboaf, J. A., and Klokholm, E., 1980 INTERMAG Boston to be published.
Mimura, Y., Imamura, N., and Kushiro, Y., 1976, J Appl Phys, 47:337.
Mitchell, E. N., Haukaas, H. B., Bule, H. D., and Streeper, J. B., 1964, J Appl Phys, 2604.
Mizoguchi, T., Gambino, R. J., Hammer, W. N., and Cuomo, J. J., 1977a, IEEE Trans MAG 13:1618.
Mizoguchi, T., Malozemoff, A. P., and Cox, D. E., 1978, J Appl Phys, 49:2461.
Mizoguchi, T., McGuire, T. R., Gambino, R. J., and Kirkpatrick, S., 1977, Phys Rev Lett, 38:89; Physica, 86-88B:738.
Mooij, J. H., 1973, phys stat sol (a), 17:521.
Müller, H. R., and Perthel, R., 1978, phys stat sol (b), 87:203.
Nagel, S. R., 1978, Phys Rev Lett, 41:990.
Nagel, S. R., and Tauc, J., 1975, Phys Rev Lett, 35:380.
Nishihara, Y., Katayama, T., Yamaguchi, Y., Ogawa, S., and Tsushima, T., 1978, Jap J Appl Phys, 17:1083.
Nowick, A. S., and Mader, S., 1965, IBM J Res Dev, 9:358.
Obi, Y., Fujimori, H., and Morita, H., 1977, Sc Repts RITU A, 26:214
Ogawa, A., Katayama, T., Hiramo, M., and Tsushima, T., 1976, Jap J Appl Phys supplement, 15:87.
Ogawa, A., Katayama, T., Hirano, M., and Tsushima, T., 1974, AIP Proc, 24:575.
O'Handley, R. C., 1977, in: "Amorphous Magnetism II," R. Levy and R. Hasegawa, eds., Plenum Press, New York, p. 379.
O'Handley, R. C., 1978a, Phys Rev, B18:930.
O'Handley, R. C., 1978b, Phys Rev, B18:2577.
O'Handley, R. C., 1980, this volume.

O'Handley, R. C., and Berger, L., 1978, in: "Transition Metals," M. G. J. Lee, L. M. Perz, and E. Fawcett, eds., The Inst of Phys, Bristol, p. 477.
O'Handley, R. C., and Boudreaux, D. S., 1978, phys stat sol a, 45: 607.
O'Handley, R. C., Hasegawa, R., Ray, R., and Chou, C-P., 1977, Appl Phys Lett, 29:330.
Ohkawa, F. J., and Yoshida, K. J., 1977, Phys Soc Jap, 43:1545.
Okamoto, K., Shirakawa, T., Matsushita, S., and Sakurai, Y., 1974, AIP Conf Proc, 24:113.
Orehotsky, J., and Schroder, K., 1972, J Appl Phys, 43:2413.
Panofsky, W. K. H., and Phillips, M., 1962, "Classical Electricity and Magnetism," Addison Wesley, New York, p. 143.
Pickart, S. J., 1979, Private communication, (see abstract DK11, Bull APS, 25, No 3, March 1980).
Pietrokowski, P., 1963, Rev Sci Instrum, 34:445.
Poate, J. M., 1978, "Proceedings of the AVS Symposium on Ion Implantation - New Prospects for Materials Modification," W. L. Brown, ed., AIP, New York, p. 1636.
Polk, D. E., 1970, Scripta Met, 4:117.
Polk, D. E., 1971, J Non-Cryst Sol, 5:365.
Pond, R., and Maddin, R., 1969, Trans Met Soc AIME, 245:2475.
Poon, J., and Durand, J., 1977, Phys Rev, B16:316.
Raeburn, S. J., and Aldridge, R. V., 1978, J Phys F: Met Phys, 8: 1917.
Raj, K., Durand, J., Budnick, J. I., and Skalski, S., 1978, J Appl Phys, 49:1671.
Rao, K. V., Malmhall, R., Backstrom, G., and Bhagat, S. M., 1976, Sol St Comm, 19:193.
Rao, K. V., Malmhall, R., Backstrom, G., and Bhagat, S. M., 1978, Anals Israel Phys Soc, 2:435.
Ratajczak, H., Gscianska, I., and Szlaferek, A., 1979, in: "Proc Ninth Int Coll on Magnetic Films and Surfaces," p. 115.
Samoilovich, A., and Kon'kov, U., 1950, J Exp Theo Phys (USSR), 20:782.
Shimada, Y., and Kojima, H., 1978, J Appl Phys, 49:932.
Shirakawa, T., Nakajima, Y., Okamoto, K., Matsushita, S., and Sakurai, Y., 1976, AIP Conf Proc, 34:349.
Sloope, B. W., and Tiller, C. O., 1965, J Appl Phys, 36:3174.
Spaepen, F., and Turnbull, D., 1976, in: "Rapidly Quenched Metals," N. J. Grant and B. C. Giessen, eds., MIT Press, Cambridge, p. 205.
Stanley, E. H., 1971, "Introduction to Phase Transitions and Critical Phenomena," Oxford Univ Press, New York, p. 47.
St. Amand, R., and Giessen, B. C., 1978, Scripta Metallurgia, 12: 1021.
Steinhardt, P., Alben, R., and Weaire, D., 1974, J Non-Cryst Sol, 15:199.
Stobiecki, T., 1978, Thin Solid Films, 15:197.

Stobiecki, T., and Hoffmann, H., 1980, J of Mag and Mag Mat, ICM, Munich, 1979, to be published.
Taylor, R. C., 1974, J Vac Sci Technol, 11:1148.
Taylor, R. C., McGuire, T. R., Coey, J. M. D., and Gangulee, A., 1978, J Appl Phys, 49:2885.
Taylor, R. C., and Gangulee, A., 1980, to be published.
Tsuei, C. C., and Duwez, P., 1966, J Appl Phys, 37:435.
Turnbull, D., and Cohen, M. H., 1958, J Chem Phys, 29:1049.
Volkenshtein, N. V., and Fedorov, G. V., 1960, Sov Phys JETP, 11:48.
von Molnar, S., Guy, C. N., Gambino, R. J., and McGuire, T. R., 1980 J of Mag and Mag Mat, ICM, Munich, 1979, to be published.
Wang, R., Merz, M. D., and Brimhall, J. L., 1978, Scripta Met, 12: 1037.
Waseda, Y., Kazaki, O., and Masumoto, T., 1977, Jour Mat Sc, 12:1927
Whyman, P. J., and Aldridge, R. V., 1974, J Phys F:Met Phys, 4:L6.
Whyman, P. J., and Aldridge, R. V., 1975, J Phys F:Met Phys, 4:2176.
Willens, R., 1973, private communication cited by B. G. Bagley in: "Amorphous Magnetism," H. O. Hooper and A. M. de Graaf, Plenum Press, New York, p. 148.
Williams, F. C., and Mitchell, E. N., 1968, Jap J Appl Phys, 7:739.
Ziman, J., 1979, "Models of Disorder," Cambridge University Press, New York.

THE HALL EFFECT IN LIQUID METALS

L. E. Ballentine

Simon Fraser University
Burnaby, B.C., Canada

1. INTRODUCTION

The Hall effect in liquid metals is usually described by contrasting it with the effect in crystalline metals. In solid metals the Hall coefficient may be of either sign, and it is often temperature dependent. But upon melting the Hall coefficient becomes independent of temperature with a magnitude close to the classical value of

$$R_H^\circ = 1/(nec) \qquad (1)$$

Here $e = -|e|$ is the electronic charge and n is the number of conduction electrons per unit volume. At the time of the first review paper on electronic properties of liquid metals (Cusack, 1963) the experimental results were so few and uncertain that only this qualitative picture could be established.

More extensive measurements collected during the next ten years (see Busch and Güntherodt, 1974) confirmed this general picture, but also established the existence of deviations from (1) of about 10% to 30% for Tℓ, Pb, and Bi. Some results have also been obtained for liquid transition and rare earth metals, and except for Ni they are quite different from (1) in both magnitude and sign.

2. SUCCESS OF THE CLASSICAL FORMULA

The success of the classical formula (1) for so many liquid metals is easily explained. Upon melting the complex anisotropic features of the Fermi surface are destroyed, since there are no preferred directions in the liquid metal. If the electron states in a liquid metal can be characterized by an energy-momentum relation (an assumption that becomes less plausible as the strength of the scattering increases), then E(k) must be spherically symmetric but need not be parabolic. The response of the group velocity $\vec{v} = \nabla_k E/h$ to an external force $\vec{F}$ can be characterized by an effective mass tensor that has two independent components (Jan 1962),

$$\frac{dv}{dt} = \frac{1}{m_n^*} \vec{F}_n + \frac{1}{m_t^*} \vec{F}_t \tag{2}$$

where $\vec{F}_n$ and $\vec{F}_t$ are the components of $\vec{F}$ normal and tangential to the spheres of constant energy in k-space. Here

$$m_n^* = \hbar^2/E'', \text{ and } m_t^* = \hbar^2 k/E' \tag{3}$$

Only the tangential effective mass, involving the first derivative of E(k), is relevant to steady state transport properties.

The Boltzmann equation can be solved if an (energy dependent) relaxation time τ exists, yielding (Jan 1962, Ballentine 1977a) the conductivity tensor

$$\sigma_{xx} = \frac{ne^2}{m_t^*} \tau \tag{4}$$

$$\sigma_{xy} = \frac{ne^3 \tau^2}{(m_t^*)^2 c} \alpha B \tag{5}$$

to first order in the magnetic field B (in the z-direction). Thus the Hall coefficient $R_H = \sigma_{xy}/(B\sigma_{xx}^2)$ becomes

$$R_H = \frac{\alpha}{nec} \tag{6}$$

where $\alpha = +1$ if $dE/dk > 0$,
$\alpha = -1$ if $dE/dk < 0$.
There is no effective mass correction to R_H.

This ease in explaining the success of the free electron formula for the Hall effect in liquid metals later became an embarrassment, however, when the quantitative deviations from that formula proved very difficult to explain.

3. DEVIATIONS FROM THE CLASSICAL FORMULA

Because the variety of Hall coefficients in solid metals is caused by the influence of the lattice potential on the electronic states, it was natural to look to the influence of the disordered potential on the electronic states for an explanation of the more limited variety in liquid metals. In the light of the previous section, this effect would be associated with the fact an energy-momentum relation E(k) does not really exist for electrons in a liquid (Ballentine, 1975), and it would evidently be outside the scope of the Boltzmann transport equation.

For this reason the current correlation (Kubo) formula for the conductivity tensor and the force correlation formula for the resistivity tensor were applied to liquid metals. The problem has proven to be very difficult. Those limited results that have been obtained were reviewed by Ballentine (1977a), and they have not led to any explanation of measured values of Hall coefficients. Indeed the theory of the Hall effect in a disordered potential remains a challenging problem.

However there is a kind of scattering that can be treated by the Boltzmann equation and which was not taken into account in Section 2. Most textbooks assume equality of the direct and inverse collision probabilities,

$$W(\vec{k},\vec{k}') = W(\vec{k}',\vec{k}) \tag{7}$$

and indeed this is true in the first Born approximation. But (7) is not generally true and the antisymmetric part of the scattering probability, called the <u>skew-scattering</u> probability, can not be taken into account in the relaxation time, moreover it contributes directly to the Hall effect. The symmetry (7), which does not always hold, should not be confused with time-reversal invariance,

$$W(\vec{k},\vec{k}') = W(-\vec{k}',-\vec{k}) \tag{8}$$

which is always true.

A general form for the skew-scattering probability is

$$\vec{a} \cdot \hat{k}' \times \hat{k}\, W_s(\theta,\varepsilon) \tag{9}$$

where $\hat{k}$ and $\hat{k}'$ are unit vectors in the directions of the initial and finite momenta, W_s is an arbitrary function of the scattering angle and energy, and $\vec{a}$ must be an axial vector so that the entire expression is scalar. Possible choices for $\vec{a}$ include the electron spin $\vec{\sigma}$, the orbital angular momentum $\vec{L}$, and the magnetic field $\vec{B}$. Each of these corresponds to a skew-scattering mechanism, and they are discussed in the following sections.

4. SPIN-ORBIT SCATTERING - THE SIMPLE METALS

The spin-orbit interaction potential has momentum space matrix elements proportional to $\vec{\sigma} \cdot \vec{k}' \times \vec{k}$. The scattering probability is the absolute square of the scattering amplitude, and the cross term between the spin-orbit amplitude and the ordinary scattering amplitude yields a form of skew-scattering with the electron spin as the axial vector in (9). The scattering rate, without spin flip, from $\vec{k}$ to $\vec{k}'$ (per unit solid angle of $\vec{k}'$) can be written as a sum of symmetric and antisymmetric parts,

$$W(\vec{k}',\vec{k}) = W_o(\theta,\varepsilon) + \hat{\sigma}_z \cdot \hat{k}' \times \hat{k} \;\; W_s(\theta,\varepsilon) \tag{10}$$

Here $\hat{\sigma}_z$ is a unit vector in the $\pm$ z direction (+ for spin up, - for spin down), with the magnetic field B determining the z-axis. There is no skew-scattering accomplanied by spin flip because the ordinary scattering amplitude has no spin flip component, and skew-scattering arises only as a cross term.

The Boltzmann equation for the electron distribution $f(\vec{k})$ is

$$\vec{F} \cdot \frac{\partial f(\vec{k})}{\hbar \partial \vec{k}} = \int \left[W(\vec{k},\vec{k}')f(\vec{k}') - W(\vec{k}',\vec{k})f(\vec{k}) \right] d\Omega_{\hat{k}'} \; ,$$

where $\vec{F} = e(\vec{E} + \vec{v} \times \vec{B}/c)$ is the external force and $e = -|e|$ is the electronic charge. This can be solved to first order in $\vec{E}$ by the familiar ansatz, $f(\vec{k}) = f°(k) + \hat{k} \cdot \vec{A}$, where the vector $\vec{A}$ may depend on energy and spin, but is independent of the direction of $\hat{k}$. By this substitution the Boltzmann equation is reduced to

$$\frac{e\hbar}{m} k \frac{\partial f°}{\partial \varepsilon_k} \vec{E} = \vec{A} \times \left(\frac{e}{mc} \vec{B} + \frac{\hat{\sigma}_z}{\tau_s}\right) - \frac{\vec{A}}{\tau_o} \tag{11}$$

The normal relaxation time τ_o and the skew-scattering time τ_s are given by

$$\frac{1}{\tau_o} = 2\pi \int_{-1}^{1} W_o(\theta,\varepsilon) \; (1-\cos\theta) \; d(\cos\theta) \tag{12}$$

$$\frac{1}{\tau_s} = \frac{2\pi}{3} \int_{-1}^{1} W_s(\theta,\varepsilon) \; [1-P_2(\cos\theta)] d(\cos\theta) \tag{13}$$

where P_2 is a Legendre polynomial. We see in (11) that the effect of skew-scattering is the same (for a given spin orientation) as that of the magnetic field, and hence it will modify the Hall coefficient.

The Hall coefficient at low magnetic fields (i.e. cyclotron frequency $\omega_c << \tau_o^{-1}$) is (Ballentine and Huberman, 1977)

$$R_H = R_H^o \left\{1-r^2+\frac{\hbar}{\varepsilon_F\tau_s}\left(\frac{\chi_p}{\chi_p^o}\right)\left[\frac{3}{4}(1+r^2) + \gamma_o - \frac{1}{2}(1-r^2)\gamma_s\right]\right\} \tag{14}$$

The notation here is $\varepsilon_F = \hbar^2 k_F{}^2/2m$, $r = \tau_o/\tau_s$, and

$$\gamma_o = \left.\frac{\partial \ln \tau_o}{\partial \ln \varepsilon}\right|_{\varepsilon = \varepsilon_F}, \quad \gamma_s = \left.\frac{\partial \ln \tau_s}{\partial \ln \varepsilon}\right|_{\varepsilon = \varepsilon_F} \tag{15}$$

Because the skew-scattering term in (11) has opposite effect on electrons of opposite spin, the net effect is proportional to the spin polarization. This has been taken to be the free electron value enhanced by the ratio of the spin susceptibilities for interacting and non-interacting electrons, $(\chi_p/\chi_p{}^o)$. The logarithmic derivatives γ_o and γ_s arise because spin-up and spin-down electrons have different kinetic energies, and hence slightly different values of τ_o and τ_s. The corrections involving r^2 are usually small, and may be comparable to the omitted effect of non-diagonal elements of the spin density matrix (Huberman and Ballentine, 1978).

The scattering probability has been calculated from the muffin-tin model of a liquid metal (Dreirach et al., 1972), which consists of spherical atoms embedded in a constant intersticial potential. The scattering amplitude of each atom is expressed in terms of phase shifts, with the spin-orbit interaction treated to first order. The total scattering amplitude is obtained by adding the scattering amplitudes of all atoms with the appropriate relative phases. This yields the skew-scattering function in (10) to be

$$W_s(\theta,\varepsilon_F) = \frac{4n_i}{\hbar} S(q) \frac{k_F}{\kappa} U_s(\theta) \tag{16}$$

where n_i is the number of atoms per unit volume, $\hbar^2 k_F/2m = \varepsilon_F$ is the kinetic energy of the electron relative to the bottom of the conduction band, $\hbar^2\kappa^2/2m = E$ is the energy relative to the intersticial potential, $S(q)$ is the liquid structure factor, $q = 2k_F \sin(\theta/2)$ is the momentum transfer, and

$$U_s(\theta) = -\sum_{\ell\,\ell'} (2\ell+1)(2\ell'+1)\sin\delta_\ell \sin(2\delta_{\ell'}-\delta_\ell) \Lambda_{\ell'} P_\ell(\cos\theta) \frac{dP_{\ell'}(\cos\theta)}{d(\cos\theta)} . \tag{17}$$

The spin-orbit coupling parameter is

$$\Lambda_\ell = \frac{\hbar^2}{4m^2c^2}\int_o^a \frac{1}{r}\frac{dV_{MT}}{dr} |R_\ell(r)|^2 r^2 dr \tag{18}$$

where a is the muffin-tin radius beyond which $V_{MT}(r)$ vanishes.

The wavefunction satisfy scattering boundary conditions,

$$R_\ell(r) = \cos\delta_\ell\ j_\ell(\kappa r) - \sin\delta_\ell\ n_\ell(\kappa r) \quad , r \geq a \tag{19}$$

As is well know, the factor (1-cosθ) in (12) causes backward scattering to dominate the electrical resistivity. The skew-scattering rate, on the other hand, is dominated by scattering angles near 90° because of the factor $(1-P_2)$ in (13). For non-transition metals the only non-negligible parameters in (17) are δ_o, δ_1 and Λ_1. The qualitative form of $U_s(\theta)$ for Bi or Pb is shown in Fig. 1. The magnitude of the skew-scattering rate τ_s^{-1} depends critically on the proximity of the peak of the structure factor to $\theta = \pi/2$ and to the zero of $U_s(\theta)$. Therefore Fig. 1 provides a qualitative explanation for the fact that the deviation of R_H from the classical value is significant for Bi (Z = 5 conduction electrons per atom) and Pb (Z = 4), but becomes negligible for Hg (Z = 2) and Au (Z = 1), even though the intrinsic strength of the spin-orbit interaction is comparable in these elements.

The theoretical values (Ballentine and Huberman, 1977) of the deviation of R_H, calculated by (14), from the classical value are

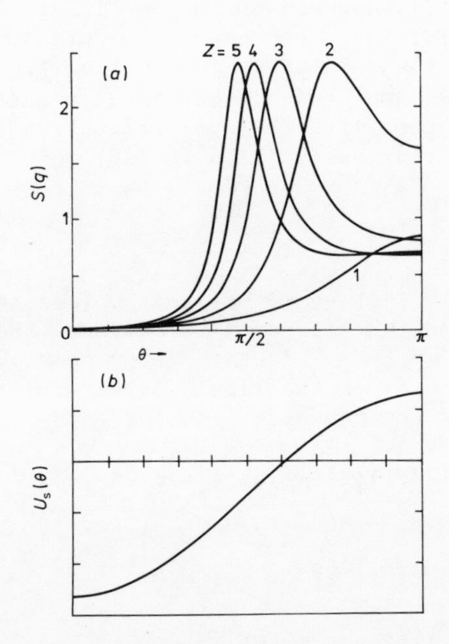

Fig. 1. (a) Percus-Yevick hard-sphere structure factor versus scattering angle for Z = 1, 2, 3, 4 or 5 conduction electrons per atom; (b) Skew-scattering function defined in equation (17), (arbitrary units). (From Ballentine and Huberman, 1977.)

Table 1.

$R_H/R_H^\circ - 1$

	Sb	Au	Hg	Tℓ	Pb	Bi
theory	-0.008	-0.0008		-0.084	-0.26	-0.10
expt.	+0.14	0.0	0.0	-0.24	-0.27	-0.31

compared with experiment in Table 1.

Discrepancies between theory and experiment are attributable to uncertainties in the parameters δ_ℓ and Λ_ℓ, and in the measurements. The quoted experimental values are selected from a number of measurements that do not always agree well among themselves (see Busch and Güntherodt, 1974). The calculated phase shifts for Hg yield such an exceptionally bad value for the resistivity that no meaningful calculation could be done. On the other hand, we predict that it is the measured value for Sb that is in error. The main features of the Hall effect in simple liquid metals are accounted for by this theory, including the fact that no significant deviations from the classical value occur for the many simple liquid metals not listed in the table.

No systematic study has been made of the Hall effect in liquid alloys, but several alloys based on Hg or Bi exhibit significant deviations from a linear interpolation between the pure components. (Hg-In: Cusack et al., 1963, 1964; Andreev and Regel, 1965; Güntherodt et al., 1966. Hg-Na, -Cd, -Zn, -In, -Tℓ, -Pb, -Bi, -Sn: Davies and Leach, 1973. Bi-Cd, -In, -Sn: Shiota and Tamaki, 1977.) The deviation is always a reduction in the magnitude of R_H. In several of these cases both pure components obey the classical formula (1). Parabolic deviations from linearity are qualitatively consistent with our theory. The skew-scattering probability in such cases would arise from the product of the spin-orbit amplitude of the heavy atom and the ordinary scattering amplitude of the lighter atom, yielding roughly a c(1-c) dependence on composition. No quantitative calculations have been performed because of the lack of partial structure factor data for these alloys.

5. EXCHANGE SCATTERING - TRANSITION METALS

In contrast to the simple metals, the Hall coefficients of liquid transition metals differ greatly from the classical formula in both magnitude and sign. Because the conduction electron density in these metals is uncertain and model dependent, it is not convenient to relate R_H to R_H°. Instead we shall express R_H in terms of $R_1 = \Omega_o/ec$, which is the classical Hall coefficient for a metal having one free electron per atom. (Ω_o = volume per atom.) The Hall coefficients of liquid Fe, Co, and Ni (from Fig. 23 of Busch and Güntherodt, 1974) are $R_H/|R_1|$ = 4.5, 1.5, -1.5.

The theory of the previous section can be applied to liquid transition metals. The important parameters in (17) are now δ_o, δ_2, and Λ_2, and hence the shape of $U_s(\theta)$ is different from Fig. 1. Because the normalization of the wavefunction is fixed outside of the atom (19), its amplitude inside the atom becomes very large at a resonance (virtual bound state). Therefore Λ_2 is enhanced by the d-resonance in transition metals. Nevertheless, spin-orbit scattering yields a contribution to R_H of only about 1% R_1.

However, the Hartree-Fock exchange operator yields another form of skew-scattering that is an order of magnitude stronger in transition metals than is spin-orbit scattering. The exchange operator has the form

$$V_{ex}(\vec{r},\vec{r}\,') = -\sum_i u_i(\vec{r})u_i{}^*(\vec{r}\,')\frac{e^2}{|\vec{r}-\vec{r}\,'|} \tag{20}$$

the sum being over occupied states. The contribution of V_{ex} to the scattering amplitude has been calculated in the distorted wave Born approximation for the case in which $u_i(\vec{r})$ is an eigenfunction of orbital angular momentum (Ballentine and Huberman, 1980). For a complete shell V_{ex} is spherically symmetric and yields no skew-scattering. A partially filled shell yields a skew-scattering probability of the form (9) with $\vec{a}$ being the net orbital angular momentum of the occupied states.

The magnitude of the exchange skew-scattering is proportional to the orbital polarization of the electrons in a particular spin state. This orbital polarization is caused by the spin-orbit interaction, which has the form $V_{so} = \lambda\,\vec{\sigma}\cdot\vec{L}$ within an atomic sphere. Let $|n\nu\rangle$ be an energy eigenvector in the absence of spin-orbit interaction, and let $|\psi_{n\nu}\rangle$ be the corresponding eigenvector to first order in V_{so}. Here $\nu = \pm 1$ is a spin quantum number and n represents all other quantum numbers. Then to first order in V_{so} we have

$$\langle\psi_{n\nu}|L_z|\psi_{n\nu}\rangle = 2\lambda\nu \sum_{n'\neq n} \frac{|\langle n\nu|L_z|n'\nu\rangle|^2}{E_{n\nu}-E_{n'\nu}} \tag{21}$$

Clearly this quantity will be negative near the bottom of a band, positive near the top of a band, and its sum over a full band vanishes. The orbital polarization of electrons of spin ν is the sum of (21) over occupied states,

$$<L_z>_\nu = \nu K^{(\nu)} \tag{22}$$

The skew-scattering time τ_s can now be calculated from (13), (16), and (17) as in the previous section, with the sole modification that the parameter Λ_ℓ now becomes

$$\Lambda_\ell = \Lambda_\ell^{so} + K^{(\nu)} \Lambda_\ell^{ex} \tag{23}$$

where Λ_ℓ^{so} is given by (18). The detailed calculation of Λ_ℓ^{ex} is given by Ballentine and Huberman (1980); it involves the wavefunctions of the occupied states within an atomic sphere.

Unfortunately a direct calculation of $K^{(\nu)}$ from (21) is not practical for a liquid metal, and it was necessary to estimate its magnitude from certain data pertaining to the solid state. However, it is evident from (21) that orbital polarization will be most important in partially filled narrow bands, such as the d bands of Fe and Co, and that it will be less important in wide band metals.

In the absence of any reliable calculations of the electronic states in liquid transition metals, it was necessary to rely on solid state bandstructure calculations to estimate the number of conduction electrons per atom. The qualitative features of the bandstructure of a transition metal in the Fe series have been described by Stearns (1978). There is an s-like band, which is completely filled and irrelevant for transport processes. Above it are several d bands, which may be subdivided into the narrow quasi-localized bands, whose contribution to transport processes is small and will be neglected, and a parabolic band containing a fraction of an electron per atom that is assumed to dominate the electrical conduction. Ballentine and Huberman (1980) estimated the effective number of conduction electrons per atom in Fe, Co, and Ni to be Z = 0.4, 0.5, and 0.5.

The only non-negligible parameters determining the skew-scattering probability in (17) are δ_o, δ_2 and Λ_2. Thus $U_s(\theta)$ is very nearly an odd function of $\cos\theta$, antisymmetric about $\theta = \pi/2$. If the structure factor S(q) were constant for $0 \le q \le 2k_F$ (as is nearly the case for the small values of k_F that we have here), the skew-scattering rate (13) would vanish. Thus it is only the non-constant part of S(q) (roughly the q^2 term) that contributes to the Hall effect in liquid transition metals, and the structure factor data is more uncertain in this region than elsewhere.

The susceptibility enhancement factor (χ_p/χ_p°) in (14) refers to only the conduction electrons. We can not use the observed susceptibility because most of it comes from the quasi-localized

Table 2

$R_H/|R_1|$

	Theory	Experiment
Fe	(-2.5 + 0.43)= -2.1	4.5
Co	(-2.0 + 1.4) = -0.6	1.5
Ni	(-2.0 + 0.1) = -1.9	-1.5

electrons, which are much more numerous than the conduction electrons. A reasonable value can be obtained by assuming that each group of electrons contributes to the total susceptibility in proportion to its density of states at the Fermi energy. Then the susceptibility enhancement factor for conduction electrons is found to be 10.3, 34 and 3.5 for Fe, Co and Ni.

The Hall coefficient (14) is the sum of a spin-independent part and a spin-dependent part, the latter being related to the susceptibility. Since the two parts can, in principle, be experimentally distinguished, the results in Table 2 are given in the form $R_H = (a+b)$, where the second term (b) is the spin-dependent part.

Only for Ni can these results be considered satisfactory. Güntherodt et al. (1977) have exploited the large temperature dependence of the susceptibility of liquid Co to separate the two parts of the Hall effect, obtaining a spin-independent part of $-2.0|R_1|$. A similar value was obtained for amorphous Co by Bergmann (1977), who used a strong magnetic field to saturate the magnetization and hence separate the two parts of R_H. Thus it appears that the estimated number of conduction electrons per atom for Co (Z = 0.5) is correct, but the calculated value of the spin-dependent part of R_H is about three times too small in this case.

Ballentine and Huberman (1980) also treated a model of local magnetic order, in which the majority and minority spin bands are split by the ferromagnetic exchange energy. The bulk magnetization vanishes at the Curie temperature in this model because of continuous fluctuations in the direction of the local magnetization while its magnitude remains nearly constant. Theoretical (Korenman et al., 1977) and experimental (Weber et al., 1977) evidence suggest that this model of itinerant electron magnetism may hold even into the liquid state. Although the theory of the Hall effect requires considerable modification for this model, the results are (fortuitously) rather similar to those discussed above, and so they will not be reviewed in detail.

The conclusion is that although exchange scattering contributes significantly to the Hall effect in liquid transition metals, it is not strong enough to account for the observed Hall coefficients of liquid Fe and Co.

6. MAGNETIC FIELD DEPENDENT SCATTERING

Chambers (1973) pointed out that the effect of a magnetic field on the orbital motion of an electron should modify the scattering amplitude. To first order in the field B, the eigenvalues of the partial wave equation are modified by the Zeeman energy,

$$E = \frac{\hbar^2 k_m^{\ 2}}{2M} + m\mu_B B \tag{24}$$

where $\hbar m$ is the angular momentum parallel to $\vec{B}$, and $\mu_B = |e|\hbar/(2Mc)$. Since the total energy E is the same for all partial waves, the wave vector $k = k_m$ must depend on m. Since the phase shift is a function of k, we have an m-dependent phase shift,

$$\delta_{\ell m} = \delta_\ell(k_m) \tag{25}$$

The scattering amplitude becomes

$$f(\vec{k}',\vec{k}) = 4\pi \sum_\ell \sum_m \frac{\sin\delta_{\ell m}\, e^{i\delta_{\ell m}}}{k_m} Y_\ell^{\ n}(\hat{k}')Y_\ell^{\ m*}(\hat{k}) \tag{26}$$

Expansion of this expression to first order in B yields a skew-scattering term of the form (9) with $\vec{a} = \vec{B}$.

Unlike the previous two forms of skew-scattering, which act oppositely on each spin state and so contribute to the spin-dependent part of the Hall coefficient, this mechanism contributes to the spin-independent part. The parameter that governs the strength of this skew-scattering mechanism is the energy derivative of the phase shift, thus it should be most important for transition metals, where the Fermi energy lies near a d-resonance.

No detailed study of this skew-scattering mechanism has yet been carried out. An estimate of its magnitude indicates that it would reduce the magnitude of the spin-independent Hall effect in Co by 7%, yielding (compare Table 2) $R_H/|R_1| = (-1.86 + 1.4) = -0.46$. This does not significantly alter the discrepancy between theory and experiment.

In the original paper Chambers attempted to calculate the contribution of this mechanism to Hall effect of Fe impurities in liquid Ge, however, he neglected the interference between the skew-scattering amplitude of Fe atoms and the symmetric scattering amplitude of the host atoms, which is likely as important as the effect that he considered.

Although the theory sketched above is simple and intuitive, it is subject to the resolution of fundamental theoretical questions about the theory of scattering in a magnetic field. The usual asymptotic state, consisting of an incident plane wave and an outgoing spherical scattered wave, is not valid in the presence of a magnetic field, and the entire scattering theory needs to be reconstructed from first principles. A preliminary sketch of such a theory by this author (using similar methods to those of Goldberger and Watson, 1964) has confirmed the result (26). On the other hand, a different approach by Huberman and Overhauser (to be published) led to the conclusion that the scattering amplitude is not modified to first order in B. This fundamental theoretical problem remains open.

7. SUMMARY AND OUTLOOK

There are two main kinds of physical mechanisms that may be responsible for non-classical Hall effects.

The first is the modification of the dynamical character of the electron states by the electron-ion potential. In a crystal this leads to Bloch waves, anisotropic Fermi surfaces, electron-like and hole-like states, and the wide range of Hall coefficients that are observed in crystalline metals and semiconductors. At present the study of the modification of electron states by the disordered potential has not led to any explanation of non-classical Hall effects in liquid metals.

The second kind consists of the various skew-scattering mechanisms. These are present in both the solid and liquid states, but are usually less important in the solid state than the first kind of mechanism. Skew-scattering has provided an explanation of non-classical Hall coefficients in liquid non-transition metals, but seems to be inadequate to explain the Hall effect in liquid transition metals.

Theoretical treatments of skew-scattering in liquid metals have thus far assumed the conduction electrons to be in freely propagating states. This assumption is in doubt for transition metals. Ballentine (1977b) gave an argument suggesting that it might be valid for Ni but not for Fe. Raeburn and Aldridge (1978) obtained a positive value for the spin-independent Hall coefficient of amorphous films of Fe. Amorphous metals are usually similar to liquid metals in their electronic properties. That result is incompatible with our theory of the Hall effect, so if it were confirmed and shown to apply also to liquid Fe, it would constitute very strong evidence for the need to consider the influence on transport properties of the modification of electron states by the liquid potential. In other words, we shall have to study the first and second kinds of mechanisms together in order to understand liquid Fe.

REFERENCES

Andreev, A. A., and Regel, A. R., 1965, Fiz. Tverd. Tela., 7:2567.
Ballentine, L. E., 1975, Adv. Chem. Phys., 31:263.
Ballentine, L. E., 1977a, Inst. Phys. Conf., 30:188.
Ballentine, L. E., 1977b, Inst. Phys. Conf., 30:319.
Ballentine, L. E., and Huberman, M., 1977, J. Phys. C: Solid State Phys., 10:4991.
Ballentine, L. E., and Huberman, M., 1980, J. Phys. C: Solid State Phys., (in press).
Bergmann, G., 1977, Phys. Lett., 60A:245.
Busch, G., and Güntherodt, H.-J., 1974, Solid State Physics, 29:235.
Chambers, W. G., 1973, J. Phys. C: Solid State Phys., 6:2441.
Cusack, N. E., 1963, Reports on Progress in Physics, XXVI:361.
Cusack, N., and Kendall, P., 1963, Phil. Mag., 8:157.
Cusack, N., Kendall, P., and Fielder, M., 1964, Phil. Mag. 10:871.
Davies, H. A., and Leach, J. S. L., 1973, The Properties of Liquid Metals, in "Proceedings of 2nd Int. Conf. held at Tokyo, 1972," S. Takeuchi, ed., Taylor and Francis, London.
Dreirach, O., Evans, R., Güntherodt, H.-J., and Künzi, H.-U., 1972, J. Phys. F: Metal Phys., 2:709.
Goldberger, M. L., and Watson, K. M., 1964, "Collision Theory," Wiley, New York.
Güntherodt, H.-J., Meuth, A., and Tièche, Y., 1966, Phys. Kondens. Mater., 5:392.
Güntherodt, H.-J., Künzi, H. U., Laird, M., Müller, R., Oberle, R., and Rudin, H., 1977, Inst. Phys. Conf., 30:342.
Huberman, M., and Ballentine, L. E., 1978, Can. J. Phys., 56:704.
Jan, J.-P., 1962, Amer. J. Phys., 30:497.
Korenman, V., Murray, J. L., and Prange, R. E., 1977, Phys. Rev. B., 16:4032; 16:4048; 16:4058.
Raeburn, S. J., and Aldridge, R. V., 1978, J. Phys. F: Metal Phys., 8:1917.
Shiota, I., and Tamaki, S., 1977, J. Phys. F: Metal Phys., 7:2361.
Stearns, M. B., 1978, Physics Today, 31-4:34.
Weber, M., Steeb, S., and Knoll, W., 1977, Phys. Lett., 61A:78.

HALL EFFECT IN LIQUID METALS: EXPERIMENTAL RESULTS

H. U. Künzi and H.-J. Güntherodt

Institut für Physik, Universität Basel
Klingelbergstrasse 82
CH-4056 Basel, Switzerland

1. INTRODUCTION

1.1 Why Do We Study Liquid Metals?

The development of quantum mechanics at the beginning of this century gave rise to a well-established and fundamental understanding of crystalline solids. One of the basic facts that simplified the theoretical problem was the fundamental property of the crystal lattice, namely its periodicity. This allowed one of calculate approximate yet realistic solutions of the Schrödinger equation. The theory of non-periodic structures is much more difficult and detailed calculations have not yet been performed on a large scale. Since it was not clear to what extent the periodicity itself was directly responsible for the physical properties of crystals, direct experimental studies of non-periodic structures became of the utmost importance. An intense interest in such studies on metallic materials started about 20 years ago. For several reasons liquid metals proved to be excellent materials for such investigations. Above the normal melting temperature liquid metals are in stable thermodynamic equilibrium. They have well-defined atomic structures and can be studied in the form of alloys as well as pure liquid elements. As most liquid metals are completely miscible their alloys can be studied over the entire concentration range. Because of phase diagram restrictions this is not possible for most crystalline alloy systems. Furthermore samples of any desired quantity can easily be prepared.

Another group of disordered metals, the amorphous and glassy metals which also were known to exist at that time had, at least in the initial stages of these investigations, been used rather scarcely for such studies. The reason for this was that only very few alloys were known to form glassy metals and the preparation of good samples was a difficult task. This situation has, however, completely changed. The preparation techniques have been considerably improved and a great number of alloys are now known to form glasses over quite extended concentration ranges and we have a much better understanding of their atomic structure. This development means that it is now possible to extend studies of the influence of the non-periodicity on physical properties of metals to low temperatures and study phenomena not observable in the liquid state (e.g. ferromagnetism and superconductivity).

Other properties, however, such as electronic transport, on which we are focusing here, do not seem to be systematically different between the liquid and the amorphous state. Many of the results obtained on liquid metals have been duplicated in amorphous alloys. This means that the results already obtained in the liquid state form a useful basis for understanding metallic glasses; the latter have many important technological applications.

1.2 Significance of Hall-effect Measurements in Liquid Metals

A fundamental characteristic of a crystalline material and one which determines many of its properties, is its band structure E(k). A full experimental determination of this quantity is, however, not yet possible. In a metal or semiconductor details of the Fermi surface which contains a good deal of important information on E(k), can be obtained using a variety of experimental measurements. Rather fundamental information on the Fermi-surface comes from the observation of quantum oscillatory phenomena such as the de Hass-van Alphen- and Shubmikov-de Haas-effect. The experimental observation of these effects, however, requires very low temperatures and cannot therefore be carried out in a high temperature solid and certainly not in metals in the liquid state. The lowest known melting points are -38.9°C for Hg and -48°C for the eutectic $Cs_{77}K_{23}$ alloy.

In the liquid state there is, in general, no band structure E(k) since k is no longer a good quantum number, nevertheless one would like to probe the electronic structure. Properties which are sensitive to the electronic structure and which can be studied include the Hall-effect and other electronic transport properties (electrical resistivity and thermopower), various photoemission spectra and paramagnetic properties. Experience has shown that much valuable information on the electronic structure of liquid metals and alloys can be obtained through measurements of the Hall-effect.

1.3 Historical Remarks

The first successful attempts to measure the Hall-effect in liquid metals were as early as 1901 by Des Coudres (1) in Göttingen. His primary interest was not the electronic structure of metals, he rather described a simple electrical circuit based on the Hall-effect which could be used for the rectification of ac-currents for electrical power measurements as well as for the determination of the Hall-coefficient. In order to check whether in liquid metals magnetohydrodynamic effects due to the interaction of the measuring current with the applied field do not give rise to "unforeseen inconveniences" he measured the Hall-coefficient of liquid mercury and a concentrated bismuth amalgam (Table 1). In view of the small Hall-coefficients of these materials and the limited sensitivity of his measuring equipment, his results were quite reasonable. Subsequent authors of this time did not obtain comparable results. In fact, Rausch von Traubenberg in 1905 (2) and Nielsen 1924 (3) who both studied liquid bismuth and Fenninger in 1914 (4) who tried to measure mercury did not observe any effect above the resolution of their instruments for measurements in the liquid state. Consequently it was assumed that due to their ideal isotropy, molten metals do not show galvanomagnetic effects (Gerlach 1928) (5) and that in solid metals the Hall-effect is mainly due to their crystallinity. A theory which was thought to account for the apparent non-existence of the Hall-effect in liquid metals was put forward by Eldrige in 1923 (6).

This view rapidly disappeared when important progress was made in the theory of electrons in solids by Sommerfeld (7) (free electron model) and by Peierls (8) who applied the ideas of Bloch (9) on the theory of metallic conductivity to the galvanomagnetic properties.

The work of Peierls explained for the first time the well-known fact that in certain crystalline metals, e.g. Zn, the Hall-coefficient can have a positive sign. Mott and Jones (10) and Fröhlich (11) suggested that for liquid metals the free electron model should be applicable. They noted that in a microscopically isotropic material the energy of an electron E(k) can only depend on the magnitude of the vector $|k|$ and not on its direction. Consequently the Fermi surface has to be a sphere.

Indeed, subsequent experimental studies by Zahn (12) and Kikoin and Fakidow (13) with presumably more sensitive instrumentation gave reproducible results for the Hall-coefficient of a liquid Na-K alloy and, morevoer, their results were in agreement with the prediction of the free electron model.

The current large interest in liquid metals dates back to about 1960. In this year Kendall and Cusack (14) and independently

Table 1. Early Measurements of the Hall-coefficient in Liquid Metals.

	$R_H(10^{-11}m^3/As)$	Year	Reference	
Hg	-5.2	1901	Des Coudres	1
Bi-Hg	-12.4	1901	Des Coudres	1
Bi	> -2500.	1905	Rausch v. Traubenberg	2
Hg	0.2	1914	Fenninger	4
Bi	2.5	1924	Nielsen	3
$Na_{50}K_{50}$	-42.	1930	Zahn	12
Hg	0.2	1931	Kikoin Fakidow	13
$Na_{50}K_{50}$	-44.	1931	Kikoin Fakidow	13
Rb	-42.	1948	Fakidov	23
Hg	-7.46	1960	Kendall, Cusack	14
Hg	-7.3	1960	Tièche	15
Hg	-7.5	1961	Cusack, Kendall	24
Hg	-8.8	1961	Wilson	25

Tièche (15) were able to determine reliable values for the Hall-coefficient in liquid mercury. At about the same time major progress was also being made in the understanding of electronic conduction in liquid simple metals. Ziman (16) proposed a theory based on the nearly free electron model. In the following years this was carefully tested. Hall-effect studies on various liquid metals which furnished information on the number of conduction electrons per atom, an important parameter in this theory, were an important contribution for the acceptance of the theory. Further support for the free electron picture in liquid simple metals was also obtained from studies in liquid alloys of simple metals by Busch and Güntherodt (17) and others.

Since the measuring and handling of liquid metals at high temperatures is difficult little was known about the properties of liquid transition metals, lanthanides and actinides until relatively recently. Progress in the technology of chemically highly inert ceramic materials meant that suitable measuring cells could be constructed so more and more results on these groups of metals have become available. The Hall-effect has been studied in the liquid light rare earth metals, uranium and in a great number of liquid transition metal alloys (Güntherodt and Künzi (18)). More recently measurements on pure Co and Ni by Müller (19) were reported.

Although the Hall-effect data obtained on these more complicated metals was indeed surprising, it is probably fair to say that it has not yet made a great impact on the understanding of the electronic structure in these materials. The most remarkable and, in fact, unexpected result is that many of these metals have a positive Hall-coefficient in the liquid state. This clearly indicates that the free electron model is inadequate and as a consequence, it is not possible to determine the number of conduction electrons in these more complicated metals. Contrary to these findings the Hall-coefficient in liquid semiconductors was always found to be negative even though many of these materials have positive coefficients in the crystalline state. (20,21)

In spite of various theoretical efforts to tackle the problem of the Hall-effect in the liquid transition and rare earth metals (see e.g. Ballentine (22)) a fully satisfying explanation of the experimental results is still lacking.

2. PHYSICAL METHODS OF INVESTIGATION

The study of the Hall-effect in liquid conductors is difficult from the standpoint of the experimental technique. The Hall-field at a given current density and at practical magnetic fields is about 10^6 times smaller than the electric field associated with the electrical resistivity. For liquid metals in particular, magnetohydrodynamic circulation currents have to be prevented by keeping the electric current and the magnetic field as low as possible. This makes it necessary to use electrical instruments capable of measuring signals as low as 1 nV. In addition a number of thermoelectric and thermomagnetic side effects have to be removed.

Further problems come from the corrosive properties of liquid metals at high temperatures. Suitable ceramics for constructing measuring cells and materials to make electrical contacts with the molten metal cannot always be found. Above about 1000°C the mechanical stability of the cell may put severe limitations on the accuracy of the measurements if the cell is not properly designed.

The following sections are intended to provide a survey of some of the major problems rather than a detailed discussion of any particular point. Details can be found in many articles and review papers (18,19,26-39).

2.1 Disturbing Effects

In a typical Hall-effect measurement we apply a magnetic field B_z in the z-direction and pass a current j_x along the x-direction. The usual isothermic Hall-coefficient R_H is defined from the electric field (Hall-field) E_y, arising from the action of the magnetic field on the electric current

$$E_y = R_H j_x B_z$$

In order that this relation properly determines the isothermal Hall-coefficient, we have to ascertain that the following subsidiary conditions are fulfilled. The current in the y-direction must be zero and the temperature gradients along the x- and y-directions must vanish (isothermal). While the first condition is easily achieved the others may be the cause of severe problems. The reason for this is that in a field B_z thermal currents w_x and w_y are coupled with the electric currents through the Nernst- and the Ettinghausen effect respectively. These heat currents finally give rise to temperature gradients which in turn produce a thermal emf in the Hall-probes. The Ettinghausen effect predicts a transverse temperature gradient dT/dy parallel to the Hall-field due to the action of the field B_z on the current j_x

$$\frac{dT}{dy} = P\ B_z\ j_x$$

P is the Ettinghausen-coefficient. The thermal emf caused by this gradient adds to the Hall-signal.

Similarly the Nernst-effect gives rise to a longitudinal temperature gradient dT/dx which in turn gives rise to a transverse electric field through the Ettinghausen-Nernst-effect and a transverse temperature gradient dT/dy through the Righi-Leduc-effect. This latter temperature gradient again causes a thermal emf in the Hall-probes. These additional effects may be reduced by making Hall-probes out of the same material as the sample under investigation. The thermoelectric voltages thereby cancel out. Further ways of eliminating these unwanted contributions include ac-current and/or ac-field methods. In the case where the frequency of measurement is large compared to the reciprocal thermal relaxation time $\tau \sim b^2/\pi^2 D$, the amplitude of the temperature gradients remains small and isothermal rather than adiabatic conditions will prevail. Here b is the distance between the Hall-probes and D the thermal diffusivity of the sample.

Further problems stem from misalignment of the Hall-probes. Under real experimental conditions it is difficult to place the Hall-probes on the same equipotential of the sample current. This problem can, however, be easily solved by using an ac-field method.

Additional difficulties specific to the liquid state arise from the excitation of magnetohydrodynamic circulation currents. Since the magnetohydrodynamic equations in their most general form are rather complex little is known about the exact paths of such currents. Busch and Vogt (26), however, argue that in the case of a laminar current the absolute value of its velocity v is proportional to $|j \cdot B|$ and consequently the electric field $E_{MHD} = v \times B$ induced by these currents is proportional to $j \cdot B^2$. Consequently in order to minimize these effects, Hall-effect measurements on liquid metals should be performed at the lowest possible magnetic fields. The double frequency method again helps to eliminate such contributions since they appear at frequencies different from the Hall-signal.

2.2 Methods of Measuring the Hall-effect

Various techniques, each one having its own advantages and disadvantages, have been used to study the Hall-effect in liquid metals. We can categorize these methods according to the geometrical form of the sample. Thin plates of rectangular (usual method) or circular (Carbino-method) shapes with their surfaces perpendicular to the magnetic field lines are mostly used. With each form four possible techniques are possible depending on the choice of the time dependence of the current and field. Direct or alternating currents and fields can be used.

2.2.1 DC-current - DC field method. In this method a constant current J as well as a constant field B are used. Consequently the Hall-voltage U_H will also be constant

$$U_H = R_H \cdot \frac{J \cdot B}{d} .$$

Here d is the thickness of the rectangular measuring cell. The major disadvantage of this technique is that it is very susceptible to all the disturbing effects enumerated above. In addition, a temperature gradient due to a inhomogeneous temperature distribution in the furnace may give rise to a thermal emf much larger than the Hall-effect. In principle this effect and other disturbing voltages can be eliminated by using four measurements obtained at different polarities of field and current (13). This procedure is, however, time consuming. Measurements on liquid metals using this technique have been performed by Cusack et al. (30), Enderby (28) and Langeheine and Mayer (40).

2.2.2 DC-current - AC-field method. An alternating field $B = B_o \sin \Omega t$ will give rise to a Hall-effect with the same time dependence. Effects brought about by non-equipotential positioning of the Hall-probes V_ρ remain time independent and can be filtered out. Thermomagnetic and thermoelectric contributions to the

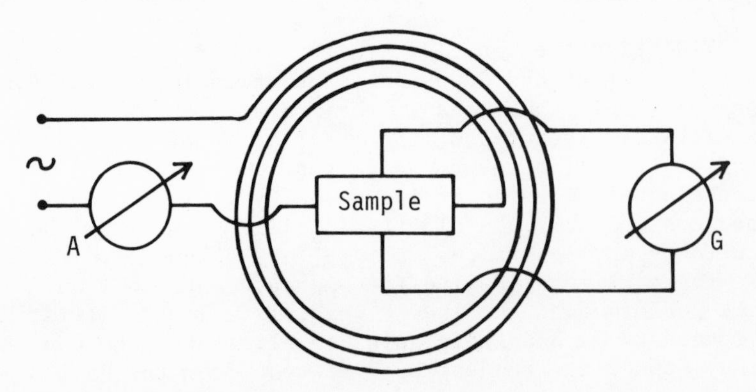

Fig. 1. Des Coudres method to measure the Hall-effect (1).

Hall-voltage can be sufficiently reduced by using frequencies above the thermal relaxation time of the sample. High frequencies, however, may cause large inductive pick-up signals in the Hall-probes. Although these signals are of the same frequency as the Hall-emf they differ in phase by $\pi/2$ and can be rejected with a phase-sensitive detector.

2.2.3 <u>AC-current - DC-field method</u>. With this choice of current and field the Hall-voltage has the same frequency and is in phase with the current. Inductive pick-up signals are greatly reduced and high frequencies help to reduce problems caused by magnetohydrodynamic movements of the liquid conductors. A major disadvantage of this method is that the misalignment voltage V_ρ is in phase with the Hall-signal. A separation of these two signals is possible by making two measurements with inversed field polarities or by compensating V_ρ. Instabilities in V_ρ, however, strongly affect the accuracy of this method. Measurements on liquid metals using this method have been performed by Wilson (25).

2.2.4 <u>AC-current - AC-field method</u>. This is now the most popular method. It was pioneered by Des Coudres (1). As shown in the circuit in Fig. 1, he passed the same current through the sample and the field coil. The Hall-signal then appears as a pulsating dc-voltage

$$V_H = R_H \frac{I_o B_o \sin^2 \omega t}{d}$$

This particular technique does not seem to be in use nowadays.

The adaptation suggested by Russel and Wahlig (41) which uses two different frequencies has proved to be more advantageous. In this method the Hall-signal appears at the sum and at the difference frequency. All the spurious signals, except the thermomagnetic ones, appear at other frequencies and can be filtered out. The thermomagnetic disturbances can be controlled as mentioned in the dc-current - ac-field method.

Although this method has considerable advantages over the others care has to be taken in the choice of the electronic equipment necessary for filtering and amplifying. Since signals at the original current and field frequencies are inevitably picked up by the Hall-probes, non-linearities in the electronic system may give rise to intermodulation noise having the same frequency as the Hall-signal. In order to minimize such effects the original parasitic signals have to be carefully compensated and in addition a phase sensitive detection system should be used (31).

This method has been used for most studies of liquid metals. Its development is well documented in the literature (1,17,27,29, 33,35,38,42).

Figure 2 shows the Block diagram of a recently improved version (43) of the ac-field — dc-current experimental set-up previously described by Busch and Tièche (27) and Güntherodt and Busch (17). For economic reasons the field current is supplied directly from the main circuit (line) which has a frequency of $\Omega = 50$ Hz. The frequencies $(\omega \pm \Omega)$ of the Hall-signal should be kept low enough to minimize inductive and capacitive pick-up signals. Here ω is the frequency of the current. But on the other hand the frequencies should be above the thermal relaxation τ of the transverse thermomagnetic effects (i.e. $(\omega \pm \Omega)\tau >> 1$). Furthermore, they should be well separated from the harmonics of the field and current. 80 Hz for ω seems to be a good choice. The Hall-signals then appear at 30 and 120 Hz.

$$U_H = R_H \frac{I_o B_o}{2d} (\cos(\omega-\Omega)t - \cos(\omega+\Omega)t)$$

Both signals can be measured and for all materials studied so far give the same results.

The sample current (80 Hz) is generated in an oscillator, amplified and fed through an isolating transformer into the sample. Current densities of up to $2A/mm^2$ can be used.

The Hall-voltage is measured by a compensation method. This eliminates the necessity of measuring the field and current separately. For this purpose a reference signal, which is strictly proportional to the sample current, and the magnetic field is generated in an electronic multiplier. The two inputs of the multiplier are connected to a pick-up coil, placed coaxially to the field coil, and to the sample current circuit. A previously set

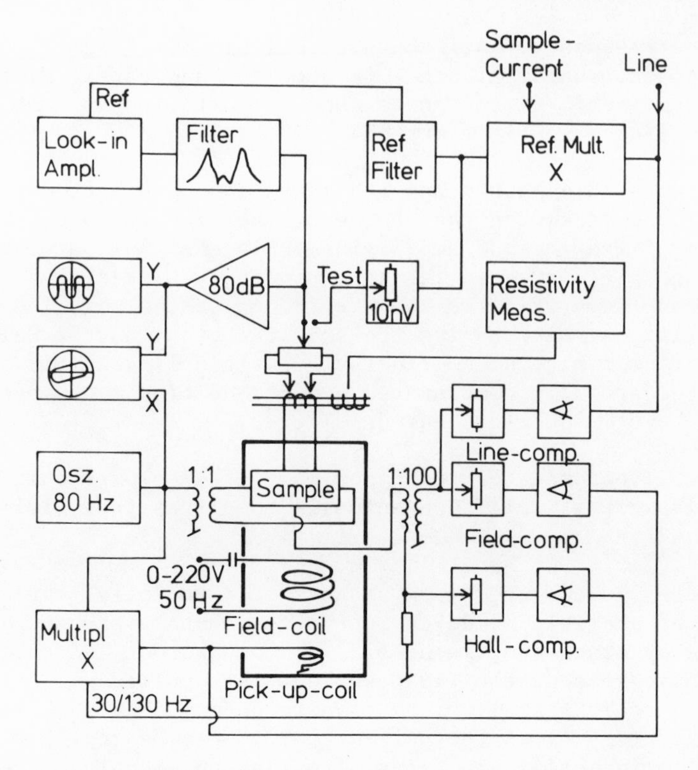

Fig. 2 Modern double AC method for Hall-effect measurements

and fixed phase shifter brings the multiplier output into the proper compensating phase (180°) with respect to the Hall-signal. The phase shifted signal is then attenuated by a potentiometer and fed into the Hall-measuring circuit. After full compensation of the Hall-signal the potentiometer reading times the calibration factor of the measuring cell gives the Hall-coefficient directly.

Before discussing the zero detecting circuit it is essential to mention that spurious signals in the measuring circuit due to inductive pick-up from the alternating field and the line (50 Hz) as well as from the misalignment voltage (80 Hz) are almost always present. These signals may be orders of magnitude larger than the Hall-effect itself. It is therefore essential, in order to prevent subsequent intermodulation effects, to buck these signals as well as possible. Three compensating circuits are therefore integrated in the set-up. The field-bucking circuit is fed from the pick-up coil, and the line bucking circuit which compensates signals not proportional to the field amplitude, is connected to a reduced line signal. Both of these bucking circuits consist of a phase shifter and a potentiometer to allow for compensation in amplitude and phase. The misalignment voltage is compensated with a potentiometer branched in the usual five point probe method. This also allows the simultaneous measurement of the electrical resistivity using the usual four point probe method. In studies on liquid metals the simultaneous measurement of the electrical resistivity has proven to be of great value. Apart from the purely scientific aspect the sample resistance may also serve to monitor the state of the sample. Incompletely filled measuring cells or bubbles in the sample are readily detected in this way. The bucking of the parasitic signals is monitored by two oscilloscopes branched via a 80 dB amplifier to the measuring circuit. With the present system bucking down to 1 μV is possible.

The zero detecting system consists of two narrow band filters for the two frequencies of the Hall-effect and two notch filters to reject the remaining, incompletely compensated, disturbing voltages appearing at the field- and current frequencies. At the output of this filter there is a 60 dB amplifier followed finally by a phase sensitive detector. The reference-signal is generated by an electronic multiplier fed by the field pick-up coil and the line. The multiplier is followed by a filter selecting the desired frequency either 30 or 130 Hz at which the Hall-signal should be observed.

2.2.5 Corbino method. In the circular sample of constant thickness (Corbino disc) the main current flows radially, but the Hall-effect gives it a circulating component. If this current alternates the circulating component can be detected inductively in a coaxial coil.

At low frequencies the induced signal in the coil is independent of the sample thickness. This would be a remarkable advantage if other geometrical constraints were not important. However, the mutual induction between disc and coil necessitates a highly reproducible and mechanically stable positioning of the coil with respect to the sample. This considerably complicates an accurate absolute determination of R_H. It should also be mentioned that the Corbino method measures the Hall-mobility $\mu_H = \sigma \cdot R_H$ rather than the Hall-coefficient. The conductivity of the sample has to be determined separately in order to obtain the Hall-coefficient.

Other choices using either constant or alternating fields and currents are also possible. But none of them brings major advantages for liquid metals. Results on liquid metals using the Corbino-effect have been obtained by Velly et al. (37) who used the ac-current dc-field method and Shackle (36) who used the double ac-technique.

2.2.6 Electrodeless determination of the Hall-mobility. Although a completely electrodeless determination of the Hall-coefficient is, in principle, possible, little experience has yet been acquired so its reliability and usefulness in liquid metals is uncertain. The only successful trial seems to be the one by Swenunson et al. (39). They used samples in the form of a cylindrical shell with its axis perpendicular to a strongly diverging magnetic field. The sample current flows parallel to the axis. An azimuthal Hall-current perpendicular to the sample current now appears. Since the cylindrical shell is in a non-homogeneous magnetic field, the Lorentz force driving the azimuthal current is stronger on the side where the high magnetic field acts and is in the opposite direction but weaker on the low field side. The balance gives rise to the Hall-current. In a homogeneous field the effect would, for obvious reasons, cancel out. Because the non-homogeneity of the field is important this technique requires a highly rigid positioning of the sample with respect to the field coils.

Furthermore a completely electrodeless determination of the Hall-coefficient also demands an inductive generation of the axial sample current. This implies that the liquid sample must form a closed ring (transformer secondary). The construction of such measuring cells may be relatively easy for the study of low melting point metals but for liquid transition metals, where an electrodeless method would indeed bring a major advantage, rather difficult design problems might arise.

2.3 The Measuring Cell

The measuring cell is the heart of the entire experimental set-up. Much of the success depends on its design, its mechanical stability and chemical compatibility with molten metals at high temperatures. Figure 3 shows the details of the cell construction and probes which have been used for many years by the present authors. The cell consists of two optically polished plates. The upper one is tightly machined to fit to the lower end of a tube which contains the liquid metal. In the lower plate a cavity of about 0.1 mm thickness is ground and holes serving for the elecrical contacts are drilled. A capillary (Ø = 0.5 mm) through the upper plate connects the cavity with the liquid metal reservoir. As probes we used various conducting materials such as graphite, vitreous carbon, Mo, W, Fe and refractory metal carbides. The choice depends on the material used to construct the cell and the metal under investigation. For many of the simple metals cells

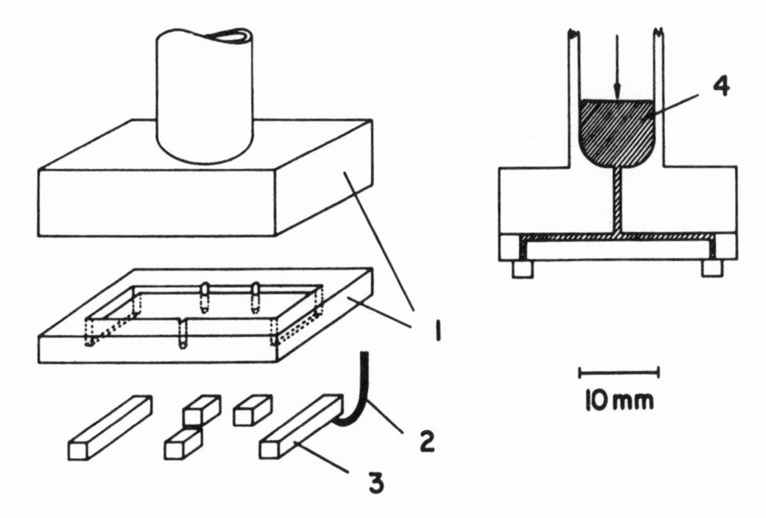

Fig. 3 Hall-effect measuring cell 1 for solid Hall-probes 3. Hall-probes are kept in position by tungsten hooks 2 and springs (not shown). 4 is the liquid metal reservoir.

machined out of quartz can be used successfully up to 1200°C. Above this temperature quartz becomes too soft. For higher temperatures and chemically more aggressive metals such as Al and some of the transition metals and their alloys, cells made out of sintered alumina proved to be satisfactory. A ceramic material far better than alumina is beryllia. This even allows the measurement of some of the light rare earth metals and alloys. Cells made out of other ceramics such as MgO, ThO_2, ZrO_2. BN and Si_3N_4 were also used on some occasions.

A major problem was encountered in the study of the Hall-effect in liquid Fe, Co and Ni. Although alumina cells proved to be sufficiently resistant in contact with the molten metal no successful combination of materials for a corrosion free contact between metal and probe or cell and probe could be found. The only solution of this problem seems to use liquid probes as shown in Fig. 4. Instead of putting solid probes on the holes in the lower plate of the cell, ceramic tubes of 5 cm length and 2 mm inner diameter were tightly machined to fit into the holes. The lower end of these tubes were closed with tightly fitting tungsten bars entering about 2 cm into the tube.

The metal that had flown down from the cavity into the tubes thus established the electrical contact with the liquid film in the cell. During measurements the temperature of the tungsten bars was always kept below the melting point of the metal under consideration. The metal in the tubes therefore solidified in the vicinity of the tungsten bars and the solid liquid interface was confined to be above the bars, thus preventing its rapid dissolution.

The only geometrical dimension which is required for the absolute determination of the Hall-coefficient is the cavity thickness. This is easily determined by a direct measurement or by calibration with a liquid metal of known Hall-coefficient. Mercury seems to have a well established value for this purpose. The latter procedure has the advantage that the cell can be checked for its tightness at the same time.

Other cell designs suitable for the investigation of liquid conductors have been described by Greenfield (29), Cusack et al. (30), Shuchannek and Minghetti (32) and Langeheine (40). The construction of cells for the Corbino-method is discussed in detail by Shackle (36) and Velley et al. (37).

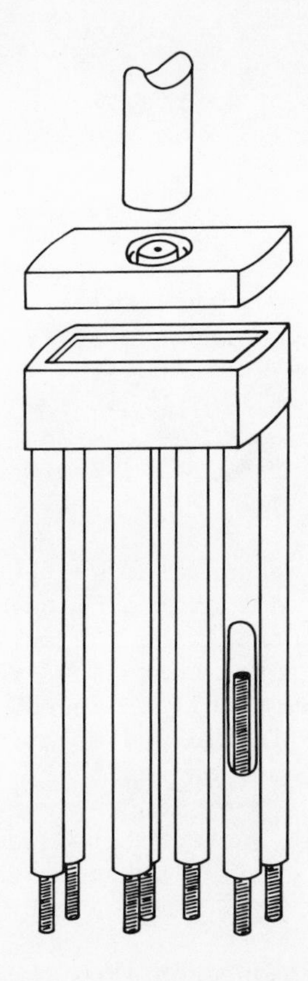

Fig. 4 Hall-effect measuring cell with liquid probes. This cell was used to measure the Hall-coefficient in liquid Ni and Co.

3. EXPERIMENTAL RESULTS

3.1 Simple Metals

With the exception of the alkaline earth metals the Hall-coefficients of liquid simple metals are now well established. Table 2 gives a summary of the more recent data. The most significant point is that the experimentally measured results are equal to the values predicted by the free electron theory

$$R_A = -\frac{1}{n|e|} = -\frac{A}{|e|N_L n_A D} .$$

n is the number of conduction electrons per unit volume, A is the atomic weight, N_L Avogadro's number and D the density. The number of conduction electrons per atom n_A is equal to the number of valence electrons or the group number in the periodic table. The alkali and noble metals have one, Zn two, Ga three, Sn four and Sb five conduction electrons per atom in the liquid state.

In Fig. 5 the Hall-coefficient and the electrical resistivity of Li (44) as a function of temperature are shown. At the melting point the Hall-coefficient changes by less than 2% whereas the electrical resistivity roughly doubles. The latter reflects the strong scattering power of the disordered structure rather than a change in the number of conduction electrons as is manifested by the Hall-effect. In the liquid state the Hall-coefficient follows

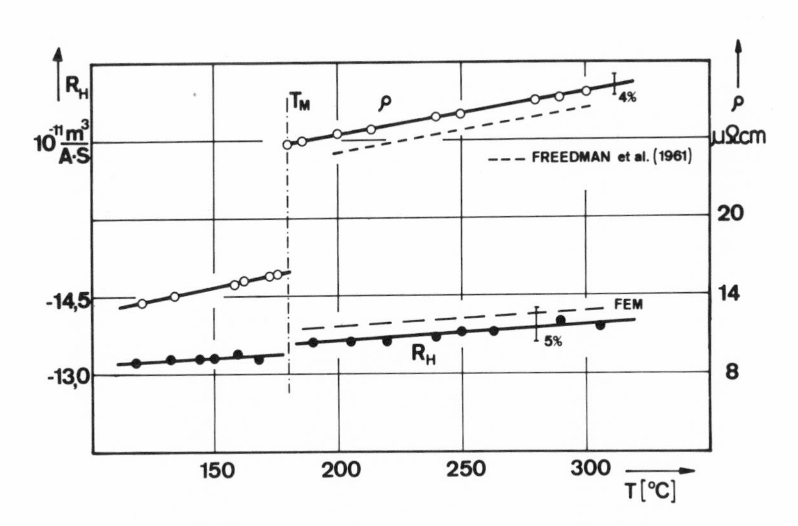

Fig. 5 Hall-coefficient and electrical resistivity of Li. FEM = free electron value.

Table 2. Hall-coefficients of liquid simple metals. R/R_o gives the measured Hall-coefficient R relative to its free electron value R_o. T indicates the range of measurement. All the data have been obtained by either the usual DC-current DC-field method or the double AC-method except the results obtained by Wilson (25) and Velly et al. (37). They used the AC-current DC-field method (indicated by AC*).

Metal	R_o ($10^{-11}m^3/As$)	R/R_o	T°C	Method	Reference
Li	13.6	0.98	180-320	AC	44
Na	25.5	0.98	98-400	AC	27
		1.00	98-110	DC	40
K	48.9	1.00	63.7-70	DC	40
Rb	59.1	1.00	38.7-45	DC	40
Cs	76.8	1.00±0.03	34-100	AC	54
		1.00	28.7-40	DC	50
Cu	8.25	1.00	1083-1150	AC	17
Ag	12.0	1.02	960-1100	AC	17
		0.97	-	AC	36
Au	11.8	1.00	1063-1150	AC	17
Zn	5.1	1.00	420-600	AC	27
		1.01	420-650	DC	28
		1.01	420-500	AC	29
		1.00	-	AC	36
Cd	7.25	0.96	321-420	DC	55
		0.98	321-500	AC	27
		1.04	321-650	DC	28
		0.99	321-400	AC	29
Hg	7.6	0.98	-30-100	DC	24
		0.96	20-300	AC	27
		1.22	20-200	AC*	25
		1.04	20	DC	28
		0.99	30-210	AC	29
		1.00	-	DC	56
		1.20	-	AC	31
		1.00	-	AC	50
		1.00	-	AC	36
		1.00	25	AC	35
		1.07	-39.5-20	AC*	37
Al	3.9	1.00	660-850	AC	17

Table 2. (continued)

Metal	$R_o(10^{-11}m^3/As)$	R/R_o	T°C	Method	Reference
Ga	3.95	0.96	35-110	DC	57
		0.99	30-600	AC	27
		0.97	35	AC	29
		1.00	-	DC	56
		1.04	-	AC	50
		1.00	-	DC	33
		1.02	30-50	AC	37
In	5.65	0.98	156-350	DC	55
		0.80	175-207	AC*	25
		1.04	156-500	DC	28
		0.93	156-320	AC	29
		0.95	-	AC	50
		1.00	156-700	AC	17
		0.94	-	DC	58
		0.92	156-180	AC	35
Tl	6.27	0.76	320-450	AC	29
Ge	3.40	1.00	900-1010	AC	27
Sn	4.42	0.98	230-425	DC	55
		1.07	250-310	AC*	25
		1.00	250-320	AC	29
		1.00	-	DC	56
		1.00	-	AC	50
		1.00	230-800	AC	17
		1.00	-	DC	33
Pb	5.05	0.38	400-800	AC	27
		0.88	330-550	DC	28
		0.73	340-500	AC	29
		0.88	350	AC	36
Sb	3.87	1.14	630-900	AC	27
		1.07	-	AC	42
Bi	4.30	0.95	271-425	DC	55
		0.60	271-800	AC	27
		0.69	285-330	AC	29
		0.68	-	AC	52
		0.96	-	AC	42

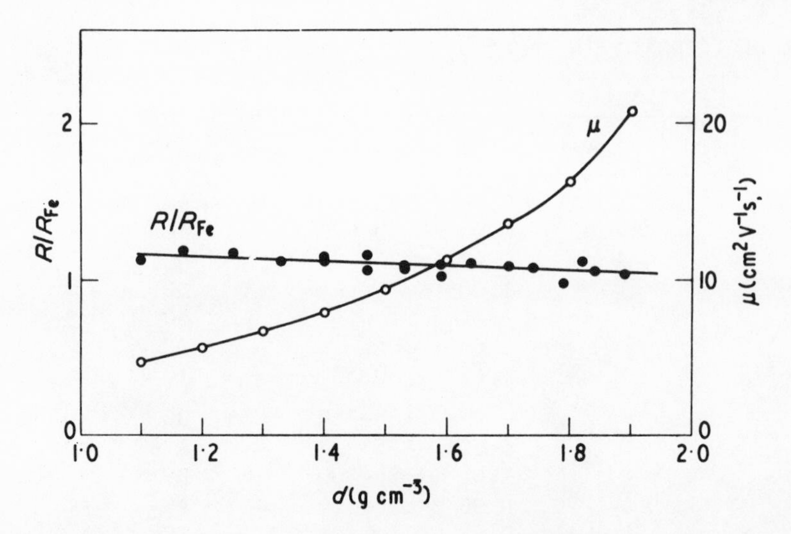

Fig. 6 Hall-mobility μ and Hall-coefficient R relative to the free electron value for Cs at 100 atm. as a function of density. (45).

the free electron value, assuming $n_A = 1$, well within the experimental accuracy. The slight increase with temperature is due to the temperature dependence of the mass density of Li. The other

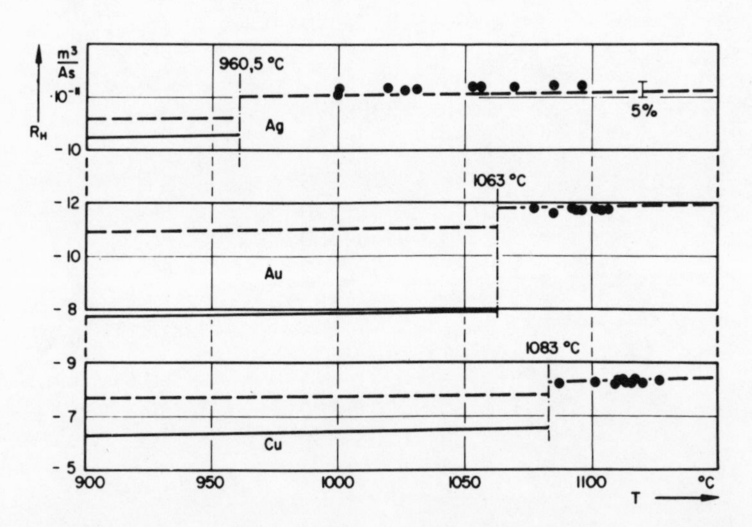

Fig. 7 Hall-coefficients of solid and liquid noble metals Ag, Au and Cu. The dashed line represents the free electron values. The experimental values are given by the full line for the solid state and the dots for the liquid state.

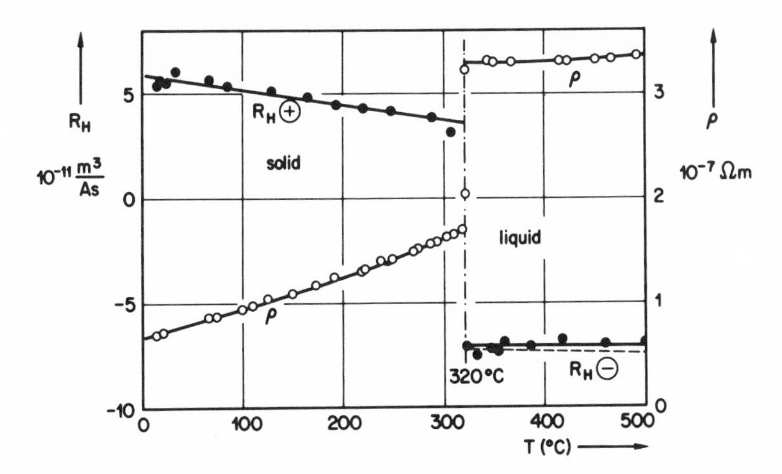

Fig. 8 The Hall-coefficient and the electrical resistivity of solid and liquid Cd. The free electron values are indicated by a dashed line. Note that the Hall-coefficient changes sign from positive to negative at the melting point (27).

alkali metals and the noble metals also contribute one conduction electron per atom. Langeheine and Mayer (40) in their measurements on solid alkali metals observed a small field dependence in the Hall-coefficient due to a non-isotropic contribution to the conductivity tensor. Above the melting temperature this field dependence disappears since the liquid is isotropic.

A rather interesting and important experiment has been performed by Even and Freyland (58). They studied the Hall-effect of expanded liquid Cs in the metallic region (subcritical conditions) in the temperature range 100-1400°C and at a pressure of 100 atm. Their results are shown in Fig. 6 plotted as a function of density. Surprisingly the Hall-coefficient takes its free electron value throughout the whole range even though the conductivity decreases by a factor of 10 in the range of measurements.

Figure 7 shows the Hall-coefficients of the liquid noble metals. The experimental values (points) closely follow the theoretical prediction (dashed line) assuming one free electron per atom (17). In the solid state deviations are noticeably larger and indicate that hole-like charge carriers contribute to the electronic transport. This is in agreement with a model for the Fermi surface due to Ziman (46). In solid Cd (Fig. 8) such deviations are much larger than in the noble metals and in fact give rise to a positive sign for the Hall-coefficient. But upon melting, the Hall-coefficient changes sign and indicates that in liquid Cd we have two conductions electrons per atom. Similar results have been

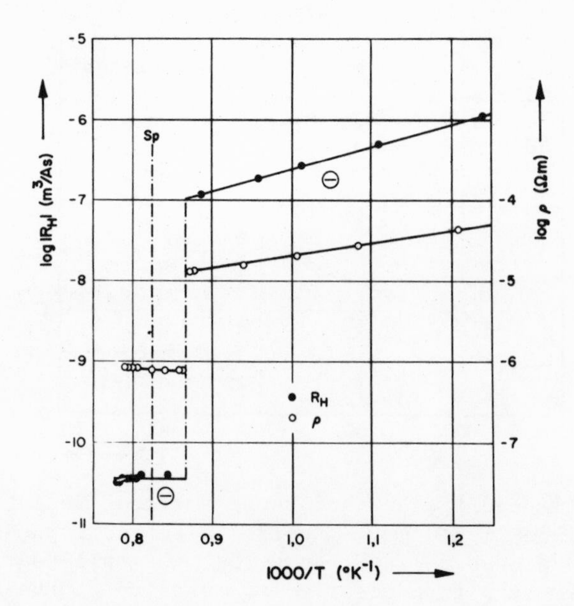

Fig. 9 The Hall-coefficient and the electrical resistivity of solid and liquid Ge. Note that the Hall-coefficient changes by four orders of magnitude at the freezing point. Subcooling has been observed (27).

observed in Hg and Zn.

The free electron theory also gives the correct results for the three and four valent metals Al, Ga, In and Sn. Of particular interest is the case of Ge (Fig. 9). In the solid state Ge is well-known for its semiconducting properties. It exhibits a large electrical resistivity and a Hall-coefficient which is orders of magnitude larger in its absolute value than is typical for metals. Above the melting point, however, Ge becomes metallic, its resistivity decreases to a value comparable to that of liquid Sn and the Hall-coefficient indicates that the four valence electrons of Ge become free conduction electrons. Liquid Si is expected to show equivalent behavior. The electrical resistivity does indeed decrease to metallic values but the Hall-coefficient has not yet been measured due to experimental difficulties.

The only liquid simple metals that are known to show substantial deviations from the free electron value are Tl, Pb and Bi. For liquid Pb it should also be mentioned that the various experimental results are in rather poor agreement with each other (see Table 2). Various theoretical models taking into account anomalous contributions (not directly caused by the Lorentz-force) have been applied to these metals in order to explain the observed deviations. For a review see Ballentine (22).

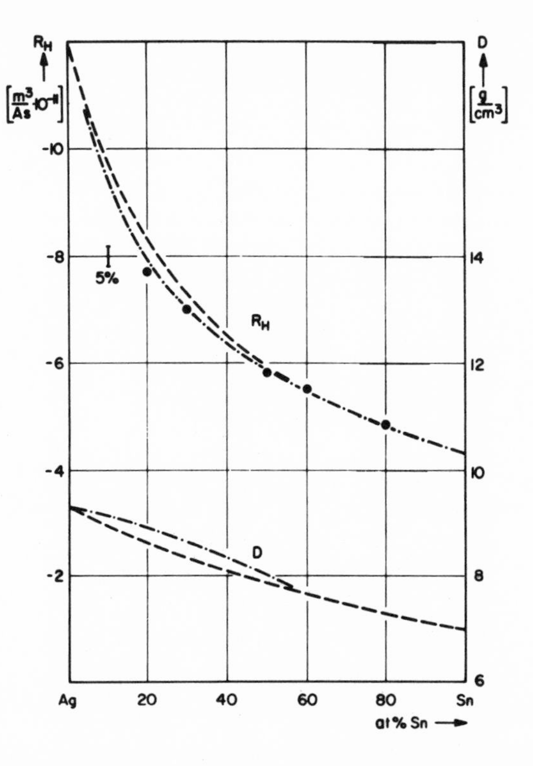

Fig. 10 The Hall-coefficient and the mass density of liquid Ag-Sn alloys. The dashed lines represent weighted mean densities (lower curve) and corresponding free electron values (upper curve). The dashed-dot line represents measured densities (47) and corresponding free electron values. The experimental values are indictaed by the full dots.

3.2 Alloys of Simple Metals

A great number of experimental investigations have been carried out on binary alloys of simple metals. In order to properly interpret the data in terms of the number of conduction electrons per atom, experimentally determined mass densities are required. It is not always sufficient to calculate the density of the alloys by assuming that the atomic volumes of the constituent elements remain concentration independent (Vegard's law). In many alloys a non negligible concentration dependence of the partial volume can be observed. If this is properly taken into account it is found that the mean number of conduction electrons per atom varies linearly with the atomic concentration. Figure 10 shows this for the case of liquid Ag-Sn alloys. The lower dashed line represents the density obtained from Vegard's law and the dashed dot line the measured densities from Sauermann and Metzger (47). The Hall

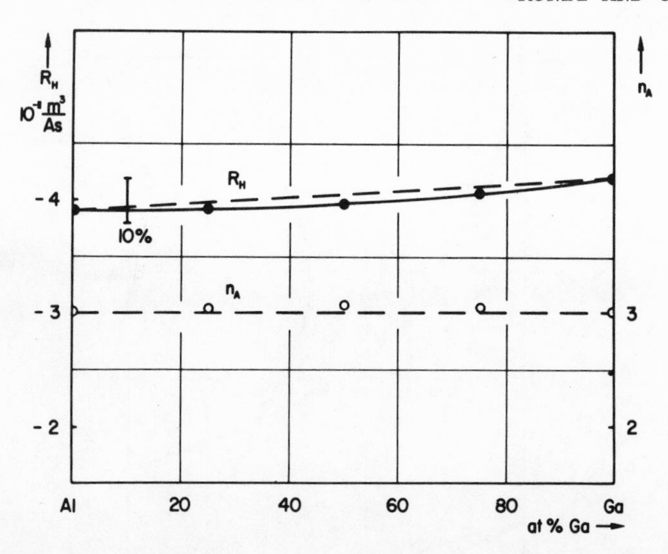

Fig. 11 The Hall-coefficient and the number of conduction electrons per atom n_A of liquid Al-Ga alloys. The dashed line represents the free electron value.

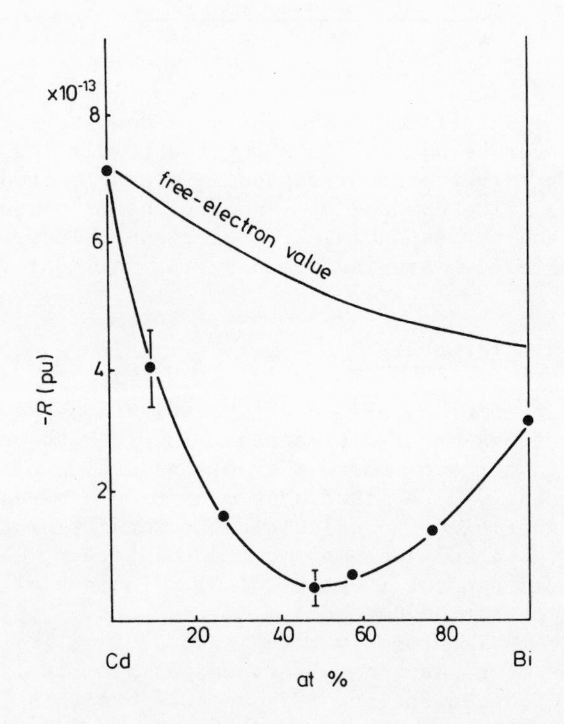

Fig. 12 Concentration dependence of the Hall-coefficients of liquid Bi-Cd alloys. (52)

coefficients were calculated using both curves. In both cases it was assumed that the number of conduction electrons per atom varies linearly with the atomic concentration (one for pure Ag and four for pure Sn). The experimental values (full dots) are in good agreement with the curve based on the measured densities. Similar deviations of the measured Hall-coefficients from the values calculated assuming Vegard's law have been observed in Au-In and Au-Sn (17) and in Cu alloys (42). For these alloys experimentally determined densities are not yet available.

Figure 11 shows the Hall-coefficient and number of conduction electrons in liquid AlGa alloys. In this system both components belong to the same group of the periodic table. The measured Hall-coefficient agrees with the prediction of the free electron model. The number of conduction electrons remains three independent of concentration.

Deviations of the Hall-coefficient from the free electron value have been reported for some liquid mercury alloys (48-51) whereas the pure constituents take on the appropriate free electron values. Alloys of Bi with Cd, In and Sn have been studied by Shiota and Tamaki (52). These exhibit substantial deviations from the free electron prediction. Their results on Cd-Bi alloys, which show the largest deviations, are reproduced in Fig. 12. The equiatomic alloys show almost a zero Hall-coefficient. It is also surprising that very small additions of Bi to Cd, which as a pure metal exhibits free electron behavior, produce substantial deviations. This is different to the behavior which is found in Pd-Sn alloys. In Pb-Sn alloys (53) the Hall-coefficients are in agreement with the free electron prediction on the Sn rich side but on the Pb rich side deviations become more significant as the concentration of Pb increases. Fig. 13.

Contrary to these findings in Bi and Pb based alloys, Takeuchi and Marukami (42) observed almost perfect agreement with the free electron model in Cu-Bi alloys over the entire concentration range including pure Bi. These results raise the obvious question as to whether the reported deviations are indeed real or whether they are due to insufficiently characterized samples (purity, dissolved gases, bubbles etc.)? As we stressed earlier it should be noted (Table 2) that for pure Pb and Bi in particular we find considerable disagreement between the results of different authors. Nevertheless, all the deviations reported so far in simple metals or alloys are in the same direction. They indicate a larger number of conduction electrons if interpreted with the free electron model.

3.3 Liquid Transition Metals and Alloys

As mentioned previously Hall-effect studies on pure liquid

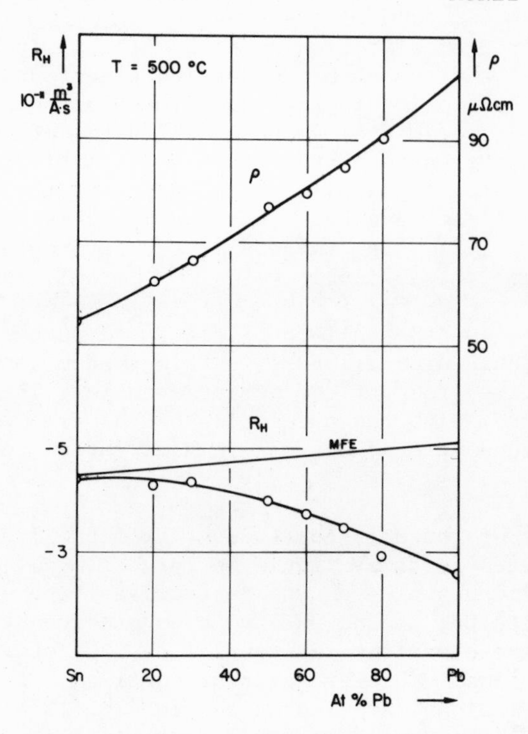

Fig. 13 Hall coefficients R_H and electrical resistivity ρ of liquid Pb-Sn alloys. MFE represents the free electron value and the experimental concentration dependence.

transition metals are extremely difficult due to the high melting temperatures and the corrosive properties of these materials. The only results which are available seem to be those for Ni and Co. (19) Somewhat more is known about the alloys of transition metals with simple metals.

Figure 14 shows the Hall-coefficient and electrical resistivity of Ni. Above the Curie temperature the Hall-coefficient is strongly temperature dependent and above about 500°C stays constant up to the melting point where it jumps to a slightly more negative value. The electrical resistivity also clearly jumps at this temperature. The Hall-coefficient, if interpreted within the free electron model, yields about 0.6 conduction electrons per Ni atom in the liquid state. This number is consistent with results obtained from magnetic data on Ni in the solid and liquid state. Although Co and Ni are neighbouring elements in the periodic table, the results obtained for Co (Fig. 15) are far more complicated and difficult to understand than those for liquid Ni. The Hall-coefficient of Co above the Curie temperature and in the liquid state is positive and strongly temperature dependent. This temperature dependence

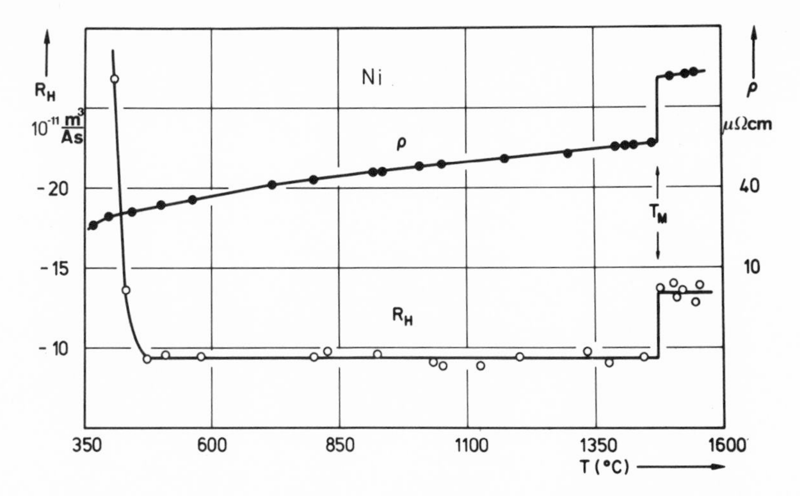

Fig. 14 Temperature dependence of the electrical resistivity and Hall-coefficient of solid and liquid Ni.

is closely related to the behavior of the magnetic susceptibility. Indeed it is well-known that in ferromagnetic materials additional anomalous contributions to the Hall-coefficient, which are proportional to the magnetization rather than the applied field, become important. It is also known that these contributions may persist above the Curie temperature in the paramagnetic range. For a recent review on the Hall-effect in ferromagnetic materials see Hurd (70) and the relevant chapters in this book. It is important to note

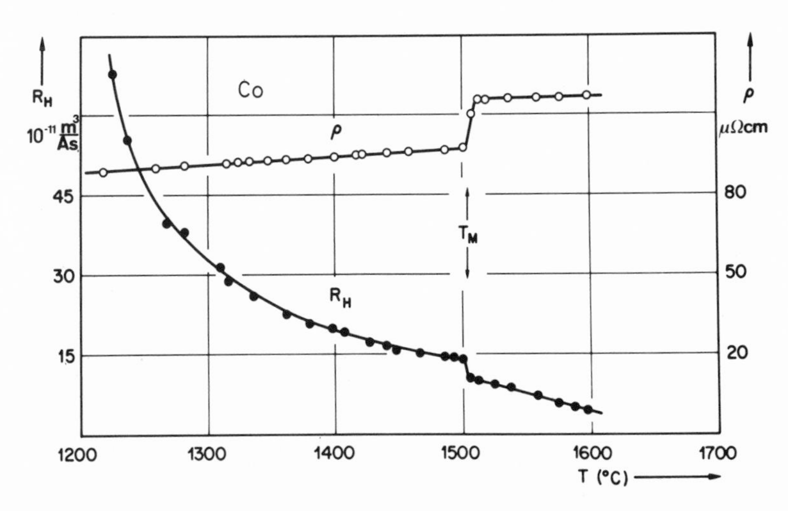

Fig. 15 Temperature dependence of the Hall-coefficient and electrical resistivity of solid and liquid Co.

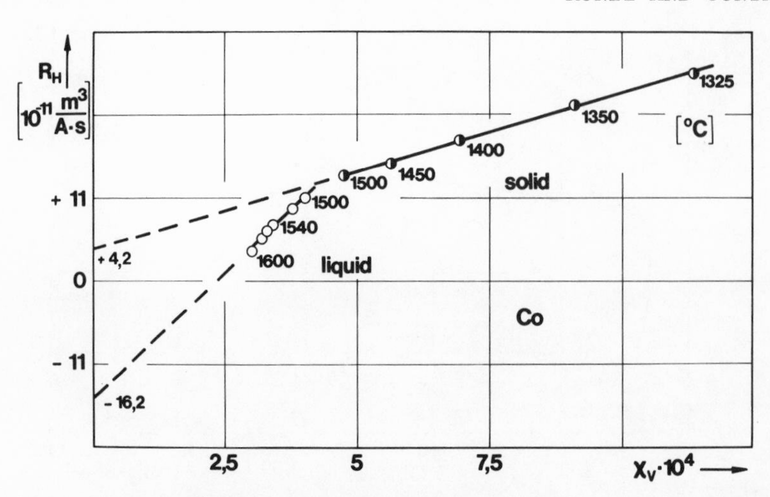

Fig. 16 Hall-coefficient of solid and liquid Co in the paramagnetic region as a function of the magnetic susceptibility.

that the Hall-field consists of the normal contribution (R_o) proportional to the field $B = \mu_o(H+M)$ and the anomalous or spontaneous part (R_1) proportional to the magnetization.

$$E_H = (R_oB + R_1\mu_oM)j$$

A separation of the two contributions is possible at fields above the magnetic saturation. In the paramagnetic range, however, $M = \chi H$ and the total effect becomes proportional to B.

$$E_H = (R_o + R_1\frac{\chi}{1+\chi})\ Bj \cong (R_o + R_1\chi)Bj$$

A single Hall-coefficient $R_H^* = R_o + R_1\chi$ arising from the two different contributions will be measured. A separation into the two components is possible if the magnetic susceptibility χ is known as a function of temperature and, in addition, R_o and R_1 do not depend strongly on temperature. In this case a plot of R_H^* vs. χ gives R_1 from the slope and R_o as the intercept of the extrapolated line at $\chi = 0$. Figure 16 shows this plot for Co. It follows that R_o is positive for solid Co near the melting point and becomes negative in the liquid state. The spontaneous Hall coefficient R_1 is larger for the liquid than for the solid. The value obtained for R_o from the extrapolation of liquid data is $-16.2\text{x}10^{-11}m^3/As$ and corresponds to 0.5 electrons per atom if interpreted according to the free electron model.

Since the total Hall-coefficient of these liquid metals can be positive and the normal contribution implies a fractional number of conduction electrons per atom it is of interest to study the

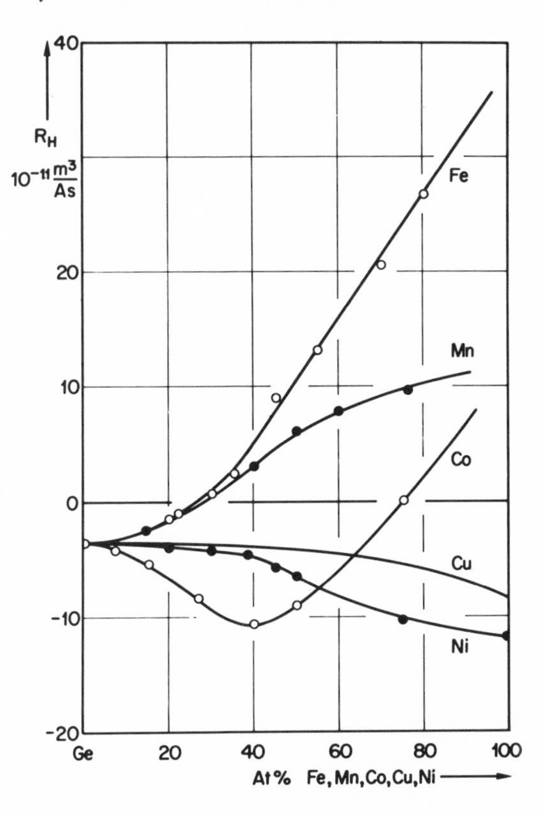

Fig. 17 Concentration dependence of the Hall-coefficients in liquid Mn-, Fe-, Co- and Ni-Ge alloys. For comparison the calculated free electron values for Cu-Ge alloys are plotted.

evolution of these particular features of the Hall-effect in alloys as function of concentration. It is convenient to examine alloys with a simple metal whose electronic properties seem to be well understood. For this purpose a number of liquid transition metal alloys with Au and Ge have been studied. The results are shown in Fig. 17 for the Ge alloys and in Fig. 18 for Au Alloys. In both types of alloys it was found that, with the exception of the Ni alloys, the Hall-coefficient changes its sign as a function of concentration. A noticeable temperature dependence within the range of measurements up to about 1200°C was detected only in Mn-Ge alloys near the equiatomic concentration (see Fig. 19). A separation into R_o and R_1, as was done for Co, using the susceptibility measurements of Güntherodt and Meier (59) therefore gives no accurate results. From the observed results it becomes evident that in Co-, Fe- and Mn-Ge and in Fe- and Co-Au alloys, deviations from the free electron model arise even at very dilute concentration of the

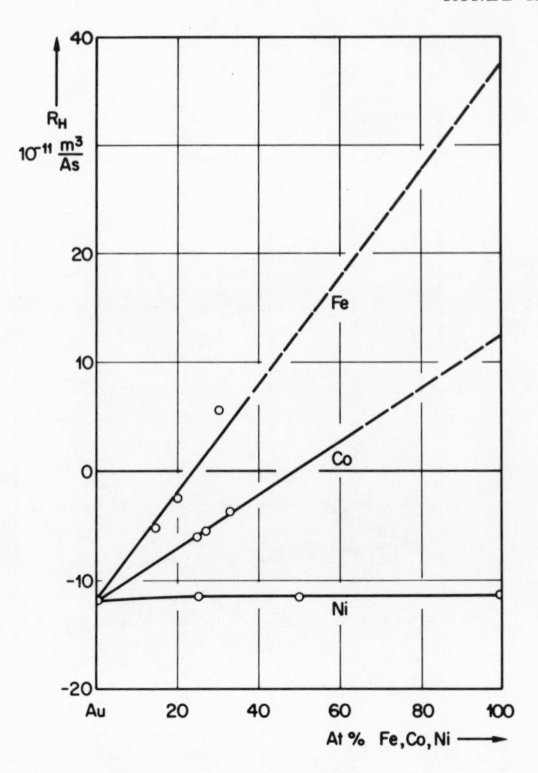

Fig. 18 Hall-coefficients of liquid Au-Fe, Au-Co and Au-Ni alloys.

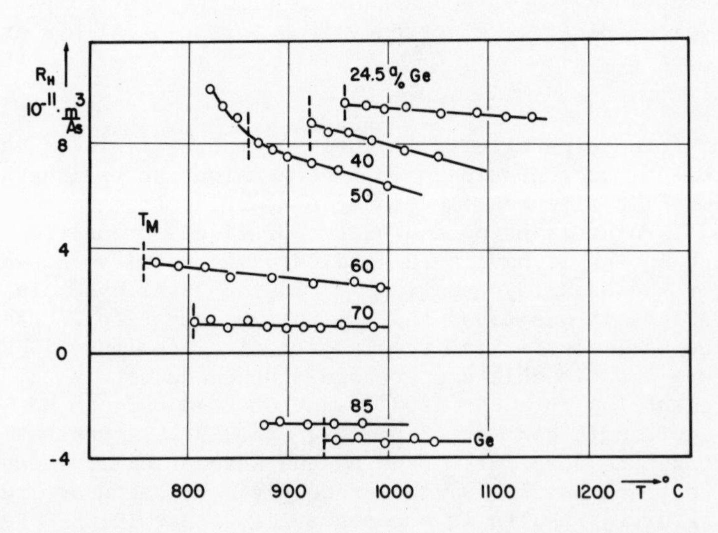

Fig. 19 Temperature dependence of the Hall-coefficients in liquid Mn-Ge alloys. T_M indicates the freezing temperature.

transition metal. For a clearer comparison the calculated free electron values for Cu-Ge are also drawn in Fig. 17. This curve also shows what we might roughly expect the normal contribution to the Hall-coefficient in these alloys to look like. The Ni alloys behave differently from the other alloys; they follow the free electron model more closely. In fact up to about 40% Ni in Ge the free electron model seems to be obeyed, assuming four electrons for Ge and zero charge carriers for Ni. It is well-known from resistivity measurements that in these alloys appreciable charge transfer may take place (18). Such a charge transfer may occur if the empty d-states which are expected to be localized on the transition metal are fully or partially filled up by s-like conduction electrons.

For Ni the number of empty d-states corresponds to the number of s-like conduction electrons. Assuming that in Ge rich alloys all the empty d-states are filled up, the effective contribution of charge carriers from Ni is then zero and we only have to count the electrons coming from Ge. Furthermore, we may argue that the full d-band does not contribute to the electronic transport. The Hall-coefficient in the alloy can then be attributed to the diluted number of conduction electrons coming from Ge.

This rather simple picture appears to be valid up to 40% Ni. Above this concentration the experimental results indicate that

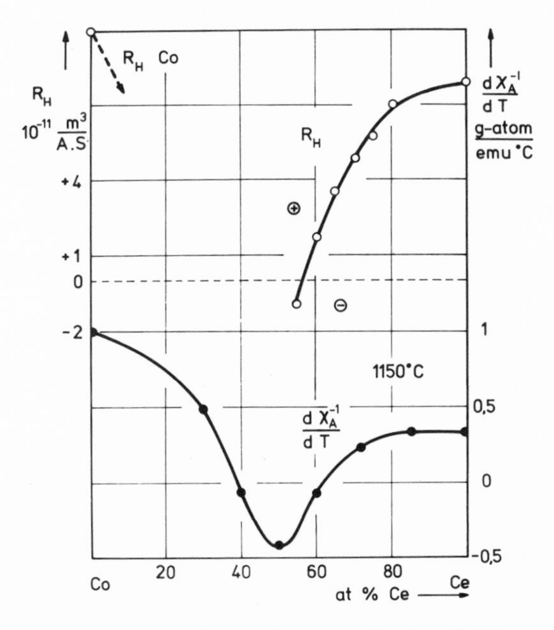

Fig. 20 The concentration dependence of the Hall-coefficient and the temperature coefficient of the reciprocal magnetic susceptibility of liquid Co-Ce alloys.

additional contributions to the Hall-coefficient become important. Such contributions might be due to the influence of an incompletely filled d-band. This view is supported by the observation that at this particular concentration the electrical resistivity starts to increase much faster with concentration than is the case for the dilute alloys.

These additional contributions seem, however, to be rather complicated in origin, since the free electron model would yield an even smaller electron concentration than we would expect from the dilution of the Ge electrons. This might occur if some of the conduction electrons became localized in covalent bonds; this seems rather unlikely in the liquid state. Furthermore a two band model does not seem to account for such behavior if realistic assumptions are made concerning the partial conductivities. Moreover, the Ge rich alloys below 80% Ni are not magnetic in the sense that they do not show a Curie-Weiss like behavior. Their susceptibilities increase with increasing temperature (58). The same is also true for liquid Co-Ge below 55% Co and for Fe-Ge below 30% Fe but deviations from the free electron model become apparent long before these concentrations are reached.

The experimental results obtained on liquid Co-Ce alloys (Fig. 20) seem to be even more puzzling. Starting from pure Ce the initially positive Hall-coefficient decreases and changes sign at 55% Ce. Since liquid Co also has a positive Hall-coefficient another "zero Hall-effect alloy" should exist on the Co rich side. Contrary to the Ge alloys discussed above, the correlation with the magnetic susceptibility (60) is more pronounced. The change of sign in the Hall-effect occurs at about the concentration where the alloys become non-magnetic as shown in Fig. 20 where the temperature coefficient of the reciprocal susceptibility $d\chi^{-1}/dT$ is plotted.

3.4 Liquid Lanthanides and Actinides

Among the metals of this group only the Hall-coefficients of the light rare earth elements La, Ce, Pr, Nd and of U have been measured. As shown in Table 3 all these liquids have positive Hall-coefficients. No results are available for the heavy rare earth

Table 3. Hall-coefficient for Liquid Rare Earth Metals and Liquid Uranium near their Melting Points. (18,61,62).

	La	Ce	Pr	Nd	U
$R_H (10^{-11} m^3/As)$	+6.15	+7.8	+9.3	+11.7	+3.88

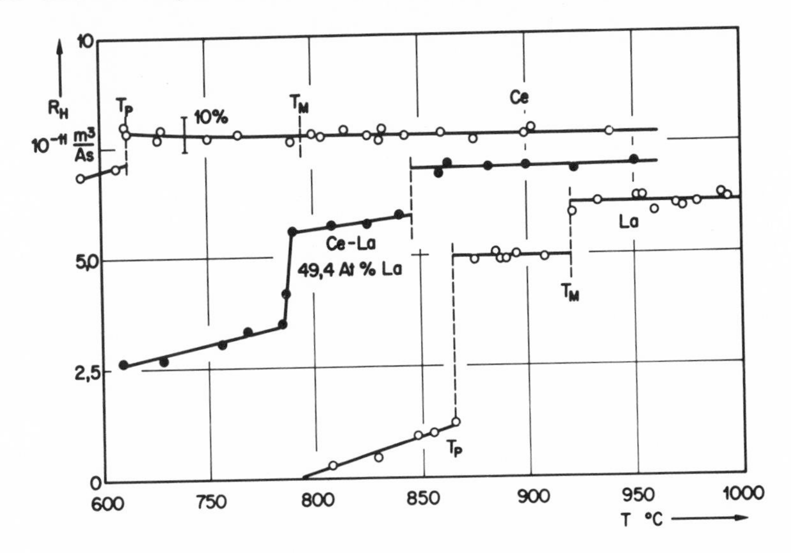

Fig. 21 Hall-coefficients of pure Ce and La and a La-Ce alloy as a function of temperature. T_P indicates the fcc-bcc phase transition and T_M the melting point.

metals and the transuranes. Measurements on alloys of these metals are also rather scarce.

Figure 21 shows the Hall-coefficients of La, Ce and a La-Ce alloy near the melting point. (18) All these metals have positive Hall-coefficients at high temperatures in the solid state as well as in the liquid state. The Hall-coefficient of Ce does not change at the melting point. A change has been observed at the temperature T_P of the phase transition from fcc to the bcc phase which occurs in most of the rare earth metals shortly before they melt. The other measurements in Fig. 21 show changes of about the same magnitude at T_P and T_M.

Figure 22 shows the Hall-coefficient and the electrical resistivity of Ce over a larger temperature range. At lower temperatures the Hall-coefficient of Ce shows a temperature dependence similar to that observed in Co. An attempt at separation into a normal and spontaneous part was, however, not successful since the total Hall-effect plotted versus the magnetic susceptibility did not give a linear relationship. This can occur if either R_o or R_1 are temperature dependent.

As in Ce, the Hall-effect of U does not change at the melting point (Fig. 23). Furthermore the Hall-coefficient of solid U does not change at the various phase transitions which it undergoes. Within the experimental accuracy the Hall-coefficient remains

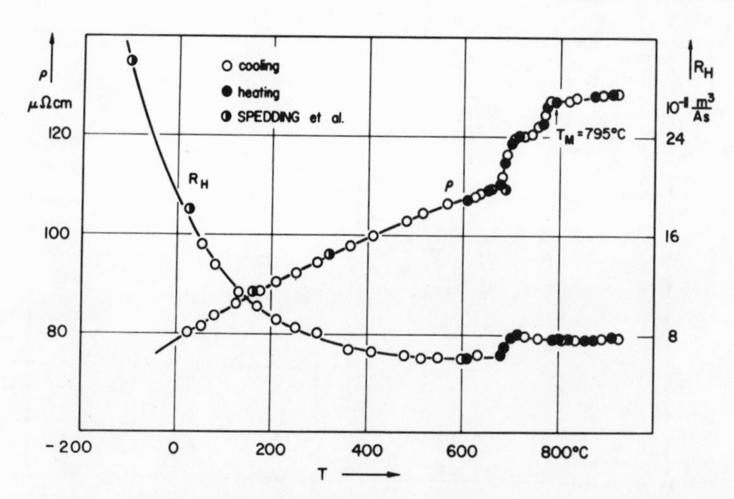

Fig. 22 The Hall-coefficient and the electrical resistivity of Ce. The full dots indicate values for heating and empty dots for cooling. (62)

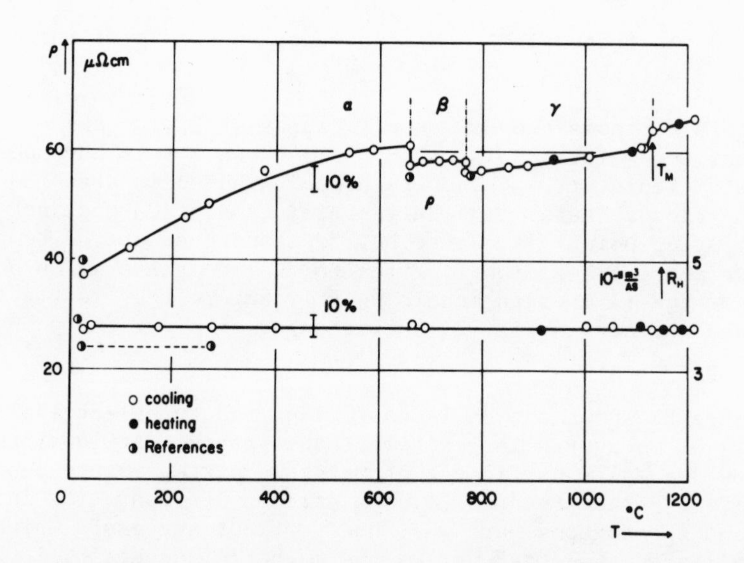

Fig. 23 The Hall-coefficient and the electrical resistivity of U. α, β and γ indicate the different solid phases and T_M the melting point. The data are compared with values of the Hall-coefficient in the solid state obtained by Boescholen and Huiszoon (67) and T. G. Berlincourt (68) and with resistivity values obtained by Pascal et al. (69).

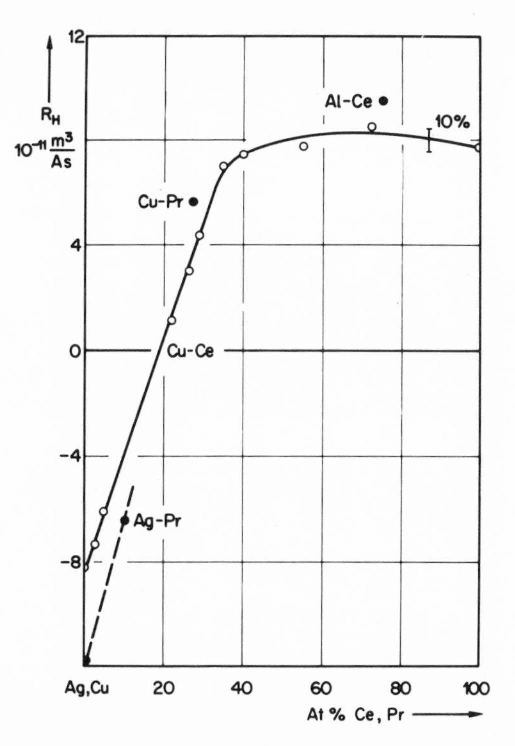

Fig. 24 Hall-coefficients of liquid Cu-Ce alloys (open dots) as a function of concentration and results of three other rare earth metal alloys (full dots).

essentially constant from -200°C to liquid state temperatures. The electrical resistivity (upper curve) clearly changes at the phase transitions and at the melting point. It should also be mentioned that by comparison with the liquid transition and rare earth metals, the value of the electrical resistivity of U is rather low and is in fact comparable with that of liquid Sn.

Figure 24 shows the concentration dependence of the Hall-coefficient in liquid Cu-Ce alloys. (18) Starting from pure Cu the Hall-coefficient increases drastically and changes sign from negative to positive near 20 at.% Ce. In Ce-rich alloys the Hall-coefficient remains practically concentration independent. Values for three other alloys are also indicated in the figure (full dots). Apart from the data given in Figs. 20 and 21 these seem to be the only results for alloys of the rare earth metals in the liquid state.

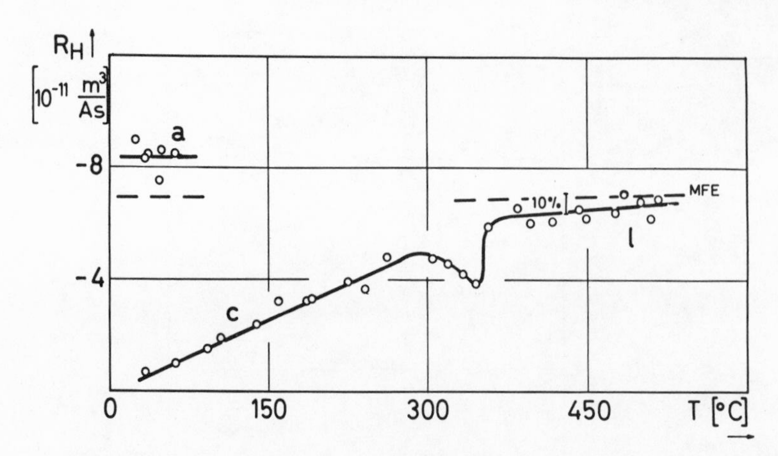

Fig. 25 Hall-coefficient of an amorphous, a liquid and a crystalline $Mg_{70}Zn_{30}$ alloy; dashed line indicates the free electron values for the Hall-coefficient.

3.5 Hall-effect in Metallic Glasses

This final chapter is not intended to constitute a comprehensive review of the work done in this interesting field. We would rather like to indicate some important similarities between these two different states of matter and, in particular, compare results on glassy alloys with those from the same alloys in the liquid state.

It is now well established that the atomic structure of metallic glasses closely resembles the structure in the molten state. Radial distribution functions measured in the two states exhibit only small differences. As a consequence we might also expect that, apart from specific low temperature effects, the electronic transport properties of the glass and the liquid should be rather similar. similar.

Figure 25 shows the results obtained from a $Mn_{70}Zn_{30}$ alloy in the amorphous, crystalline and liquid state (63). In the liquid state the Hall-coefficient is close to its free electron value whereas in the amorphous state a small deviation seems to exist. In the crystalline state the Hall-coefficient is strongly temperature dependent. Similar results have also been observed in Pd-Si alloys (64). A comparison between the two states becomes more difficult and the interpretation more uncertain in those cases where the glassy lloy is strongly magnetic or even ferromagnetic.

The electrical resistivity also shows a strong correlation between the two states. (63,65,66) It has been found that for many of the liquid alloys the resistivity versus temperature curve can simply be extrapolated to obtain the resistivity of the corresponding amorphous alloy. This also holds for compositions where the temperature coefficient of the electrical resistivity in the liquid state is negative.

4. CONCLUSIONS AND OUTLOOK

The determination and understanding of the Hall-coefficients in liquid metals has been an area of intensive investigation for the past 20 years. To the casual observer this may appear a relatively short period of time in view of the fact that the Hall-effect itself has been known for a century. Earlier experimenters, however, did not have the highly sophisticated and sensitive electronic instruments which we now have at our disposal. Although the art of handling liquid metals at high temperatures has undergone a considerable development in the past decade, further progress is necessary on this front, rather than in the electronic part, before we can learn more about the liquid transition and rare earth metals.

The most outstanding result in this field is certainly the remarkable agreement between the measured Hall-coefficients in liquid simple metals and the free electron value. This is clearly one of the major sources of the good theoretical understanding we now have of the electronic structure of these metals.

Further theoretical work and further experimental measurements on well characterized samples are necessary to achieve a better understanding and more consistent data for those cases where the Hall-coefficient deviates from its free electron value or shows anomalous contributions. This is undoubtedly a difficult task but one which should be sufficiently rewarding to warrant the large effort which is required.

ACKNOWLEDGEMENTS

We would like to express our sincere gratitude to Dr. R. Müller, Dr. M. Müller, Dr. M. Liard and Miss R. Oberle. Some of the more recent data are based on their thesis work. We also would like to thank H. R. Hidber and A. Nassenstein for designing and constructing the new double frequency experimental set-up. The authors are very much indebted to Dr. R. Evans for many stimulating discussions and for carefully reading the manuscript.

Financial support from the Swiss National Science Foundation and the Research Center of Alusuisse is gratefully acknowledged.

REFERENCES

1. T. Des Coudres, Phys. Z. 2:586 (1901).
2. H. Rausch v. Traubenberg, Ann. Phys. 17:109 (1905).
3. W. M. Nielsen, Phys. Rev. 23:302 (1924).
4. W. N. Fenninger, Phil. Mag. 27:109 (1914).
5. W. Gerlach, in: "Handbuch der Physik XIII," Geiger und Scheel, ed., Berlin (1928), p. 259.
6. J. A. Eldridge, Phys. Rev. 21:131 (1923).
7. A. Sommerfield, Z. f. Physik 47:1 (1928).
8. R. Peierls, Z. f. Physik 53:255 (1929).
9. F. Bloch, Z. f. Physik 52:555 (1928).
10. N. F. Mott and E. H. Jones, "The Theory of the Properties of Metals and Alloys," (1936).
11. H. Fröhlich, "Elektronentheorie der Metalle," (1936).
12. H. Zahn, Die Naturwissenschaften 40:848 (1930).
13. J. Kokoin and I. Fakodow, Z. Phys. 71:393 (1931).
14. P. W. Kendall and N. E. Cusack, Phil. Mag. 5:100 (1960).
15. Y. Tièche, Helv. Phys. Acta 33:963 (1960).
16. J. M. Ziman, Phil. Mag. 6:1013 (1961).
17. H.-J. Güntherodt and G. Busch, Phys. kondens. Materie 6:325 (1967).
18. H.-J. Güntherodt and H. U. Künzi, Phys. kondens. Materie 16:117 (1973).
19. R. Müller, Thesis, University of Basel.
20. N. F. Mott and E. A. Davies, "Electronic processes in non-crystalline materials," Clarendon Press, Oxford (1971).
21. R. S. Allgaier, Phys. Rev. 185:227 (1969).
22. L. E. Ballentine, this book.
23. I. G. Fakidov, Doklady Akad. Nauk SSSR 63:123 (1948).
24. N. Cusack and P. Kendall, Phil. Mag. 6:419 (1961).
25. E. G. Wilson, Phil. Mag. 7:989 (1962).
26. G. Busch and O. Vogt, Helv. Phys. Acta 27:24 (1954).
27. G. Busch and Y. Tièche, Phys. kondens. Materie 1:78 (1963).
28. J. E. Enderby, Proc. Phys. Soc. 81:772 (1963).
29. A. J. Greenfeld, Phys. Rev. 135:1589 (1964).
30. N. E. Cusack, J. E. Enderby, P. W. Kendall and Y. Tièche, J. Sci. Instr. 42:226 (1965).
31. R. G. Suchannek, Rev. Sci. Instr. 37:589 (1966).
32. R. G. Suchannek, L. Minghetti and S. Naiditch, Rev. Sci. Instr. 37:782 (1966).
33. V. A. Alekseev, A. A. Andreev and Yu. F. Ryzhkov, Zavodskaya Laboratoriya 35:691 (1969); Indus. Lab. 35:829 (1969).
34. H. L. McKinzie and D. S. Tannhauser, J. Appl. Phys. 40:4954 (1969).
35. J. C. Perron, Rev. de Physique Appl. 5:611 (1979).
36. P. W. Shackle, Phil. Mag. 21:987 (1970).
37. J. P. Velly, A. M. Martin and E. J. Picard, Phys. kondens. Materie 15:36 (1972).
38. H. U. Tschirner, R. F. Wolf and M. Wobst, Wiss. Z. d. Techn.

Hochsch. Karl-Marx-Stadt 17:335 (1975).
39. R. D. Swenumson, U. Even and J. C. Thompson, Rev. Sci. Instr. 49:519 (1978).
40. L. Langeheine and H. Mayer, Z. Physik 249:386 (1972).
41. B. R. Russel and C. Wahlig, Rev. Sci. Instr. 21:1028 (1950).
42. S. Takeuchi and K. Murakami, Sci. Rep. RITU A25:73 (1974).
43. H. R. Hidber and A. Nassenstein, in preparation.
44. H.-J. Güntherodt, H. U. Künzi and R. Müller, Phys. Lett. 54A, 155 (1975).
45. U. Even and W. Freyland, J. Phys. F: Metal Phys. 5:L104 (1975).
46. J. M. Ziman, Advan. Phys. 10:1 (1961).
47. I. Sauermann and G. Metzger, Z. Phys. Chem. 216:37 (1961).
48. N. E. Cusack and P. W. Kendall, Phil. Mag. 8:157 (1963); 10:871 (1964).
49. H.-J. Güntherodt, A. Menth and Y. Tièche, Phys. kondens. Materie 5:392 (1966).
50. A. A. Andreev and A. R. Regel, Fiz. Tverd. Tela 7:2567 (1965) (Sov. Phys.-Solid State 7:2076 (1966)); Fiz. Tverd. Tela 8:3681 (1966) (Sov. Phys.-Solid State 8:455 (1967)).
51. H. A. Davies, J. S. Llewelyn Leach and P. H. Draper, Phil. Mag. 23:1163 (1971).
52. I. Shiota and S. Tamaki, J. Phys. F: Metal Phys. 7:2361 (1977).
53. H.-J. Güntherodt, H. U. Künzi and R. Müller, Helv. Phys. Acta 46:403 (1973).
54. P. W. Kendall, J. Nucl. Mater. 35:41 (1970).
55. S. Tackeuchi and H. Endo, Trans. Jap. Inst. of Metals 2:243 (1961).
56. Y. I. Dutchak, P. O. Stetskiv and J. P. Klyus, Sov. Phys.-Solid State 8:455 (1966).
57. N. E. Cusack, P. W. Kendall and A. S. Marwaha, Phil. Mag. 7:1745 (1962).
58. M. Benkirane and J. Robert, Compt Rend B264:1584 (1967).
59. H.-J. Güntherodt and H. A. Meier, Phys. kondens. Materie 16:25 (1973).
60. L. Schlapbach, Phys. condens. Matter 18:189 (1974).
61. G. Busch, H.-J. Güntherodt and H. U. Künzi, Phys. Lett. A32:376 (1970).
62. G. Busch, H.-J. Güntherodt, H. U. Künzi and L. Schlapbach, Phys. Lett. A31:191 (1970).
63. R. Oberle, H. U. Künzi, H.-J. Güntherodt, B. G. Giessen, in preparation.
64. H.-J. Güntherodt, H. U. Künzi, M. Liard, M. Müller, R. Müller and C. C. Tsuei, in: "Amorphous Magnetism II," R. A. Levy and R. Hasegawa, eds., Plenum Press, New York (1977), p. 257.
65. H.-J. Güntherodt, H. U. Künzi, M. Liard, R. Müller, R. Oberle and H. Rudin, in: "Proc. 3rd Int. Conf. on Liquid Metals Bristol," (1976), p. 342.
66. H.-J. Güntherodt and H. U. Künzi, in: "Metallic Glasses," J. J. Gilman, ed., American Society for Metals, Metal Park, Ohio (1977), p. 247.

67. F. Boescholen and C. Guiszoon, Physica 23:704 (1957).
68. T. G. Berlincourt, Phys. Rev. 114:969 (1959).
69. J. Pascal, J. Morin and P. Lacombe, C. R. Acad. Sci. 256:4899 (1963).
70. C. M. Hurd, "Hall Effect in Metals and Alloys," Plenum Press, New York (1972).

EXPERIMENTAL HALL EFFECT DATA FOR A SMALL-POLARON SEMICONDUCTOR

P. Nagels

Materials Science Department, S.C.K./C.E.N.
B-2400 MOL (Belgium)

ABSTRACT

The current state of small-polaron theory is briefly reviewed. Experimental evidence for the existence of small polarons in reduced single crystals of lithium niobate, based on electrical conductivity, thermopower and Hall effect measurements, is presented. The transport coefficients of this material exhibit all the typical features of hopping motion of small polarons in the non-adiabatic regime. The drift mobility μ_D is very low and manifests a thermally-activated temperature dependence. The Hall mobility μ_H is greater than the drift mobility and increases exponentially with temperature dependences of μ_D and μ_H are in accord with that derived by Holstein and Friedman for small-polaron hopping.

1. INTRODUCTION

About a decade ago, the physics of crystalline low-mobility semiconductors attracted much attention both from experimental and theoretical side. Typical examples of this group of materials are the transition metal oxides, such as NiO, CoO and MnO. Most of these materials are large bandgap semiconductors and hence, are good insulators. Appreciable conductivity can be achieved by introducing a deviation from the stoichiometric composition or by incorporating at substitutional positions foreign ions with valences different from that of the original metal ion. A well-known example is the doping of NiO by Li^+ ions. For each Li^+ ion on a lattice site a neighboring Ni^{2+} is transformed into a Ni^{3+} ion. Much effort was put in the measurement of the basic transport coefficients commonly used to characterize a semiconductor. Due to its small magnitude the experimental determination of the Hall coefficient R_H was found to be extremely difficult. The Hall mobility μ_H, given by the product $R_H \times \sigma$, is typically less than $10^{-4}\ m^2V^{-1}s^{-1}$ in the transition metal oxides. The conductivity mobility, more often called the drift mobility μ_D, was also estimated to be small. These low mobilities lead to a mean free path of the charge carriers comparable with the interatomic distances. It was realized that such a situation might question the applicability of the classical band theory, originally developed for a broad band semiconductor. The nature of the electronic conduction process in the oxides remained a subject of discussion for a long time, but even now, the understanding is far from being complete. Electronic conduction in crystalline oxides can occur by motion of free charge carriers in a band or by a diffusion-like motion called "hopping" process. It was suggested that an electron in an ionic crystal can be trapped by the polarization of the surrounding lattice induced by the carrier itself. If the interaction between the electron and the lattice is strong enough, a quasi-particle, called a "polaron", will be formed and at high temperatures this will move by a hopping process. An extensive theoretical literature was published concerning small-polaron formation and its motion in a crystalline solid. On the basis of the extension of the lattice distortion a distinction was made between a "large" and a "small" polaron. Although the possibility of small-polaron formation in certain low-mobility semiconductors, especially the transition metal oxides, was accepted from the theoretical viewpoint, experimental studies did not always yield a decisive test for the occurrence of hopping conduction in these crystalline materials. The typical case is that of NiO on which extensive experimental results have been gathered in the past. The interpretation of the data has provoked much controversy. For a long time this material was considered by many authors to be a nice example of a small-polaron semiconductor. Later more reliable experimental facts seemed to be incompatible with the theoretical predictions of the small-polaron model. For analogous materials, such as CoO, Fe_2O_3, TiO_2 and $BaTiO_3$, the experimental data did not allow to draw definite conclusions with regard to their conduction process.

The main purpose of this article is to present some results, made in our laboratory, on the electronic properties of single-crystalline $LiNbO_3$. This study includes measurements of d.c. electrical conductivity, Hall effect and thermopower as a function of temperature obtained on $LiNbO_3$ of different oxygen deficiencies. It will be shown that $LiNbO_3$ undoubtedly exhibits all the features characteristic of small-polaron hopping. We shall discuss in detail the behavior of the Hall coefficient and the Hall mobility and demonstrate their great importance in the understanding of the conduction mechanism. The section devoted to the description and the analysis of the experimental data obtained on $LiNbO_3$ will be preceded by a short review on the small-polaron motion in crystalline solids. This part will be primarily concerned with the theoretically predicted behavior of the Hall coefficient, Hall and drift mobility of the small polaron in the hopping regime. We shall restrict ourselves to the fundamental characteristics of these transport coefficients.

2. SOME THEORETICAL ASPECTS

The purpose of this section is to present some basic features of the transport properties of small polarons. Extensive reviews on the subject of small-polaron theory have been published by Emin[1] and by Austin and Mott.[2] The essential idea of small-polaron formation is that, if the electron-lattice interaction is sufficiently strong, a charge carrier may get self-trapped at a particular atomic site. The unit built up by the trapped excess carrier and its induced lattice deformation is called a polaron, and more specifically a small polaron if the lattice deformation extends over a distance of the order of the interatomic spacing. The mobility of the small polaron will be much smaller compared to that of a free electron, since the particle must carry its induced lattice deformation with it as it moves through the lattice. The occurrence of a small value of the drift mobility μ_D in a crystalline or amorphous material is often used as a first criterion for showing the existence of small polarons. In his original paper on small-polaron theory, Holstein[3] showed that a small polaron in an ideal crystal may move via two distinct processes. This can be understood on the basis of the following arguments. At low temperatures the small-polaron states overlap sufficiently to form a polaron band and the small polaron moves by a Bloch-type band motion. However, the width of the polaron band rapidly decreases with temperature and at characteristic temperature, which was estimated to be in the vicinity of half the Debye temperature, the polaron-band width becomes less than the uncertainty in energy due to the finite lifetime of the polaron states. The small polaron becomes localized and can only move by hopping from one atomic site to a neighboring one. The physical picture for the hopping process is based on occasional dynamic fluctuations of the lattice which at a given moment may produce equal distortions at the occupied and a neighboring site. The electronic energy of the two adjacent sites will then become momentarily equal and the carrier will have a certain probability of tunnelling. The momentary occurrence of equal energy was called a "coincidence event" by Holstein. The transition probability during a coincidence event was studied in two limiting cases, called the adiabatic and non-adiabatic regimes. In the adiabatic regime the electron can follow the motion of the lattice and the carrier will possess a high probability of hopping to the adjacent site during a coincidence event. In the non-adiabatic regime the electron cannot follow the lattice vibrations and the time required for an electron to hop is large compared to the duration of a coincidence event. In this case many coincidence events will occur before the the carrier hops to a neighboring site. The transition probability per unit time will be much smaller in the non-adiabatic regime than in the adiabatic one.

The theory of the drift mobility of the small polaron in the non-adiabatic approximation was developed by Holstein.[3] In the low temperature band motion the drift mobility is mainly determined by the phonon scattering and, hence, it decreases with increasing temperature. This conduction process will not be discussed in further detail, because, as far as we know, it has not been observed experimentally in crystalline low-mobility semiconductors. At high temperature, where hopping conduction dominates, Holstein derived the following approximate formula for the drift mobility in the non-adiabatic case:

$$\mu_D = \frac{3ea^2}{2\hbar}\frac{J^2}{kT}\left(\frac{\pi}{4W_H kT}\right)^{\frac{1}{2}} \exp\left(-\frac{W_H}{kT}\right) \quad (1)$$

Here, a is the distance between neighboring hopping sites, J is the electronic transfer integral and W_H is the hopping activation energy. The other parameters have their usual meaning. The parameter J is a measure of the overlapping of the wave functions.

If the drift mobility is plotted as:

$$\mu_D T^{3/2} \propto \exp\left(-\frac{W_H}{kT}\right)$$

then the activation energy for hopping can be deduced from the slope of this line. The principal result of Holstein's calculation is the exponential increase of the drift mobility with increasing temperature, but it should be emphasized that equation (1) is a high-temperature classical limit of a more general expression, which predicts a continuous lowering in slope towards lower temperatures. The thermally-activated behavior of the drift mobility is one of the essential features of small-polaron theory. It should be noted that the pre-exponential term, which varies at $T^{-3/2}$, will become predominant at the temperature where kT becomes of the order of W_H.

The adiabatic process of small-polaron hopping has been studied by Emin and Holstein.[4] They arrive at the result that:

$$\mu_D = \frac{3ea^2}{2}\left(\frac{\omega_o}{2\pi kT}\right) \exp\left[-\left(\frac{W_H - J}{kT}\right)\right] \quad (2)$$

Here, ω_o is the longitudinal optical phonon frequency. The activation energy of the drift mobility is lower than the non-adiabatic counterpart by an amount J. However, this will not produce a major different in the temperature dependence of μ_D since J is usually much smaller than W_H.

The calculation of the Hall mobility of the small polaron has proved to be more difficult. According to the theory of Friedman and Holstein[5] the mechanism of the Hall effect is based on the occurrence of coincidence events which involve the initially occupied site and at least two final sites. They found that the Hall mobility is extremely sensitive to the geometric arrangement of the sites involved in the hopping process. For the case of an equilateral triangular lattice where each atom belongs to a group of three mutually nearest neighbors, Friedman and Holstein arrive at the following expression for the non-adiabatic hopping regime:

$$\mu_H = \frac{ea^2}{2h} J \left(\frac{\pi}{4kTW_H}\right)^{\frac{1}{2}} \exp\left(-\frac{1}{3}\frac{W_H}{kT}\right) \tag{3}$$

Thus, in the classical high-temperature limit, the temperature dependence of the Hall mobility is represented by:

$$\mu_H \propto T^{-\frac{1}{2}} \exp\left(-\frac{1}{3}\frac{W_H}{kT}\right)$$

An important feature of this theory is that the Hall mobility is much larger than the drift mobility and increases exponentially with temperature. The activation energy of μ_H is only one-third of that associated with the drift mobility. At sufficiently high temperatures, i.e. when $kT \simeq W_H/3$, the Hall mobility reaches a maximum and then decreases with increasing T, due to the decrease of the pre-exponential term with increasing temperature. For the triangular lattice the Hall coefficient is readily found from the relation:

$$R_H = -\frac{1}{nqc}\left(\frac{\mu_H}{\mu_D}\right)$$

where n is the number of charge carriers (electrons) per unit volume. From eqns. (1) and (3) it follows that the ratio μ_H/μ_D is given by:

$$\frac{\mu_H}{\mu_D} = \frac{kT}{3J} \exp\left(\frac{2}{3}\frac{W_H}{kT}\right)$$

so that:

$$R_H = -\frac{1}{nqc}\frac{kT}{3J} \exp\left(\frac{2}{3}\frac{W_H}{kT}\right) \tag{4}$$

The Hall coefficient for motion in a three-site network is greater than its normal value for a broad band semiconductor ($R_H = -1/nqc$); the ratio μ_H/μ_D is a decreasing function of temperature.

The Hall effect for the four-site non-adiabatic case has been calculated by Emin.[6] For the square lattice in the high-temperature limit the Hall mobility is of the form:

$$\mu_H \propto \left(\frac{\hbar\omega_o}{kT}\right)^{3/2} \exp\left(-\frac{1}{3}\frac{W_H}{kT}\right) \tag{5}$$

while the drift mobility manifests a temperature dependence of the form:

$$\mu_D \propto \left(\frac{\hbar\omega_o}{kT}\right)^{\frac{1}{2}} \exp\left(-\frac{W_H}{kT}\right) \tag{6}$$

It follows that the temperature dependence of the Hall mobility is much more effected by the temperature dependence of its pre-exponential term than in the three-site case. The Hall mobility for the cubic geometry exhibits a broad maximum and decreases, as indicated by Emin, above a temperature equal to $2W_H/9k$. It is concluded, therefore, that the absence of a thermally-activated Hall mobility does not exclude the possibility of small-polaron conduction in a solid. The temperature dependence of the Hall mobility in the three- and four-site configuration as calculated in the non-adiabatic theory for the choice of parameters $a = 4$ Å. $J = \hbar\omega_o = 6.4 \times 10^{21}$ J(0.04 eV) and $W_H = 5\hbar\omega_o$ is illustrated in Fig. 1. In the high temperature range the drift mobility still manifests a thermally-activated behavior.

The Hall mobility for the three-site hopping model in the adiabatic regime has been calculated by Emin and Holstein.[4] They have shown that for high temperatures:

$$\mu_H = \frac{1}{2}ea^2\left(\frac{\omega_o}{2\pi kT}\right) f(T) \exp\left[-\left(\frac{1}{3}\frac{W_H - J}{kT}\right)\right] \tag{7}$$

where $f(T) = 1$ for $J << kT$ and $f(T) = kT/J$ for $J >> kT$. Figure 2 presents a plot of μ_H, calculated using the non-adiabatic and adiabatic approach, versus $\hbar\omega_o/kT$ for the set of parameters $a = 4$ Å, $J = \hbar\omega_o = 6.4 \times 10^{-21}$ J(0.04 eV) and $W_H = 5\hbar\omega_o$. Using eqns. (2) and (7), the Hall coefficient is found to be:

$$R_H = -\frac{1}{nqc}\left(\frac{f(T)}{3}\right) \exp\left(\frac{2}{3}\frac{W_H}{kT}\right) \tag{8}$$

An important feature of the adiabatic theory is that the activation

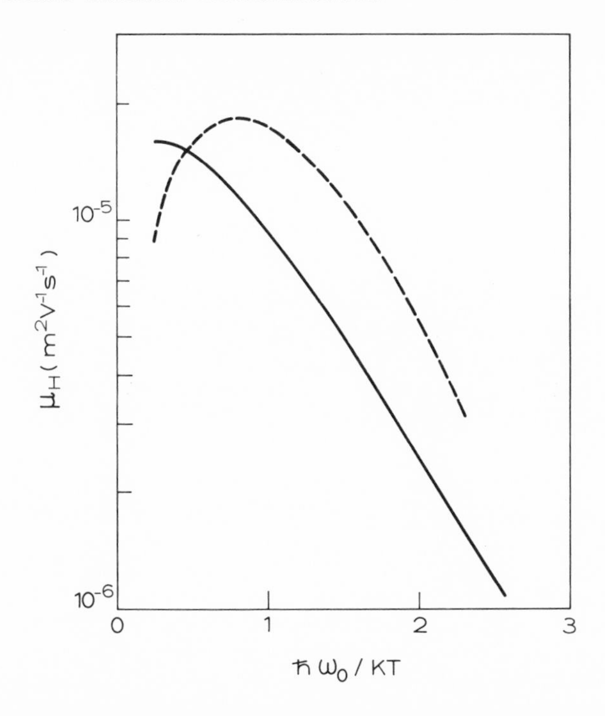

Fig. 1. The Hall mobility versus $\hbar\omega_o/kT$ in a traingular (full line) and a cubic lattice structure (dotted line) as calculated in the non-adiabatic theory for the parameters a = 4 Å, $J = \hbar\omega_o = 6.4 \times 10^{21}$ J and $W_H = 5\hbar\omega_o$ (after Ref. [5] and [6]).

energy of μ_H is less than that of the non-adiabatic case by an amount J. For low values of the activation energy, i.e. when $J \simeq W_H/3$, the Hall mobility is a continuously decreasing function of temperature. Moreover, μ_H may become smaller than μ_D at high temperatures. Thus, as in the four-site non-adiabatic case, the absence of an activated Hall mobility does not in itself rule out small-polaron motion.

An important result of Holstein's theory is concerned with the sign of the Hall coefficient.[7] Indeed, the author concluded that the sign of the Hall effect for holes is the same as that for electrons for the three-site model and opposite for the four-site case. In other words, the theory predicts a sign anomaly of R_H when hole-like polarons move by hopping solids characterized by a three-site geometry.

A further extension of the small-polaron hopping theory was made by taking into account the role of vibrational dispersion on the hopping motion of small polarons. Let us first mention that

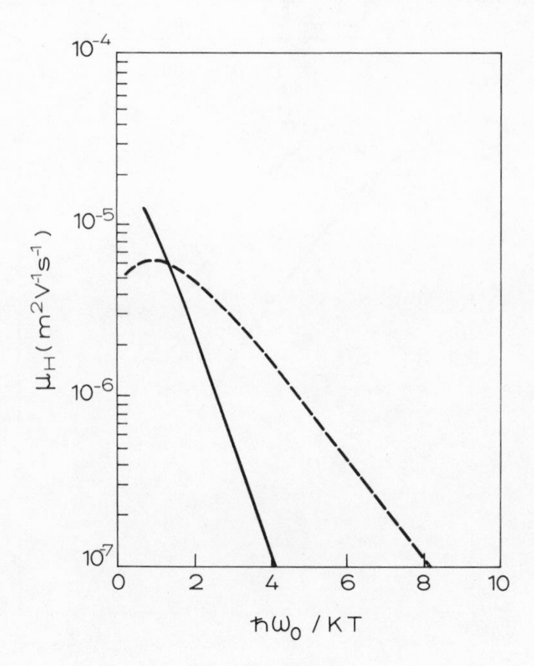

Fig. 2. The Hall mobility versus $h\omega_o/kT$ in a triangular lattice structure as calculated in the non-adiabatic (full line) and adiabatic approximation (dotted line) for the parameters $a = 4$ Å, $J = \hbar\omega_o = 6.4 \times 10^{21}$ J and $W_H = 5\hbar\omega_o$ (after Ref. [4] and [5]).

in all previous studies it was assumed that eac hop occurs independently of its previous hops. In the uncorrelated hopping regime the mean time between two successive small-polaron hops is much greater than the time associated with the relaxation of the lattice following a small polaron hop. Emin[8] considered now the case of correlated small-polaron hopping, applicable to intermediate-mobility materials ($\mu_D > 10^{-5}$ $m^2V^{-1}s^{-1}$) or stated in another way, to materials having a rather large J value. In this situation the carrier has a substantial probability of hopping either to a nearest neighbor or back to its original site before the lattic relaxes. Emin's theoretical predictions are that in the correlated case the hopping activation energy will be substantially reduced from that of the uncorrelated hopping motion. The drift mobility goes through a broad maximum and then diminishes slowly with increasing temperature as a consequence of the decrease of the pre-exponential term. The important result is that the drift mobility in this regime need not manifest a clear thermally activated behavior. In the same paper Emin demonstrated that the Hall mobility also will be affected by lattice relaxation effects. In particular, he showed that the activation energy of the Hall mobility will be smaller than that obtained for the uncorrelated situation and, moreover, that it will be smaller than the drift mobility activation energy. The latter feature was already encountered in the previous studies.

In conclusion, we have today adequate theories of small-polaron motion applicable for different situations. In general, these theories lead to temperature dependence of the drift and Hall mobility which are quite different from the dependence for ordinary band conduction. The experimental situation is not as favorable since until now one has found only few materials which clearly show the characteristic small-polaron properties.

3. EXPERIMENTAL VERIFICATION OF THE SMALL-POLARON MODEL: $LiNbO_3$

Lithium niobate has been the subject of considerable interest because of its piezoelectric, electro-optical and non-linear optical properties. At present, the acoustical and optical applications of this material are numerous. The physical properties are largely influenced by compositional variations of $LiNbO_3$. Much work has been devoted, therefore, to the improvement of the material quality. $LiNbO_3$ is available now under the form of large single crystals of good optical quality and is transparent throughout the visible and the near infrared (from 0.35 to 5 μm). Very pure and stoichiometric $LiNbO_3$ is a water-clear crystal. Heating the crystals at high temperature in a reducing atmosphere results in black $LiNbO_3$. Subsequent annealing in an oxygen atmosphere transforms the black crystals again in transparent ones. For a detailed review of the chemistry and physics of $LiNbO_3$ the reader is referred to an article by Räuber.[9]

The structure of $LiNbO_3$ at room temperature belongs to the rhombohedral space group R3c. This form is ferroelectric with an unusually high Curie temperature of 1410 K. The basic net of the structure is formed by six equidistant plane layers of oxygen per unit cell. The oxygen arrangement can be described by a distorted hexagonal close packing. There are six octahedral interstices lying between each pair of oxygen layers. The first and fouth contain Nb, the second and fifth are empty, the third and sixth contain Li. In the rhombohedral classes there are several choice of the unit cell and directions of axes. For most physical applications a hexagonal cell is preferred, with all axes orthogonal. The unit cell parameters for hexagonal axes of $LiNbO_3$ are: a_H = 5.150 Å and c_H = 13.867 Å. There are six formula units per unit cell.

The composition of a lithium niobate crystal pulled from a stoichiometric melt is not completely described by the formula $LiNbO_3$. Nassau and Lines[10] have shown that it contains a slight deficiency of lithium ions compensated by an excess of niobium ions. Doping of $LiNbO_3$ by foreign metal ions is easy because the structure is able to adjust for fairly large amounts of additions. Crystals doped by transition metal ions and rare earth metal ions have been grown and have been extensively investigated for their use in holographic applications.

Pure $LiNbO_3$ is a good insulator. The optical band gap is reported to be 5.9×10^{-19} J (3.7 eV) at room temperature.[11] Redfield and Burke[12] studied the shift of the optical absorption edge in the temperature range between 10 and 667 K. The edge is remarkably broad and curved even at low temperatures, which may be explained by the large amount of defects introduced by the deviation from the stoichiometric composition. The temperature coefficient of the optical gap is equal to about 2.1×10^{-22} J K^{-1} (1.3×10^{-3} eV K^{-1}) in the range between 300 and 667 K.

Now we shall discuss the electrical properties of lithium niobate. After describing some earlier experimental results, which were mainly concerned with one transport coefficient only, we shall treat in detail some unpublished experimental data obtained by us on reduced $LiNbO_3$ single crystals. These results will greatly help in elucidating the nature of the conduction mechanism in this matrial. A first study of the electrical conductivity at high temperatures (400-1000°C) was performed by Shapiro et al.[13] on stoichiometric $LiNbO_3$. The material was highly ohmic, showing a conductivity of the order of 5×10^{-4} Ω^{-1} m^{-1} at 400°C. The authors observed a decrease in slope of the conductivity around 600°C, which they related to the small discontinuous reduction in the lattice parameter along the c-axis occurring between 610 and 620°C. In order to study the nature of the structural defects in reduced $LiNbO_3$, Bergmann[14] measured the electrical conductivity of single crystals under various pressures of oxygen below 10^5 P and at temperatures up to about 1400 K. The electrical conductivity was found proportional to $P_{O_2}^{-\frac{1}{4}}$ which can be explained by a defect equilibrium involving singly ionized oxygen vacancies. Oxygen in the atmosphere is in equilibrium with these defects according to the following reaction:

$$O^{2-} \rightleftharpoons e_o^- + e^- + \tfrac{1}{2}O_2(g)$$

where e_o^- is an electron at an O^{2-} site and e^- a free electron. This process leads, therefore, to the formation of free electrons in reduced $LiNbO_3$.

A similar but more extensive study of the electrical conductivity as a function of temperature and oxygen pressure has been published by Jorgensen and Bartlett.[15] These authors confirmed the dependence of the conductivity on $P_{O_2}^{-\frac{1}{4}}$ for oxygen pressures less than approximately 10^{-1} P. In addition to the electrical conductivity, they also measured the thermoelectric power which was found to be negative at low oxygen pressures. By determining electrical transport numbers with the help of an emf method, Jordonsen and Bartlett

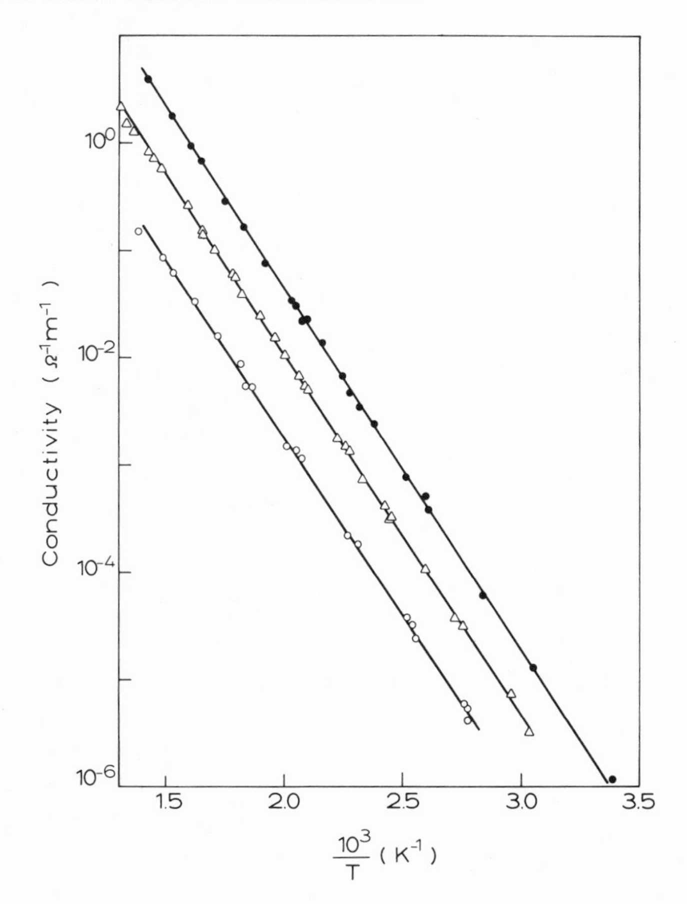

Fig. 3. Electrical conductivity as a function of reciprocal temperature for three $LiNbO_3$ single crystals reduced at 1010°C under different oxygen pressures (P_{O_2} = 10^{-5} (o), 10^{-9} (Δ) and 10^{-13} P (●), respectively).

could further differentiate between ionic and electronic conduction. At 1000 K, the electrical conductivity was found to be purely ionic at 10^5 P of oxygen and purely electronic at low oxygen partial pressures.

Bollmann and Gernand[16] studied the optical absorption, electrical conductivity and photoconductivity of $LiNbO_3$ after thermal treatments in different atmosphere. Reduction produces two distinct absorption bands situated at 480 and 1200 nm. The crystals were photosensitive in the energy ranges of both bands. In a later paper Bollmann[17] proposed that the following reaction should take place during reduction:

$$2Nb^{5+} + O^{2-} \rightleftharpoons 2Nb^{4+} + O_v^{2-} + \tfrac{1}{2}O_2$$

Free electrons are produced by the ionization of Nb^{4+} ions according to the mechanism:

$$Nb^{4+} \rightleftharpoons Nb^{5+} + e^-$$

Nb^{4+} ions are considered as the reason for the intense black coloration of the reduced crystals.

All previous studies, mainly based on electrical conductivity measurements, were carried out with the intention to get some insight in the defect structure of reduced $LiNbO_3$. The authors were not able, however, to unfold the electronic conduction mechanism in this material. An investigation of transport coefficients other than the electrical conductivity is necessary for a complete understanding of this mechanism. In this view the present author performed measurements of d.c. electrical conductivity, thermopower and Hall effect as a function of temperature on $LiNbO_3$ single crystals reduced at various low oxygen pressures. The next part of this section will be devoted to a detailed description and analysis of these results.

We shall first briefly mention some details of the experimental procedures used in this study. All measurements were made on samples cut from Czochralski grown single crystals (kindly supplied by Métallurgie Hoboken, Belgium). The samples were reduced by heating in streaming gas mixtures of CO and CO_2 at a temperature of 1010°C for 24 hrs. The oxygen partial pressure during the reduction was adjusted by using three different CO_2/CO ratios equal to 99/1, 50/50 and 1/99. This yields very low P_{O_2} values amounting to 10^{-5}, 10^{-9} and 10^{-13} P, respectively. The four-point probe method of Van der Pauw[18] was used to measure the d.c. electrical conductivity and the Hall coefficient. Four gold contacts were alloyed at the corners of rectangular plane-parallel platelets, typically 7x7x0.5 mm^3 in size. Thermopower data were obtained on small bars on which at each end a chromel-constantan thermocouple was fixed by point-welding the junction of the thin wires on a gold contact. The reliability of the results was checked by reversing several times the direction of the temperature gradient. Figure 3 shows the temperature dependence of the d.c. electrical conductivity of three $LiNbO_{3-y}$ samples, having different oxygen deficiencies. The measurements, which were carried out in a gas stream of the appropriate CO_2/CO mixture, were limited to an upper temperature of about 750 K in order to avoid changes in the defect concentrations. In this way the crystal always remains in the state of the frozen-in high temperature defect equilibrium. Data taken during a second thermal cycle did coincide completely with those obtained during the first

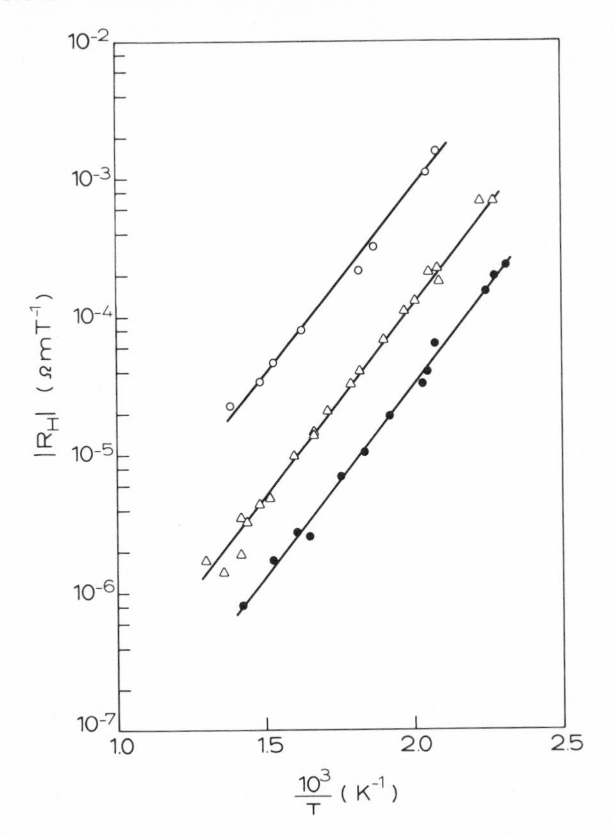

Fig. 4. The absolute value of the Hall coefficient as a function of reciprocal temperature for the same $LiNbO_3$ single crystals as in Fig. 3.

one. A.c. conductivity measurements at a fixed frequency of 10 KHz were also performed as a function of temperature. Within the limits of experimental error the a.c. conductivities did not differ from the d.c. values, which indicates that the conduction is purely electronic. It can be seen from Fig. 3 that the electrical conductivity of the three samples varies exponentially with temperature in the whole temperature range of measurement. The curves are running exactly parallel, the activation energy characterizing each curve being equal to 1.07×10^{-19} J (0.67 eV). Earlier reported values of 1.09×10^{-19} J (0.68 eV),[19] 1.12×10^{-19} J (0.70 eV)[17] and 1.15×10^{-19} J (0.72 eV)[14] are in good agreement with the one found here.

The Hall coefficient data obtained between 430 and 750 K are shown in Fig. 4. The Hall coefficients show a negative sign and appear to depend exponentially on temperature. The Hall coefficients

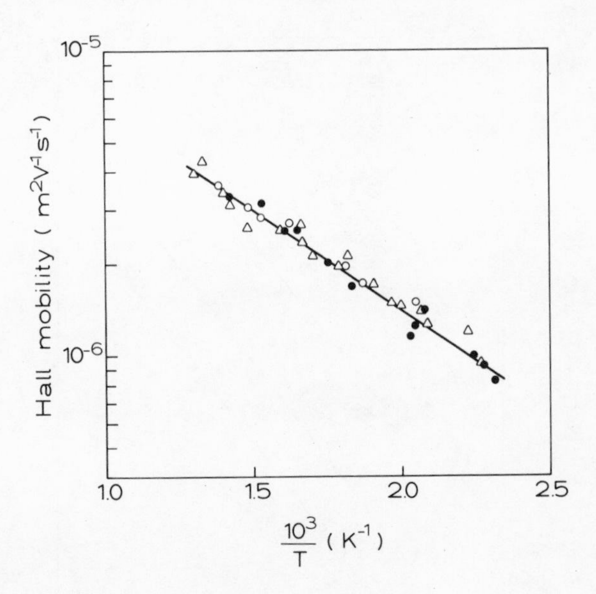

Fig. 5. Hall mobility as a function of reciprocal temperature for the same $LiNbO_3$ single crystals as in Fig. 3.

show a negative sign and appear to depend exponentially on temperature. The Hall coefficient ratio of samples reduced under the two extreme conditions (partial oxygen pressure equal to 10^{-5} and 10^{-13} P, respectively) amounts to a factor of 30. The activation energy deduced from the slope of the R_H curves (8.5×10^{-20} J (0.53 eV)) is less than that found for the conductivity.

In Fig. 5 the Hall mobility is plotted versus the reciprocal temperature. The Hall mobility is found to be very low (μ_H = 8.5×10^{-7} $m^2v^{-1}s^{-1}$ at 430 K) and completely independent of the defect concentration. Its temperature dependence can be described as thermally activated with an activation energy of 2.08×10^{-20} J (0.13 eV).

The thermopower was measured on only one sample showing the highest conductivity. The results of these measurements are presented in Fig. 6. The values were always negative indicating conduction by electrons in support of the electron donor defect model. The thermopower varies linearly with 1/T in the temperature range of measurement (325 - 560 K). The activation energy calculated from the slope of the straight line is equal to 4.0×10^{-20} J (0.25 eV), which is 6.7×10^{-20} J (0.42 eV) less than that activating the electrical conductivity.

The most striking features of the results are: 1. The difference in activation energy between the conductivity and the

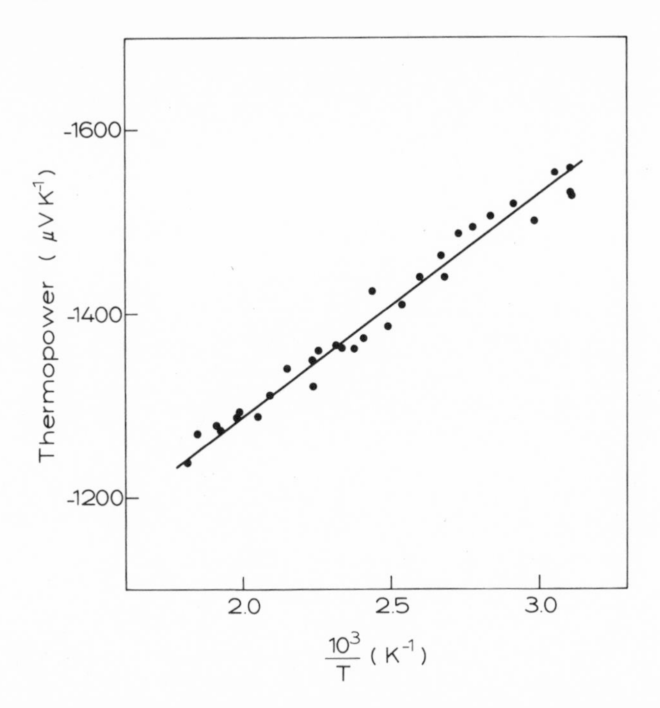

Fig. 6. Thermopower as a function of reciprocal temperature for a $LiNbO_3$ single crystal reduced at 1010°C under an oxygen pressure of 10^{-13} P.

thermopower. According to the general relation $\sigma = nq\mu_D$, the variation of the conductivity with temperature includes both the variation of the number of charge carriers and of the mobility. On the contrary, the thermopower is directly related to the number of charge carriers only. The difference in slope between that of the conductivity (E_σ) and of the thermopower (E_S) can therefore be attributed to the activation energy of the drift mobility W_H. From this it follows that the drift mobility of electrons in $LiNbO_3$ increases exponentially with increasing temperature. 2. The Hall mobility is very low and shows a similar behavior as the drift mobility, i.e. μ_H manifests a thermally-activated temperature dependence. An important point in the further discussion of the results is the comparison of the activation energies associated with μ_D and μ_H. If we assume at first instance that the conductivity and Hall mobility can be represented by pure exponentials (neglecting hereby possible temperature-dependent pre-exponential terms), as is shown in Figs. 3 and 5, then the activation energy of the drift mobility is equal to $W_H = E_\sigma - E_S = 6.7 \times 10^{-20}$ J (0.42 eV), which is approximately three times as high as that of the Hall mobility (2.08×10^{-20} J (0.13 eV)).

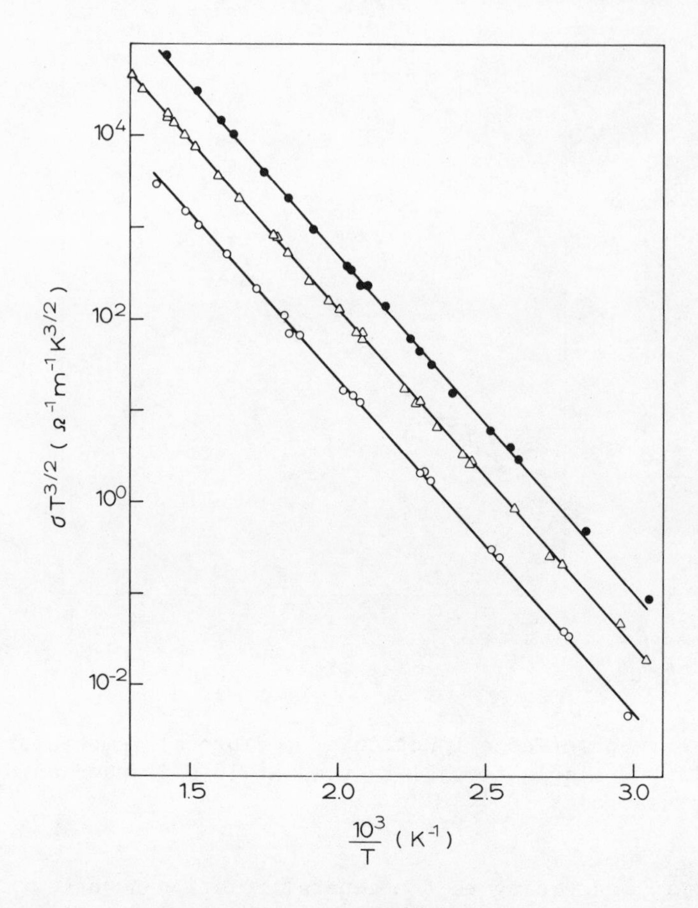

Fig. 7. Conductivity plotted as $\sigma T^{3/2}$ versus 1/T for the same $LiNbO_3$ single crystals as in Fig. 3.

The observed temperature dependence of the drift and Hall mobility together with the small magnitude of μ_H give strong evidence for the existence of small polarons in $LiNbO_3$. These small polarons would be formed by trapping of electrons at niobium cations. This results in a transformation of a number of Nb^{5+} ions into Nb^{4+} ions. The relatively high temperatures ($T > \theta/2$) at which the measurements were carried out would lead one to expect that the transport occurs by a hopping mechanism in which an electron jumps from a Nb^{4+} ion to a neighboring Nb^{5+} ion. The Debye temperature of stoichiometric $LiNbO_3$ deduced from heat capacity measurements at very low temperatures[21] and from elastic constant data[22] is equal to approximately 560 K.

As discussed in the previous section, two different regimes can be considered for the small-polaron transfer: non-adiabatic

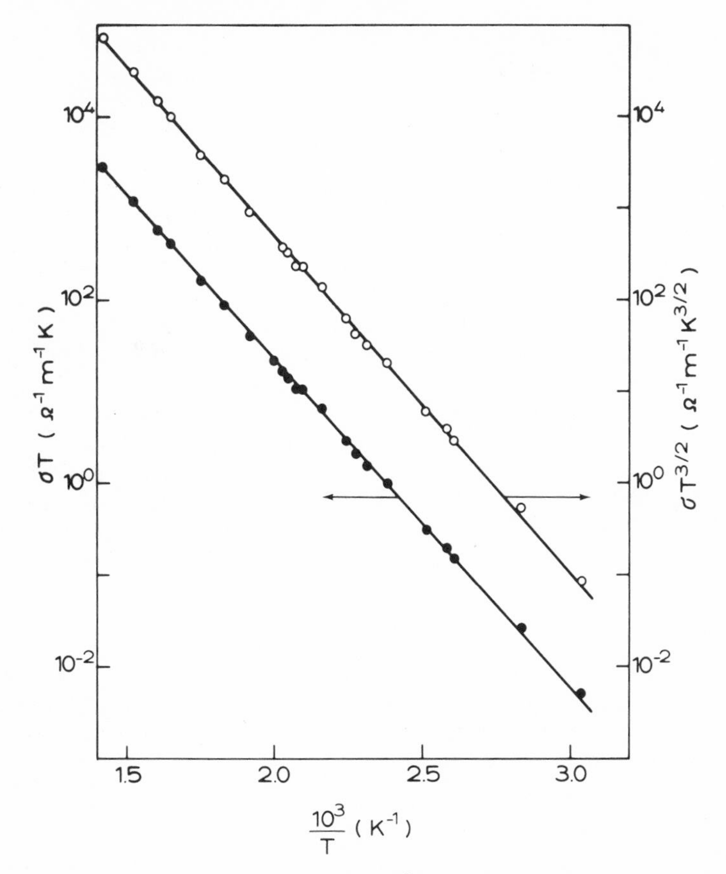

Fig. 8. Conductivity plotted as $\sigma T^{3/2}$ (upper line) and σT (lower line) versus 1/T for a $LiNbO_3$ single crystal reduced at 1010°C under an oxygen pressure of 10^{-13} P.

and adiabatic hopping. In Holstein's non-adiabatic theory, it is assumed that the charge carrier responds sufficiently slowly to the vibrational motion of the lattice so that it is not able to follow the lattice motion. In this case, the electron will not always tunnel to a neighboring site when two neighboring sites have reached equal energy at a given moment. In the non-adiabatic approximation the conductivity is expressed by a relation of the form (see eqn. 1):

$$\sigma \propto T^{-3/2} \exp\left(-\frac{W_H}{kT}\right)$$

Figure 7 presents such plots of $\sigma T^{3/2}$ versus the reciprocal temperature for the three $LiNbO_3$ crystals. They exhibit a good linear behavior, in accordance with the Holstein's relation. The activation energy deduced from the slope of the straight lines is found to be 1.18×10^{-19} J (0.74 eV).

In the adiabatic approximation, in which the charge carrier is able to adjust instantaneously to the vibrational motion, the conductivity is also thermally activated but possesses a temperature-dependent pre-exponential factor which is different from that of the non-adiabatic formula. This pre-exponential term varies as T^{-1} (see eqn. 2) rather than as $T^{-3/2}$, valid for the non-adiabatic case. In Fig. 8 the experimental conductivity data obtained on the crystal reduced in a 99% CO/1% CO_2 atmosphere are plotted in two different ways: the upper curve represents a $\sigma T^{3/2}$ vs 1/T plot, the lower curve a σT vs 1/T plot. The experimental data can be approximated by straight lines equally well in either one of both plots, so that the inclusion of the pre-exponential term does not help to make a distinction between the two regimes. The conductivity possesses such a high activation energy that the temperature-dependent pre-exponential factor does not play an important role in determining the overall temperature dependence. The activation energy calculated from the slope of the σT vs 1/T curve is 1.14×10^{-19} J (0.715 eV).

As indicated by Holstein[1] in his original paper, the non-adiabatic small-polaron theory may be applied under the condition that the electron transfer integral J lies below a maximum value given by:

$$J_{max} = \left(\frac{2W_H kT}{\pi}\right)^{\frac{1}{4}} \left(\frac{\hbar\omega_o}{\pi}\right)^{\frac{1}{2}} .$$

Here, ω_o is an average optical phonon frequency which can be estimated from the relationship $\hbar\omega_o = k\theta$. In the case of $LiNbO_3$ we have $\theta = 560$ K and $k\theta = 7.7 \times 10^{-21}$ J (0.048 eV). The use of an average value is a very rough approximation, especially if one takes into account that the lattice dynamics of $LiNbO_3$ are characterized by a rather large number of phonon branches. The activation energy for hopping W_H can be obtained from the difference in slopes between the conductivity and thermopower curves. This yields $W_H = (1.18 - 0.40) \times 10^{-19}$ J $= 0.78 \times 10^{-19}$ J (0.74 − 0.25 = 0.49 eV). At 500 K, the right-hand term $(2W_H kT/\pi)^{\frac{1}{4}} (\hbar\omega_o/\pi)^{\frac{1}{2}}$ takes the value 6.7×10^{-21} J (0.42 eV).

In order to estimate J we can use the experimental value of the Hall mobility and insert this in the appropriate expression of μ_H for the non-adiabatic case. In the preceding section, we have emphasized that the Hall mobility depends on the local geometry of the atomic sites. The theory leads to different results for a three- or four-site configuration. The hopping centers in reduced $LiNbO_3$ are niobium ions. In our experiments the electrons are

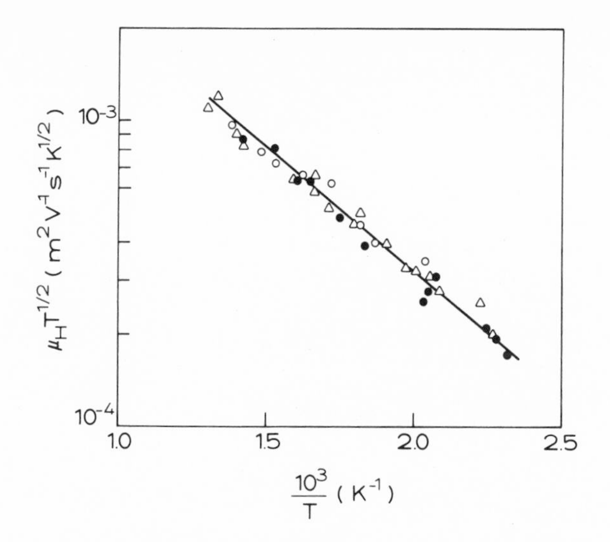

Fig. 9. Hall mobility plotted as $\mu_H\ T^{\frac{1}{2}}$ versus 1/T for the same $LiNbO_3$ single crystals as in Fig. 3.

driven in the direction of the c-axis under the influence of the applied magnetic field. One has to consider, therefore, the positions of the Nb^{5+} ions with respect to each other in the successive planes along this axis. An examination of the structure of $LiNbO_3$ (see ref. [9]) built up by hexagonally packed oxygen layers shows that the stacking sequence of the Nb^{5+} ions in these planes favors a transition via a triangular configuration.

It may be remembered that, according to Friedman and Holstein's theory, the Hall mobility in a hexagonal lattice is given by the expression:

$$\mu_H = \frac{ea^2}{\hbar} J \left(\frac{\pi}{16kTW_H}\right)^{\frac{1}{2}} \exp\left(-\frac{W_H}{3kT}\right)$$

where a is the hopping distance.

Figure 9 shows a plot of $\mu_H T^{\frac{1}{2}}$ versus the reciprocal temperature obtained with the experimental Hall mobility data of the three non-stoichiometric $LiNbO_3$ crystals. The activation energy calculated from the slope of the straight line is 2.6 x 10^{-20} J (0.16 eV), which, within the limits of accuracy of the measurements, is exactly equal to one-third of the activation energy associated with the drift mobility. Indeed, as mentioned earlier, W_H is found to be equal to 7.8 x 10^{-20} J (0.49 eV). The difference in activation energy between the drift and Hall mobility and more specifically the 1/3 ratio is one of the essential predictions of Friedman and Holstein's theory. The experimental finding of the mobility ratio lends considerable support for the applicability of the non-adiabati approximation.

The formula of the Hall mobility includes the hopping distance "a". In a regular hexagonal lattice, as considered in Holstein's theory, this distance corresponds to that of two neighboring atomic sites. In $LiNbO_3$ the shortest interatomic Nb-Nb distance is 3.765 Å but this value will not be used in the calculation for the following reason. The triangle, which most probably will be involved in the transition from the initial to the final state, is composed of Nb ions lying at unequal distances. In general, the transition probability is determined by the longest distance over which an electron has to jump. For electron transfer along the c-axis in $LiNbO_3$ this distance is of the order of 6 Å.

Substituting the experimentally found Hall mobility (μ_H = $1.6 \times 10^{-6}\ m^2V^{-1}s^{-1}$ at 500 K) into the above expression, one finds that the electron transfer integral J is equal to 4.6×10^{-21} J (0.029 eV). Hence, it follows that the criterion for the applicability of the non-adiabatic small-polaron theory, i.e.:

$$J < \left(\frac{2W_H kT}{\pi}\right)^{\frac{1}{4}} \left(\frac{\hbar\omega_o}{\pi}\right)^{\frac{1}{2}}$$

is fulfilled.

It may be remarked here that the Hall mobility formula used to calculate the transfer integral is a simplified expression which holds to a very good approximation for high temperatures ($kT \gg \hbar\omega_o$, i.e. for $T > 560$ K). A same calculation of the left and right-hand terms of the inequality at the highest temperature of the Hall mobility measurements (715 K) yields $J = 5.3 \times 10^{-21}$ J (0.033 eV) and $(2W_H kT/\pi)\ (\hbar\omega_o/\pi) = 7.4 \times 10^{-21}$ J (0.046 eV), which again satisfies the condition required for the validity of the non-adiabatic approach.

If the adiabatic model were applicable an estimate of the transfer integral could also be made using the experimental σ and S data. According to the theoretical predictions of the adiabatic approximation, the drift mobility is thermally activated with an activation energy equal to $W_H - J$. The Hall mobility possesses a smaller activation energy given by $W_H/3 - J$ and may, therefore, display a milder temperature dependence. From the slope of the σT vs 1/T plot (see Fig. 8) an activation energy of 1.14×10^{-19} J (0.715 eV) is deduced. After substracting the energy for carrier creation (4.0×10^{-20} J), $W_H - J$ is found equal to 7.4×10^{-20} J (0.465 eV). The Hall mobility, when plotted as log μ_H versus 1/T (valid for

J > kT) is characterized by an activation energy $W_H/3 - J$ = 2.1×10^{-20} J (0.13 eV). From these two values one obtains W_H = 8.0×10^{-20} J (0.50 eV) and $J = 5.6 \times 10^{-21}$ J (0.035 eV). For J < kT, μ_H varies as $T^{-1} \exp[-(W_H/3)kT]$. This case cannot be considered here since it would give a negative value for J.

In addition, assuming the adiabatic case to be valid, the transfer integral could also be estimated from the pre-exponential term of the Hall mobility formula (eqn. 7):

$$\mu_H = \frac{1}{2} ea^2 \left(\frac{\omega_o}{2\pi kT}\right) f(T) \exp\left[-\left(\frac{W_H}{3} - J\right)/kT\right]$$

where $f(T) = kT/J$ for J > kT, yielding a temperature-independent pre-exponent.

At 500 K, we obtain $J = 1.9 \times 10^{-21}$ J (0.012 eV), which is inconsistent with the condition J > kT and, moreover, gives a too small value to satisfy the inequality expressed above.

From the above discussion it may be concluded that the data obtained on $LiNbO_3$ are best interpreted on the basis of the non-adiabatic small-polaron theory. It can also directly be shown that the general condition for the existence of small polarons, i.e.:

$$J << 4 W_H$$

is fulfilled. Indeed, we found $J = 3.5 \times 10^{-21}$ J (0.029 eV) at 500 K and $W_H = 7.8 \times 10^{-20}$ J (0.49 eV).

We shall discuss now some other parameters relevant to small-polaron formation. In a first approximation the polaron hopping energy is half the polaron binding energy W_p:

$$W_H = \frac{1}{2} W_p$$

Using simple electrostatics, Mott[22] showed that in polar lattices W_p is given by:

$$W_p = \frac{e^2}{2\kappa_p r_p}$$

where r_p is the polaron radius and $\kappa_p^{-1} = (\kappa_\infty^{-1} - \kappa_o^{-1})$; here, κ_∞ and κ_o are the high frequency and static dielectric constants.

Bogomolov et al.[23] estimated the polaron radius to be:

$$r_p = \frac{1}{2} \left(\frac{\pi}{6} N\right)^{1/3}$$

where N is the number of sites per unit volume.

The high and low frequency dielectric constants of $LiNbO_3$ diffe largely: $\kappa_\infty = 80$ and $\kappa = 5.3$.[24,25] This gives $\kappa_p = 5.68$. The polaron radius calculated from the above formula is 1.5 Å, which yields $W_p = 1.34 \times 10^{-19}$ J (0.84 eV). This is in rather good agreement with the experimental value $W_p = 1.57 \times 10^{-19}$ J (0.98 eV), deduced from $W_p = 2W_H$.

The parameter which determines the strength of the electron-phonon interaction is the coupling constant γ. In Holstein's theory γ is defined as:

$$\gamma = \frac{W_p}{\hbar\omega_o}$$

From this relation we deduce that $\gamma = 20$. This very large value is a convincing argument that strong coupling theory applies.

We shall turn now to an evaluation of the drift mobility from the d.c. conductivity data. If one type of charge carrier dominates the transport, the conductivity is written as the produce of the carrier density n, the charge of the carrier q and the mobility μ_D : $\sigma = nq\mu_D$.

To estimate the carrier density two ways will be followed. A first method to separate mobility from carrier concentration is the use of the thermoelectric power. In general, the thermoelectric power for electrons is given by the expression:

$$S = -\frac{1}{qT} (E_F - AkT) \tag{9}$$

where AkT describes the transport of kinetic energy by a charge carrier. If the carriers are small polarons in the hopping regime, then the number of charge carriers can be found from:

$$b = N \exp\left(-\frac{E_F}{kT}\right) \tag{10}$$

where N is the density of sites per unit volume on which a carrier can sit. With the help of this equation the thermopower can be written as:

$$S = \frac{k}{e} \left(\frac{n}{N} + A\right) \tag{11}$$

The kinetic term has been discussed by a number of authors.[26-29] In the simple model in which the charge carrier hops between sites which interact under equal strength with the lattice, there is no transfer of vibrational energy associated with the small-polaron hop. Under this condition the transport term in eqn. (11) tends to zero and the thermopower is directly related to the carrier density.

Let us turn now to experimental situation in $LiNbO_3$ (see Fig. 6). The intercept of the thermopower at $1/T = 0$ is found to be equal to 765 $\mu V\ K^{-1}$, which yields $A = 9.2$. Emin[30] argued that there may be a number of more complicated models for which a non-zero transport term may arise. Values approaching 10 may be envisioned for A on this basis. Such a situation may occur e.g. when the effective electron-lattice interaction at the initial and final sites is not the same, so that the energy levels associated with these sites are unequal. In this case a fraction of the hopping activation energy, determined by the non-uniformity of induced polarization, will be transferred during a hop. When the disorder energy increases, the transport term tends to increase in magnitude and may become very large.

In order to calculate the density of carriers from eqn. (11) one has to know the number of available hopping sites "N". As discussed earlier, $LiNbO_3$ reduced in CO/CO_2 atmosphere contains oxygen vacancies and reduced Nb ions (Nb^{4+}). Conduction occurs by hopping of an electron between a Nb^{4+} ion and a Nb^{5+} ion. Hence, N may be taken equal to the number of remaining Nb^{5+} ions per unit volume, which gives $N = 1.87 \times 10^{28}\ m^{-3}$. The carrier density determined from the thermopower data is plotted in Fig. 10 as a function of the reciprocal temperature. In the temperature range of measurements, n varies from $2.5 \times 10^{24}\ m^{-3}$ at 325 K to $1 \times 10^{26}\ m^{-3}$ at 560 K. At 560 K, the highest temperature of the S measurements, the donor centers are not completely ionized. This is also evidenced by the behavior of the conductivity which in the $\sigma T^{3/2}$ vs $1/T$ plot exhibits a constant slope up to 715 K, the upper limit reached in the σ measurements. At 715 K, the carrier density will be equal to $3.2 \times 10^{26}\ m^{-3}$. Bergmann[14] measured the electrical conductivity of $LiNbO_3$ crystals reduced in different CO/CO_2 mixtures up to temperatures of about 1400 K. The data can be represented by straight lines with a slope of 1.15×10^{-19} J (0.77 eV) up to about 900 K. At this temperature the curves start deviating from the linear behavior, which may mark the onset of the exhaustion range. At 900 K, the carrier density in our sample will amount to approximately $7 \times 10^{26}\ m^{-3}$. From optical absorption measurements Clark et al.[11] estimated the number of Nb^{4+} ions to be of the order of 10^{27} ions m^{-3} in reduced $LiNbO_3$ crystals having a conductivity of 2×10^{-5} $\Omega^{-1}m^{-1}$ at room temperature. This result confirms the high number of donor centers predicted by our electrical measurements.

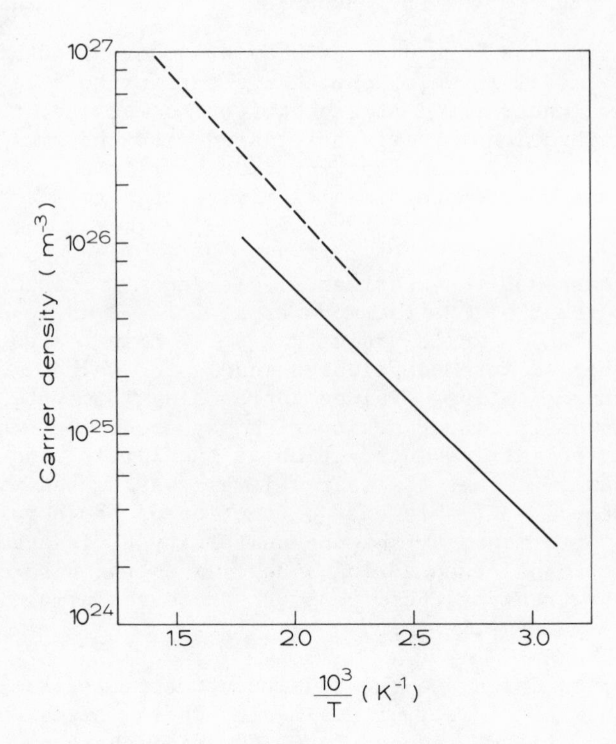

Fig. 10. Charge carrier density of $LiNbO_{3-y}$ versus 1/T, as calculated from the thermopower data (full line) and from the Hall coefficient data (dotted line) shown in Figs. 4 and 7.

A second way to determine the carrier density is to start from the Hall coefficient R_H, which in the non-adiabatic case and for a charge transfer involving three sites is given by (eqn. 4):

$$R_H = - \frac{1}{nqc} \frac{\mu_H}{\mu_D}$$

$$= - \frac{1}{nqc} \frac{kT}{3J} \exp \left(\frac{2}{3} \frac{W_H}{kT}\right)$$

With the experimental value of $R_H = 3.1 \times 10^{-5}$ Ωm T^{-1} (at 500 K) and taking $J = 4.8 \times 10^{-21}$ J (0.03 eV) (the value deduced from the Hall mobility) we find $n = 1.6 \times 10^{26}$ m^{-3} for the sample on which both the the thermopower and Hall coefficient were measured. At the same temperature the carrier density calculated from the thermopower amounts to 5.7×10^{25} m^{-3}, which is a factor of three lower. The agreement between both values is rather good. For comparison, the carrier density derived from the Hall coefficient data is also represented in Fig. 8 (dotted line).

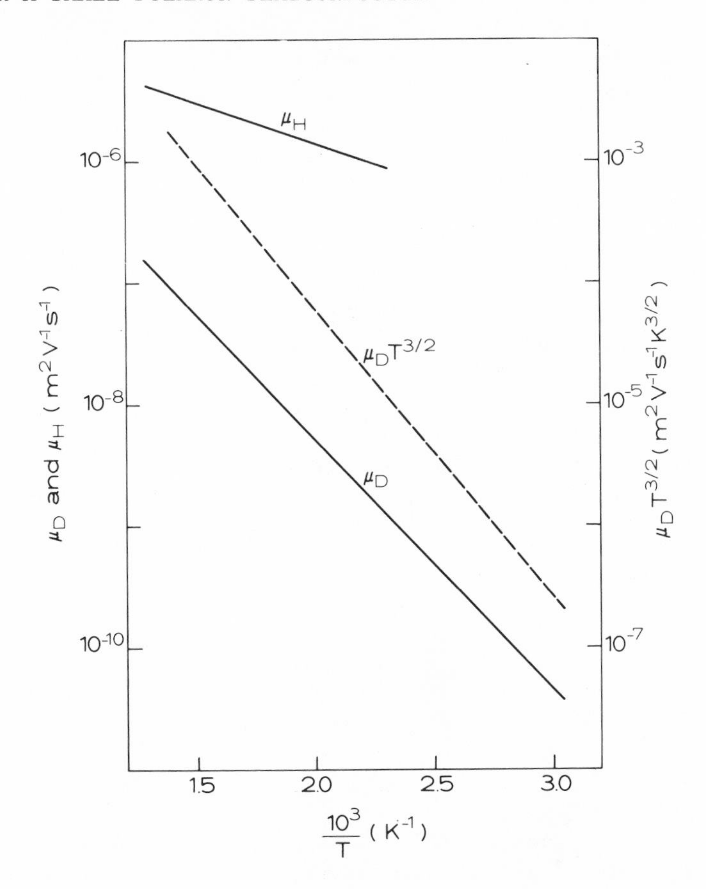

Fig. 11. Temperature dependence of the Hall mobility μ_H and the drift mobility μ_D for the same $LiNbO_3$ single crystals as in Fig. 3. The drift mobility is also plotted as $\mu_D\ T^{3/2}$ versus 1/T (dotted line).

We can proceed now to a calculation of the drift mobility by using the conductivity and carrier density data. In Fig. 11 the drift mobility, derived from the σ and S results, is shown as a function of the reciprocal temperature. The thermopower was taken as a measure for the charge carrier concentration, since it seemed to us a more reliable parameter than the Hall coefficient. In the same figure we have plotted the experimentally measured Hall mobility. At 500 K, the drift mobility is found to be 5.0×10^{-9} $m^2v^{-1}s^{-1}$, whereas the Hall mobility is much larger than μ_D and amounts to 1.4×10^{-6} $m^2v^{-1}s^{-1}$. This leads to a μ_H/μ_D ratio equal to 280. It may be recalled that the drift mobility in the non-adiabatic approximation is given by (eqn. 1):

$$\mu_D = \frac{3ea^2}{2\hbar}\frac{J^2}{kT}\left(\frac{\pi}{4W_H kT}\right)^{\frac{1}{2}} \exp\left(-\frac{W_H}{kT}\right)$$

Due to the temperature of the pre-exponential term one should have to plot log $\mu_D T^{3/2}$ versus 1/T, if one wants to obtain the correct value of the hopping energy W_H. Such a plot is also shown in Fig. 11.

As has been pointed out previously, the much higher value of μ_H compared to that of μ_D is one of the principal features of the non-adiabatic small-polaron theory for the case of an electron moving in a triangular array. Furthermore, the very low value of μ_D provides convincing evidence for the existence of small polarons in $LiNbO_3$. Indeed, the upper limit for the drift mobility in the uncorrelated small-polaron theory is given by:[31]

$$\mu_D << \frac{ea^2}{kT}(\Delta\nu)\left(\frac{kT}{2W_H}\right)^{2/3}$$

where $\Delta\nu$ is the dispersion of the optical phonons.

This inequality can be written as:

$$\mu_D << \left[\frac{ea^2}{h}\left(\frac{h\nu_o}{kT}\right)\right]\left[\left(\frac{\Delta\nu}{\nu_o}\right)\left(\frac{kT}{2W_H}\right)^{2/3}\right]$$

Chowdhury et al.[32] determined the dispersion of several phonon branches in $LiNbO_3$ by inelastic neutron scattering. From this work we can deduce that for the high-frequency optical mode at ν_o = 7.15 x 10^{12} Hz the ratio $\Delta\nu/\nu_o$ is of the order of 0.2. Taking a = 6 Å, the first term within the brackets is equal to about 10^{-4} $m^2v^{-1}s^{-1}$ at 400 K. The second term is of the order of 0.1. Then it follows that the drift mobility must be smaller than 10^{-5} $m^2v^{-1}s^{-1}$, condition which is amply fulfilled in the case of $LiNbO_3$.

We conclude that the results obtained on reduced $LiNbO_3$ can be interpreted in a straightforward way by the non-adiabatic small-polaron theory. In our opinion, this material provides the clearest case of small-polaron hopping conduction reported until now. The main freatures which support this view can be summarized as follows:

1. The drift mobility is very low and manifests a thermally-activated temperature dependence, in accord with Holstein's theory.
2. The Hall mobility is greater than the drift mobility and increases exponentially with temperature, the associated activation energy being exactly one-third of that of the drift

mobility.

3. The sign of both the thermopower and Hall coefficient is negative, indicating that electrons dominate the conduction process. According to Holstein's prediction, the sign of R_H is normal when electrons move in a triangular geometry.
4. The charge carrier density calculated from the thermopower and the Hall coefficient, using the appropriate formulae for hopping conduction, are in reasonable good agreement.

REFERENCES

1. D. Emin, Aspects of the Theory of Small Polarons in Disordered Materials, in: "Electronic and Structural Properties of Amorphous Semiconductors," P. G. Le Comber and J. Mort, eds., Academic Press, New York (1973), p. 261.
2. I. G. Austin and N. F. Mott, Adv. Phys. 18:41 (1969).
3. T. Holstein, Ann. Phys. (N.Y.) 8:343 (1954).
4. D. Emin and T. Holstein, Ann. Phys. (N.Y.) 53:439 (1969).
5. L. Friedman and T. Holstein, Ann. Phys. (N.Y.) 21:494 (1963).
6. D. Emin, Ann. Phys. (N.Y.) 64:336 (1971).
7. T. Holstein, Phil. Mag. 27:225 (1973).
8. D. Emin, Phys. Rev. B3:1321 (1971).
9. A. Räuber, Chemistry and Physics of Lithium Niobate, in: "Current Topics in Materials Science, Vol. 1," E. Kaldis, ed., North-Holland, Amsterdam (1978), p. 481.
10. K. Nassau and M. E. Lines, J. Appl. Phys. 41:533 (1970).
11. M. G. Clark, F. J. Disalvo, A. M. Glass and G. E. Peterson, J. Chem. Phys. 59:6029 (1973).
12. D. Redfield and W. J. Burke, J. Appl. Phys. 45:4566 (1974).
13. Z. I. Shapireo, S. A. Fedulov, Yu. N. Venevtsev and L. G. Rigerman, Sov. Phys. Cryst. 10:725 (1966).
14. G. Bergmann, Solid State Commun. 6:77 (1968).
15. P. Jorgensen and R. Barlett, J. Phys. Chem. Solids 30:2639 (1969).
16. W. Bollmann and M. Gernard, Phys. stat. sol. (a) 9:301 (1972).
17. W. Bollmann, Phys. stat. sol. (a) 40:83 (1977).
18. L. J. Van der Pauw, Philips Res. Rep. 13:1 (1958).
19. V. K. Lapshin and A. P. Rumyanzev, "Materialy vseroyusn. konf. fizika dielektr. i perspekt. ego rasvit," Leningrad (1973), Vol. 1, p. 145.
20. A. M. Glass and M. E. Lines, Phys. Rev. B13:180 (1976).
21. G. E. Peterson, J. R. Carruthers and A. Carnevale, J. Chem. Phys. 53:2436 (1976).
22. N. F. Mott, J. Non-Cryst. Solids 1:1 (1968).
23. V. N. Bogomolov, E. K. Kudinov and Yu. A. Firsov, Soviet Phys.-Solid State 9:2502 (1968).
24. K. Nassau, H. J. Levinstein and G. M. Loiacona, J. Phys. Chem. Solids 27:989 (1966).
25. E. Bernal, G. D. Chen and T. C. Lee, Phys. Letters 21:259 (1966).

26. G. L. Sewell, Phys. Rev. 129:597 (1963).
27. K. D. Schotte, Z. Phys. 196:393 (1966).
28. A. L. Efros, Soviet Phys.-Solid State 9:901 (1967).
29. M. I. Klinger, Rep. Progr. Phys. 30:225 (1968).
30. D. Emin, Electrical Transport in Semiconducting Non-Crystalline Solids, in: "Physics of Structurally Disordered Solids," S. S. Mitra, ed., Plenum Press, New York (1976), p. 461.
31. D. Emin, Phys. Rev. Letters 25:1751 (1970).
32. M. R. Chowdhury, G. E. Peckham, R. T. Ross and D. H. Saunderson, J. Phys. C.: Solid State Phys. 7:L99 (1974).

THE HALL EFFECT IN HOPPING CONDUCTION*

David Emin

Solid State Theory Division-5151
Sandia National Laboratories†
Albuquerque, New Mexico 87185

ABSTRACT

The Hall mobility associated with hopping motion differs qualitatively from that attributed to the notion of itinerant carriers. Particularly striking is the occurrence of Hall-effect sign anomalies: electrons and holes which circulate in a magnetic field in the sense of positive and negative charges, respectively. A concise overview of the theory of the small-polaron Hall effect is presented with emphasis being placed on the physical original of the small polaron's distinctive properties.

I. INTRODUCTION

The Hall effect associated with electronic hopping motion differs qualitatively from that which is associated with the quasifree motion of itinerant charges. A primary reason for the difference is that itinerant carriers are usually associated with deBroglie wavelengths far in excess of an interatomic separation. For this reason a charge in a high-mobility semiconductor, such as crystalline silicon, germanium, or gallium arsenide, may be viewed as free-electron-like and be treated within the effective-mass approximation. Then, as is well known, in a magnetic field electrons are deflected in the sense of negatively charged positively massed free particles while holes are deflected in the sense of positively charged and

*Supported by the U. S. Department of Energy, DOE, under Contract DE-AC04-76-DP00789.
†A U. S. Department of Energy facility.

positively massed free particles. Furthermore, within the effective-mass approximation the magnitudes of the respective responses of a charge to electric and magnetic fields, as measured by the drift and Hall mobilities, are nearly equal to one another. In other words, the Hall factor, the ratio of the Hall mobility to drift mobility is typically of the order of unity. It is for this reason that a Hall coefficient determination in tandem with a conductivity measurement can be utilized to estimate the quasifree carrier's mobility and density. However, in instances for which the effective-mass approach becomes inappropriate these well-known and oft-exploited results cannot be utilized.

In instances in which the effective-mass scheme cannot be adopted the Hall effect is more complicated yet more sensitive probe of systems wherein the local atomic geometry and the nature of the local electronic orbitals play significant roles. Two situations that lie outside the domain of the quasifree approach are 1) conduction involving energy bands which are narrower than the thermal energy, kT, and 2) situations which are best described as hopping conduction. In the first instance the quasifree picture fails because the deBroglie wavelength of a particle is reduced to atomic dimensions. In the second situation, rather than viewing a charge carrier as moving freely through a material where its motion is occasionally impeded by a scattering event, the charge is treated basically as remaining confined within some limited spatial region with its motion proceeding via spasmodic transfers between different locales. In contrast to the quasifree situation, in these instances the Hall and drift mobilities are often found to be quite different from one another. In fact one even finds circumstances in which the sign of the Hall effect is anomalous. That is, an excess electron can be deflected by a magnetic field in the sense of a positively charged and positively mass free particle. Analogously, the circulation of a hole in a magnetic field can take place in the sense of a free electron. The predominant theme is that when the coherent motion of a particle is destroyed its Hall effect may differ qualitatively from that of a free particle. Furthermore, in magnitude, temperature dependence, and even in sign the Hall mobility may be quite distinct from the drift mobility.

In this article attention will be primarily directed toward situations in which charge carriers hop (make phonon-assisted transitions) between severely localized states within which the charge is strongly coupled to the atomic motion. In these situations the hopping motion may be designated as being small-polaronic; the adjective "small" indicates a well localized charge and the term "polaronic" signifies that there is a substantial interaction of the carrier with the atomic displacements. The scheme of this article is as follows. In section II the basic notions of the Hall effect will be briefly reviewed and the principal results of the scattering studies enumerated. Section III provides a brief description of the physics of small-polaron hopping motion, with some additional commentary contrasting this strong-coupling situation to that of (weak-coupling) hopping conduction among shallow impurities. Section IV addresses the origin and the procedure for calculating the Hall effect in a hopping situation. The magnitude and temperature dependence of the Hall mobility are the subjects of Section V. The calculation of the sign of the Hall effect is then addressed in Section VI. The article concludes with a very brief synopsis of some experimental results.

II. SCATTERING APPROACH TO ITINERANT MOTION

The fundamental idea of a Hall effect experiment is to measure the steady-state deflection of a current passing through a material when a small homogeneous magnetic field is applied perpendicular to the current flow. A schematic representation of a Hall effect experiment is shown in Fig. 1. Here a slab of isotropic material is connected in a circuit such that a dc current density, j_x, flows in the x-direction under the influence of an external electric field, E_x, applied in the x-direction. In the steady state, with the application of the magnetic field, H, in the z-direction a transverse electric field, E_y, the Hall field, is established in the y-direction as a result of the current being constrained to flow in only the x-direction. In particular, for a single species of carrier the transverse current associated with the Hall field, $nq\mu_D E_y$, must offset the transverse component associated with the magnetic deflection of the current flowing in the x-direction, $nq\mu_H(-\mu_D E_x H)$, yielding $E_y/E_x H = \mu_H$, an expression for the Hall mobility. The quantity which is obtained in a Hall effect experiment is the Hall coefficient:

$$R_H \equiv E_y/j_x H = \mu_H/nq\mu_D. \tag{1}$$

With measurements of the conductivity, σ, and the Hall coefficient one obtains the Hall mobility.

$$\begin{aligned}\mu_H &= (nq\mu_D)R_H \\ &= \sigma R_H.\end{aligned} \tag{2}$$

The scattering approach to the calculation of the Hall mobility associated with the motion of carriers of a given energy band of a semiconductor involves simply solving the Boltzmann equation. One then finds that, for the arrangement depicted in Fig. 1, the Hall mobility for electrons ($q = -|e|$) is

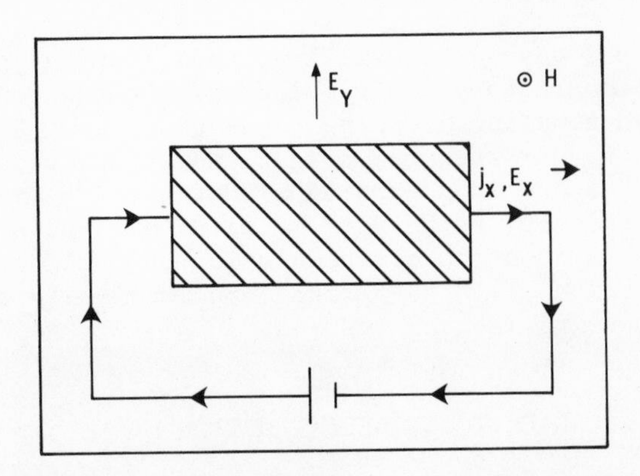

Fig. 1. A schematic representation of a Hall effect experiment. E_y is shown directed in the positive y-direction, as it is for positively charged and positively mass free particles. If the particles are negatively charged the direction of E_y is reversed. The magnetic field is directed upward out of the plane of the paper.

$$\mu_H = \frac{-|e| \sum_{\underset{\sim}{k}} v_y(\underset{\sim}{k}) \left| \frac{v_y(\underset{\sim}{k})}{M_{xx}(\underset{\sim}{k})} - \frac{v_x(\underset{\sim}{k})}{M_{xy}(\underset{\sim}{k})} \right| \tau^2(k) \exp[-E(\underset{\sim}{k})/\kappa T]}{\sum_{\underset{\sim}{k}} v_x^2(\underset{\sim}{k}) \tau(\underset{\sim}{k}) \exp[-E(\underset{\sim}{k})\kappa T]}. \quad (3)$$

where $v_x(\underset{\sim}{k})$ and $v_y(\underset{\sim}{k})$ are components of the velocity vector, $M_{xx}(\underset{\sim}{k})$ and $M_{xy}(\underset{\sim}{k})$ are components of the effective-mass tensor, $E(\underset{\sim}{k})$ is the carrier energy, and $\tau(\underset{\sim}{k})$ is the collision time, each quantity being associated with a state labeled by wavevector $\underset{\sim}{k}$. For hole motion in this band one must reverse the sign of both the charge carrier ($q = |e|$) and the electronic energy appearing explicitly and implicitly in the quantities of Eq. (2). When the width of the relevant energy band is very much greater than the thermal energy, κT, the effective-mass approximation is customarily employed; namely, $E(\underset{\sim}{k})$ is expanded about a band extremum to second order in the components of $\underset{\sim}{k}$. If, for simplicity, one also takes the scattering time to be independent of $\underset{\sim}{k}$ $[\tau(\underset{\sim}{k}) = \tau]$, then evaluation of Eq. (1) for an isotropic system yields

$$\mu_H = \frac{q\tau}{|M^*|} = \mu_D, \quad (4)$$

where M^* is the diagonal element of the effective-mass tensor evaluated at the appropriate band extremum. If the restriction that $\tau(\underset{\sim}{k})$ be a constant is removed the Hall and drift mobilities generally only differ from one another by a factor of the order of unity which depends upon the details of the scattering, i.e., the Hall factor. Thus when the transport primarily involves states for which the effective-mass approximation is valid the Hall and drift mobilities are essentially equal to one another. The sign of the Hall coefficient then provides the sign of the carrier, and the magnitude of the Hall coefficient provides a direct estimate of the carrier density: $n \approx 1/qR_H$.

The complementary situation occurs in the case of transport in bands which are narrow compared with the thermal energy. In this instance, states throughout the energy band have a significant probability of occupancy. Concomitantly the deBroglie wavelength is comparable to an interatomic intercell separation. The Hall mobility will then depend on the values of the velocity vector, effective-mass tensor, energy, and scattering time throughout the entire band. The Hall mobility may therefore differ substantially from the drift mobility. Indeed, since the signs of the terms in the $\underset{\sim}{k}$-summation of the numerator of Eq. (3) are not uniformly positive one sees that the sign of the Hall coefficient need not be identical with the sign of the carrier. The sign of the Hall mobility depends on the dispersion relation, $E(\underset{\sim}{k})$, and on the weighting assigned to different states within the $\underset{\sim}{k}$-summation by the scattering time, $\tau(\underset{\sim}{k})$. It is important to note that the dispersion relation (and the quantities derived from it) depend on the structure of the material and on the nature and relative orientations of the local orbitals between which the carrier transfers. Hence beyond the effective-mass situation the Hall coefficient generally depends in sign as well as in magnitude and temperature dependence on the details of the system under study. These features have been explicitly demonstrated for transport within, narrow bands in the constant collision-time approximation, $\tau(\underset{\sim}{k}) = \tau$. The results are quoted and discussed in Ref. (2).

One aspect of the narrow-band results is however especially worthy of note. This comment concerns those situations for which evaluation of the numerator of Eq. (3) in the narrow-band limit yields an expression proportional to an odd number of transfer integrals. In these instances the Hall mobility associated with an electron added to an otherwise empty band is identical in magnitude, temperature dependence and <u>sign</u> to that of the band when it is shy one electron of being filled—the band contains a single "hole." The equality of sign arises because the transformation from a representation in terms of electrons to a representation in terms of holes involves reversals of both the sign of the charge ($q \rightarrow -q$) and the signs of the transfer integrals.[3] Thus when the numerator of Eq. (3) is proportional to an odd number of transfer integrals in addition to the charge, the transformation between the excess-electron and excess-hole situations involves an even number of sign changes — there is no net change of sign.

III. SMALL-POLARON HOPPING CONDUCTION

In some situations it is appropriate to view charge transport in terms of hopping motion rather than within a simple scattering picture. In particular, one typically employs this point of view in describing both (1) the tunneling of an electronic carrier among a sufficiently low density of unoccupied defect states and (2) the intrinsic tunneling of a small polaron. In the latter situation, the transfer of the carrier occurs primarily between adjacent orbitals.

Since the strength of the interaction of a carrier with the atomic vibrations varies inversely as some model-dependent power of the characteristic radius of the electronic state,[4-6] small polarons and carriers residing in deep well-localized defects generally interact strongly with the atomic vibrations. In other words, severely localized states are strongly coupled to the atomic displacements while shallow (weakly localized) states are only weakly coupled to the atomic vibrations. In the case of a strongly coupled system the alterations of the equilibrium positions of the atoms surrounding the carrier are very large compared with the amplitude for zero-point atomic motions. For weakly coupled states the converse is true. Charge transport involving at least one state which is strongly coupled to the lattice is qualitatively different from that of the weak-coupling situation. In the strong-coupling case transport typically involves the emission and absorption of many phonons, while in weak-coupling circumstances processes involving a minimal number of phonons predominate.[5,6]

The muliphonon hopping that is characteristic of hopping in the strong-coupling regime can be described in terms of tunneling of both the electronic carrier and the atomic displacement pattern associated with the carrier's occupancy of a given site. To understand this, first note that associated with the carrier occupying a site the equilibrium positions of the atoms about it are displaced. Thus even for a carrier to move between geometrically equivalent sites the atoms about both the initial and final sites of the transition must alter their positions. At the lowest of temperatures, where the thermal agitation is minimal, this typically involves tunneling of the atoms of the solid between positions associated with the carrier's occupancy of the initial site to positions appropriate to the carrier's occupancy of the final site. Thus, at low temperatures, in addition to the tunneling of the electronic carrier between initial and final sites, there is tunneling of the atoms of the solid. This tunneling is associated with the overlap of the atomic wavefunctions of initial and final atomic configurations. This overlap decreases as the electron-lattice coupling strength increases, since the disparity between initial and final atomic configurations concomitantly increases. Thus, for a strong-coupled situation the net hopping rate can be extremely

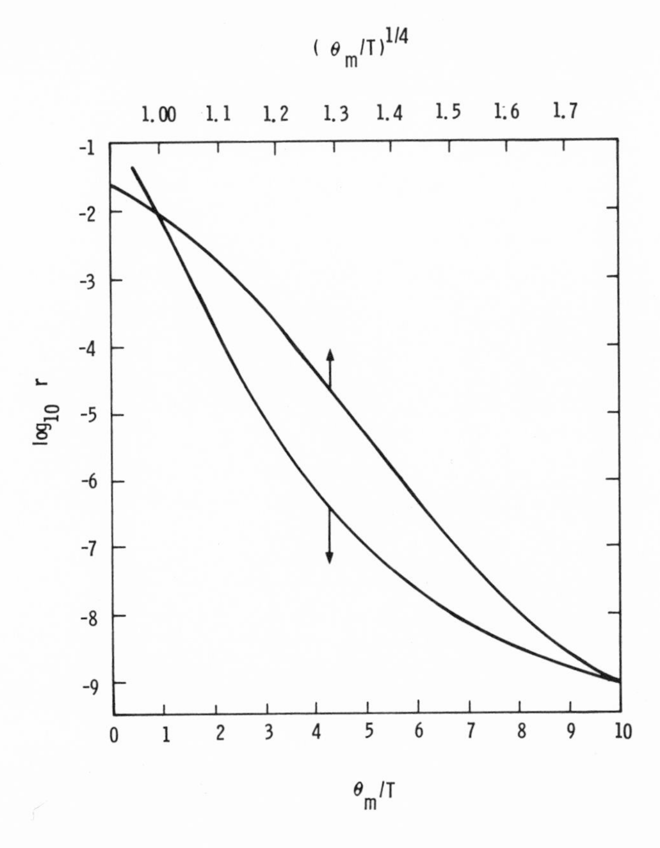

Fig. 2. The logarithm of the small-polaron jump rate is plotted against reciprocal temperature in units of the temperature corresponding to the maximum-energy phonon with which the carrier interacts appreciably, θ_m. The curve is also plotted against the fourth root of θ_m/T. See Ref. (6) for details.

small at low temperatures.

With increased temperature there is an increased probability that the atoms will assume configurations for which there is larger overlap between atomic configurations associated with the electronic carrier respectively occupying initial and final sites. Thus, as illustrated in Fig. 2, the phonon-assisted hopping rate rises with temperature as the atomic mismatch associated with the predominate atomic tunneling process is reduced. Furthermore, the non-Arrhenius behavior shown in Fig. 2 results when there is a distribution of suitable atomic configurations of sufficiently different energies so that they become thermally accessible over a range of temperatures.[6]

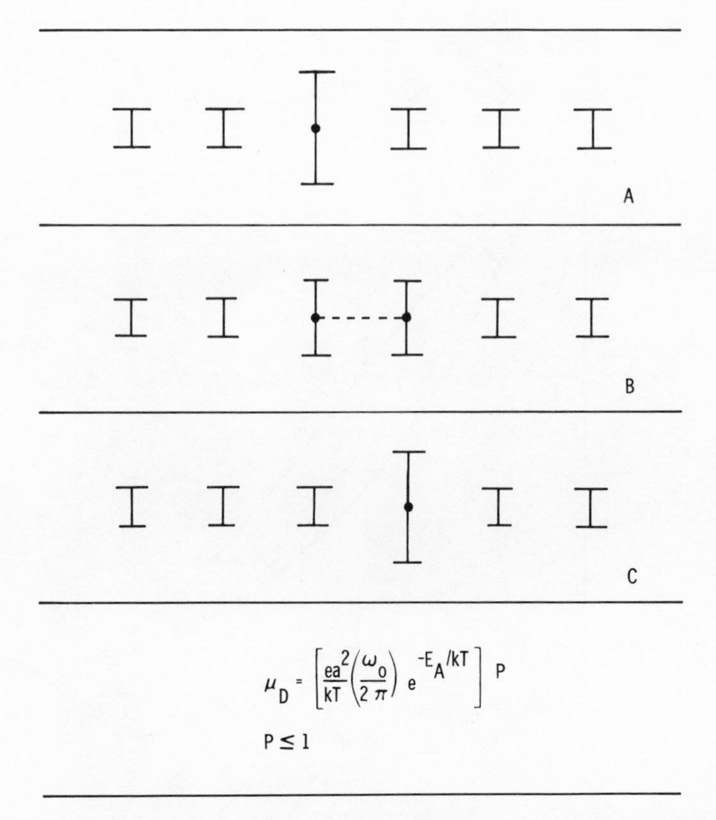

Fig. 3. A schematic representation of the small-polaron jump process in the high-temperature semiclassical regime. Nonessential "thermal" displacements as well as vibrational dispersion effects are suppressed for clarity.

At sufficiently high temperatures the predominate hopping transitions occur when the atoms assume positions which are consistent with the carrier being shared between initial and final sites.[7] Then there is a momentary degeneracy of the electronic energy levels of initial and final sites. In such a configuration, termed a coincidence configuration, the electronic charge can move between sites without any atomic tunneling. The hopping rate is activated in this high-temperature regime. The activation energy, E_A, is simply the minimum strain energy which must be supplied to the equilibrated system to form a coincidence configuration. In Fig. 3 the three states of the jump process are illustrated for a six-site line of deformable molecules. Here each molecule is characterized by a single deformation coordinate of vibration frequency ω_o which is represented in the figure by the length of the vertical line associated with each molecule. As illustrated, at equilibrium an occupied molecule, signified by the presence of the dot, is expanded. A hop occurs when, as represented in Fig. 3B, the molecular deformation is such that the carrier can equally reside on either of two sites. The hopping rate is given by the rate at which such coincidences occur,

$$R_{co} = (\omega_o/2\pi)\exp(-E_A/\kappa T), \tag{5}$$

multiplied by the probability, P, that the carrier will avail itself of the opportunity to hop when a coincidence occurs. The factor P depends on the product of the "duration of the coincidence" and the transfer rate for the electronic carrier between degenerate states. If this product is sufficiently large (much greater than unity) the carrier can adiabatically adjust to the atomic motion; P is then essentially unity and the hops are termed adiabatic. Otherwise P is much less than unity, the carrier cannot always respond to the establishment of a coincidence, and the hopping is termed nonadiabatic. The formula of Fig. 3 results from relating the jump rate, R, to the diffusion constant, D, via the relation

$$D = Ra^2, \tag{6}$$

where a is the intersite separation, and then relating the diffusion constant to the drift mobility using the Einstein relation,

$$\mu_D = (e/\kappa T)D. \tag{7}$$

In the weak-coupling situation, as for hopping between shallow impurities, the results are quite different.[4] Multiphonon hopping is improbable.[6] Then at low (cryogenic) temperatures an individual hop is simply activated with an energy equal to the difference between the site energies of final and initial sites for a hop upward in energy.[4,6] The individual hop downward in energy is temperature independent. At higher temperatures the hopping rate rises with temperature in a faster than simply activated manner.[5]

IV. THE ORIGIN OF THE HALL EFFECT

In considering the Hall effect associated with hopping transport, one is tempted to conclude that there is either no Hall effect or that it is exceptionally small. The first idea results from viewing the hopping process as simply a two-site problem. As such there is no clear mechanism by which an applied magnetic field can affect the direction of current resulting from intersite hopping motion. Indeed, as will be discussed in this section, it is essential in addressing the Hall effect to consider elemental jump processes which involve more than two sites. However, the involvement of higher order (more than two-site) jump processes does not

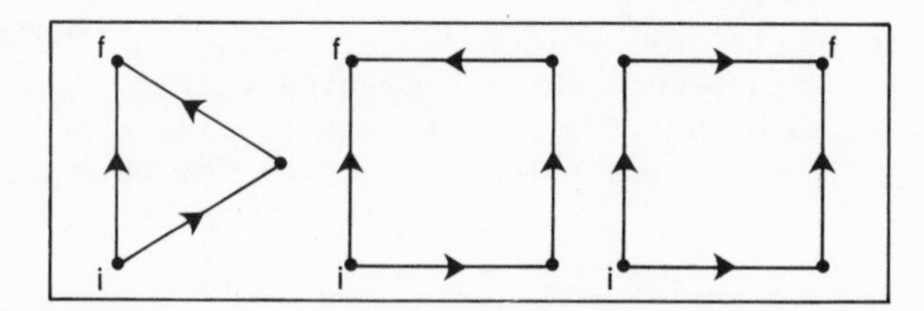

Fig. 4. Some transfer processes which link initial (i) and final (f) sites and interfere to give rise to a Hall effect are illustrated for triangular and square geometrics.

necessarily lead to small Hall mobilities. Indeed, the converse is usually the case; the Hall mobility due to hopping motion is usually very much larger than the drift mobility. In this section the processes which produce a Hall effect in hopping situations will be succinctly described.

With the application of a magnetic field the local atomic-like orbitals in a tight-binding representation garner magnetic-field dependences which, to terms linear in the magnetic field strength, may be simply expressed as additional phase factors. As a result the transfer integral linking any pair of sites then also contains a magnetic-field dependent phase factor.[8,9] Since the elementary hopping rate associated with only a pair of sites involves the absolute square of the relevant transfer integral no linear magnetic-field dependence of the jump rate results. However, if one considers higher order contributions to the jump rate which involve additional sites a linear magnetic-field dependence of the jump rate results from interference between different transfer paths. The diagrams at the left and center of Fig. 4 illustrate this situation for three- and four-site processes respectively.[9,10] If the local orbitals are s-like the Hall mobility associated with a given process is simply proportional to the area subtended by the (three or more) sites involved in the two interfering processes. When the orbitals are not s-like the pertinent area may differ substantially from the subtended area.[11]

Studies of the small-polaron Hall effect have centered on the semiclassical regime where the predominate charge transport takes place without atomic tunneling. Here, as discussed in the previous section, an elementary hop is described in terms of the occurrence of coincidence events during which a carrier can equally well reside at either of two sites. For the Hall effect in the nonadiabatic semiclassical regime it has been demonstrated that a triple coincidence is required.[9,10] Most generally the carrier must be presented with a choice of final sites in order to have its direction of motion influenced by the magnetic field.[12] Since the Hall mobility measures the carrier's response to a unit Lorentz force (proportional to μ_D) the Hall mobility involves the ratio of the magnetic-field dependent and independent portions of the total jump rate. Thus the activation energy of the Hall mobility is the difference of the energy associated with a triple coincidence, ε_3, and that due to a two-site coincidence, ε_2:[9-12]

$$\mu_H \propto e^{-(\varepsilon_3-\varepsilon_2)/\kappa T}, \tag{8}$$

where generally $\varepsilon_3 \geq \varepsilon_2$.[12] In a special model, the narrow-band limit of the molecular crystal model, one finds that $\varepsilon_3 = 2\varepsilon_2/3$ so that that $E_A^{Hall} = \varepsilon_3-\varepsilon_2 = \varepsilon_2/3 = E_A/3$.[9] However it must be emphasized that the simple relationship, $E_A^{Hall} = E_A/3$ applies to a specific model and is not a general feature of small-polaronic hopping.

V. HALL MOBILITY TEMPERATURE DEPENDENCE

Studies of the small-polaron Hall mobility have been carried out within a semiclassical perturbative approach (corresponding to nonadiabatic hopping motion) for both triangular[9] and square lattice structures.[10] In addition, the Hall effect of the triangular lattice has been calculated within the adiabatic approach.[12] In these calculations only those interference processes represented in Fig. 4, involving the minimum number of sites commensurate with the lattice structure, are considered. All of the calculations indicate that the Hall mobility posseses a smaller activation energy than the drift mobility. In Fig. 5 three pairs of plots of the drift and Hall mobilities against reciprocal temperature are displayed. It is seen that for some choices of the parameters the Hall mobility may even decrease with increasing temperature while the drift mobility remains essentially activated.

The tendency of the Hall mobility to fall with increasing temperature at sufficiently high temperature is related to the temperature dependence of the prefactor of the Hall mobility expression.[11,12] As the number of members of the basic loop is increased the prefactor garners additional factors of $|J|/\kappa T$, where J is the intersite transfer integral.[2,10] These factors arise when those sites for indirect transfer between pairs of (the three) coincident sites. Thus, as shown in Fig. 6, even for a sizeable drift mobility activation energy, $E_A = 0.3$ eV (with $E_A^{Hall} = 0.1$ eV) the Hall mobility will display a rather mild temperature dependence at high temperatures if the number of elements in the basic loop, n, is sufficiently large. Obviously the significance of this effect decreases as the magnitude of the Hall-mobility activation energy increases.

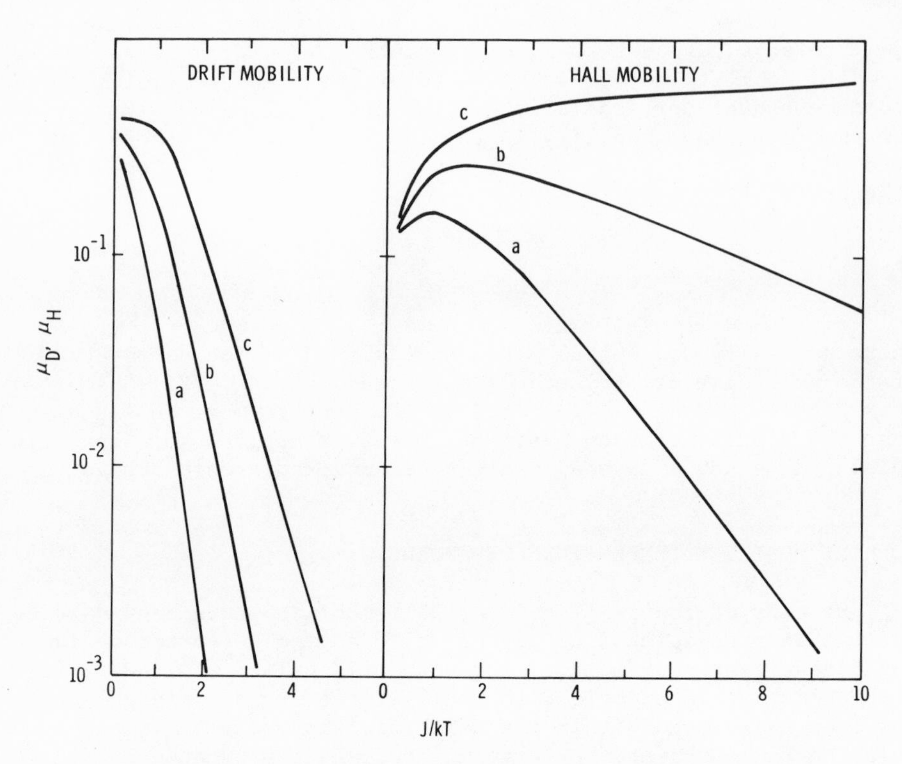

Fig. 5. The absolute values of the Hall and drift mobilities associated with the adiabatic hopping of an electron-like small polaron within a triangular lattice structure are plotted versus the ratio of the magnitude of the transfer integral, J, to the thermal energy. See Ref. (12).

VI. SIGN OF THE HALL EFFECT

The most significant aspect of the occurrence of Hall-effect sign anomalies is that their occurrence inidcates that the predominate mobile charges are not itinerant free-electron-like carriers.[2,11] In particular, Hall-effect sign anomalies can result when the predominate transport is of the hopping variety. As described previously, in the hopping situation the Hall effect arises from interference processes involving the transfer of a carrier between pairs of sites of a closed path which includes the initial and final sites. As a result the sign of the Hall effect depends upon the product of electronic transfer integrals which link local orbitals about the closed loop. In particular, with the major effect of the magnetic field being contained in the Peierls contribution to the magnetic phase factors, the sign of the Hall coefficient is given by[11]

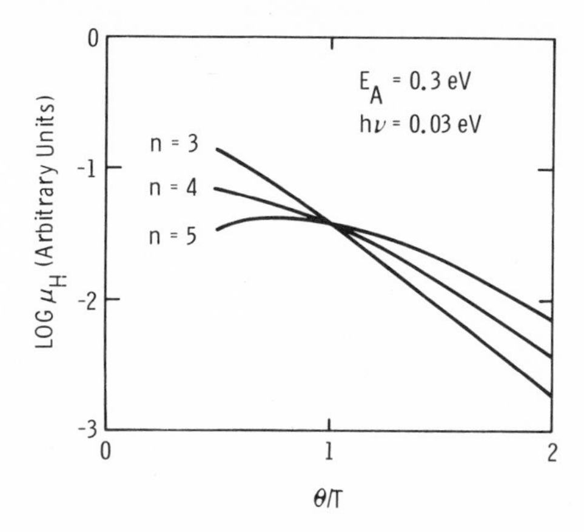

Fig. 6. The logarithm of the small-polaron Hall mobility is plotted against reciprocal temperature measured in units of the phonon temperature, θ, for interference processes involving three, four and five sites. The parameters are chosen to represent the situation in amorphous arsenic.

$$\mathrm{sign}(R_H) = \mathrm{sign}(\varepsilon^{n+1} \prod_{j=1}^{n} J^i_{j,j+1})$$

where $\varepsilon < 0$ is the carrier is an electron, $\varepsilon > 0$ if the carrier is a "hole", and $J^i_{j,j+1}$ is the field-free electronic transfer integral of the ith band linking sites j and j+1; here the site labeled as n+1 refers to the same site as that labeled by 1.

For processes involving an odd number of orbitals (n being odd) the sign of the Hall coefficient depends only on the sign of the product of transfer integrals. This is because, in this circumstance, the factor sign (ε^{n+1}) is always positive. If however the number of orbitals is even with all the transfers being equivalent to one another, the sign of the Hall coefficient is always normal:negative for electrons and positive for holes. Thus the Hall-effect sign anomalies are associated with geometries for which the primary interference processes involve odd numbers of sites. Furthermore the sign of the Hall coefficient for odd-membered rings is determined by the sign of the product of transfer integrals which in turn depend upon the symmetry and relative orientations of the orbitals between which the carrier moves.[11]

To illustrate this aspect of the theory consider the examples of Fig. 7.[2,11] In subfigure 7a a loop composed of three antibonding orbitals (centered at the three solid dots), each linking a pair of atoms (dark squares), is depicted. Presuming that the energy which enters into the transfer integral is (as is typical) negative, each

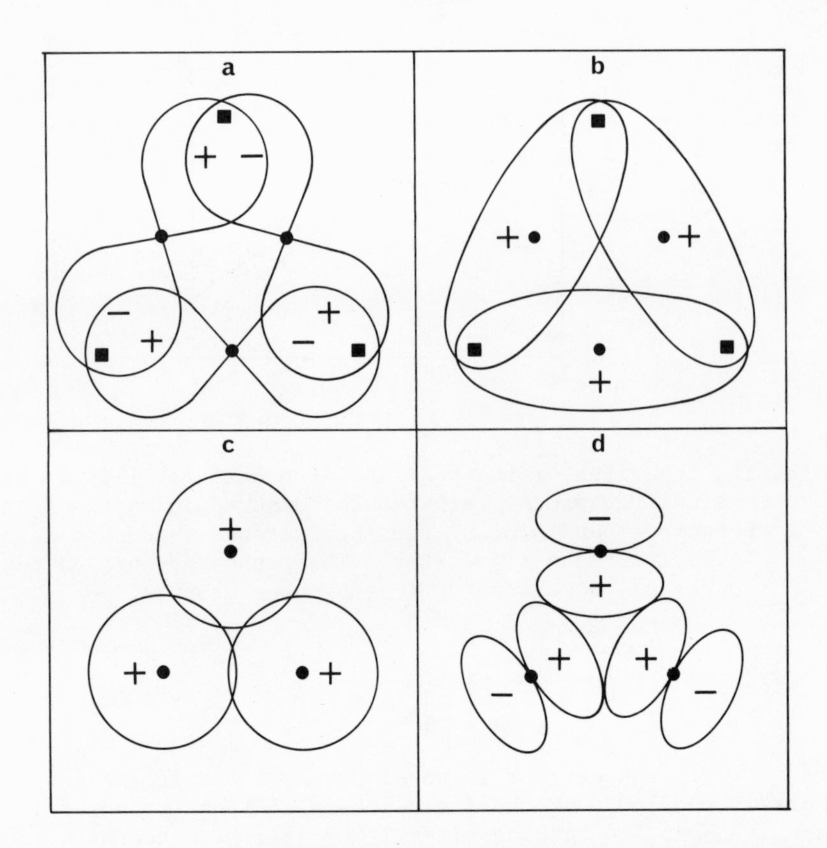

Fig. 7. A schematic representation of the orbitals involved in the examples cited in the text. The positive and negative signs denote particular (arbitrary) choices of the phases of the local orbitals. The solid dots indicate the centroids of the orbitals between which the carrier transfers and the solid squares denote the positions of the atoms of the solid in cases (a and b) in which they differ from the orbital centroids.

of the three transfer integrals associated with the overlap of the closest lobes of adjacent antibonding orbitals is positive. Then the product of the three transfer integrals about the closed loop is also positive. This means that electrons moving in a structure composed of odd-membered rings of antibonding orbitals will give rise to a positive Hall coefficient. In subfigure 7b, three bonding orbitals of a three-membered loop are shown. Here the transfer integrals and their product are negative. Thus the hopping motion of a carrier (a hole) between the bonding orbitals of an odd-membered structure generates a negative Hall coefficient. In the third portion of the figure three s-like donor orbitals are shown. Here, as in the case of the bonding orbitals, the Hall coefficient is negative. Thus the Hall effect associated with the hopping of donor electrons is expected to be negative, independent of the degree of occupancy of the donor band. Finally, in Fig. 7d a trifurcate arrangement of directed orbitals is illustrated. These can correspond to antibonding orbitals emanating from a single site or a trifurcate arrangement of p-orbitals at the base of a pyramidal structure. Presuming, as above, the usual case to prevail, with the transfer being dominated by the overlap of adjacent lobes of adjacent orbitals, the Hall coefficient will be negative.

Thus, Hall-effect sign anomalies can arise in instances of hopping conduction. The essential feature is that in these instances the carrier motion is not free-electron-like. In detail the presence of the sign anomalies depends on the atomic structure as well as on the symmetry and relative orientations of the orbitals between which the carrier transfers.

VII. SOME EXPERIMENTAL RESULTS

Measurements of the Hall effect associated with hopping motion have been almost exclusively carried out in situations in which the hopping is of the strong-coupling small-polaronic type. This has occurred despite the fact that the rather high resistivities which frequently characterize materials in which small-polarons are formed tend to make the measurements difficult to perform. Nonetheless there is a growing list of materials in which the small-polaron Hall effect has been observed. Since "Experimental Hall Effect Data for Small-polaron Semiconductors" is the subject of an article by P. Nagels in this volume, only a few general comments on this topic will be included in the present discussion.

At present there are two classes of materials for which there are extensive measurements which may be viewed as evidence of the small-polaron Hall effect. They are narrow-band-oxide crystals, of which MnO,[13] $Nb_{1-x}V_xO_2$[14] and UO_2[15] are examples, and various covalent noncrystalline solids: numerous chalcogenide glasses,[16-21] and amorphous silicon,[22] germanium,[23] and arsenic.[24] In all of these cases interpretation of other transport data supports the view of a small-polaron drift mobility, with a magnitude and temperature dependence that is in accord with the discussion of Sec. III. In these cases the Hall mobility is typically suitably small, much less than 1 cm^2/V-sec, and generally possesses a milder temperature dependence than the drift mobility.

In the simplest of models, the molecular crystal model (which treats intermolecular hopping with the presumption that a carrier only interacts with the molecule upon which it resides) in the limit of vanishing intermolecular electronic transfer, the

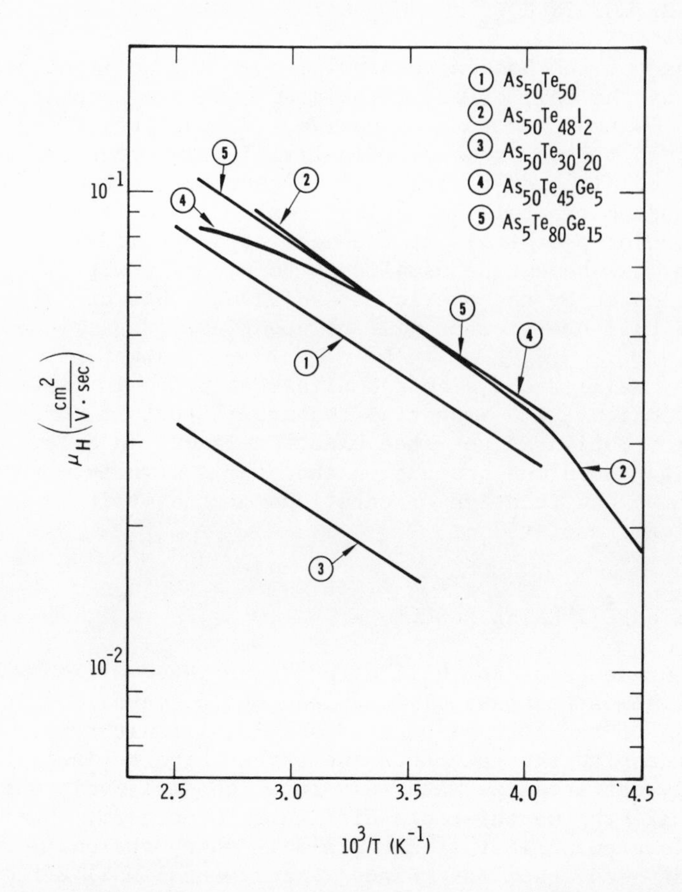

Fig. 8. The Hall mobility is plotted against reciprocal temperature for some representative chalcogenide glasses. See Ref. (16)

Hall-mobility activation energy is one-third that of the drift mobility.[9,10] Such a relationship is often observed.[14,16-21] However, it must be emphasized that beyond this oversimplified model the temperature dependences of the Hall and drift mobilities are usually more complicated.[12] Generally the temperature dependence of the Hall mobility depends upon (1) the range of the electron-lattice interaction, (2) the magnitude of the electronic transfer integrals, and (3) the nature of the local geometry. As such it is often inappropriate to emphasize agreement or disagreement with detailed aspects of the model Hall-effect calculations as being significant. What is of importance is the magnitude of the Hall mobility and the tendency of the Hall mobility to increase with temperature, albeit sometimes very mildly.

In noncrystalline solids the short-range order is often such as to yield Hall effect sign anomalies for small-polaron hopping.[2]

Appropriate sign anomalies have been observed in numerous chalcogenide glasses[6-21] as well as in amorphous silicon,[22] germanium,[23] and arsenic.[24] In addition, the magnitudes and temperature dependences of the measured Hall mobilites are consistent with the predictions of the small-polaron theory. As examples, in Fig. 8 the Hall mobilities of several chalcogenide glasses are plotted against reciprocal temperature. Here, as in crystalline small-polaronic solids, other transport measurements are also in accord with a small-polaron interpretation.[2] It is crucial to note that the observation of the Hall effect sign anomalies belie interpretation of the transport in terms of high-mobility (free-carrier-like) extended-state motion.

VIII. SUMMARY

The Hall mobility associated with small-polaron hopping motion is qualitatively different from that associated with itinerant carrier motion. Namely, the Hall mobility is exceptionally low, $\ll 1\ cm^2/V\text{-sec}$, and tends to increase with increasing temperature. Strikingly, its sign can be anomalous for appropriate arrangements of hopping sites. Unlike the situation for itinerant carriers, the mobility determined from a Hall-effect experiment usually differs substantially from the mobility which enters into the conductivity. In fact, the sign of the Hall mobility depends on both the relative positions and the nature of the orbitals between which the localized carriers move. These departures from the free-carrier-like behavior reflect the fact that the deBroglie wavelength of the carriers is comparable to an interatomic separation.

REFERENCES

1. L. Friedman, Phys. Rev. 131:2445 (1963).
2. D. Emin in: "Amorphous and Liquid Semiconductors," W. E. Spear, ed., University of Edinburgh (1977), p. 249.
3. T. Holstein, Phil. Mag. 27:225 (1973).
4. A. Miller and E. Abrahams, Phys. Rev. 120:745 (1960).
5. D. Emin, Adv. Phys. 24:305 (1975).
6. D. Emin, Phys. Rev. Lett. 32:303 (1974).
7. T. Holstein, Ann. Phys. (N.Y.) 8:343 (1959).
8. T. Holstein, Phys. Rev. 124:1329 (1961).
9. L. Friedman and T. Holstein, Ann. Phys. (N.Y.) 21:494 (1963).
10. D. Emin, Ann. Phys. (N.Y.) 64:336 (1971).
11. D. Emin, Phil. Mag. 35:1188 (1977).
12. D. Emin and T. Holstein, Ann. Phys. (N.Y.) 53:439 (1969).
13. C. Crevecoeur and H. J. de Wit, J. Phys. Chem. Solids 31:783 (1970).
14. I. K. Kristensen, J. Appl. Phys. 40:4992 (1970).
15. P. Nagels, private communication.

16. D. Emin, C. H. Seager and R. K. Quinn, Phys. Rev. Lett. 28:813 (1972).
17. A. J. Grant, T. D. Moustakas, T. Penny and K. Weiser, in: "Amorphous and Liquid Semiconductors," J. Stuke and W. Brenig, eds., Taylor and Francis, London (1974), p. 325.
18. M. Roilos and E. Mytilineou, Ref. (17), p. 319.
19. G. R. Klaffe and C. Wood, in: "Proc. Int. Conf. Phys. of Semiconductors," Int. Union Pure and Aplly. Phys., London (1976).
20. P. Nagels, R. Callaerts and M. Benayer, Ref. (17), p. 867.
21. W. F. Peck and J. F. Dewald, J. Electrochem. Soc. 111:561 (1964).
22. P. G. LeComber, D. J. Jones and W. E. Spear, Phil. Mag. 35:1173 (1977).
23. E. Mytilineou and E. A. Davis, Ref. (2), p. 632.

HALL EFFECT AND THE BEAUTY AND CHALLENGES OF SCIENCE

A. C. Beer

Battelle Columbus Laboratories
505 King Avenue
Columbus, Ohio 43201

I. INTRODUCTION

It is a great honor to be asked to participate in this commemorative symposium on the centennial of the discovery of the Hall effect here at Johns Hopkins. The Hall effect is a superb example of a phenomenon which, while simple in initial concept, is really very profound in its implications and application to a variety of solids of inherently differing characteristics. The evolution of the versatility of Hall data in interpreting complex transport in solids is indeed a classic example of the beauty and challenges of science. Let us try to follow some of the major developments in this progression of theoretical and experimental achievements.

II. LORENTZ FORCE

Fundamentally the Hall effect is a manifestation of the Lorentz force on an ensemble of charged particles constrained to move in a given direction and subjected to a transverse magnetic field. The physics involved is illustrated in Figure 1 by consideration of a confined stream of free particles, each having a charge e and a steady-state velocity ν_x due to the impressed electric field E_x. A magnetic field in the z direction produces a force F_y in the y direction in accordance with Eq.(1), below, which gives the force on a charged particle subjected to an electric field $\vec{E}$ and a magnetic field $\vec{H}$:

$$\vec{F} = e\,[\vec{E} + (1/c)\vec{v} \times \vec{H}] \text{ (Gaussian units).} \tag{1}$$

However, the boundary condition imposed by the orientation of the

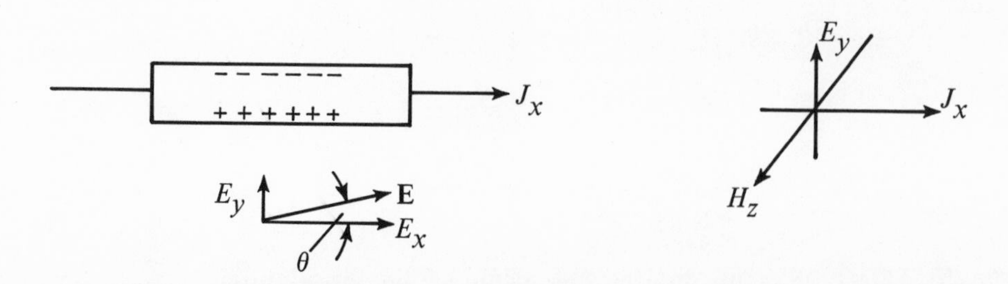

Fig. 1. Illustration of Hall field E_y due to action of magnetic field on a confined stream of positive charge carriers. For the example shown, e, J_x, and H_z are positive—as are also ν_x, E_x, and E_y. The resultant electric field $\vec{E}$ and the Hall angle θ are shown at the lower left. In the case of electrons, the Hall field is reversed for positive J_x, both ν_x and e then being negative.

specimen requires that the transverse current J_y vanish, and hence at steady state F_y will vanish. Therefore a transverse electric field E_y must build up in order to counteract the Lorentz force, as is indicated in Eq.(2):

$$F_y = 0 = E_y - (1/c)\nu_x H_z. \tag{2}$$

This transverse field is known as the Hall field, and we shall designate it by E^H. Its magnitude is given by Eq.(3), where we have also expressed the charge carrier velocity in terms of current density J_x and charge carrier concentration n:

$$E^H(\equiv E_y) = (1/c)\nu_x H_z = J_x H_z/nec, \quad \text{using } J_x = ne\nu_x. \tag{3}$$

The Hall field can be related to the current density by means of the Hall coefficient R, as noted below:

$$E^H = RJ_x H_z, \qquad R = 1/nec \quad \text{(Gaussian units*)}, \tag{4}$$

*In the Gaussian system, mechanical quantities are expressed in cgs units, electrical quantities in esu, and magnetic fields in gauss or oersteds. In the practical system, mechanical quantities are in cgs units, electrical quantities in volts and coulombs, and magnetic fields in gauss or oersteds. In the equations involving transport phenomena, conversion from the Gaussian system to the practical system is effected by replacing c by unity and replacing H by $H/10^8$, where those quantities explicitly occur. The practical units for R are cm^3/coulomb; those for mobility are cm^2/volt-sec.

$$E^H = RJ_xH_z/10^8, \qquad R = 1/ne \quad \text{(practical units*)}. \qquad (4a)$$

Our sign convention is that $e < 0$ for electrons.

Thus we see that for our simple model, the Hall coefficient is inversely proportional to the charge carrier density n, and that it is negative for electron conduction and positive for hole conduction. Finally, referring again to Fig. 1, we can define a <u>Hall angle</u> by the relation

$$\theta = \tan^{-1}(E_y/E_x). \qquad (5)$$

An important point to be made here is that the Hall effect may be regarded as the rotation of the electric field vector in the sample as a result of the applied magnetic field. More will be said later of the Hall angle.

In view of the relations we have presented thus far, some of the readers will perhaps be tempted to utter the cliché "So what could be simpler?" To which I shall respond that, as this talk progresses, it will be seen to be more appropriate to borrow instead from Shakespeare his classic phrase "How infinite in faculties!"

III. PRACTICAL CONSIDERATIONS

In perfecting the use of Hall phenomena as a tool for studying the electronic characteristics of solids, one can identify two classes of complications. The first has to do with the experimental determination of the Hall coefficient, and the second with its interpretation in terms of the electronic character of the solid.

1. Measurement Techniques

With regard to measurement of the Hall coefficient, a number of techniques are available.[1-6] The conventional procedure is to measure, by means of a potentiometer or other high-impedance device, the Hall voltage V^H across a parallelepiped in a direction normal to both $\vec{H}$ and $\vec{J}$. The arrangement is shown in Figure 2. The expression for the Hall coefficient, given in Eq.(6), follows directly from the previous equation (4a) when the geometry of the sample is taken into account:

$$R = 10^8\, V_{3,4}\, t/IH \qquad \text{(practical units)}. \qquad (6)$$

---

* See footnote on previous page.

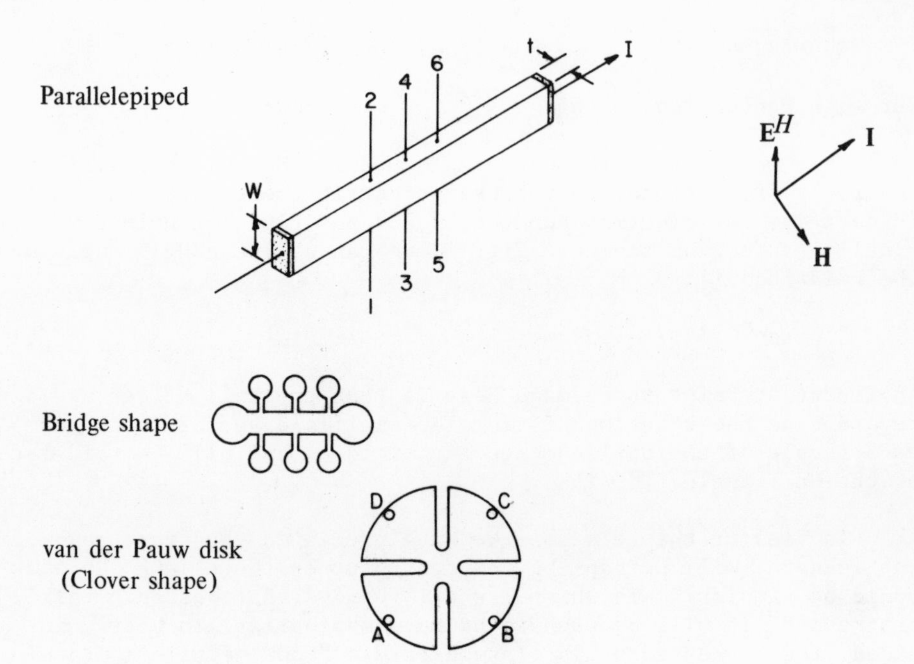

Fig. 2. Measurement samples for Hall coefficient and related effects. Top: Conventional parallelepiped. Hall probes are 3 and 4; resistivity probes are 2 and 6 and also 1 and 5. Middle: Typical form of "bridge-type" sample. Bottom: Clover-shaped design, or van der Pauw disk.

Here t is the thickness of the sample, measured along $\vec{H}$, $\vec{I}$ is the current (as contrasted to the current density, which we call $\vec{J}$), and $V_{3,4}$ is the Hall voltage V^H. Inasmuch as V^H may be the order of microvolts, extreme care must be taken to avoid extraneous voltages. An example is the misalignment voltage, caused by probes 3 and 4 not being on an equipotential plane when H = 0. A somewhat common procedure for eliminating this spurious voltage is to do measurements for opposite directions of H and take half the algebraic difference, inasmuch as the Hall voltage is odd in H. The misalignment voltage is, of course, independent of H but proportional to I. As we shall see later, however, if anisotropic solids are being studied and various galvanomagnetic effects are to be measured, then it is desirable to avoid any misalignment voltage by adjusting the position of the Hall probes to obtain a null reading when H = 0. An alternative technique involves replacing probe 4 by a sliding contact[1,5] to a resistor which is attached to the sample at two points, such as 2 and 6 in Figure 2. Disposing of misalignment voltages in this fashion allows one to use the reversal of magnetic field to separate even and odd galvanomagnetic effects.

Also, reversals of H and I can be used to eliminate unwanted voltages resulting from certain thermal effects. Such spurious voltages can arise from temperature gradients and resulting thermoelectric or thermomagnetic effects, as will be discussed later.

Connections to the sample shown in the sketch should be arranged so as to avoid, as much as possible, perturbations of the electric field in the sample. Thus the potential probes are small points. The end contacts, which short the Hall field, should be sufficiently far removed from the Hall potential probes. The latter condition can be ensured for sample length-to-width ratios of at least four.[7] Hall samples can also be cut in a bridge-shaped design,[8] as is shown in the figure. This has the advantage of providing large contact areas with a reasonably small disturbance of the electric fields in the sample. Another popular design is the van der Pauw disk.[9] A number of other arrangements are presented in the literature,[10] including methods for determining the Hall coefficient without contacting the specimen.[10a] Information on profiling techniques is also available.[10b]

2. General Transport Phenomena

The Hall effect is but one manifestation of a variety of electrical transport phenomena occurring in solids in a magnetic field, which are referred to as galvanomagnetic effects and thermomagnetic effects. To illustrate these effects, we write the phenomenological equations which relate the densities of electric and thermal currents to the gradients producing the transport (that is, electric fields and temperature gradients).[11]

$$J_i = \sum_j \sigma_{ij}(\vec{H})E_j^* + \sum_j M_{ij}(\vec{H})\frac{\partial T}{\partial x_j}, \qquad (7)$$

$$j = 1,2,3$$

$$q_i = \sum N_{ij}(\vec{H})E_j^* + \sum_j L_{ij}(\vec{H})\frac{\partial T}{\partial x_j}. \qquad (8)$$

Summations are of course carried out over the three coordinate directions, and the four coefficients in the above equations are tensor quantities, which are magnetic-field dependent. The σ_{ij} are the components of the electrical conductivity tensor, and the other coefficients are related to thermoelectric phenomena and to thermal conductivity. The field $\vec{E}^*$ is the negative gradient of the electrochemical potential, namely

$$E_j^* = -\frac{1}{e}\frac{\partial(e\phi+\zeta)}{\partial x_j} = E_j - \frac{1}{e}\frac{\partial\zeta}{\partial x_j}, \qquad e<0 \text{ for electrons.} \qquad (9)$$

In the case where there are no temperature or carrier concentration variations throughout the sample, then the chemical potential ζ is constant and $\vec{E}^*$ reduces to the electric field $\vec{E}$, that is, the negative gradient of the electrostatic potential ϕ.

Since most measurements are carried out under conditions where $\vec{J}$ is the independent variable, it is desirable to invert[12] the electrical current density equation, which process yields relations (10) and (11) below

$$E_i^* = \sum_k \rho_{ik}(\vec{H}) J_k + \sum_k \alpha_{ik}(\vec{H}) \frac{\partial T}{\partial x_k} , \qquad k = 1,2,3 \tag{10}$$

$$q_i = \sum_k \pi_{ik}(\vec{H}) J_k - \sum_k \kappa_{ik}(\vec{H}) \frac{\partial T}{\partial x_k} . \tag{11}$$

Here the respective tensors represent well-known coefficients, namely: ρ, resistivity; α, thermoelectric power (Seebeck coefficient); π, Peltier coefficient; and κ, thermal conductivity. For the sample orientation such that $J \equiv J_x$, a number of the terms in the summations drop out; and with further simplifying assumptions including $H \equiv H_z$, the preceding equations reduce to

$$E_y^* = \rho_{yx}(H) J_x + \alpha_{yx}(H) \frac{\partial T}{\partial x} + \alpha_{yy}(H) \frac{\partial T}{\partial y} , \tag{11a}$$

$$E_x^* = \rho_{xx}(H) J_x + \alpha_{xx}(H) \frac{\partial T}{\partial x} + \alpha_{xy}(H) \frac{\partial T}{\partial y} , \tag{11b}$$

$$q_x = \pi_{xx}(H) J_x - \kappa_{xx}(H) \frac{\partial T}{\partial x} - \kappa_{xy}(H) \frac{\partial T}{\partial y} , \tag{11c}$$

$$q_y = \pi_{yx}(H) J_x - \kappa_{yx}(H) \frac{\partial T}{\partial x} - \kappa_{yy}(H) \frac{\partial T}{\partial y} , \tag{11d}$$

$$J \equiv J_x, \ H \equiv H_z, \ \rho_{xz} = \rho_{yz} = \rho_{zx} = \rho_{zy} = 0.$$

With the current restricted to the x direction and the magnetic field lying along the z direction, the "further simplifying assumptions" mentioned above include the premise that all tensor components involving index z vanish. This condition is satisfied, for example, in isotropic systems or in cubic systems with coordinate axes along the cube axes. As a result of these constraints, there are no gradients in the z direction. In anisotropic media in general, however, the limitation of the gradients to two dimensions may not be possible.[13]

At this point we want to emphasize that there exists a set of general but very powerful relations formulated by Kohler and Onsager, which apply to the resistivity and thermal conductivity tensors and which connect the Peltier and Seebeck coefficients,[14] namely:

$$\rho_{ij}(\vec{H}) = \rho_{ji}(-\vec{H}), \qquad \kappa_{ij}(\vec{H}) = \kappa_{ji}(-\vec{H}), \tag{12}$$

$$\pi_{ij}(\vec{H}) = T\alpha_{ji}(-\vec{H}). \tag{13}$$

As we shall examine in more detail later, these Kohler-Onsager reciprocal relations tell us that the symmetric parts of the ρ_{ij} and the κ_{ij} tensors [Eq.(12)] must be even in H and the antisymmetric parts, odd in H.

Going back now to Eqs.(11a) to (11d), we see that in general the electric fields and temperature gradients are connected via the thermoelectric coefficients. This is an example of how life can be complicated. For by Eq.(11a) we see that a measured Hall field will in general contain contributions arising from temperature gradients and entering through the thermoelectric and thermomagnetic coefficients. Inasmuch as the usual theoretical relationships involving the Hall effect are based on isothermal conditions, we now have two choices: (a) do measurements in a manner so as to achieve reasonable approaches to isothermal conditions, or to correct for or to cancel or balance out the thermoelectric and thermomagnetic voltages. For certain cases, cancellation can be achieved through appropriate reversals of current and magnetic field. The other alternative is (b) to enlarge the theory and take account of the complications introduced through Eqs.(11a) to (11d). This approach would only be feasible for simple isotropic solids. As we shall see presently, the theory even for isothermal Hall effect can be very complicated, except for ideal materials. Nevertheless, an <u>adiabatic</u> Hall coefficient is sometimes measured. In this case, the boundary condition $\partial T/\partial y = 0$ is replaced by $q_y = 0$.[15,15a]

To shed some light on what particular properties of a solid make for large differences between isothermal and adiabatic measurements, let us consider the simple example of the resistivity at zero magnetic field. Eqs.(11b) and (11c) then yield, for the isothermal and adiabatic measurement conditions, the following relationships:

$$\text{Isothermal:} \quad \begin{cases} E_x^* = \rho J_x \\ q_x = \pi J_x \end{cases}, \ \text{grad } T = 0 \tag{14}$$

$$\text{Adiabatic:} \begin{cases} E_x^* = \rho J_x + \alpha \partial T/\partial x \\ 0 = q_x = \pi J_x - \kappa \partial T/\partial x. \end{cases} \tag{15}$$

Therefore, for a determination of resistivity under the isothermal and adiabatic conditions, the following expressions apply:

$$\rho^i \equiv (E_x^*/J_x)_{T=const} = \rho, \tag{16}$$

$$\rho^{ad} \equiv (E_x^*/J_x)_{qx=o} = \rho(1 + \pi\alpha/\kappa\rho) \equiv \rho(1 + Z^*), \tag{17}$$

where

$$Z^* \equiv \pi\alpha/\kappa\rho = \alpha^2 T/\kappa\rho \quad \text{(thermoelectric figure of merit)}. \tag{18}$$

Hence,

$$(\rho^{ad} - \rho)/\rho = Z^*. \tag{19}$$

The thermoelectric figure of merit Z^*, defined in Eq.(18), is a measure of the efficiency of the material in a thermoelectric cooling or power generation process. We thus note that it is the thermoelectric figure of merit which is an important consideration in the difference between data taken under isothermal and adiabatic conditions. It is to be pointed out that the above analysis was done for one-dimensional transport, namely resistivity. The situation for Hall effect would be more complex. Nevertheless, we can expect differences between isothermal and adiabatic Hall effects to be small for materials where $Z^* \ll 1$. For example, in materials with the diamond or zincblende crystal structure, where the lattice contribution to the thermal conductivity is relatively large, the correction to isothermal coefficients for specimens measured under adiabatic conditions is quite small—often under one percent.[15]

Going back again to option (a), we wish to point out that one can reduce temperature gradients along the sample by the use of large end-contacting blocks of copper or other material with a large thermal conductivity, by use of isothermal baths, or—in the more difficult cases—by resorting to a-c measurements.[16] An effective technique in the latter case is to use square wave currents of alternating polarity.[17] In d-c measurements, the appropriate switching in polarity of both magnetic field and sample current is useful in some instances.

Before leaving this topic, I should like to point out how thermomagnetic effects can arise on the basis of our simple free-electron model. The Lorentz force acts in principle as a velocity selector—the slow electrons being deflected less than the more energetic ones—a process which of course creates a temperature gradient in the transverse direction. This temperature difference will result in a transverse potential difference due to the Seebeck coefficient of the material. The former phenomenon is called the Ettingshausen effect. Detailed information on the various thermoelectric and thermomagnetic phenomena is available in numerous places.[17a]

3. Galvanomagnetic Effects, Anisotropic Solids

We now concentrate on isothermal conditions and discuss in more detail the effects of a magnetic field on electrical transport. In particular, we consider the major galvanomagnetic effects in the general case of anisotropic solids. For a specimen with essentially no temperature gradients nor composition variations, the distinction between the gradient of the electrochemical potential, $\vec{E}^*$, and the electric field $\vec{E}$ vanishes. Equations (11a) to (11d) now reduce to the "short form" shown below:

$$E_i = \sum_k \rho_{ik}(\vec{H}) J_k. \tag{20}$$

The words "short form" will probably cause some eyebrow raising after one looks at Eq.(21), which represents the expansion of the resistivity tensor in powers of the magnetic field.

$$\rho_{ik}(\vec{H}) = \rho^o_{ik} + \sum_\ell \rho^o_{ik\ell} H_\ell + \sum_{\ell,m} \rho^o_{ik\ell m} H_\ell H_m + \dots \tag{21}$$

The superscript zero on a tensor component indicates that it is independent of magnetic field. Even a casual glance at Eq.(21) reveals that, in spite of the neglect of terms higher than second order in H, the resistivity tensor for an anisotropic material can be a complex creature—especially if the magnetic field is not along a direction of high symmetry that can cause the coefficients of an appreciable number of terms in each summation to vanish.

We shall identify the linear terms in H in Eq.(21) with the Hall effect, those quadratic in H with magnetoresistivity, and further terms as higher order effects of these phenomena. However, such a convention does not appear universally throughout the literature. Some of the quadratic terms—i.e. those ones responsible for an electric field normal to the current—have been identified with a quadratic "Hall effect"[18] and also a planar "Hall effect".[19] The latter phenomenon exists even in isotropic solids—as we shall see later. To avoid these problems in definition, and to establish a base to see how the complications come about, we shall adopt what seems to me to be the most straightforward of the various conventions. This is to divide the resistivity tensor into antisymmetric and symmetric parts and to identify the antisymmetric part with the Hall effect, and the symmetric part with generalized magnetoresistivities. This division is written as follows:

$$\rho_{ik}(\vec{H}) = \rho^a_{ik}(\vec{H}) + \rho^s_{ik}(\vec{H}), \tag{22}$$

where

$$\rho^a_{ik}(\vec{H}) = -\rho^a_{ki}(\vec{H}); \qquad \rho^s_{ik}(\vec{H}) = \rho^s_{ki}(\vec{H}). \tag{23}$$

The components of the general Hall field E^a are then given by

$$E^a_i(\vec{H}) = \sum_k \rho^a_{ik}(\vec{H}) J_k. \tag{24}$$

Now an expression involving an antisymmetric tensor such as Eq.(24) can be written in terms of a vector cross product, as was done by Casimer[20] and is shown in Eq.(25):

$$\vec{E}^a(\vec{H}) = \vec{R}(\vec{H}) \times \vec{J}. \tag{25}$$

The quantity $\vec{R}(\vec{H})$ is called the Hall vector. We note, in view of the Kohler-Onsager relations, [Eq.(12)], that $\vec{R}(\vec{H})$ is odd in $\vec{H}$. It likewise follows that the $\rho^s_{ik}(\vec{H})$, which we identify with magnetoresistivity, are even in $\vec{H}$.

$$\vec{R}(\vec{H}) = -\vec{R}(-\vec{H}), \qquad \rho^s_{ik}(\vec{H}) = \rho^s_{ik}(-\vec{H}). \tag{26}$$

The components of the Hall vector can be expressed in terms of the antisymmetric components of the resistivity tensor as follows:[20a]

$$R_k(\vec{H}) = \frac{1}{2}\sum_{\ell,m} \varepsilon_{k\ell m}\, \rho^a_{m\ell}(H). \tag{27}$$

The $\varepsilon_{k\ell m}$ are the permutation tensors, which are ± 1 when $k\ell m$ are all different and zero otherwise. They are positive when the indices are in standard cyclic order, negative for the reverse case. The factor 1/2 occurs because the tensor $\rho^a_{m\ell}$ is summed over its antisymmetric indices.

Looking again at Eq.(25), we make the startling discovery that while the Hall field is always perpendicular to $\vec{J}$, it is not necessarily perpendicular to $\vec{H}$. We thus need to distinguish between the conventional Hall field (i.e. the transverse Hall field), which is perpendicular to both $\vec{J}$ and $\vec{H}$, and another component of $\vec{E}^a(\vec{H})$ which is perpendicular to both $\vec{J}$ and the transverse Hall field.

The phenomenon in this latter case is called the longitudinal Hall effect. The two contributions to the general Hall field, and their components along unit vectors in the appropriate directions, are shown below:

$$\vec{E}^a(\vec{H}) = \vec{E}^H(\vec{H}) + \vec{E}^{H,L}(\vec{H}), \tag{28}$$

$$= E^H(\vec{H})\frac{\vec{H}\times\vec{J}}{|\vec{H}\times\vec{J}|} + E^{H,L}(\vec{H})\frac{\vec{J}\times(\vec{H}\times\vec{J})}{|\vec{J}\times(\vec{H}\times\vec{J})|}. \tag{29}$$

Explicit expressions for the conventional and longitudinal Hall fields can be put in the following useful form:

$$\text{Transverse:}\quad \vec{E}^H = \frac{\vec{E}^a(\vec{H})\cdot\vec{H}\times\vec{J}}{|\vec{H}\times\vec{J}|}\,\frac{\vec{H}\times\vec{J}}{|\vec{H}\times\vec{J}|}, \tag{30}$$

$$\text{Longitudinal:}\quad \vec{E}^{H,L} = \frac{\vec{E}^a(\vec{H})\cdot\vec{J}\times(\vec{H}\times\vec{J})}{|\vec{J}\times(\vec{H}\times\vec{J})|}\,\frac{\vec{J}\times(\vec{H}\times\vec{J})}{|\vec{J}\times(\vec{H}\times\vec{J})|}. \tag{31}$$

In view of the identity

$$\vec{J}\times(\vec{H}\times\vec{J}) \equiv J^2\vec{H} - \vec{J}\cdot\vec{H}\,\vec{J}, \tag{32}$$

we see that if $\vec{H}$ is perpendicular to $\vec{J}$, then the longitudinal Hall field will lie along $\vec{H}$.

It is of obvious interest to determine the principal components of the Hall vector. To do this, we use Eqs.(25) and (29) to obtain

$$\vec{R}(\vec{H})\times\vec{J} = \frac{E^H(\vec{H})}{|\vec{H}\times\vec{J}|}\vec{H}\times\vec{J} + \frac{E^{H,L}(\vec{H})}{|\vec{J}\times(\vec{H}\times\vec{J})|}(\vec{J}\times\vec{H})\times\vec{J}. \tag{33}$$

From which it follows at once that

$$\vec{R}(\vec{H}) = \frac{E^H(\vec{H})}{|\vec{H}\times\vec{J}|}\vec{H} + \frac{E^{H,L}}{|\vec{J}\times(\vec{H}\times\vec{J})}\vec{J}\times\vec{H}. \tag{34}$$

Thus we see that for the conventional Hall effect, the Hall vector lies along $\vec{H}$; for the longitudinal Hall effect, it lies along $\vec{J}\times\vec{H}$. We are now in a position to express the Hall vector, $\vec{R}^T(\vec{H})$, for the conventional (transverse) Hall effect in terms of the usual Hall coefficient $R(\vec{H})$. We have, from Eq.(34),

$$\vec{R}^T(\vec{H}) = \frac{E^H(\vec{H})}{|\vec{H}\times\vec{J}|}\vec{H} = R(\vec{H})\vec{H}, \text{ where } R(\vec{H}) \equiv \frac{E^H(\vec{H})}{|\vec{H}\times\vec{J}|}. \tag{35}$$

Thus the familiar Hall field becomes

$$\vec{E}^H(\vec{H}) = R(\vec{H})\vec{H}\times\vec{J}. \tag{36}$$

Of course we knew relation (36) all along—way back from Eq.(4)—but it's gratifying to have it readily spring forth from our general and somewhat esoteric derivation involving the antisymmetric components of the resistivity tensor. The longitudinal Hall effect has been studied theoretically by Grabner,[21] with experimental data for n-type germanium. He finds that it vanishes for (a) spherical energy surfaces, (b) when $\vec{H}$ is parallel to an axis of rotation of the crystal, and (c) in the limit of strong magnetic fields.

In order to examine in more detail the various galvanomagnetic phenomena which produce electric fields in directions coincident with the Hall field, we examine the expansion of the components of the resistivity tensor $\rho(\vec{H})$, as was done in Eq.(21). The expression for the electric field, through second order terms in H, is

$$E_i = \sum_k \rho^o_{ik} J_k + \sum_{k,\ell} \rho^o_{ik\ell} J_k H_\ell + \sum_{k,\ell,m} \rho^o_{ik\ell m} J_k H_\ell H_m + \dots \quad . \tag{37}$$

Here I repeat again our convention that we associate the terms that are odd in H with the Hall effect and those that are even with magnetoresistive fields. To simplify matters, we now place the current density vector $\vec{J}$ along axis 1 and the magnetic field $\vec{H}$ along axis 3, so that the expression for the conventional Hall field reduces to (38) below, and the longitudinal Hall field is given by Eq.(39).

$$E^H \equiv E_2 = \rho^o_{213} H_3 J_1 \quad , \tag{38}$$

$$H \equiv H_3, \qquad J \equiv J_1$$

$$E^{H,L} \equiv E_3 = \rho^o_{313} H_3 J_1 . \tag{39}$$

Cubic crystal with cube axes along 1,2,3:

$$\rho^o_{ik\ell} = \varepsilon_{ik\ell} \rho^o_{123} . \tag{40}$$

For a material with cubic symmetry, if $\vec{H}$ is along a cube axis then there is only one value for the Hall coefficient, as is indicated by Eq.(40). The longitudinal Hall field vanishes. However, if $\vec{H}$ is not along a cube axis, then as Grabner showed, ρ^o_{313} will not vanish so long as the energy surfaces are nonspherical.

To see how additions to the Hall fields can arise via the magnetoresistive coefficients, we consider the contributions to E_2 and E_3 from the second order terms in H of Eq.(37). We have

$$E_2 = \rho^o_{2133} H_3^{\,2} J_1 , \tag{41}$$

$$H \equiv H_3, \quad J \equiv J_1$$

$$E_3 = \rho^o_{3133} H_3^{\,2} J_1 . \tag{42}$$

The above contributions vanish if H is along a crystal axis of sufficient symmetry. For example, if H is along a cube axis in the cubic system or along the three-fold axis in a trigonal system, the tensor components ρ^o_{2133} and ρ^o_{3133} vanish. However, they are non-zero in a trigonal crystal for certain magnetic field directions that are normal to the three-fold axis. Phenomena depicted by Eq.(41) have been referred to in early literature as the quadratic "Hall effect".[18] The recent tendency is, as we have enunciated, to consider the effect as a magnetoresistive contribution inasmuch as it arises from the symmetric part of the resistivity tensor.

Let us now recapitulate a bit. We recall that by our definition the conventional Hall field is along $\vec{H} \times \vec{J}$, and is therefore perpendicular to both $\vec{J}$ and $\vec{H}$. However, it is not necessary that $\vec{H}$ be perpendicular to $\vec{J}$, just so long as it is not parallel. For simplicity, though, the experimenter usually orients the magnetic field normal to $\vec{J}$. For such case, the simplified equations (38) to (42) are applicable. There are, however, examples in the literature where the only condition imposed is that $\vec{E}^{meas}$ be normal to $\vec{J}$.[22] For such a case, magnetoresistive coefficients can contribute to $\vec{E}^{meas}$ even in isotropic solids. To see how this can come about, we again choose $\vec{J}$ along axis 1, $\vec{E}^{meas}$ along 2, but now permit $\vec{H}$ to have an arbitrary direction, as indicated by the constraints shown as Eq.(42a) below. Equation (37) then yields the results for E^{meas} ($\equiv E_2$) shown in Eqs.(43) and (44).

$$J \equiv J_1, \quad E^{meas} \equiv E_2, \quad H \equiv H(H_1, H_2, H_3), \tag{42a}$$

$$E_2^{(odd)}/J = \rho^o_{211}H_1 + \rho^o_{212}H_2 + \rho^o_{213}H_3 , \tag{43}$$

$$E_2^{(even)}/J = \rho^o_{2111}H_1^{\ 2} + \rho^o_{2122}H_2^{\ 2} + \rho^o_{2133}H_3^{\ 2} + \boxed{2\rho^o_{2112}H_1H_2} + 2\,\rho^o_{2113}H_1H_3 + 2\,\rho^o_{2123}H_2 . \tag{44}$$

These equations, which can be applied to an arbitrary crystal orientation, show how complicated the expressions can be when the currents and magnetic fields are not along directions of sufficient symmetry to cause most of the galvanomagnetic coefficients to vanish. Of course, as we noted before, if $\vec{H}$ is along a cube axis in, say, germanium then only the third term in Eq.(43), namely the conventional Hall term, is non-zero. But if the direction of $\vec{H}$ is arbitrary with respect to directions of crystal symmetry, we can get a longitudinal Hall field by means of the other terms.

With regard to contributions to E_2 from magnetoresistive coefficients, results are much more formidable, as is seen in Eq.(44).[22a] Again, for a cubic crystal with H along a cube axis only the term in the box is non-zero. In a trigonal system, we can get one or two additional terms depending upon which coordinate is aligned along the three-fold axis. We saw examples of some of these phenomena in Eqs.(41-42) earlier, where we mentioned that the effect had been called a quadratic "Hall effect" in early literature. We now concentrate on the boxed term in Eq.(44). Interestingly enough, the galvanomagnetic coefficient ρ^{o}_{2112} is non-zero even in isotropic systems. Since this coefficient describes the state of affairs when the current and both fields are confined to a single plane, the phenomenon is sometimes known as the planar "Hall effect".[19,22] It is usually expressed in terms of the angle between H and J, thus

$$E_2 = 2\,\rho^{o}_{2112} H_1 H_2 J = \rho^{o}_{2112} J H^2 \sin 2\phi, \qquad \phi \equiv \measuredangle H,J. \tag{45}$$

Some authors have referred to the planar "Hall effect" as the "pseudo Hall effect" since it arises from a magnetoresistive coefficient (i.e. the symmetric part of the resistivity tensor), and therefore according to our definition it is not a Hall effect. However, both it and the so-called quadratic "Hall effect" produce electric fields that are normal to the current. Therefore care must be exercised not to confuse such fields with the conventional Hall field. Inasmuch as these magnetoresistive contributions are even in H, they can be eliminated by taking data for both directions of magnetic field.

4. Specimen Homogeneity

A major consideration in the measurement and interpretation of Hall data is the homogeneity of the sample. This is especially important in the case of semiconductors where the densities of impurities and crystal imperfections are kept small. Thus any deviations from such a state are especially conspicuous. In addition, most semiconductors of interest have relatively large mobilities, and a large mobility accents all galvanomagnetic effects. We shall discuss the effect of inhomogeneities in more detail in the next part of this talk. We shall see there that inhomogeneities can produce an intermixing of Hall and magnetoresistivity effects and can thus cause very large magnetic field dependencies of the Hall coefficient. We also wish to point cut that the transverse magnetoresistance can really go wild, since this effect is very sensitive to any perturbation of the Hall field.[23] Although time does not permit me to include magnetoresistance in any detail in this talk on Hall effect, I would point out that it is even more sensitive than the Hall effect to scattering processes, anisotropy, band structure, inhomogeneities, etc.[24] The trouble is that it is so sensitive to so many things that results are difficult to interpret. Nevertheless, magnetoresistance measurements are an important adjunct to Hall data when a thorough study is being made of a semiconductor. With regard to inhomogeneities, it is obviously very important to discover their existence. Ways of doing this include checks of resistivity as a function of position in the bulk material with a four-point probe, comparison of data on measured samples cut from different locations, and analyses of magnetic-field dependence of Hall and other galvanomagnetic effects.[24a]

5. Summary to Part III

It is now appropriate to summarize some of the major considerations we have discussed during the first part of this talk. We showed that, although measurements of the Hall effect are relatively straightforward, there are a number of areas where care needs to be exercised. These include voltage probe misalignment problems, end-contact shorting, unwanted contributions due to thermoelectric and thermomagnetic effects, and, in anisotropic solids, contributions from magnetoresistive effects which are collinear with the Hall field but are even in H. Lastly, but always to be kept in mind, especially if unfamiliar materials are being studied, are effects of sample inhomogeneities.

IV. THEORETICAL ASPECTS

For the remaining part of this paper, I want to discuss theoretical aspects involved in the interpretation of Hall data in terms of the electronic character of the solid. We must thus make the transition from the simple free-particle treatment discussed at the beginning of this talk to the state of affairs existing in a real solid. For example, we need to take account of the distribution of velocities and the interactions of the charge carriers with impurities, defects, and lattice thermal vibrations of the solid (i.e. the *scattering*), as well as the band structure of the material. The latter consideration relates to the fact that the charge carriers in the solid are not free but exist in a potential energy field having the periodicity of the lattice. As a result of these constraints, only certain energy states, or *bands*, are allowed for the charge carriers. In addition, the relationship between the energy and the velocity of the carriers is not the simple $1/2\ mv^2$ of the free particle, but is more complex. An approximation which works satisfactorily in conventional semiconductors is to characterize the charge carriers by an *effective* mass m^*.[24b]

Taking into account most of the complexities we have mentioned, one still obtains an expression equating the Hall coefficient to the inverse of the carrier concentration for the case where conduction occurs by means of carriers of a single type, i.e. all electrons or all holes of a single effective mass. This means that conduction is in a single band. The more complex relationships for multiband conduction will be discussed later.

6. Single-Band Conduction

The expression for the Hall coefficient under the above conditions is commonly written in the form

$$R = r/ne \qquad \text{(practical units, } e < 0 \text{ for electrons).} \tag{46}$$

The Hall coefficient factor, r, depends on the nature of the scattering, the band structure, the statistics characterizing the distribution of velocities of the carriers, and the magnetic field strength. Fortunately it depends rather weakly on these factors, and its value is usually within, say, 70% of unity.

To develop some relationships involving the Hall coefficient, we consider an isothermal and isotropic solid with $\vec{H}$ along the z axis and $\vec{J}$ along the x axis. The general transport relationships, given previously as Eq.(7), then reduce to

$$J_x = \sigma_{xx}E_x + \sigma_{xy}E_y, \tag{47}$$

$$0 = -\sigma_{xy}E_x + \sigma_{xx}E_y. \tag{48}$$

$$\left\{ \begin{array}{l} H \equiv H_z \\ J \equiv J_x \\ \sigma_{ij} \equiv \sigma_{ij}(H) \end{array} \right.$$

The above relations plus the definitions of the Hall coefficient R and the conductivity σ allow us to express these quantities in terms of the components of the conductivity tensor, namely

$$R = E_y/J_xH = \rho_{xy}/H = \sigma_{xy}/(H[\sigma_{xx}^2 + \sigma_{xy}^2]), \tag{49}$$

$$\sigma = J_x/E_x = 1/\rho_{xx} = [\sigma_{xx}^2 + \sigma_{xy}^2]/\sigma_{xx}. \tag{50}$$

It is to be noted that for the practical system of units, H (in gauss) is to be divided by 10^8. From Eqs.(48)-(50), in conjunction with Eq.(5), some very useful relations can be derived:

$$E_y/E_x = \sigma_{xy}/\sigma_{xx} = R\sigma H = \tan\theta. \tag{51}$$

An important attribute of charge carriers in a solid is their mobility, defined in the simple case as their drift velocity per unit electric field. The usual mobility, or more precisely the conductivity mobility, is commonly designated by μ and is related to the conductivity σ as indicated below:

$$\sigma = ne\mu . \tag{52}$$

It is also customary to define a Hall mobility μ^H by the relation

$$\mu^H = R\sigma . \tag{53}$$

From this definition plus Eq.(51), we see that

$$R\sigma = E_y/HE_x = (\tan\theta)/H, \tag{54}$$

$$\mu^H H = \tan\theta. \tag{55}$$

Thus we see that the Hall mobility bears a simple relation to the Hall angle. By use of Eq.(46) with (52) and (53), we establish the interesting finding that the Hall coefficient factor is the ratio of the Hall and conductivity mobilities, thus

$$r = \mu^H/\mu . \tag{56}$$

For many common semiconductors, the scattering of the charge carriers can be characterized by a relaxation time τ, which in most instances is energy-dependent. In such cases the conductivity mobility can be expressed by[25]

$$\mu = (e/m^*)\langle\tau\rangle , \tag{56a}$$

where $\langle\tau\rangle$ is the average of the relaxation time over the distribution of the charge carriers in energy. The Hall coefficient factor can also be related to averages of τ. For the case of weak magnetic fields, the result assumes the simple form[25]

$$r = \langle\tau^2\rangle/\langle\tau\rangle^2, \quad \mu^H H/10^8 \ll 1. \tag{57}$$

Both Eqs.(56a) and (57) assume a constant effective mass, that is, the existence of spherical energy surfaces and parabolic bands. Spherical energy surfaces ensure an isotropic effect mass; parabolic bands (the case where the charge-carrier energy is a quadratic function of momentum or wave vector) ensure that the effective mass does not vary with energy. If the preceding conditions are not met, then the m* in Eq.(56a) is also subject to the averaging. The case of nonspherical energy surfaces will be discussed shortly. The theory for nonparabolic bands has been presented by Kołodziejczak,[25a] with detailed treatment of the case of spherical energy surfaces and nonparabolic bands—representative of the conduction band in InSb.

Going back now to Eq.(57), we see that the value of r is affected by the nature of the scattering process. It is also magnetic-field dependent, approaching unity in the strong field region. Because of this behavior, it should be possible in principle to determine r from the ratio of the Hall coefficients measured at the given field and at the strong-magnetic-field limit. In practice however, the strong field must not be so great that quantum effects become significant.[26] This means that we require that $\hbar eH \ll kT$. Here $\hbar$ is the Dirac-Planck's constant, k is Boltzmann's constant, and T is the absolute temperature. Also the perfection of the measurement sample must be extremely good, since even minor inhomogeneities can cause aggravated problems at high magnetic field strengths.

A summary of how some common scattering mechanisms affect the Hall coefficient factor[26a] is shown in Table I. All entries except the last are for classical statistics, i.e. the Fermi level ζ lies outside the conduction or valence bands. The table also includes results for an arbitrary variation of the relaxation time τ with energy, given by the parameter λ. The value $\lambda = -1/2$ applies to acoustic phonon scattering, and $\lambda = 3/2$ is for scattering by ionized impurities. If the semiconductor is highly degenerate, i.e. the Fermi level lies deep within the band, then of course Fermi-Dirac statistics must be used. In the limit of high degeneracy it is essentially only those charge carriers of energy equal to the Fermi energy that contribute to conduction. Thus we have, in essence, a constant-τ case for all simple scattering mechanisms, and r is unity. In the strong field limit, the Hall coefficient factor is unity for all cases, provided that the magnetic field is not so huge that the separation between the Landau levels becomes comparable with kT. If the latter occurs, then quantum effects must be considered. Although the table shows only values of r in the limiting cases of zero and strong magnetic field strengths, the behavior of r can be calculated for arbitrary values of $\mu^H H$. However, the results are considerably more complicated.[26b]

As we implied in our discussion following Eq.(57), nonsphericity in the energy surfaces and nonparabolic dependences of charge-carrier energy on wave vector can also affect the Hall coefficient factor. We shall not take time to discuss the latter situation, a reference thereto having already been cited.[25a] To illustrate the effect of nonsphericity in the energy surfaces of a semiconductor, we include an anisotropy factor in the previous expression for r. Thus

$$r(H) = f(\tau,H)a(H). \quad (58)$$

Here the factor $f(\tau,H)$ represents the contribution of the scattering processes to $r(H)$, and the factor $a(H)$ is the result of nonisotropic energy surfaces. At very weak magnetic fields, $f(\tau,H)$ becomes the ratio of relaxation time averages shown before in Eq.(57), and the anisotropy factor approaches a value which we call a_o. At strong fields, $a(H)$ approaches unity.

Table I. Effect of Scattering on r

(Spherical Energy Surfaces)

Type of Scattering	Degree of Degeneracy	$r_\circ$ $(\mu^H H/10^8 \ll 1)$	r_∞ $(\mu^H H/10^8 \gg 1)$*
Acoustic phonon	None ($\zeta/kT \ll 0$)	1.18 (i.e., $3\pi/8$)	1
Ionized impurity	Ditto	1.93 (i.e., $315\pi/512$)	1
Constant-τ	"	1	1
$\tau \sim \varepsilon^\lambda$	"	$[3\sqrt{\pi}/4][(2\lambda+3/2)!][\lambda+3/2)!]^{-2}$	1
Arbitrary	Highly degenerate ($\zeta/kT \gg 1$)	1	1

*But must have $heH/10^8 m^* \ll kT$ so that quantum effects are negligible.

$$r_o = \langle r^2\rangle/\langle r\rangle^2 a_o, \qquad \text{for } \mu^H H/10^8 \ll 1, \tag{58a}$$

$$r_\infty = 1, \qquad \text{for } \mu^H H/10^8 \gg 1. \tag{58b}$$

For an energy surface represented by a system of ellipsoids arranged to possess cubic symmetry, a_o can be expressed in terms of the principal effective masses characterizing the ellipsoids, as shown in Eq.(59).[27]

$$a_o = 3(1/m_1^* m_2^* + 1/m_2^* m_3^* + 1/m_3^* m_1^*)/(1/m_1^* + 1/m_2^* + 1/m_3^*)^2 \tag{59}$$

For ellipsoids of revolution (spheroids), such as are characteristic of the conduction bands of germanium and silicon, Eq.(59) reduces to (61), where K is the ratio of longitudinal to transverse mass longitudinal being that along the axis of revolution of the ellipsoid.

$$\text{For spheroids: } m_1^* = m_2^* = m_\perp^*, \quad m_3^* = m_\parallel^* \tag{60}$$

$$a_o = \frac{3K(K+2)}{(2K+1)^2}, \quad K \; \frac{m_\parallel^*}{m_\perp^*}. \tag{61}$$

A plot of a_o as a function of K is shown in Figure 3. We note that for prolate spheroids, representative of common semiconductors such as germanium and silicon, a_o lies between 0.78 and 1. For pie-shaped surfaces, however, it can get quite low. Numerical values of a_o for various bands in germanium and silicon are listed in Table II. The results for the warped heavy-mass bands were obtained from the data of Lax and co-workers.[28] For the light-mass valence bands, which are approximately spherical, a_o is essentially unity.

The reader will note that thus far I have discussed the Hall anisotropy factor only in the limit of very weak and very strong magnetic fields. At intermediate field strengths, things can be very complicated. For example, in the spheroid model, results can depend on what particular arrangement of spheroids is used to provide the cubic symmetry, i.e. (a) 3 or 6 spheroids with major axes along the [100] directions, (b) 12 spheroids along [110] directions, or (c) 4 to 8 spheroids along [111] directions.[29] Furthermore, as might be expected, moving the magnetic field vector off a direction of high symmetry can cause significantly different variations in the magnetic field dependence of the Hall coefficient. As an example of this, we present in Figure 4 data taken by Mason and co-workers on 3.18 ohm-cm germanium.[30] It is seen that moving H off the [001] direction can exert a strong influence on the magnetic-field dependence of R.

In the case of warped energy surfaces, the magnetic-field dependence of r can possess maxima when certain admixtures of

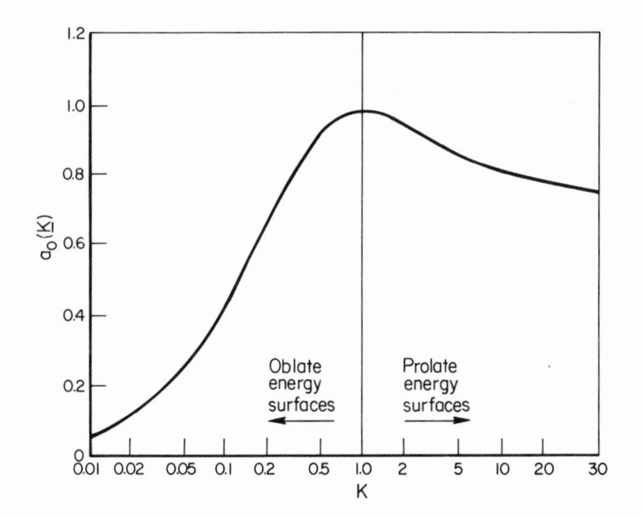

Fig. 3. Dependence of anisotropy factor in μ_o^H/μ_o on the anisotropy of the effective mass for a family of spheroidal energy surfaces arranged so as to possess cubic symmetry. The quantity K is the ratio $m_{||}^*/m_\perp^*$ (after Herring).[27]

Table II. a_o in Germanium and Silicon

Conduction band:

$$a_o = 3K(K + 2)/(2K + 1)^2$$

Ge: $K \simeq 1$, $a_o \simeq 0.79$

Si: $K \simeq 4.7$, $a_o \simeq 0.87$

Warped heavy mass valence band:

Ge: $a_o \simeq 0.71$

Si: $a_o \simeq 0.76$

Light mass valence band:

$a_o \simeq 1.$

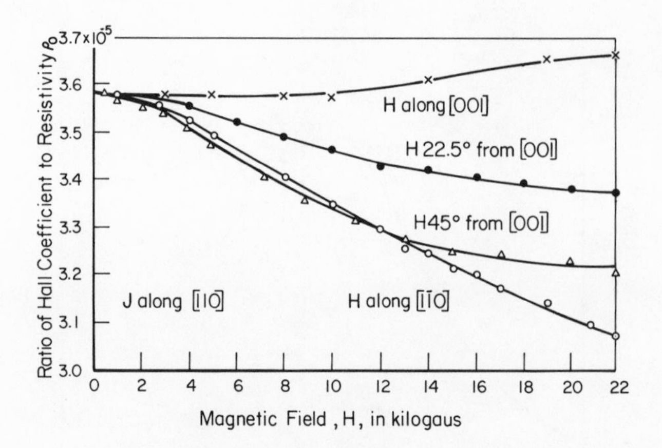

Fig. 4. Variation of $R(H)/\rho_o$ with H for four directions of the magnetic field in 3.18 ohm-cm germanium (after Mason, Hewitt, and Wick).[30]

scattering mechanisms exist,[31] as is seen in Figure 5. Here β represents the degree of admixed ionized impurity scattering, and μ_L^s is the mobility representative of lattice scattering in a spherical band. We see from the figure that the warping produces a significant lowering of the Hall coefficient factor r at weak magnetic fields. This is apparent in the curves for small β, indicative of the case where acoustic phonon scattering is strongly predominant. The admixture of increasing amounts of ionized impurity scattering introduces conspicuous changes, eventually producing striking maxima. With regard to Figure 5, I would point out that the theoretical development on which the curves shown there are based involves approximations which could stand improvement. Nevertheless, the maxima are real, as is borne out by experimental data to be shown later. I present the results to illustrate the sensitivity of the Hall effect to complexities in scattering and warping in band structure.

A general discussion of the effect of warped energy surfaces on μ^H/μ has been given by Shockley,[32] who presents a helpful physical explanation for the lowering of μ^H/μ. He points out how, for electrons in certain states in solids with re-entrant energy surfaces, the ratio μ^H/μ can become negative. Anisotropy factors have been calculated by several investigators for a variety of planar-faced energy surfaces. Values of a_o are 1/2 for a cube[33-35] and 2/3 for an octahedron,[34,35] with approaches toward unity as the number of faces is increased.[36] Effects of the rounding of edges and corners have also been considered.[37]

It is to be noted that for simplicity in presentation I have chosen to deal separately with the influence on the Hall coefficient

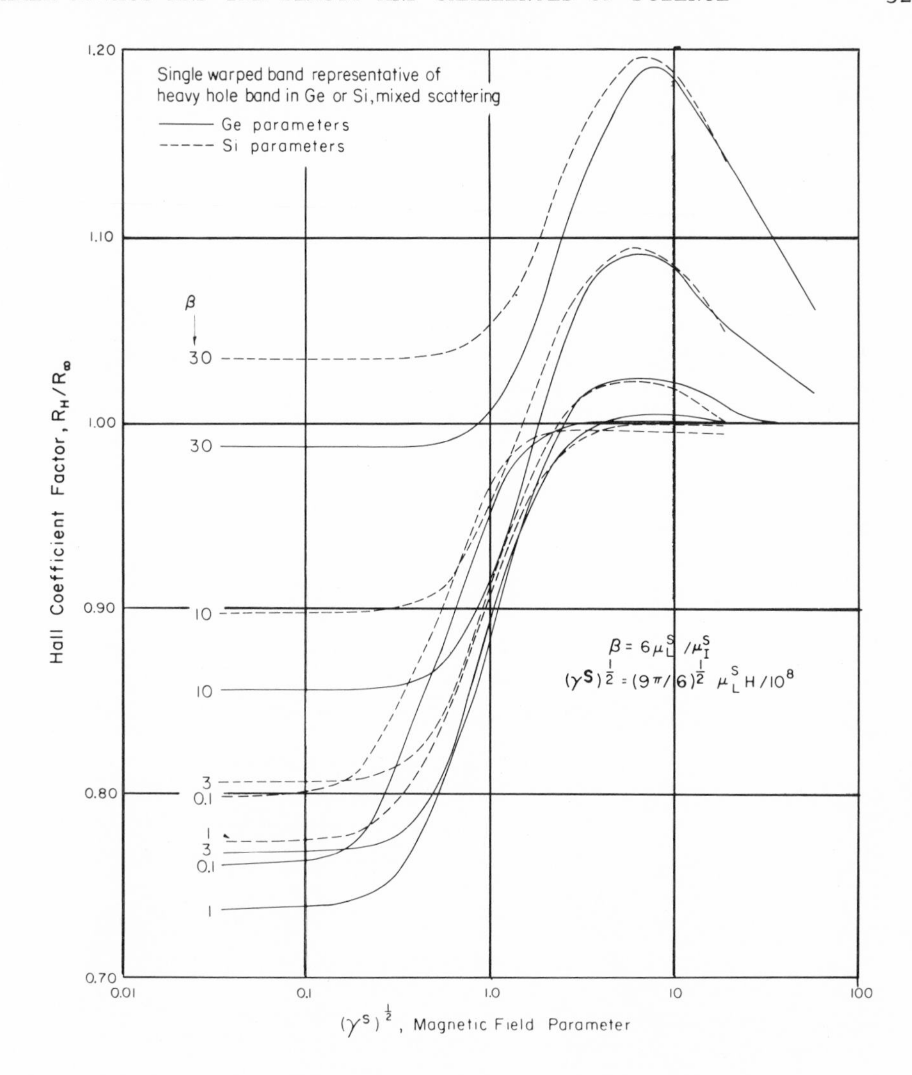

Fig. 5. Hall coefficient factor for a single warped band. The solid lines are for parameters representative of the heavy-mass band in germanium, as determined from cyclotron resonance studies. The dashed lines are applied to silicon. The parameter β measures the degree of ionized impurity scattering (after Beer and Willardson).[31]

factor of such entities as nature of the charge-carrier scattering, energy band anisotropy, and multiband effects (the latter to be discussed in Section 7). However, the same basic consideration underlies all of these cases—namely a situation where certain charge carriers possess different Hall angles. For this reason, Allgaier has referred to r as a mixing factor,[36] since it is the mixing of charge carriers with different Hall angles which causes r to differ from unity.

In concluding this section, I want to emphasize that careful studies of Hall effect can tell us something about the predominant scattering mechanisms in the semiconductor, and also they can supply information about any anistropies in band structure. In obtaining information about scattering, one needs especially to pay attention to variations in Hall coefficient factor with temperature and with magnetic field; for studies on band structure, the variations with magnetic field and with sample orientation are significant.

7. Multiband Conduction

Let us now add another complication—the case where conduction occurs in two or more bands. Then every component of the conductivity tensor receives contributions from each of the bands, and these must be summed. To simplify the equations, we consider the case of an isotropic solid (or cubic crystal with the coordinate axes along the cube directions) and place the current in the x direction and the magnetic field in the z direction. Then the application of the simplified transport equations [(47) and (48)] shown before, yields

$$J_x = [\sum_i \sigma_{xx}^{(i)}(H)]\, E_x + [\Sigma \sigma_{xy}^{(i)}(H)]\, E_y, \qquad H \equiv H_z,\; J \equiv J_x \tag{62}$$

$$0 = [\sum_i - \sigma_{xy}^{(i)}(H)]E_x + [\Sigma \sigma_{xx}^{(i)}(H)]E_y,$$

where $\sigma_{ik}^{(i)}$ refers to band i.

For the summations over the various bands contributing to the conduction process, σ_{xx} is always positive and, for non-pathological bands (the case of certain re-entrant energy surfaces has been discussed by Shockey),[32] σ_{xy} is negative for electrons and positive for holes. We consider again only the limiting cases of the Hall coefficient for very weak and very strong magnetic fields. For electrons and holes of concentrations n and p respectively, the results are[38]

$$R_o = -\frac{r_{on}\mu_n^2 n - r_{op}\mu_p^2 p}{|e|(\mu_n n + \mu_p p)^2}, \quad R_\infty = -\frac{1}{|e|(n-p)}, n \neq p. \qquad \begin{matrix} n,\text{electrons} \\ p,\text{holes} \end{matrix} \tag{63,64}$$

In the case of light and heavy holes, as is representative of the valence bands of a large number of semiconductors, the expressions for the Hall coefficient are

$$R_o = \frac{r_{o1}\mu_1^2 p_1 + r_{o2}\mu_2^2 p_2}{|e|(\mu_1 p_1 + \mu_2 p_2)^2}, \quad R = \frac{1}{|e|(p_1+p_2)}. \quad \begin{matrix} 1,\text{heavy holes} \\ 2,\text{light holes} \end{matrix} \quad (65,66)$$

The interesting observation, seen in both Eqs.(63) and (65), is that in the expression for the weak-field Hall coefficient the carrier densities are weighted by the squares of the respective mobilities. Thus a small concentration of minority carriers can exert a large effect if their mobilities are large. As the magnetic field is increased, however, the weighting factors become less controlling (this can be seen if the above sets of equations are developed as a function of magnetic field[39] rather than in the simplified forms representative of merely the low- and the high-field limits), and the mobility weighting factors are absent altogether in the high-field limit [as is seen in Eqs.(64) and (66)].

Concentrating on Eq.(63) for a moment, we see that a strongly p-type semiconductor with p=6400n will actually have a negative Hall coefficient R_o if $\mu_e > 80\mu_h$. As the magnetic field is increased, the Hall coefficient will cross over to positive values as approximation (64) begins to apply. An excellent example of this behavior[39a] is provided by the data on InSb shown in Figure 6. Because of the relatively high electron mobility, the specimen must be cooled to quite a low temperature before the extrinsic p-type Hall coefficient is attained. Actually, InSb has both heavy and light holes, but the electron mobility exceeds that of even the light holes. In semiconductors of this type, the Hall coefficient crossover moves to higher temperatures (i.e. larger values of electron density) as the magnetic field is increased. This is the behavior we mentioned above in connection with the transition from Eq.(63) to (64), and it is caused by the fact that the high-mobility carriers (which are closer to the strong-field approximation region) are electrons. Therefore an increase in magnetic field requires a greater relative density of electrons (hence a higher temperature) to maintain the Hall coefficient null.

An opposite situation to that discussed above is found in p-type germanium[39b] (see Fig. 7). Here it is the mobility of the light holes which is greater, by a significant amount, than that of the electrons. The Hall coefficient crossover therefore moves to lower temperatures (i.e. a larger relative density of light holes as compared to electrons) as the magnetic field is increased. This behavior is quite noticeable in the data presented in Figure 7. The dashed lines shown in the lower part of the figure represent a calculation for the case where there are no light holes, and therefore

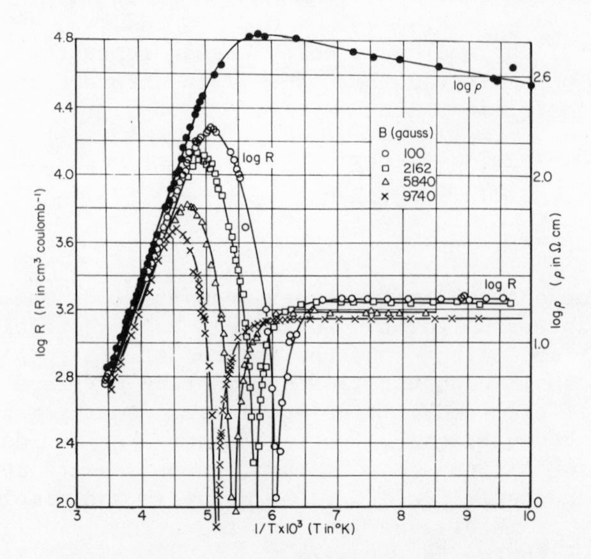

Fig. 6. Temperature dependence of the Hall coefficient of a p-type InSb speciment ($p \simeq 4 \times 10^{15}$ cm^{-3}) at different magnetic field strengths. The resistivity at zero field is also shown (after Howarth, Jones, and Putley).[39a]

the electron mobility is the larger (since it significantly exceeds that of the heavy holes). This was the situation in the InSb example. We also note in the figure that when the magnetic field becomes so large that all carriers approach the strong-field region, both curves assume similar slopes. Calling attention again to Fig. 6, I wish to point out that Hall data of the type shown can provide us with simple relationships involving carrier densities and electron-hole mobility ratios. Of value, besides the crossover temperature, is the ratio of the Hall maximum to the value at the extrinsic plateau. For a simple model with only electrons and one type of hole and a single acceptor level of density N_A and negligible activation energy, the carrier concentrations can be expressed by

$$p = n + N_A, \quad \text{or} \quad \frac{p}{N_A} = \frac{n}{N_A} + 1 = X + 1, \quad \text{where } X \equiv n/N_A \tag{67}$$

$$n = XN_A, \qquad p = (X + 1)N_A. \tag{68}$$

For simplicity, we assume that the Hall coefficient factors are the same for each band so that, with the use of relations (68), Eq.(63) can be put in the form

$$R_o = -\frac{r_o}{|e|N_A}\frac{(c^2-1)X - 1}{[(c+1)X+1]^2}, \quad c \equiv \mu_n/\mu_p, \quad r_{on} = r_{op} \equiv r_o. \tag{69}$$

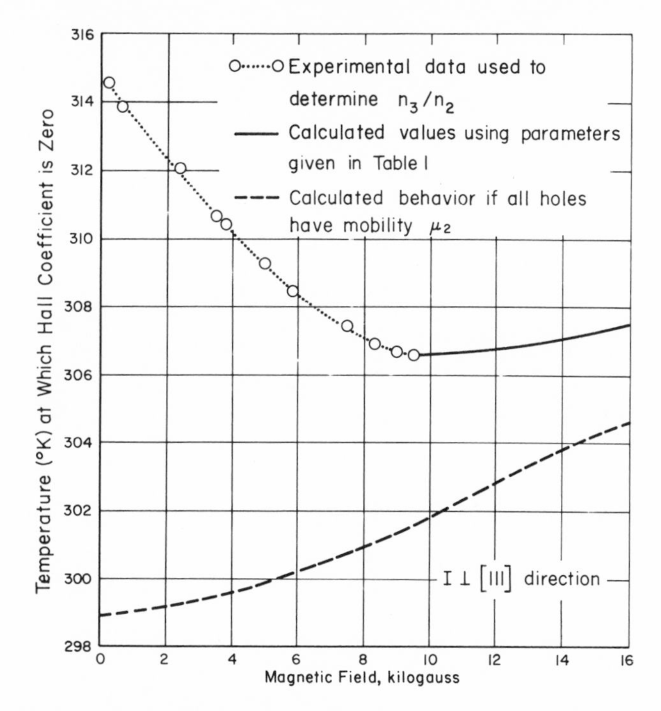

Fig. 7. Experimental data and calculated results of Hall coefficient crossover temperatues in p-type germanium ($p \simeq 4\times10^{13}$ cm^{-3}) as a function of magnetic field (after Willardson, Harman, and Beer).[39b]

It follows at once that at the Hall coefficient null

$$X = 1/(c^2-1), \quad \text{i.e. } n^{null} = N_A/(c^2-1). \tag{70}$$

With use of elementary calculus, we find that at the Hall coefficient maximum

$$X = 1/(c-1), \quad \text{i.e. } n^{max} = N_A/(c - 1). \tag{71}$$

One additional expression—simple, yet very useful—follows from the above equations. This is the ratio between the values of the weak-field Hall coefficient at its maximum and at its extrinsic plateau, namely

$$R_o^{max}/R_o^{ext} = - (c - 1)^2/4c. \tag{72}$$

This relationship depends only on the mobility ratio of the electrons and holes.

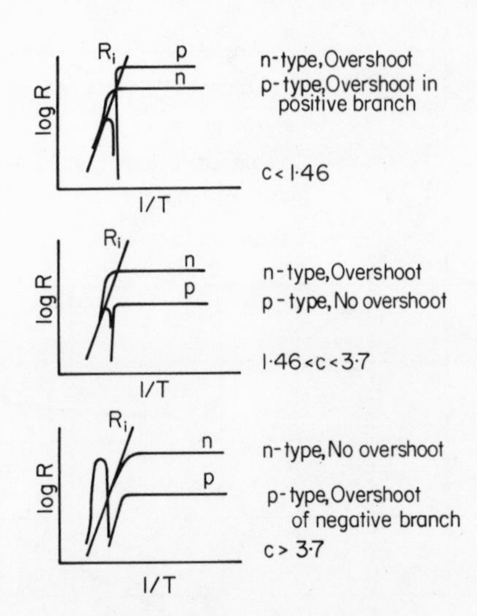

Fig. 8. Hall effect in mixed electron-hole conduction showing effect of different values of the electron-hole mobility ratio, c (after Putley).[40]

In concluding the discussion on p-type material, I cannot emphasize too strongly the importance—particularly for studies done on unfamiliar materials—of measuring magnetic-field dependences and temperature variations of the Hall effect. Think how ridiculous a result could ensue if one applied the simple formula $R = 1/pe$ when data were taken at magnetic fields which placed the material in the vicinity of the crossovers shown in Fig. 6!

The mobility ratio also influences the nature of the temperature dependence of the Hall coefficient in n-type material in the transition region between the extrinsic plateau and the intrinsic slope. This behavior has been studied by Putley.[40] and his results are shown in Figure 8. We see that the effects of the different mobility ratios are quite noticeable.

Although the crossovers are not involved, conspicuous two-band effects can be observed in extrinsic p-type semiconductors when light and heavy hole bands exist, as is the case, for example, in germanium, silicon, and common III-V compound semiconductors. Figure 9 shows the excursion of the Hall coefficient in extrinsic p-type germanium as the magnetic field is varied from the weak-field to the strong-field regions[39b]—Eq.(65) to Eq.(66). Note that for measurements done at constant magnetic fields of several kilogauss or less, the Hall coefficient shows a large temperature variation which does not exist if the data are taken at around ten kilogauss. This apparently anomalous temperature variation is more

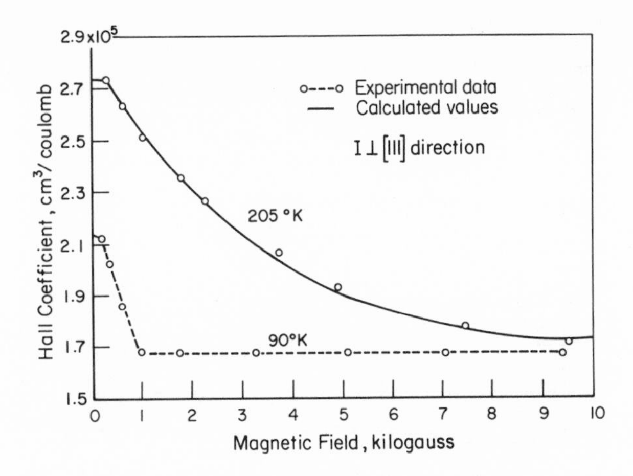

Fig. 9. Hall coefficient as a function of magnetic field at two temperatures for a p-type germanium specimen (after Willardson, Harman, and Beer).[39b]

explicitly revealed[40a] in the data plotted as a function of temperature in Figure 10. It is important to realize that in this case the naive application of the simple Hall effect relation $R = 1/pe$ would lead one to think that the hole density decreased substantially with increasing temperature—a thoroughly puzzling state of affairs. Actually what happens is that the temperature increase moves the material more toward the weak-field region so that the effect of the light holes becomes more controlling. But, you may say, the magnetic field was kept constant. The point to be emphasized is that it is the product μH which determines the existence of the weak- or strong-field region. Hence you may religiously maintain H constant, but when the mobility changes with temperature, as it does in most semiconductors, a temperature excursion will include the effects of a magnetic field variation. In contrast to the conduction resulting from the light- and heavy-hole bands, the transport due to the electrons—where only a single band is involved—shows normal behavior for $R\sigma/\mu$, according to the lower curve of Figure 10. Thus once again we see compelling reasons for including magnetic field dependences in any measurement of Hall effect.

Two-band effects can also be important in a number of extrinsic semiconductors where the bands in question are separated in energy but where carriers are activated into the higher-lying band at the measurement temperature. Then the carrier-concentration terms in Eqs.(65) and (66) are temperature-dependent. Common examples are found in a number of compound semiconductors, such as

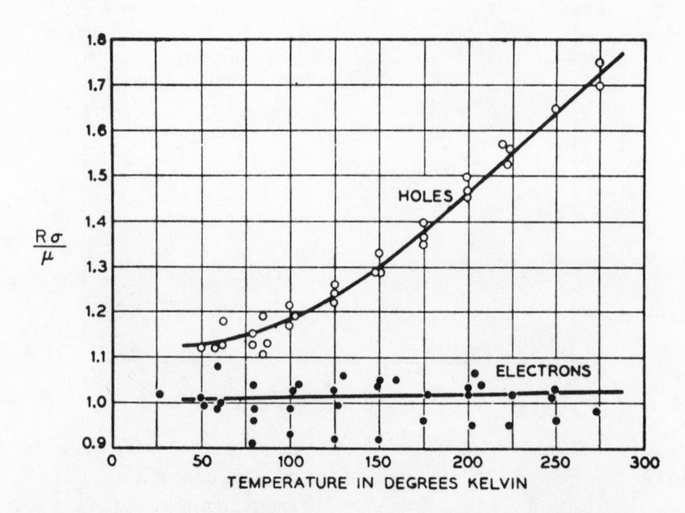

Fig. 10. Temperature dependence of $R\sigma/\mu$ for different specimens of n- and p-type germanium (after Morin).[40a]

the conduction bands in GaSb and, at high temperatures, GaAs. Note that in the III-V materials, this effect is in addition to the influence of the light- and heavy-hole bands, which are degenerate at the center of the Brillouin zone,[40b] and therefore there is no activation energy involving the light and heavy holes. Detailed analyse of Hall effect as a function of temperature in materials having conduction bands such as those mentioned above, have provided information on the separation in energy of the bands in question and on the ratio of the mobilities in those bands.[41,41a] Because of the separation in energy of the bands in question, it is readily possible that the charge carriers in one band may be degenerate (i.e. require the use of Fermi-Dirac statistics), while those in the other band can be described by the classical statistics.[41a,41b]

8. Two-Band Conduction with Warping and Mixed Scattering

Lest even at this point, one or two of the more blasé of my readers might become complacent, let me show some weird effects that can occur in the magnetic-field dependence of the Hall coefficient when we have two-band conduction involving a light-hole band and a warped heavy-hole band, each with different admixtures of acoustic phonon and ionized impurity scattering.[41c] Time and space will not allow much discussion—just a look at some graphs. Figure 11 shows p-type germanium with an interesting minimum. But let us not be partial to minima. We can take band characteristics representative of silicon and get some pretty maxima, as seen in

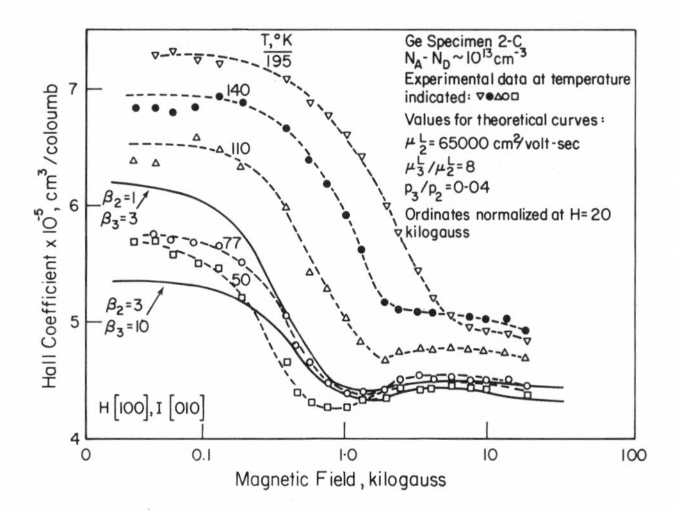

Fig. 11. Magnetic field dependence of Hall coefficient in p-type germanium for various temperatures. Theoretical curves (solid lines) are based on μ_2^L = 65,000 cm^2/volt-sec, representative of the mobility at 77°K and on normalization of ordinates at 20 kgauss for the β_2 = 1 curve to fit the experimental curves, the curve of β_2 = 3 being slightly depressed for clarity. Subscripts 2 and 3 refer to heavy- and light-mass bands respectively (after Beer).[41c]

Figure 12. A similar type of behavior was found to occur in type IIB diamond and in AlSb. Figure 13 presents such data. The resemblance to silicon is quite striking.

9. Inhomogeneities in the Measurement Sample

Finally, I want to devote a little time to show how inhomogeneities in the measurement sample can produce some disturbing results. However, there is only time to state that certain inhomogeneities can produce an admixture of Hall and magnetoresistivity effects—thereby causing large magnetic-field dependences for the Hall coefficient. This behavior is illustrated by a step-function model—the measurement parallelepiped having a discontinuity in Hall coefficient and resistivity. For $x < 0$, the respective values are denoted by R_1 and ρ_1; for $x > 0$, they are R_2 and ρ_2. These values are those of the material in the applied magnetic field, although in many cases R may vary only slightly with field or be constant. We shall omit details of derivation[42] and merely present the final expression for the Hall coefficient as measured across potential probes placed at $x = 0$, which defines

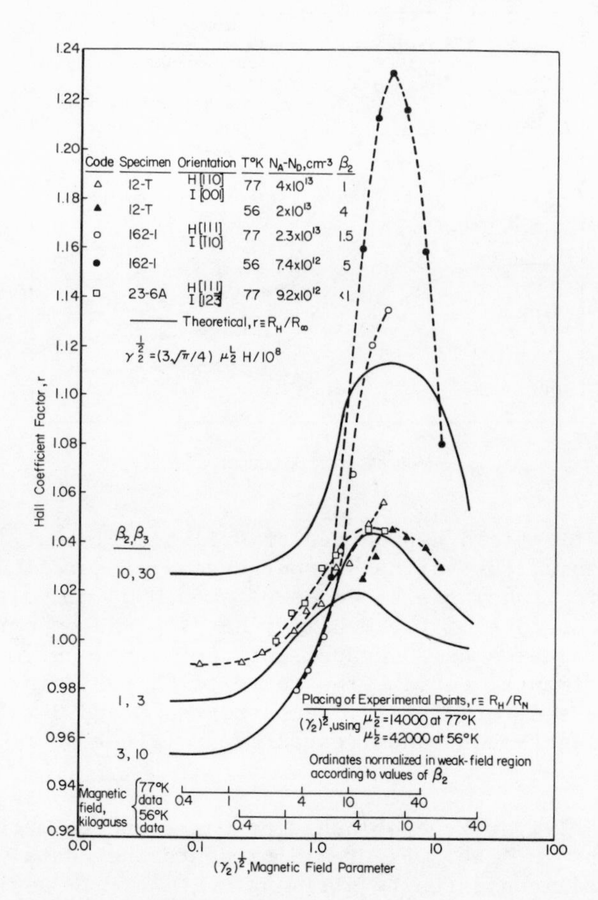

Fig. 12. Experimental data (dashed lines) indicating magnetic-field variation of Hall coefficient factor in extrinsic p-type silicon for two temperatures. Theoretical predictions for different amounts of ionized impurity scattering are shown by solid lines (after Beer).[41c]

the plane of discontintuity in R and ρ. The result is[42]

$$R_{x=0}^{meas} = (R_1\rho_2 + R_2\rho_1)/(\rho_1 + \rho_2). \tag{73}$$

Inasmuch as ρ can have a large variation with H, we see that this admixture of Hall and magnetoresistive effects may produce a large magnetic-field variation in the measured value of R, even if R_1 and R_2 should be essentially independent of H. The magnitude of the measured Hall coefficient may increase or decrease with magnetic field. For it to increase, the following condition must be

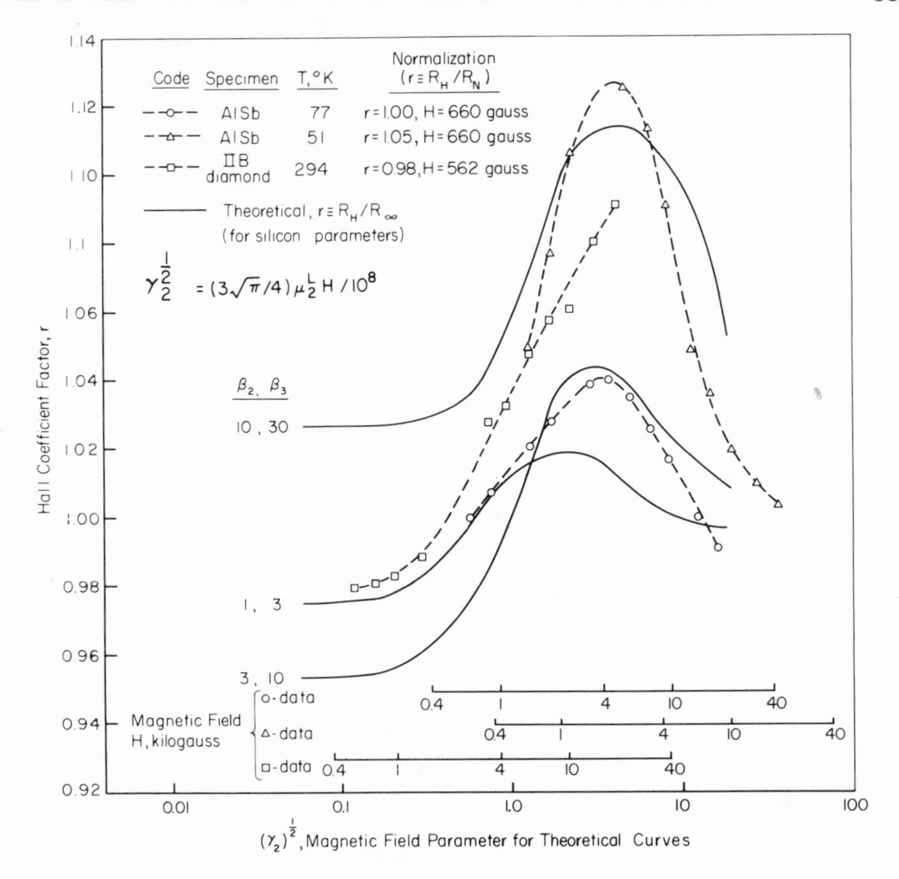

Fig. 13. Experimental data (dashed lines) showing magnetic-field dependence of Hall coefficient factor for p-type diamond and AlSb. Superimposed are solid-line theoretical curves (cf. Fig. 12) representative of silicon (after Beer).[41c]

satisfied:[42]

$$\frac{1}{R}\frac{\partial R}{\partial H} = \frac{\rho_1\rho_2}{\rho_1+\rho_2}\,\frac{R_2-R_1}{R_2\rho_1+R_1\rho_2}\left(\frac{1}{\rho_1}\frac{\partial\rho_1}{\partial H}-\frac{1}{\rho_2}\frac{\partial\rho_2}{\partial H}\right), \tag{74}$$

where $R \equiv R^{meas}_{x=0}$ of Eq.(73). In deriving the above relationship, we have neglected any changes in Hall coefficient with magnetic field in each material, since these changes are usually quite small compared to the resistive changes. Equation (74) tells us that the measured Hall coefficient will increase with magnetic field if the inherent magnetoresistance, $(1/\rho)\partial\rho/\partial H$, is larger in the region

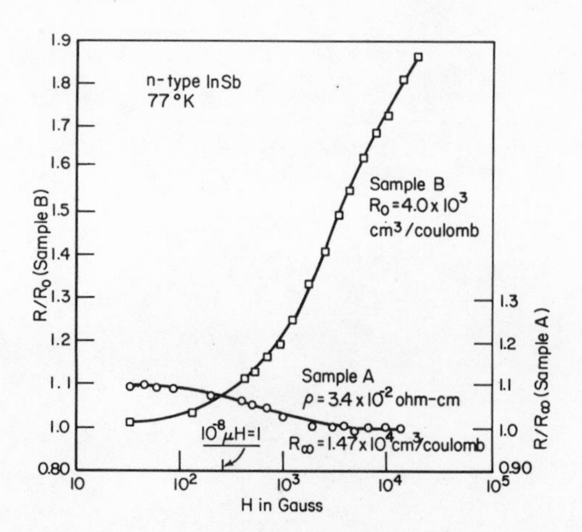

Fig. 14. Magnetic-field dependence of the normalized measured Hall coefficient in two high-purity n-type InSb samples. Sample A is homogeneous, but sample B exhibits a change in electron density of a factor of 10 along its length. Carrier concentrations are: sample A: 4.3×10^{14} cm^{-3}; sample B, 5×10^{14} to 5×10^{15} cm^{-3} (after Bate, Bell, and Beer).[42]

having the smaller inherent Hall coefficient.

An illustration of the type of behavior predicted by Eq.(73) is provided by two measurement samples of InSb—one of which was quite uniform, while the other exhibited a variation in carrier density ranging from $5x10^{14}$ to $5x10^{15}$ cm^{-3}. The data are shown in Figure 14. It is strinkingly evident that the presence of inhomogeneities can completely mask the normal variation in Hall coefficient factor with magnetic field.

Again we stress that much caution is necessary in interpreting Hall data taken on specimens of unknown character. One can obtain large distortions in apparent Hall mobility—either on the low or high side—depending on the nature of the inhomogeneities, placement of Hall and current probes, etc. An example of lowered mobility is seen in Figure 15 for the step-function resistivity variation. Mobility enhancement was studied by Wolfe and Stillman,[43,43a] who showed that multiple metallic inclusions can cause such behavior. The high apparent mobility is most commonly observed in compound semiconductors which are grown under metal-rich conditions. Temperature dependences of mobility as determined from the $R\sigma$ product for anomalous and normal GaAs samples were measured by Wolfe and co-workers, and are shown in Figure 16. Note that the ordinate is a logarithmic scale, so that the mobility augmentation is fairly large. Imagine the disappointment of a materials-preparation researcher who might first be led to believe that he had produced

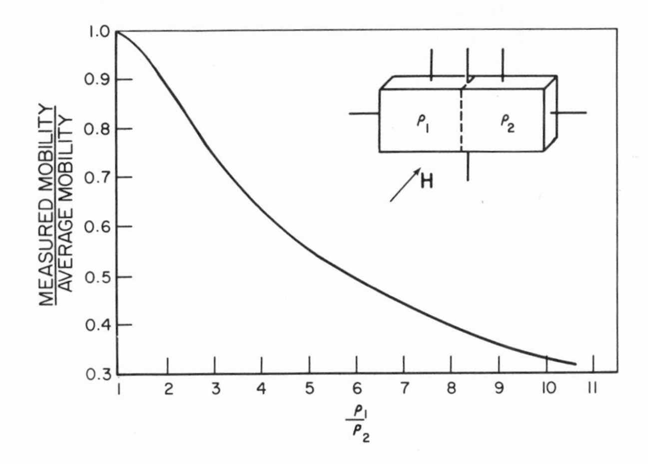

Fig. 15. Apparent Hall mobility as a function of the ratio of resistivities in a step-function model. Hall probes are placed on the plane of discontinuity. The normalization factor is the average of μ_1^H and μ_2^H (after Bate, Bell, and Beer).[42]

some super-duper once-in-a-lifetime material! I think it is abundantly clear that meaningful analyses of Hall data require checks on the homogeneity of the sample. Methods for doing this were mentioned earlier (Section 4).

10. Materials and Situations Requiring Special Treatment

Although we have covered a rather large number of complex situations that can be encountered in using Hall effect in a broad spectrum of practical cases, there are numerous types of materials and situations which require special treatment that has not been considered here. For example, certain metals have Fermi surfaces reminiscent of some kinds of modern sculpture. If open orbits are encountered, the behavior of the Hall coefficient with magnetic field and the interpretation of the Hall data are vastly complicated.[44] In ferromagnetic materials, an additional phenomenon comes into being—namely the "extraordinary Hall effect"—which is related to the magnetization.[45,46] At very strong magnetic fields in semiconductors, quantum effects appear[47] when the Landau level spacing becomes important—a situation which appears when $\hbar eH/m^*$ becomes larger than kT. Another consideration is hot carrier effects, which assume importance at strong electric fields when the current may no longer vary linearly with field.[48]

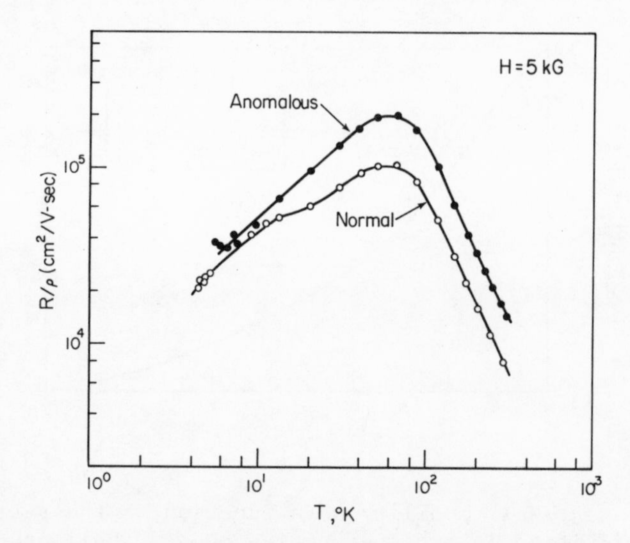

Fig. 16. Apparent mobility enhancement in GaAs. The anomalous specimen is believed to contain metallic inclusions, presumably gallium precipitates (after Wolfe et al.).[43a]

Finally, I would point out that the transport theory we have presented here is based on concepts which, in the simplest interpretation, may be related to the concept of a charge-carrier relaxation time or mean free path—quantities which are proportional to mobility.[49] There are materials, however, with very low mobilities. For such materials, I think it is apparent that our conventional transport theory will be in trouble when the mean free path becomes less than the interatomic distance.[50] A related consideration deals with the fact that conventional transport is treated by classical theory, based on the Boltzmann equation. Its validity requires that $\hbar/\tau$ be significantly greater than the Fermi energy.[49] For materials where some of the above conditions are not fulfilled, transport theory has been developed which is based on a hopping process.[50] Types of amorphous semiconductors are in the category requiring special treatment.

Some of the materials and situations mentioned above will be discussed later by other speakers at this conference.

V. CLOSING REMARKS

In closing, I would remark that we have seen how the Hall effect has far deeper substance and more arcane subtleties than one might ever guess from the simple free-electron relationships based on the Lorentz force. We have seen complexities and unexpected confrontations in measurement, in theory, and in interpretation; perturbation of sought-after fields by contacts and specimen geometries, by ubiquitous thermal effects, and by higher order galvanomagnetic coefficients—along with the unique influences of material anisotropies, mixed scattering processes, light and heavy charge carriers, warped bands, and specimen imperfections. We have seen how all these matters conspire to unfold a drama and portrait revealing ever-changing facets and challenges. Yet what a wealth of information can be extracted through diligent effort and study. But after all, is this not an example of the kaleidoscope of beauty and color that springs forth in science?

If now we had glasses of champagne, we would drink a toast to Edwin Hall. We would do it in the words of Shakespeare "How infinite in faculties" and in the poetry of Keats "A thing of beauty is a joy for ever."

ACKNOWLEDGMENT

The author wishes to thank Dr. Sherwood Fawcett of the Battelle Memorial Institute for his assistance in providing the conditions that made it possible to complete a work having the scope and magnitude of the present paper.

REFERENCES

1. H. Fritzsche, Methods of Experimental Physics in: "Solid State Physics," L. Marton, ed., Academic Press, New York (1959), Vol. 6, part B, pp. 145-160.
2. W. C. Dunlap, "An Introduction to Semiconductors," Wiley, New York (1957), pp. 178-194.
3. J.-P. Jan, in: "Solid State Physics," F. Seitz and D. Turnbull, eds., Academic Press, New York (1957), Vol. 5, pp. 17-24.
4. A. C. Beer, "Galvanomagnetic Effects in Semiconductors," Academic Press, New York (1963)., pp. 54-68.
5. E. H. Putley, "The Hall Effect and Related Phenomena," Butterworths, London (1960), pp. 23-55.
6. C. M. Hurd, "The Hall Effect in Metals and Alloys," Plenum Press, New York (1972), pp. 183-200.
7. See, for example p. 186 of Ref. 2; p. 20 of Ref. 3; pp. 57-59 of Ref. 4.
8. See, for example, P. P. Debye and E. M. Conwell, Phys. Rev. 93:693 (1954).
9. L. J. van der Pauw, Philips Research Repts. 13:1 (1958); 16:187 (1961).
10. Reference 3, pp. 22-24; Ref. 4, pp. 55, 56, and 66; Ref. 6, pp. 183-195.

10a. Reference 6, pp. 195-200.

10b. See, for example: T. Ambridge and C. J. Allen, Electronics Letters (Great Britain) 15:648 (1979).
11. Reference 4, p. 3.
12. See, for example, p. 7 of Ref. 4.
13. More discussion of this point can be found in Ref. 4, pp. 16 and 52.
14. For background information, see Ref. 4, pp. 6-9.
15. V. A. Johnson and F. M. Shipley, Phys. Rev. 90:523 (1953).
15a. Ref. 5, pp. 82-84.
16. See, for example, pp. 153, 154 of Ref. 1; pp. 183-185 of Ref. 2 pp. 21, 22 of Ref. 3; pp. 66, 346 of Ref. 4; pp. 49-53 of Ref. 5; pp. 193-195 of Ref. 6.
17. T. M. Dauphinee and E. Mooser, Rev. Sci. Instr. 26:660 (1955); discussed also in Ref. 3 (p. 22) and Ref. 4 (p. 66).
17a. See, for example, W. W. Scanlon, pp. 166-170 of Ref. 1; p. 191 of Ref. 2; pp. 9-11 of Ref. 3; pp. 66-68 of Ref. 4; pp. 26-32, 82-90 of Ref. 5; pp. 185-187, 190 of Ref. 6.
18. See, for example, M. Kohler, Ann. Physik 20:891 (1934); D. Shoenberg, Proc. Cambridge Phil. Soc. 31:271 (1935).
19. C. Goldberg and R. E. Davis, Phys. Rev. 94:1121 (1954); K. M. Koch, Z. Naturforsch. 10a:496 (1955); or see p. 69 of Ref. 4
20. H. B. G. Casimer and A. N. Gerritsen, Physica 8:1107 (1941); H. B. G. Casimer, Revs. Modern Phys. 17:343 (1945).
20a. Page 42 of Ref. 4.
21. L. Grabner, Phys. Rev. 117:689 (1960).
22. See, for example, L. P. Kao and E. Katz, Phys. and Chem. Solids 6:223 (1958); p. 69 of Ref. 4.
22a. See also, for example, W. M. Bullis, Phys. Rev. 109:292 (1958); pp. 86-91 of Ref. 4.
23. See, for example, pp. 58-65 of Ref. 4.
24. Specific examples are found in Ref. 4, pp. 142, 226-264. For comprehensive analyses of germanium-type semiconductors, consult E. G. S. Paige, The Electrical Conductivity of Germanium, in: "Progress in Semiconductors," A. Gibson and R. Burgess, eds., Wiley, New York (1964), Vol. 8, pp. 48-50. For inhomogeneity effects, see pp. 315-321 of Ref. 4. On p. 317, the equation and the figure apply to the case where the inherent magnetoresistance is small compared to the inhomogeneity effect. Otherwise, a factor ρ/ρ_o should multiply the brackets in Eq.(27.22) and a factor ρ_o/ρ, the ordinate in Fig. 50. For further details, consult R. T. Bate and A. C. Beer, J. Appl. Phys. 32:800 (1961).
24a. For more details, consult pp. 34-37 of Ref. 1; pp. 179, 187-190 of Ref. 2; pp. 56, 316-323 of Ref. 4.
24b. See, for example: F. Seitz, "The Modern Theory of Solids," McGraw-Hill, New York (1940), pp. 141, 316-319; R. A. Smith, "Wave Mechanics of Crystalline Solids," Chapman and Hall, London (1963), pp. 124-127; J. M. Ziman, "Principles of the Theory of Solids," Cambridge Univ. Press, Cambridge (1964), pp. 157-161.

25. See, for example, pp. 99-101 of Ref. 4.
25a. J. Kołodziejczak, Acta Phys. Polonica 20:379 (1961).
26. Further details can be found in Ref. 4, p. 332.
26a. Derivations of these results can be found in Ref. 4, pp. 109, 121, 124-126.
26b. Ref. 4, pp. 142-127.
27. C. Herring, Bell System Tech. J. 34:237 (1955).
28. R. Dexter, H. Zeiger, and B. Lax, Phys. Rev. 104:637 (1956); p. 184 of Ref. 4.
29. See, for example, M. Shibuya, Phys. Rev. 95:1385 (1954); pp. 247-251 of Ref. 4.
30. W. Mason, W. Hewitt, and R. Wick, J. Appl. Phys. 24:166 (1953).
31. A. C. Beer and R. K. Willardson, Phys. Rev. 110:1286 (1958); pp. 198-200 of Ref. 4.
32. William Shockley, "Electrons and Holes in Semiconductors," Van Nostrand, New York (1950), pp. 338-341.
33. C. Goldberg, E. Adams, and R. Davis, Phys. Rev. 105:865 (1957).
34. H. Miyazawa, "Proc. Intern. Conf. Phys. Semicond., Exeter, 1962," Inst. of Phys. and Phys. Soc., London (1962), p. 636.
35. R. S. Allgaier, Phys. Rev. 158:699 (1967).
36. R. S. Allgaier, Phys. Rev. 165:775 (1968).
37. R. S. Allgaier, Phys. Rev. B 2:3869 (1970); P. H. Cowley and R. S. Allgaier, Phil. Mag. 29:111 (1974); R. S. Allgaier and R. Perl, Phys. Rev. B 2:877 (1970).
38. See, for example, Harvey Brooks, in: "Advances in Electronics and Electron Physics," L. Marton, ed., Academic Press, New York (1955), Vol. VII, pp. 132 and 133; p. 148 of Ref. 1; pp. 148-167 of Ref. 4.
39. See, for example, Eq.(17.11) in Ref. 4.
39a. D. Howarth, R. Jones, and E. Putley, Proc. Phys. Soc. B70:124 (1957).
39b. R. K. Willardson, T. C. Harman, and A. C. Beer, Phys. Rev. 96:1512 (1954); also p. 165-168 of Ref. 4.
40. Ref. 5, pp. 115-119.
40a. F. J. Morin, Phys. Rev. 93:62 (1954).
40b. See, for example, Donald Long, "Energy Bands in Semiconductors," Interscience, New York (1968), pp. 100-109.
41. See, for example, L. W. Aukerman and R. K. Willardson, J. Appl. Phys. 31:939 (1960); R. D. Baxter, F. J. Reid, and A. C. Beer, Phys. Rev. 162:718 (1967).
41a. R. S. Allgaier, J. Appl. Phys. 36:2429 (1965).
41b. R. S. Allgaier and B. B. Houston, Jr., J. Appl. Phys. 37:302 (1966).
41c. A. C. Beer, J. Phys. Chem. Solids 8:507 (1959).
42. R. T. Bate, J. C. Bell, and A. C. Beer, J. Appl. Phys. 32:806 (1961).
43. C. M. Wolfe and G. E. Stillman, in: "Semiconductors and Semimetals," R. K. Willardson and A. C. Beer, eds., Academic Press, New York (1975), Vol. 10, pp. 175-220.

43a. C. Wolfe, G. Stillman, D. Spears, D. Hill, and F. Williams, J. Appl. Phys. 44:732 (1973).
44. See pp. 56-66 of Ref. 6.
45. A. W. Smith and R. W. Sears, Phys. Rev. 34:1466 (1929); E. M. Pugh, Phys. Rev. 36:1503 (1930); E. M. Pugh and N. Rostoker, Revs. Modern Phys. 25:151 (1953).
46. See also pp. 74-82 of Ref. 3; pp. 153-182 of Ref. 6.
47. See, for example, pp. 332-340 of Ref. 4.
48. E. M. Conwell, "High Field Transport in Semiconductors," Academic Press, New York (1967); also pp. 379-381 of Ref. 4.
49. See pp. 365 and 366 of Ref. 4. On p. 365, the factor m^*/m_o in Eq.(31.2) should be inverted.
50. Pages 372-379 of Ref. 4.

THE HALL EFFECT IN HEAVILY DOPED SEMICONDUCTORS

Donald Long and O. N. Tufte

Honeywell Corporate Material Sciences Center
Bloomington, Minnesota 55420

ABSTRACT

The properties of the Hall effect in relatively pure, lightly-doped semiconductors are now generally well understood.[1] However, the Hall effect and mobility in heavily-doped semiconductors, which have metal-like behavior, have been studied much less experimentally and are still not fully understood theoretically. This paper reviews the Hall effect and related properties of heavily-doped semiconductors for the cases of both uniform and non-uniform doping. The unexplained experimental results remain a challenge to theory, but the empirical knowledge of the Hall effect in heavily-doped semiconductors is useful in solid-state device technology.

1. INTRODUCTION

The Hall effect, along with the electrical resistivity, have been and continue to be the key parameters used in investigations of the basic electrical conduction processes in semiconductor materials. In this paper, we discuss the results of the Hall effect and resistivity studies of the electrical conduction and carrier scattering mechanisms in a variety of semiconductor materials uniformly doped with high concentrations of known impurities. Surprisingly, the electrical conduction processes are even today not well understood in heavily doped semiconductors. The application of the Hall effect to the analysis of impurity concentration and distribution in non-uniformly doped semiconductors has proven to be a key tool in the investigation of diffusion and ion implantation processes. The application of the Hall effect to non-uniformly doped semiconductors is discussed along with examples of results in areas of importance for electronic devices and sensors.

The reason for the importance of the Hall effect in the investigation of electrical conduction mechanisms comes from the simple relation between the Hall Coefficient, R and the free carrier concentration in the solid, n given by

$$R = A \frac{1}{ne}, \tag{1}$$

where e is the charge of the carrier. The parameter A is called the "Hall factor" which depends in detail on statistics of the carriers and their scattering but is always very close to unity. For heavily doped materials that are statistically fully degenerate, the factor is unity. The electrical resistivity ρ, is defined by

$$\rho = \frac{1}{ne\mu}, \tag{2}$$

where μ is the conductivity mobility, i.e., the carrier velocity per unit of applied electric field. The ratio of the Hall coefficient and the resistivity is the Hall mobility, given by

$$\mu_H = \frac{R}{\rho}, \tag{3}$$

where the Hall factor is assumed to be unity.

The Hall mobility is a key parameter in the investigation of electrical conduction mechanisms in solids. It is directly related to the microscope scattering of charge carriers by defects or lattice atoms by the relation

$$\mu_H = e \frac{\tau}{m^*} , \tag{4}$$

where τ is the scattering time and m^* the carrier effective mass. Theoretical analyses of scattering processes in solids leads to the calculation of τ, which allows the predictions to be compared with experiments through Eq. 4.

We need to define the meaning of a "heavily-doped" semiconductor and distinguish it from a "lightly-doped" semiconductor for the purposes of this paper. The distinction is illustrated in Fig. 1. In Fig. 1a, a typical lightly-doped n-type semiconductor is represented; the discrete donor impurity energy levels lie within the energy gap, below the conduction band edge. In Fig. 1b, on the other hand, the doping is so heavy that the donor impurity energy levels are merged with the conduction band and indistinguishable from it; this merging occurs because with heavy doping, the impurities are near each other and their electron wave-functions overlap

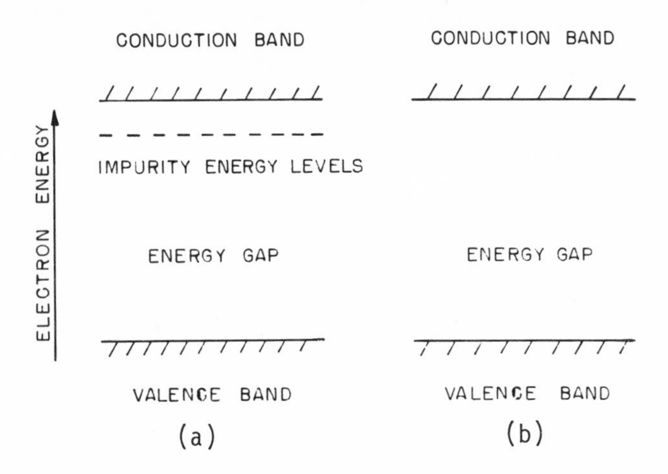

Fig. 1. Electronic energy level diagrams for (a) a typical lightly-doped semiconductor (n-type in this example) with discrete donor levels in the energy gap, separated from the conduction band, and (b) a heavily-doped semiconductor, in which the impurity energy levels are merged with and indistinguishable from the band.

to form a continuum with the conduction band.

The material of Fig. 1a has typical semiconductor-like properties. The Hall coefficient is very large at low temperatures because nearly all the electrons are in the impurity energy levels, and thus not free for conduction. As the temperature rises, the Hall coefficient decreases rapidly (exponentially) as electrons are thermally excited into the conduction band. Finally, at high temperatures all the donor-derived electrons are in the conduction band, the carrier concentration varies no more with temperature, and the Hall coefficient has a constant and relatively low value.

The heavily-doped n-type material of Fig. 1b behaves like a metal rather than like a typical semiconductor. This is because all the donor-derived electrons are in the conduction band at all temperatures, as in a metal. The Hall coefficient in this case is independent of temperature, except for the possibility of a minor temperature dependence of the Hall factor.

2. UNIFORMLY-DOPED SEMICONDUCTORS

In this section, we will consider the Hall effect and related properties in semiconductors which are uniformly doped. By uniformly doped we mean that the semiconductors are doped with

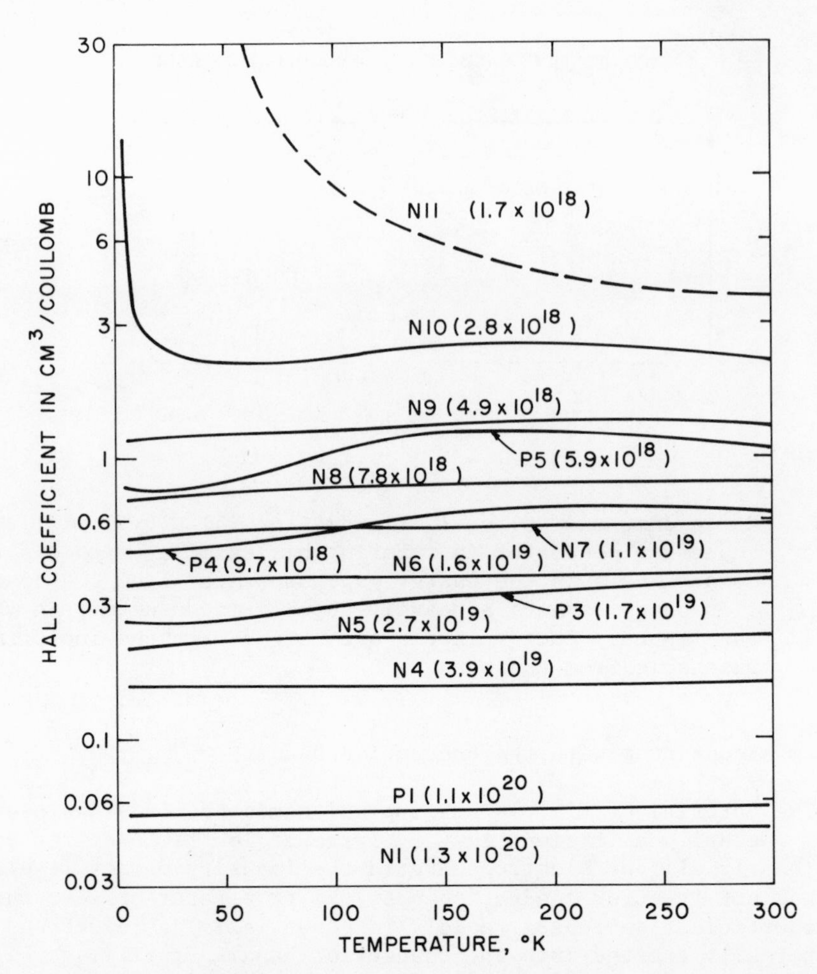

Fig. 2. Hall effect vs temperature in heavily-doped n-type (N1, etc.) and p-type (P1, etc.) Si samples.[2] Carrier concentrations in cm^{-3} are in parenthesis.

electrically effective impurities during the growth of single-crystal material, so that the impurity dopants are uniformly distributed throughout the volume of the crystal. And we will concentrate on heavily-doped material in which dopant concentrations lie in a range from nearly 10^{17} to over 10^{20} cm^{-3}. Of course, these are still relatively small impurity concentrations compared to other classes of materials; e.g., there are 5 x 10^{22} cm^{-3} Si atoms in a Si crystal, so that even a 10^{20} cm^{-3} dopant concentration in Si would represent only 0.2% impurity content.

Nearly all of the large body of research done on the Hall effect in uniformly-doped semiconductors has involved lightly-doped

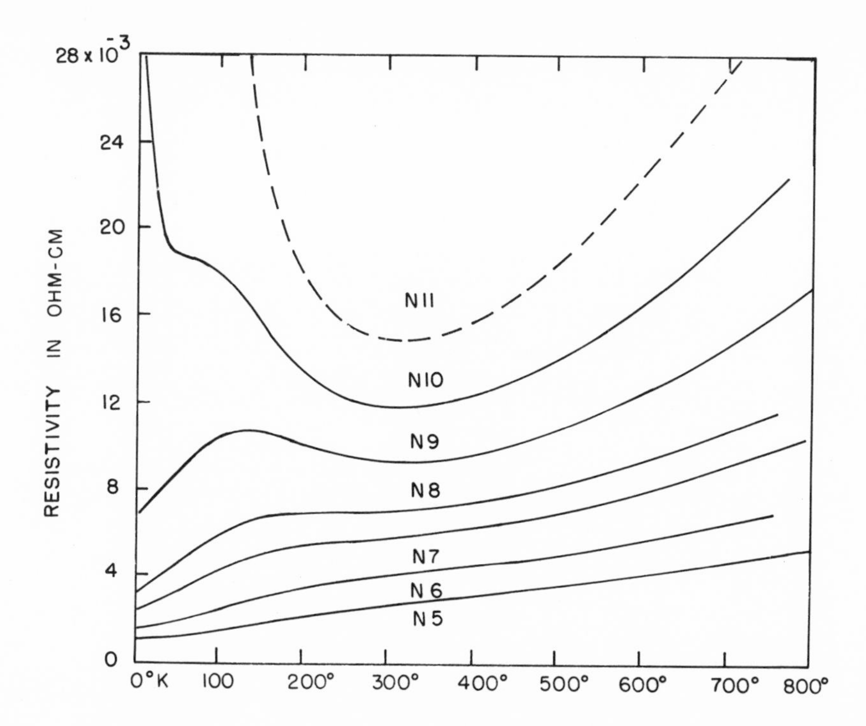

Fig. 3. Resistivity vs temperature in the n-type Si samples of Fig. 2.

material, and the characteristics of the Hall effect, resistivity, and mobility in such material are now quite well understood.[1] The research emphasis on lightly-doped material has been due to the fact that this material is of primary importance in solid-state device technology. On the other hand, relatively little research has been done on heavily-doped semiconductors.

Let us now examine what is known experimentally about the Hall effect, resistivity, and mobility in uniformly heavily-doped semiconductors.[2-5] First consider Si, the best-known and generally best-understood semiconductor material due to its widespread use in solid-state electronics. Figure 2 shows Hall coefficient vs temperature curves for both n- and p-type samples of doping concentrations ranging from 1.7 x 10^{18} cm^{-3} in sample N11 up to 1.3 x 10^{20} cm^{-3} in N1. Several characteristics of these curves are noteworthy. First of all, the Hall coefficient is smaller the higher the doping concentration, as expected; these Hall effect curves have, of course, been used to deduce the concentrations in the samples. Sample N11 shows a rather strong temperature dependence of the Hall coefficient, which increases with decreasing temperature; this is evidence of semiconductor-like behavior, corresponding to the energy

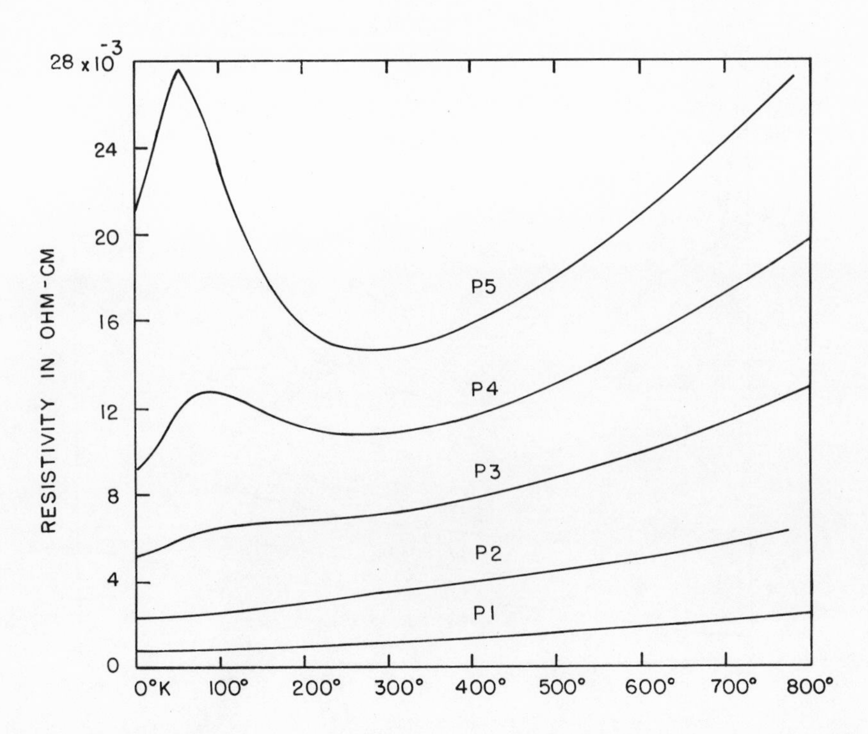

Fig. 4. Resistivity vs temperature in the p-type Si samples of Fig. 2.

level model of Fig. 1a. Thus there are still discrete impurity energy levels separated from the conduction band in sample N11. However, in sample N10, which contains 2.8×10^{18} cm^{-3} impurities, less than twice as many as in N11, the Hall coefficient vs temperature curve is very different. Then in all the other n-type samples, the Hall coefficient is nearly independent of temperature, meaning that the carrier concentration must be independent of temperature, corresponding to the energy level model of Fig. 1b; this is what is observed also in typical metals. Thus a conversion of n-type Si from semiconductor-like to metal-like behavior occurs over a narrow range of doping concentration, at about 3×10^{18} cm^{-3}. This abrupt transition has been studied and explained by Mott[6] and others and is called the "Mott transition"; it occurs in materials other than Si also.

The p-type Si curves in Fig. 2 show more temperature variation even at heavy doping levels than the n-type curves. These variations must be due to a rather weak temperature dependence of the Hall factor, rather than to any carrier concentration dependence on temperature, since the temperature variations are not

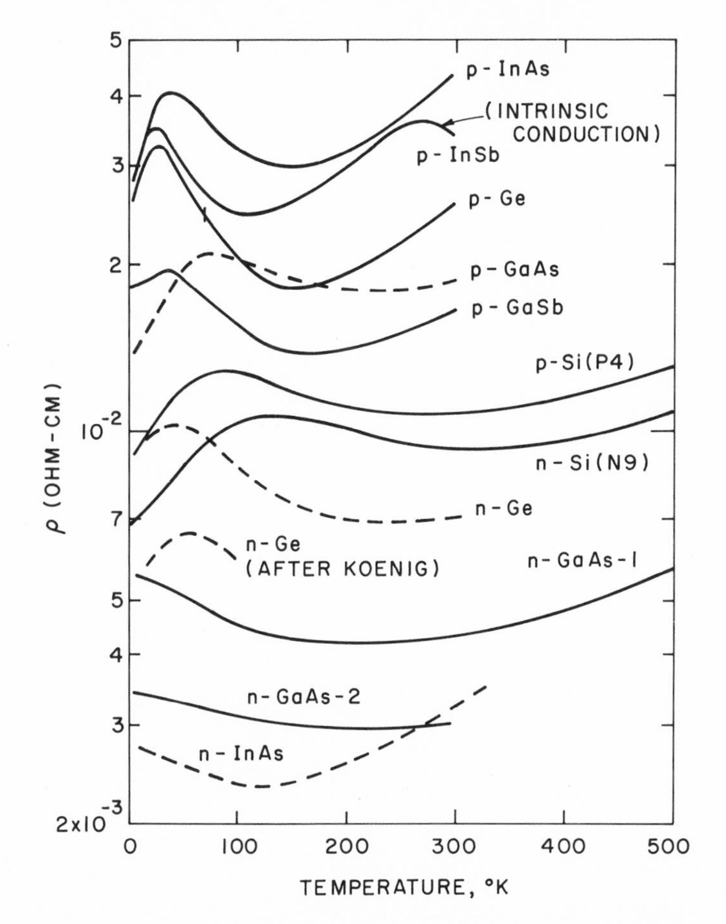

Fig. 5. Resistivity vs temperature in samples of a number of different semiconductors, both n-type and p-type.[3,4]

semiconductor-like. Sample P5 has a dopant concentration of 5.9×10^{18} cm^{-3} and P1 of 1.1×10^{20} cm^{-3}. The Mott transition in p-type Si occurs at a doping concentration lower than that of sample P5.

Now let us confine our interest to doping levels higher than that of the Mott transition. At these higher levels, the carrier concentration is constant (independent of temperature), as deduced from the Hall effect. Figures 3 and 4 show resistivity vs temperature curves for the n-type and p-type samples of Fig. 2. We immediately note an interesting pattern in these curves. A pronounced resistivity maximum appears at low temperatures in the less heavily doped samples, and as the doping increases, the maximum becomes less pronounced and moves to high temperatures, finally being absent in

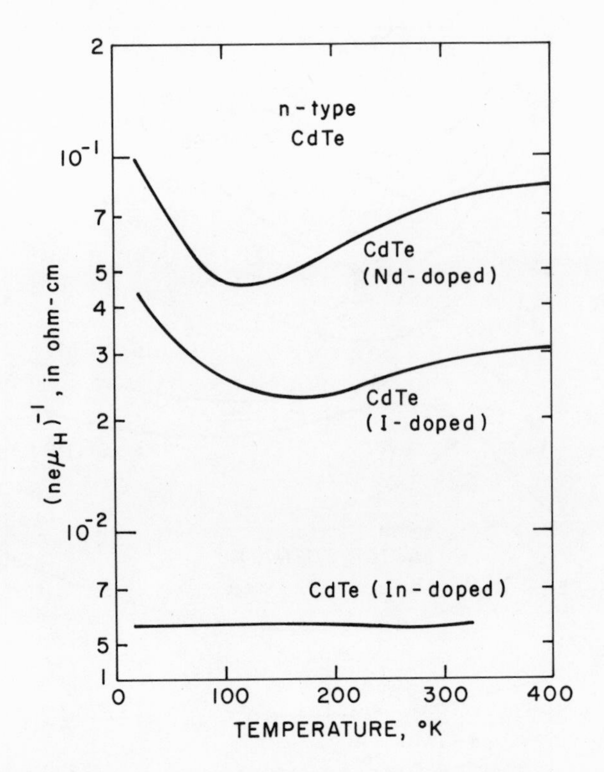

Fig. 6. Resistivity vs temperature in three samples of heavily-doped n-type CdTe.[5]

the most heavily doped samples. This resistivity maximum must correspond to a minimum in the mobility vs temperature, since the Hall effect has shown the carrier concentration to be constant.

Let us consider next the Hall effect and resistivity in other semiconductors.[3-5] We find that Ge, InSb, InAs, GaAs, GaSb, and CdTe all exhibit Hall coefficient vs temperature curves which are flat (except for minor variations like the p-type Si samples of Fig. 2) at concentrations above the Mott transition. Thus these other semiconductors are metal-like above the Mott transition also. Figures 5, 6, and 7 show resistivity vs temperature curves for several samples of these six additional semiconductors. Again we see resistivity maxima like those in Si; however, no such maxima occur in n-type InAs, GaAs, GaSb, and CdTe samples. The absence of maxima and the resistivity curve shapes in these n-type samples are significant. Figure 7 shows even heavily-doped n-type GaSb samples ranging from 5.9×10^{16} cm^{-3} carrier concentration in sample A to 1.54 to 10^{18} cm^{-3} in sample G; even over this wide doping range no maxima are seen, but the least heavily doped samples have shallow

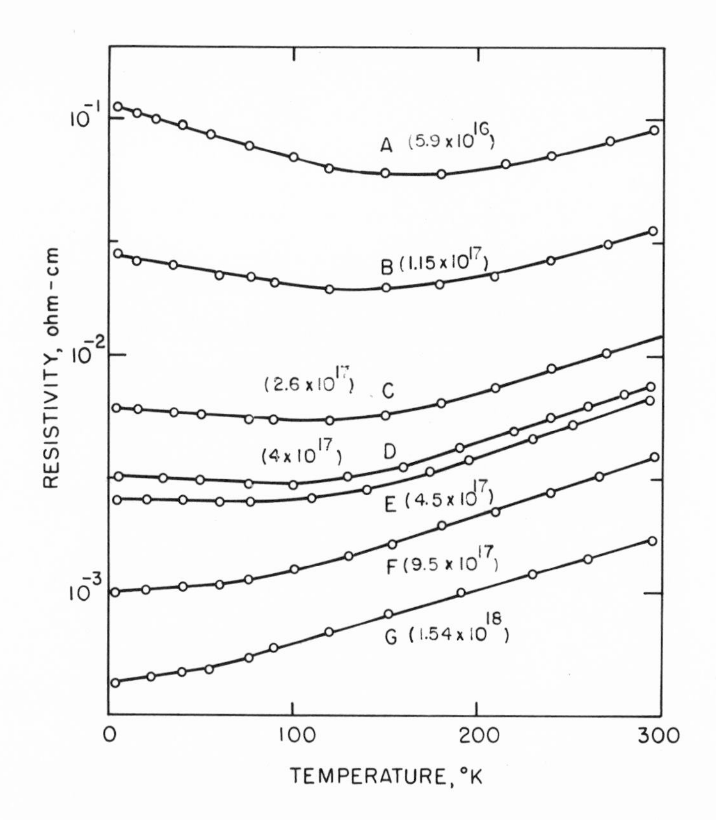

Fig. 7. Resistivity vs temperature in heavily-doped n-type GaSb samples.[4] Carrier concentrations in cm^{-3} are in parentheses.

minima, while the most heavily doped samples have resistivity vs temperature curve shapes like that of any ordinary metal (Cu, Au, etc.).

We have found in fact from the experiments on a number of different semiconductors of both n- and p-type that there is a perfect correlation as follows:

ρ vs T curve shape	Structure of band edge
• resistivity maximum with properties described above for Si	• complex ("many-valley" or degenerate)

ρ vs T curve shape	Structure of band edge
• no maximum, but shallow minimum with hint of trend toward maximum if "negative temperature" were possible	• single k=0 conduction band minimum

See Fig. 8 for definition of the band edge structures. The above correlation has not yet been fully explained theoretically.

In 1963-1965, we proposed and discussed a model in which the resistivity maximum is caused by a "resonant" scattering of the

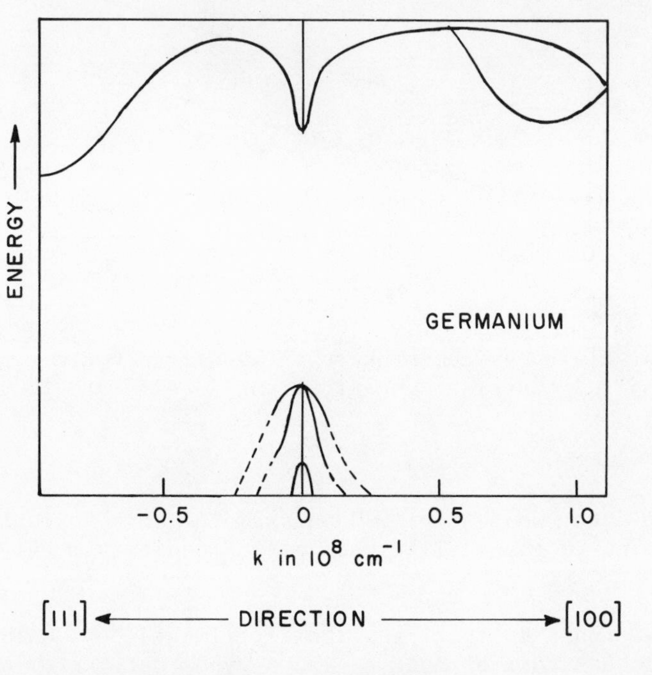

Fig. 8. Energy band structure of Ge as an example; the lowest conduction band minima lie along <111> axes in k-space and give a "many-valley" (actually 4-valley) conduction band edge. In Si, the lowest conduction band minima are those on the <100> axes, giving another "many-valley" conduction band edge. The k=0 minimum lies lowest in the conduction bands of InAs, InSb, GaAs, GaSb, and CdTe. The valence band maxima of all the p-type semiconductors studied are like that of Ge, being degenerate at k=0 and having non-spherical surfaces of constant energy of the current carriers (holes).

current carriers by the ionized impurities,[2-4] leading to a lower mobility and, therefore, a higher resistivity; the resistivity goes through a maximum at the temperature of greatest resonance. The n-type semiconductors in which a resistivity maximum does not occur show a movement toward resonance as the temperature drops, but 0°K is reached before the scattering parameters can achieve a resonant condition. The resonant scattering model is an intuitively appealing explanation but may not be fully provable by theory; we were not able to formulate a rigorous, quantitative theory.

Subsequently, other researchers have continued to study these anomalous electrical properties of heavily-doped semiconductors, concentrating on n-type Ge in most of the work. One can convert n-type Ge from a "many-valley" band edge to a single k=0 conduction band by applying a uniaxial stress to destroy the crystal symmetry and lower one valley below the others; the "single-valley" Ge then has a ρ vs T curve shape like the other single-valley semiconductors (n-GaAs, n-InSb, etc.). Most of this research has been done by Katz,[7] by Krieger,[8] and by Sosaki and coworkers.[9] Various theoretical hypotheses other than resonance impurity scattering have been proposed, but there is not yet an adequate theory.

Regardless of the correct theoretical explanation, the fact is that Hall effect and related measurements have shown a systematic pattern in the mobility vs temperature behavior of heavily-doped semiconductors which is perfectly correlated qualitatively with the energy band edge structure of the semiconductor. This pattern begs for a satisfactory, comprehensive theoretical explanation; it remains an open question and a challenge to theorists.

3. NON-UNIFORMLY DOPED SEMICONDUCTORS

The Hall effect has also proven to be a very effective tool in the evaluation of impurity densities and distributions in heavily doped semiconductors having non-uniform distributions of impurities. The non-uniform doping is typically introduced by diffusion or ion implantation and results in a doping profile that is uniform in the plane of the surface and is non-uniform perpendicular to the surface. Both of these are key processes in semiconductor device fabrication and it is essential to know the actual impurity concentrations and profiles in order to compare device characteristics with analytical models.

The physics of the Hall effect in a material that is non-uniformly doped perpendicular to the surface can be seen by considering a simple two-layer model, as shown in Figure 9. Each layer is assumed to be uniformly doped and characterized by a free carrier concentration n, equal to the impurity concentration (all impurities —fully ionized), a carrier mobility μ, and thickness t. The

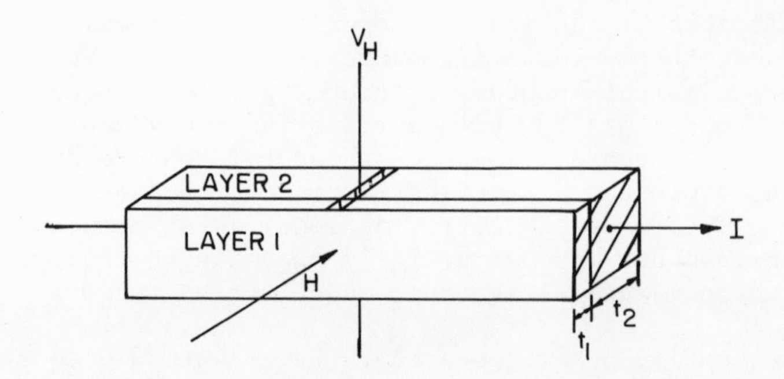

Fig. 9. Arrangement considered for the analysis and measurement of the Hall effect in a material consisting of two layers.

current applied to the sample is divided between the two layers, in direct proportion to their conductivity and in the presence of a magnetic field, a Hall voltage given by $\frac{R_{1,2}H_{1,2}I_{1,2}}{t_{1,2}}$ will be generated in each layer. The top and bottom surfaces, where the Hall voltage is measured, are equipotential surfaces, so the difference in Hall voltages will give rise to a circulating current between the two layers. The measured Hall coefficient for this simple two-layer case is given by[10]

$$R = \frac{et(n_1\mu_1{}^2t_1 + n_2\mu_2{}^2t_2)}{[|e|(n_1\mu_1t_1 + n_2\mu_2t_2)]^2} , \tag{1}$$

where e is the charge of the current carriers. For multi-layer structures, this expression can be extended by the addition of terms to Eq. 1. In the limit of a continuously varying ionized impurity profile, the Hall coefficient is given by

$$R = \frac{te\int_o^t n(x)\mu^2(x)dx}{\left[|e|\int_o^t n(x)\mu(x)dx\right]^2} . \tag{2}$$

An early application of the Hall effect in non-uniformly doped silicon was in the evaluation of the doping levels of diffused layers in silicon.[11] In this case, the impurity concentration N,

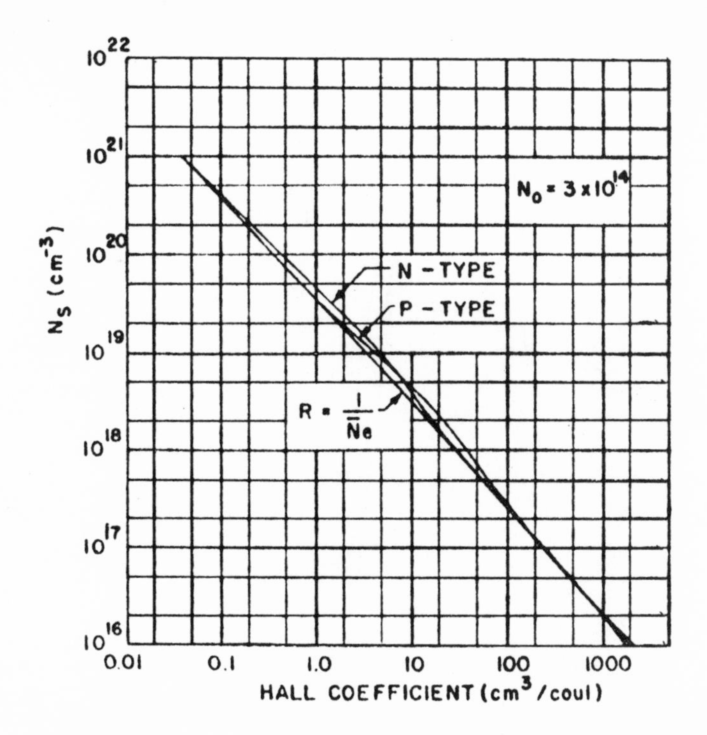

Fig. 10. Comparison of the surface concentration N_s vs average Hall coefficient curves for n- and p-type diffused layers on silicon along with the mobility independent Hall coefficient curve.[11] (Reprinted by permission of the publisher, The Electrochemical Society, Inc.)

normal to the sample surface is predetermined by the diffusion conditions and for a constant impurity source during diffusion is given by the complementary error function distributions, i.e.,

$$N(x) = N_s(1-\text{erf}\,\frac{x}{2\sqrt{D\tau}} - N_o) \tag{3}$$

where N_o is the bulk impurity concentration, N_s the surface concentration, x the distance from the surface and $D\tau$ the diffusion coefficient-time product. The diffused layer is characterized by a thickness t, where the concentration of diffused impurities equals the background concentration N_o, and if they are of opposite type, a p-n junction is formed. The surface impurity concentration can be directly related to the measured Hall coefficient of the diffused layer through Eq. 2. The unique advantage of the Hall effect for measurement of N_s, is that the results are very nearly independent of the variation in the carrier mobility with doping level. This is shown in Fig. 10, where the relation between N_s and R has

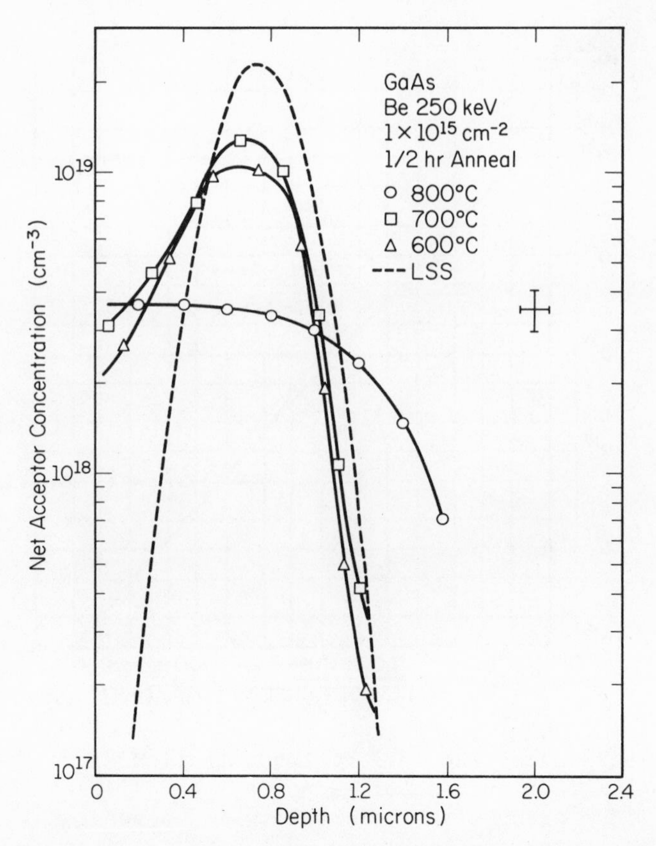

Fig. 11. Net acceptor concentration profile obtained from differential Hall effect measurements on GaAs implanted with beryllium. (Ref. 13).

been numerically calculated from Eq. 2 for n- and p-type diffused layers assuming specific variations in μ with doping level, and for the case of no mobility variation with doping.[11] Clearly, the Hall coefficient is very nearly a measure of the average carrier density and only a weak function of the mobility. Thus, the accuracy of the N_s values obtained by the Hall effect method can be much greater than the accuracy of the mobility values. This is not true for the determination of N_s by conductivity measurements, which is most commonly used. This makes the Hall effect method particularly useful for diffusion studies on less well known semiconductor materials and where a high degree of accuracy is desired in silicon or germanium.

The analysis of the Hall effect in diffused layers also clearly showed the potential for using diffused layers in Hall effect sensors. The Hall coefficient is approximately eight times larger than the coefficient corresponding to the surface concentration of the layer for all concentrations. Since the Hall voltage is inversely proportional to the layer thickness, the thin layers resulting from diffusion processes result in large voltages.

Another important application of the Hall effect in non-uniformly doped materials is in the actual measurement of impurity profiles in diffused or ion-implanted layers.[12] Neither the depth distribution of ionized impurities nor the influence of lattice disorder on mobility is generally known so the Hall effect is required. In this case, a measurement of the Hall coefficient is made for the layers and then a small portion of the layer is removed by sectioning techniques (etching). The Hall coefficient is then remeasured and the Hall coefficient, and hence carrier concentration, can be determined from Eq. 1. This technique has been applied extensively to the analysis of diffused and ion-implanted layers. An example of the results obtained by this method is shown in Fig. 11, which shows the carrier concentration profile obtained by ion implantation of beryllium in GaAs[13] as a function of the subsequent annealing temperature. The carrier mobility in the implanted layer can also be determined as a function of doping level (implant depth). Direct measurement of the free carrier concentration profile, and consequently the ionized impurity profile within ion-implanted and diffused layers is essentially in both the quantitative analysis of these processes and the understanding of the physics of electronic devices made with these processes.

ACKNOWLEDGEMENTS

We acknowledge the close collaboration of P. W. Chapman, R. J. Hager, E. L. Stelzer, and J. D. Zook in the Honeywell research on heavily-doped semiconductors, without which this paper would not be possible.

REFERENCES

1. E. H. Putley, "The Hall Effect and Related Phenomena," Butterworths, London (1960).
2. P. W. Chapman, O. N. Tufte, J. D. Zook, and D. Long, J. Appl. Phys. 34:3291 (1963).
3. D. Long, J. D. Zook, P. W. Chapman, and O. N. Tufte, Solid State Comm. 2:191 (1964).
4. D. Long and R. J. Hager, J. Appl. Phys. 36:3436 (1965).
5. D. Long and R. J. Hager (unpublished results for n-type CdTe).
6. See, for example, N. F. Mott, Physics Today 31:42 (1978), and references therein.
7. M. J. Katz, Phys. Rev. 140:A1323 (1965); M. J. Katz, S. H. Koenig and A. A. Lopez, Phys. Rev. Lett. 15:828 (1965).

8. J. B. Krieger et al., Phys. Rev. B3:1262 (1971).
9. W. Sosaki, Prog. Theor. Phys. Suppl. (Japan) 57:225 (1975); see also references therein.
10. R. L. Petritz, Phys. Rev. 110:1254 (1958).
11. O. N. Tufte, Jour. Electrochem. Soc. 109:235 (1962).
12. J. W. Mayer, O. J. Marsh, G. A. Shifrin, and R. Baron, Can. J. Phys. 45:4073 (1967).
13. W. V. McLevige, M. J. Helix, K. V. Vardyanathan, and B. G. Streetman, J. Appl. Phys. 48:3342 (1977).

ELECTRON CORRELATION AND ACTIVATED HALL MOBILITY

C. J. Adkins

Cavendish Laboratory
Cambridge, CB3 OHE, U. K.

ABSTRACT

Experiments with inversion layers show clearly that the Hall mobility becomes activated at low temperatures when the carrier concentration is low. After a brief account of the inversion layer system, the observed behaviour of the Hall effect and the conductivity is described. Neither is consistent with independent particle mobility-edge and percolation models. It is shown that electron correlation must be important at low carrier densities, and the electron liquid model is described, according to which, when correlation is dominant, the carriers become localized in the Wigner and flow past the background disorder like a viscous liquid. It is suggested that correlation must always dominate sufficiently close to any metal-insulator transition so that activated Hall mobility should generally be observable in low carrier density systems at low temperatures.

1. INTRODUCTION

In this paper, we are concerned with the effects of correlation on electrical transport properties. For correlation to dominate carrier dynamics, correlation energies must be larger than other relevant energies. This means that the effects of correlation are seen most clearly in experiments carried out at low temperatures on systems containing a low density of carriers. The clearest evidence for the failure of independent particle models comes from measurements on inversion layers, so the discussion is presented in relation to that system.

Inversion layers provide an ideal experimental system for the investigation of the fundamental physics of electrical transport at low carrier densities. They are unique in that the concentration of charge carriers may be varied directly and easily over a very large range in a single device. Also, the carrier concentration is known independently of Hall measurements. Transport occurs in a region subject to disorder so that the first carriers to enter the inversion layer are localized at low temperatures and the system exhibits a classic metal-insulator transition as their concentration is changed. This aspect—the metal-insulator transition—has initiated many recent studies of transport properties near the threshold of metallic conduction. These have been reviewed in some detail in an earlier publication (Adkins 1978a).

In this paper, we shall give a brief description of the inversion layer system and review some of the experimental results obtained with it. We shall show that they are not consistent with independent particle models but that they can be explained by assuming that correlation dominates carrier dynamics. This limit is described by the electron liquid model according to which, when correlation dominates, the carriers become localized in the Wigner sense to form an electron fluid which flows past disorder like a classical viscous liquid. Although we are primarily concerned with the Hall effect, this cannot be interpreted without reference to the basic mechanism of conduction, so the discussion covers both aspects of charge transport: conductivity and the Hall effect.

2. THE INVERSION LAYER SYSTEM

An inversion layer is formed at a semiconductor surface when a sufficiently large electrostatic field is applied normal to the surface to induce minority carriers at the surface. Figure 1 illustrates the situation when an electron inversion layer is formed on a p-type semiconductor. The electrostatic field causes a drop in electron potential towards the surface, holes are repelled, and, when the band bending is sufficient, the bottom of the conduction band near the surface falls below the Fermi level to form a narrow, nearly triangular, potential well in which electrons are bound. The well is so narrow that the energy separation of the lowest levels for motion perpendicular to the surface is typically about 20 meV and at low temperatures all carriers are in the lowest level. Motion parallel to the surface is essentially unquantized, however, so that, at low temperatures, the electrons should form a strictly two-dimensional Fermi gas. The inversion layer is electrically isolated from the bulk semiconductor by the intervening depletion layer, the region from which majority carriers are excluded by the penetration of the electrostatic field.

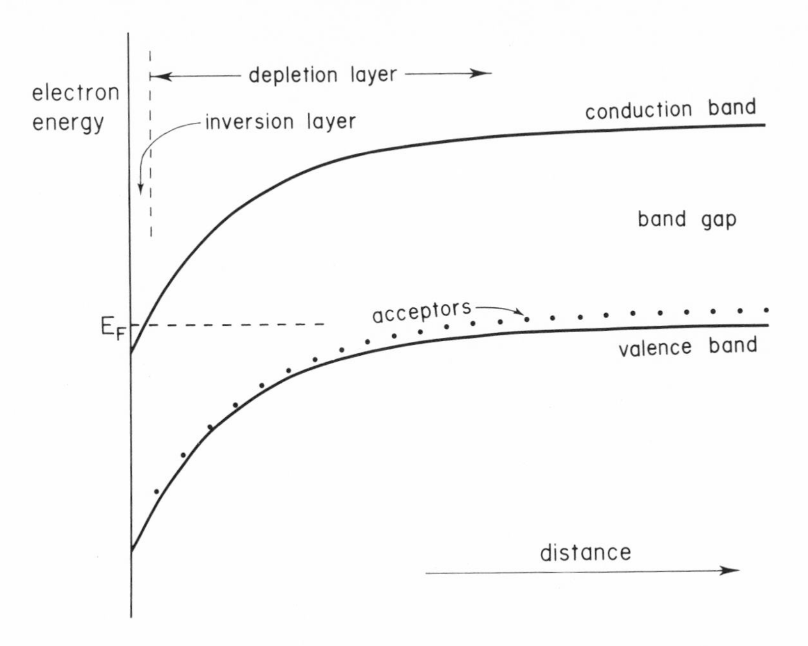

Fig. 1. Formation of an electron inversion layer at the surface of a p-type semiconductor.

Inversion layers are generally produced and controlled in insulated gate field effect transistors, the essential elements of which are shown in Figure 2 for an n-channel device. The inversion layer is induced by applying a positive potential to a thin metal

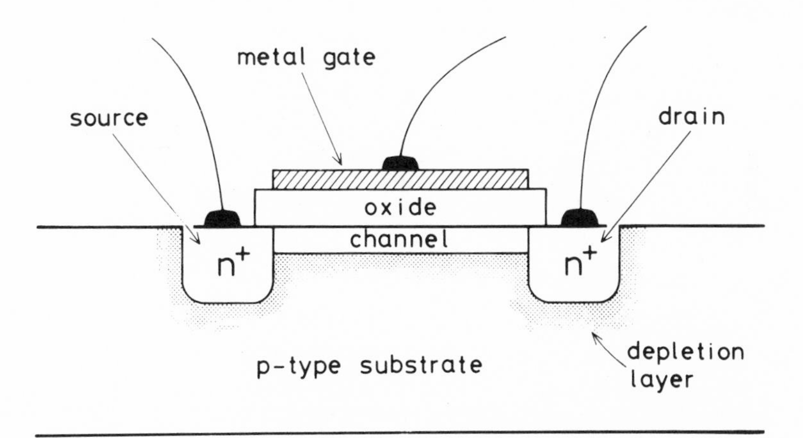

Fig. 2. The elements of an n-channel insulated gate field effect transistor.

film, the gate electrode, which is separated from the semiconductor by a thin insulating layer of silicon dioxide. Electrical connection is made to the inversion layer via heavily doped n-type regions called the source and drain. Since the induced charge is directly related to the gate voltage through the electrostatics of the device, carrier concentrations are known directly. In most other experimental systems, carrier concentrations have to be inferred from the Hall effect. The fact that carrier concentrations are known independently of the Hall effect is a unique advantage of the inversion layer system.

Although devices are normally manufactured so as to minimize disorder at the semiconductor/oxide interface, disorder is always present, primarily as a result of charges associated with defects at the interface and in the oxide, which is a glass. The disorder results in a potential which varies irregularly with position in the inversion layer. This random potential merely acts as a source of scattering at high carrier desnities, but, if only a few carriers are present, it causes them to become localized. The disorder therefore plays a vital role in determining the electrical properties of inversion layers at low temperatures near the threshold of conduction.

3. THE INDEPENDENT PARTICLE MOBILITY EDGE MODEL

According to the independent particle mobility edge model developed by Mott (1966, 1967), disorder should smear the edges of

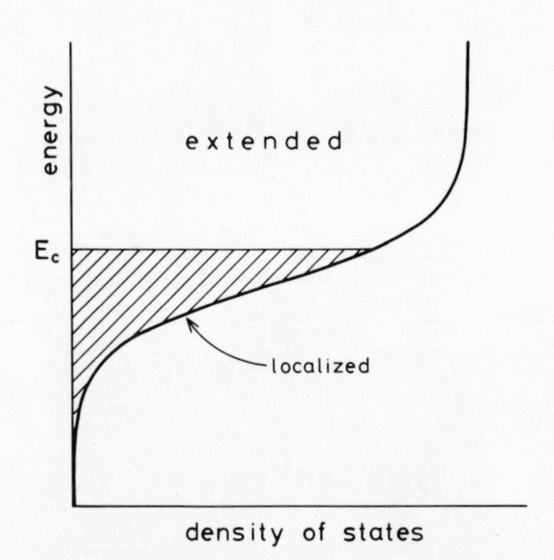

Fig. 3. The density of states at the bottom of a two-dimensional conduction band according to the independent particle mobility edge model.

bands and, in the band tails, the states should be localized (Figure 3). The localized states are separated from the extended by a mobility edge at energy E_c which, according to the computer studies of two-dimensional systems by Licciardello and Thouless (1975), should be where the density of states has risen to about three-quarters of the unperturbed band value. Localization is strongest deep in the tail, the size of the localized states increasing as the energy rises towards E_c. The mobility edge is at the energy where the localization length rises rapidly to a macroscopic magnitude. (Abrahams et al. 1979)

The conductivity predicted by this model depends on temperature and the position of the Fermi level E_F as follows:

(a) $E_F \geq E_c$ The conductivity is metallic, that is, independent of temperature at low temperatures. As the carrier concentration is reduced so that E_F approaches E_c from above, the conductivity falls as scattering becomes more important. When $E_F = E_c$, it reaches a minimum value, corresponding to an effective mean free path roughly equal to an inter-site spacing. In two dimensions the minimum metallic conductivity is predicted to be a universal constant with a value

$$\sigma_{mm} \approx 0.12\ e^2/\hbar = 3 \times 10^{-5}\ S$$

(Mott et al. 1975; Licciardello and Thouless 1975). At T = 0, the conductivity should fall very rapidly to zero as E_F falls below E_c, all carriers then being in localized states.

(b) $E_F < E_c$ If the temperature is not too low, conduction is possible by thermal activation of carriers into extended states at and above E_c. The conductivity should then be given by

$$\sigma = \sigma_{mm} \exp(-W/kT)$$

where

$$W = E_c - E_F$$

and the Boltzmann factor is an approximation for the Fermi function in the limit $kT < W$. The prefactor is σ_{mm}, regardless of carrier concentration, since the mobility at E_c is fixed, independent of state occupation, by the criterion determining the position of E_c. When $kT << W$, excitation to E_c becomes negligible and the dominant conduction process becomes variable range hopping (tunnelling) between localized states which gives a temperature dependent activation energy and a conductivity varying as $\sigma = \sigma_o \exp(-A/T^{1/3})$, where σ_o and A are constants. Eventually, as $T \to 0$, $\sigma \to 0$. The conductivity predicted by the independent particle mobility edge

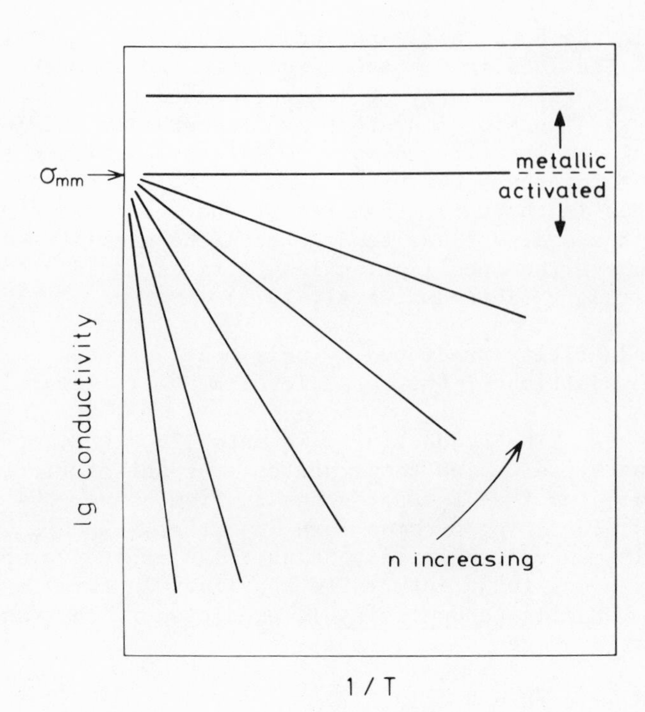

Fig. 4. The behaviour of the conductivity near the threshold of conduction as predicted by the independent particle mobility edge model. n is the density of carriers and σ_{mm} the theoretical minimum metallic conductivity.

model in the metallic and 1/T régimes is illustrated in Figure 4.

The Hall effect predicted by the mobility edge model depends on temperature and position of the Fermi level as follows:

(a) $E_F > E_c$ A fixed number of carriers will be immobilized in localized states in the band tail. The rest will be free, so that the Hall constant, R_H, will vary with carrier concentration n according to

$$1/R_H = (n - n_{loc})\, e,$$

where n_{loc} is the density of localized carriers. This relationship should be independent of temperature, except as E_F approaches E_c from above, when thermal excitation out of localized states can reduce the number of immobile carriers. As long as $E_F > E_c$, the Hall mobility $\mu_H \equiv \sigma R_H$ should equal the conductivity mobility μ given by $\sigma = (n - n_{loc})\, e\mu$.

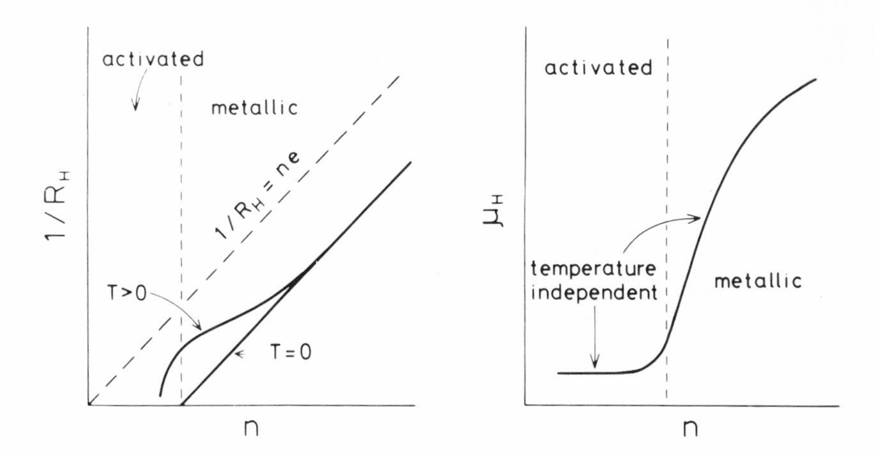

Fig. 5. The behaviour of the Hall constant and the Hall mobility near threshold as predicted by the independent particle mobility edge model.

(b) $E_F < E_c$ In the 1/T régime, the Hall constant should give the density of carriers contributing to conductivity, namely, those in extended states:

$$1/R_H = n_{ext}\, e.$$

In this regime n_{ext} is activated, so R_H should be also. The Hall mobility, on the other hand, should correspond to carriers moving at the mobility edge and should therefore become independent of both temperature and carrier concentration. The Hall predictions are summarized in Figure 5.

4. PERCOLATION MODELS

The basic assumption of percolation models (Arnold 1976) is that the potential fluctuations are relatively long-range so that the channel becomes macroscopically inhomogeneous. The system is described by supposing that the disorder causes the energy at the bottom of the conduction band to vary with position. In each region, the electronic properties are taken to be the same as those of a uniform system in which the position of the band edge E_c relative to the Fermi level E_F is the same as it is locally in the disordered system. Thus, at T = 0, a region in which $E_F > E_c$ is metallic, one in which $E_F < E_c$ is insulating.

According to the percolation model, the first carriers to enter the inversion layer go to the regions of lowest potential and form metallic lakes. As long as the carrier concentration is low, the lakes are isolated from one another by the regions where $E_C > E_F$ and (except at very low temperatures when tunnelling may become important) conduction will only occur by thermal activation over the barriers separating the lakes. In this region the temperature dependence of the conductivity should resemble simple activation with an activation energy a measure of the average barrier height. As more carriers are added, the lakes become larger and the barriers lower until metallically continuous percolation paths are established at the critical metal/insulator ratio of unity (Zallen and Scher 1971). Beyond this concentration, the conductivity will remain finite to $T = 0$ but there will still be weak activation when metallic percolation paths are first established, because, at that stage, they will be tortuous and activation over barriers will significantly reduce resistance. All trace of activation will finally disappear as the metallic areas expand to fill the whole space. In the percolation model, preexponentials in the fully activated régime depend on the mobility of carriers in extended states and could vary widely from one system to another. Generally, however, because of increasing screening, they would be expected to rise as the carrier concentration is increased.

The Hall effect predicted by the percolation model is as follows: At high carrier concentrations, the Hall coefficient is dominated by the connected metallic areas and its magnitude will correspond to the average carrier concentration <u>in the metallic regions</u> (Juretschke, Landauer and Swanson 1956). Below the percolation threshold, in the $1/T$ régime, where activation <u>over</u> the barriers dominates, there will be a large contribution to the Hall effect from the barrier regions where the (thermally activated) carrier concentration is low and therefore the mean drift velocity high. R_H should therefore increase markedly and correspond to a <u>reduced</u> carrier concentration. However, if, at very low temperatures, tunnelling begins to make a significant contribution to conduction across barrier regions, the Hall effect would fall again (assuming that the Holstein (1961) mechanism for Hall effect in tunnelling would be suppressed).

These predictions, arrived at above largely by physical argument, are confirmed by detailed calculations using two-dimensional effective-medium theory. (Adkins 1979 a,b)

5. OBSERVED BEHAVIOUR NEAR THRESHOLD

5.1 Conductivity

Some conductivity results have been reported which are in agreement with the independent particle mobility edge model (Adkins

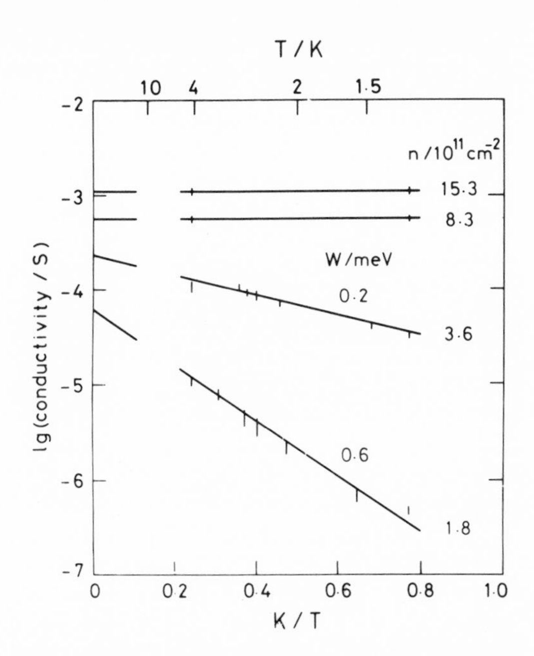

Fig. 6. Typical conductivity results near threshold. Note that (a) intercepts are greater than 3 x 10^{-5} S, the theoretical minimum metallic conductivity, and (b) they increase with carrier concentration. (Allen et al., 1975)

et al. 1976). They were obtained in samples with a small density of localized states ($\lesssim 2 \times 10^{11}$ cm^{-2}). Generally, however, the behaviour is rather different. More typical results are shown in Figure 6. We note a transition from metallic to activated conduction as the carrier concentration is reduced, activation energies increasing as the carrier density becomes smaller. However, the activated plots do not have a common intercept equal to the predicted minimum metallic conductivity but increase with carrier concentration. There are also other differences which will be discussed in more detail below. Generally it seems possible to summarize observed behaviour as follows.

I. Intercepts are never below about 1 x 10^{-5} S.

II. Intercepts increase with carrier density. Figure 7 shows the correlation of observed intercepts with carrier density. This dependence is quite general EXCEPT that

III. High oxide charge limits or suppresses the rise of intercepts with n. Charge is usually trapped in the oxide layer. When the density of trapped charges is large, the rise of intercepts is

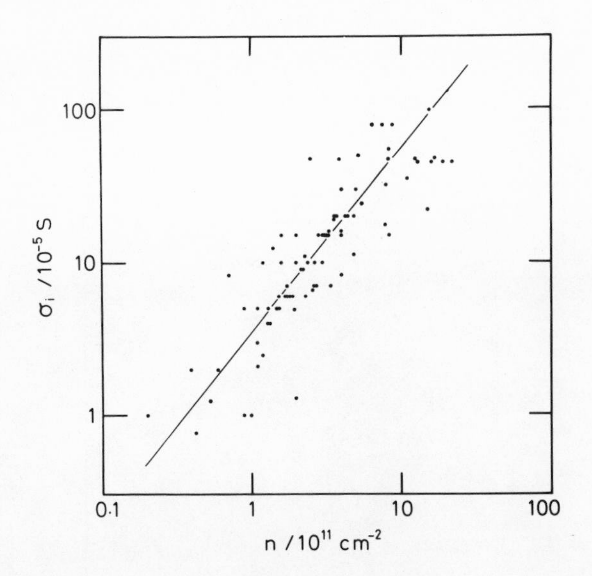

Fig. 7. A plot showing the correlation between intercepts σ_i of log σ against 1/T plots and carrier concentration n. (Data from: Allen et al., 1975; Pepper, 1977; Sjöstrand and Stiles, 1975; Tsui and Allen, 1974 and 1975; Vohralik, unpublished.)

suppressed even though n may become quite large before the onset of metallic conduction. (See Adkins 1978.)

IV. <u>When carrier densities are greater than about 3 x 10^{11} cm^{-2}, apparent densities of localized states can be impossibly large</u>. In both mobility edge and percolation models, simple activated behaviour is attributed to excitation of carriers to energies at which they can propagate freely through the system. The way in which the activation energy W varies with carrier concentration n should therefore depend on the density of available localized states at the Fermi level. Consider first the mobility edge model:

When the carrier concentration is increased by Δn, the Fermi level rises by

$$\Delta E_F = \Delta n/N(E_F)$$

where N(E) is the density of localized states (in space and energy). Corresponding to this rise in E_F, there is a decrease in the activation energy:

$$\Delta W = - \Delta E_F.$$

Δn is known from the gate-inversion layer capacitance and the change in gate voltage. ΔW is measured from the change in gradient of the plots of log σ against $1/T$. Hence, $N(E)$ is determined:

$$N(E) = - \Delta n/\Delta W.$$

Such calculations can lead to values of N(E) very much larger than the density of states in the unperturbed conduction band (a constant in two dimensions) (Adkins et al. 1976). Clearly, the density of states in a band tail cannot be greater than that in the band itself. The difference cannot be accounted for by screening of potential fluctuations, because screening would cause E_c to fall as carriers are added which would increase the observed $-\Delta W$ and decrease the calculated "densities of states". These results cannot be reconciled with an independent particle mobility edge model. Nor are they compatible with the fully activated regime of the percolation model: in the fully activated regime, the percolation model assumes that the metallic lakes are non-connected and W is a measure of the mean barrier height. When carriers are added, the rise in Fermi level in the lakes must be greater than $\Delta n/N_o$ where N_o is the unperturbed band density and Δn is the increase in average carrier density in the inversion layer. This is because carriers are concentrated into the lakes. Again, the average level of barriers could only fall through screening so that calculated "densities of states" should always be smaller than N_o. Only above the percolation threshold, in the regime of weak activation, does the percolation model predict changes of gradient which would imply densities of states greater than the unperturbed band value. However, the effective medium theory analysis indicates that the temperature dependence of the conductivity in this region should not resemble simple activation and that the range of resistance accessible by changing the temperature should be rather small (Adkins 1979 b). Again, the model does not seem to provide an adequate description of what is observed.

In some samples it is found that, as the temperature is reduced through the 1/T régime, the temperature dependence of the conductivity weakens and eventually disappears leaving a small finite constant conductivity as $T \to 0$ (Sjöstrand and Stiles 1975). This behavior suggests the survival of metallic percolation paths. However, it does not necessarily imply that the dominant conduction mechanism at higher temperatures is that described by the percolation model.

If the regime of activated conduction is interpreted in the obvious way, namely, the preexponentials are taken as a measure of the process being activated and W is the energy required, the implication of the rise of intercepts with n is that the process being activated depends on carrier concentration. If this is correct, models based on independent particle descriptions are ruled out, and we must seek an explanation for the observed behaviour involving inter-carrier correlation.

5.2 The Hall Effect

The usual behaviour of the Hall effect is also incompatible with the independent particle mobility edge model. In the metallic régime, the apparent carrier concentration, as given by $1/eR_H$, is slightly smaller than the actual carrier concentration, the difference being attributable to carriers immobilized in localized states, and the Hall mobility follows the conductivity mobility. However, as the system enters the activated régime, the apparent carrier concentration becomes equal to the total carrier concentration, that is, it appears that all carriers participate in conduction, and the Hall mobility, instead of becoming constant, becomes activated with the same activation energy as the conductivity. (Arnold 1974; Thompson 1978) Typical Hall mobility results are shown in Figure 8. The clear implication is that, in the activated régime, all carriers participate in conduction and the mobility

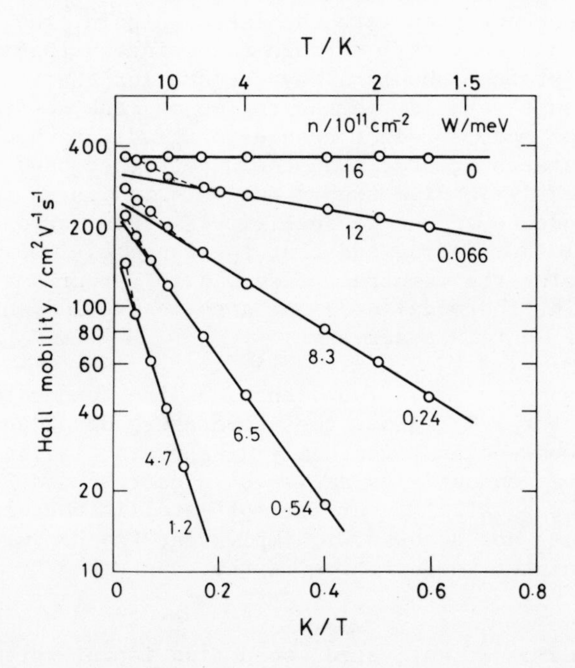

Fig. 8. Typical Hall mobility results. The Hall constant implies that all carriers participate in conduction. The Hall mobility is activated and equals the conductivity mobility if it is assumed that all carriers are mobile. (Arnold, 1974)

is activated. Such behaviour is reminiscent of classical models of viscous flow of liquids (Tabor 1968) and is embodied in the electron liquid model described in the next section.

The Hall effect is also in disagreement with the percolation model. Above the percolation threshold, when thermal activation is negligible, the apparent carrier concentration $1/eR_H$ should correspond to the actual carrier concentration in the metallic regions. It should therefore be greater than the mean concentration as given by the gate voltage and oxide capacitance by a factor equal to the inverse of the proportion of the inversion layer which is metallic. In this region, apparent carrier concentrations are never found to be greater than actual mean concentrations. This fact can only be accommodated within the percolation model by further assuming that a proportion of the carriers in the metallic regions is immobilized by the disorder. Similarly, below the percolation threshold, in the fully activated regime, the percolation model predicts a large contribution to the Hall effect from the barrier regions where the carrier concentration is low. This would give an increased Hall constant and imply a reduced apparent carrier concentration. What is observed, however, is that the apparent and actual carrier concentrations become essentially equal. Again, the behaviour is inconsistent with the percolation model.

6. CARRIER CORRELATION AND THE ELECTRON LIQUID MODEL

In the previous section it was shown that the conductivity and Hall effect, as usually observed, are in disagreement with both the mobility edge and percolation models. It was suggested that the character of the results implies that correlation may be playing an important part in determining the inversion layer properties near the conduction threshold. The electron-liquid model (Adkins 1978 a, b), which will now be described, shows the behaviour which results when correlation dominates carrier dynamics.

It is clear from an examination of relevant energies that corelation must be important near the threshold of metallic conduction. The mutual potential energy of two electrons in an inversion layer is given approximately by

$$V(r) = e^2/4\pi\varepsilon_0\bar{\varepsilon}r$$

where r is their separation and $\bar{\varepsilon}$ the mean permittivity of the silicon and the silicon dioxide (Stern 1973). Numerically this becomes

$$V(r)/\text{meV} = 5.8(n/10^{11}\text{cm}^{-2})^{\frac{1}{2}}$$

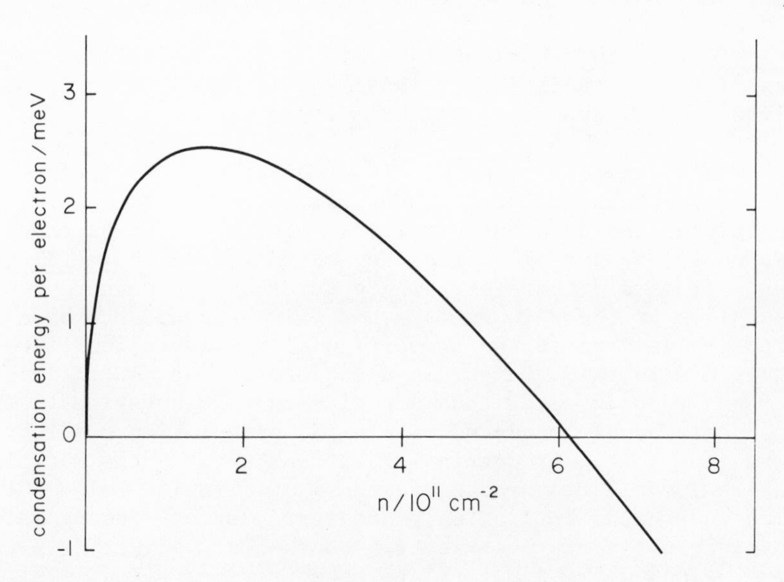

Fig. 9. Estimated condensation energy for Wigner localization of electrons at a (100) silicon surface.

so that, at typical carrier densities, V(r) is 5-20 meV. This is much greater than observed activation energies near threshold in the 1/T régime, and so it must be important there. It must also be important on the metallic side of the transition as the following argument shows.

An extreme effect of correlation in the metallic state would be the promotion of Wigner localization (Wigner, 1939). This is energetically favourable when

$$\langle KE_{ext} \rangle - KE_{loc} - \Delta U > 0$$

where $\langle KE_{ext} \rangle$ is the mean carrier kinetic energy in the normal metallic state, KE_{loc} is the carrier zero-point kinetic energy in Wigner localized states and ΔU is the change in potential energy of a carrier on Wigner localization. ΔU is given approximately by

$$\Delta U = -\beta V(r)$$

where β is a constant depending on the details of the electronic structure. In the inversion layer case, it is probably correct to put $\beta \approx 0.7$. $\langle KE_{ext} \rangle$ is known from the band density of states and the carrier density and KE_{loc} may be put equal to $\hbar^2\pi^2/2m^*r^2$. The estimated energy to be gained by Wigner localization is then shown in Figure 9, which suggests that Wigner localization is favourable below $n \approx 6 \times 10^{11}$ cm^{-2}. No account has been taken in the above estimate, however, of the effect of disorder which would be expected to enhance localization. It is thus likely that, in practice, Wigner localization may be favourable up to carrier densities exceeding 1×10^{12} cm^{-2}.

The electron liquid model seeks to describe the transport properties of Wigner localized carriers in the presence of disorder. In the absence of disorder, the carriers would form a regular lattice which would conduct by bodily movement. Disorder will act to deform and pin this lattice so that at zero temperature its structure would resemble that of a glass; but at finite temperatures, thermal activation will promote continual microscopic rearrangement so that there will be diffusion and the possibility of flow with an activated mobility, just as in classical theories of viscosity in normal fluids (Tabor 1968). Transport is therefore visualized as the viscous flow of an electron liquid past fixed pinning sites provided by the disorder. The dominance of inter-carrier forces imposes a kind of continuity condition which entrains carriers and causes essentially all to participate in conduction, but the mobility is activated. This is precisely what is implied by the conductivity and Hall results.

The following simple argument leads to an approximate expression for the conductivity of the electron fluid. We take account of correlation by supposing that each microscopic activation process results in the movement of a group of carriers. Pinning will normally be via single carriers but will be communicated to neighbours by correlation. Therefore, when thermal motions activate a jump at a pinning site, the result will be movement of a number of carriers from one configuration to a neighbouring one. Even though a group will not retain its entity from one jump to the next, a reasonably accurate expression for the conductivity should be obtained by supposing that it does. Consider, then, the diffusion of a group of z electrons as a result of thermally activated jumping. The Einstein argument for mobility gives

$$\mu = (ze)\nu R^2/4kT$$

where ze is the charge on the group, ν the jump frequency and R the jump distance. The number of such groups per area is n/z so that the conductivity is $\sigma = (n/z)(ze)\mu$, which, on substitution, becomes

$$\sigma = zne^2\nu R/4kT.$$

Activation enters through the jump rate ν. The strongest coupling to thermal motions is likely to be via the plasma modes of the electron fluid which surrounds the carrier (or carriers) in the group at which the critical pinning is taking place. Since the plasma modes (at relevant energies) have a continuous spectrum, the local configuration will change at thermal rates, kT/h, and the chance that any individual configuration will be sufficiently energetic to promote a jump introduces a Boltzmann factor. Thus

$$\nu \approx (kT/h) \exp(-W/kT)$$

where W is the height of the potential barrier. Since large movements of the group are prevented by the surrounding fluid, the most likely jump distance is an intercarrier spacing: this returns the configuration in the immediate vicinity of the pinning site to a low-energy state. We therefore put $R^2 = 1/n$. Then the conductivity becomes

$$\sigma = (ze^2/8\pi\hbar) \exp(-W/kT)$$

which is numerically

$$\sigma = 1.0 \times 10^{-5} z \exp(-W/kT) \text{ S.}$$

In this, W will depend on the degree and type of disorder and on n, reducing to zero at the transition to the metallic state.

We see that this expression is consistent with the essential features observed in the conductivity. At high carrier concentrations, the system will be metallic either because the Wigner localization criterion is no longer satisfied or because the zero point energy of carriers at pinning sites exceeds the pinning energy. Both mechanisms imply a transition to metallic conduction at carrier densities of the same order as are observed experimentally (Adkins 1978a). At lower carrier concentrations we observe activated conductivity with preexponentials from about 1×10^{-5} S upwards. The dependence of preexponentials on n is through z (which is called the <u>correlation index</u>): for a given structure of the background disorder, the number of carriers associated with each pinning site will increase with n and hence z will increase with n. However, if the oxide charge is high, the structure of the disorder becomes fine-grained, few carriers are associated with each pinning site and z remains small. When the liquid model applies, there is no connection between band densities of states and the dependence of W on n, so there is no significance in observations of high values of $\Delta n/\Delta W$.

The Hall effect is explained trivially by the electron-liquid model. Correlation causes almost all the carriers to flow as a fluid with activated mobility. The fluid is subject to forces resulting from (1) viscous drag, (2) the applied electric field, (3) the Hall electric field and (4) the motion in the magnetic field. Equating (3) and (4), which are perpendicular to the current flow, and (1) and (2), which are parallel to it, one obtains a Hall coefficient corresponding to the <u>total</u> carrier concentration and

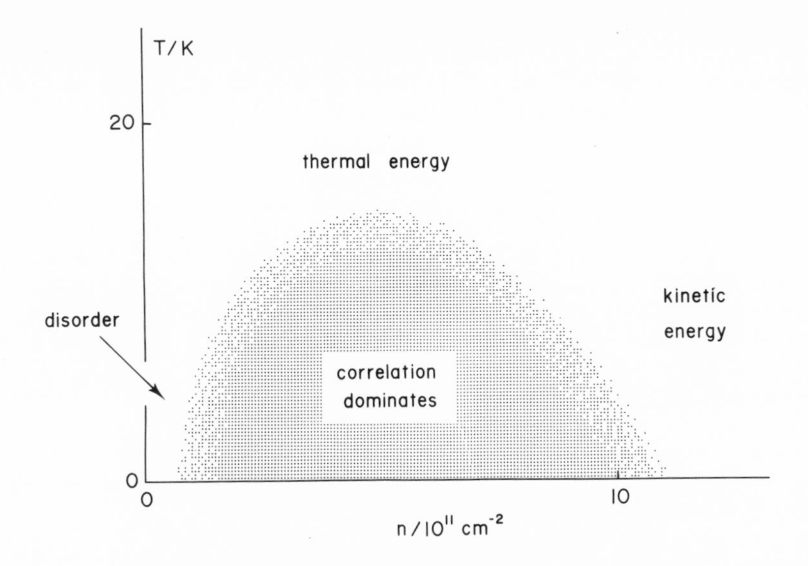

Fig. 10. A tentative phase diagram showing the region in which the electron liquid model should provide a good description of inversion layer behaviour. The independent particle model should always become valid in the limit of low carrier concentration.

a Hall mobility which is activated and identical to the conductivity mobility.

The electron liquid model is not supposed to describe the behaviour of inversion layers under all conditions but only when correlation dominates the behaviour. Figure 10 is a tentative "phase diagram" showing the region in which the electron liquid model should be a good description. At high carrier densities, the zero-point kinetic energy in localized states becomes large and Wigner localization becomes unfavourable. At high temperatures, thermal energies dominate and the gas becomes classical. At low carrier densities, the Wigner condensation energy per carrier becomes small and eventually disorder energies dominate so that carrier interaction becomes unimportant and the behaviour is adequately described by the independent particle model. (This is the region in which results agreeing in detail with the mobility edge model have been obtained. Adkins 1978a.) The system will not show any sudden change of properties at the boundary of the area in which correlation is important. Intercarrier potentials are "soft" and the effect of thermal motions and the background disorder will be to cause a gradual change of behaviour between the various regions.

(A rather different model, but one which also invokes Wigner localization, has been put forward by Katayama et al. (1977) to explain certain unusual conductance anomalies in inversion layers, but they envisage a different transport mechanism which cannot explain the observed Hall behaviour.)

7. CONCLUSIONS

This brief account of electrical transport in inversion layers at low temperatures and low carrier concentrations has shown that the behaviour which is normally observed cannot be explained in terms of current independent-particle models. An examination of relevant energies shows that, in the region of interest, correlation must play a dominant role in determining carrier dynamics. The electron liquid model seeks to describe the process of transport when correlation dominates. Correlation causes the carriers to become localized in the Wigner sense, the electron fluid so formed flowing past background disorder in a way resembling the motion of a conventional viscous fluid. The model predicts an activated Hall mobility essentially equal to the conductivity mobility, and a Hall constant corresponding to the participation of virtually all carriers in conduction.

This paper has been primarily concerned with inversion layers since the inherent advantages of this experimental system lead to particularly clearcut results. However, activated Hall mobilities and reduced Hall constants have also been seen near metal-insulator transitions in other systems (e.g. in impurity conduction: Davis and Compton, 1965). All such transitions are determined by the smooth passage through zero of some energy difference. It therefore seems highly likely that correlation will *always* be dominant sufficiently close to *any* metal-insulator transition. If this is the case, independent particle descriptions must always fail, and new models, like the electron-liquid model described here, must be developed to account for low-temperature transport properties close to the transition.

ACKNOWLEDGEMENT

I should like to thank Taylor and Francis Ltd., publishers of Philosophical Magazine, for permission to reuse material from an earlier paper (Adkins 1978b).

REFERENCES

Abrahams, E., Anderson, P. W., Licciardello, D. C., and Ramakrishnan, T. V., 1979, *Phys. Rev. Lett.*, 42:673.

Adkins, C. J., 1978a, J. Phys. C., 11:851.
Adkins, C. J., 1978b, Phil. Mag. B, 38:535.
Adkins, C. J., 1979a, J. Phys. C., 12:3389.
Adkins, C. J., 1979b, J. Phys. C., 12:3395.
Adkins, C. H., Pollitt, S., and Pepper, M., 1976, J. de Phys., 37: C4-343.
Allen, S. J., Tsui, D. C., and DeRosa, F., 1975, Phys. Rev. Lett., 35:1359.
Arnold, E., 1974, Appl. Phys. Lett., 25:705.
Arnold, E., 1976, Surf. Sci., 58:60.
Davis, E. A., and Compton, W. D., 1965, Phys. Rev. A,140:2183.
Fowler, A. B., 1975, Phys. Rev. Lett., 34:15.
Holstein, T., 1961, Phys. Rev., 124:1329.
Juretschke, H. J., Landauer, R., and Swanson, J. A., 1956, J. Appl. Phys., 27:836.
Katayama, Y., Narita, K., Skiraki, Y., Aoki, M., and Komatsubara, K. F., 1977, J. Phys. Soc. Japan, 42:1632.
Licciardello, D. C., and Thouless, D. J., 1975, J. Phys. C, 8:4157.
Mott, N. F., 1966, Phil. Mag., 13:689.
Mott, N. F., 1967, Adv. Phys., 16:49.
Mott, N. F., Pepper, M., Pollitt, S., Wallis, R. H., and Adkins, C. J., 1975, Proc. Roy. Soc. A, 345:169.
Pepper, M., 1977, Proc. Roy. Soc. A, 353:225.
Sjöstrand, M. E., and Stiles, P. J., 1975, Solid State Comm., 16:903.
Stern, F., 1973, Phys. Rev. Lett., 30:278.
Tabor, D., 1968, "Gases, Liquids and Solids," Penguin Books, pp. 230-232.
Thompson, J. P., 1978, Phys. Lett., 66A:65.
Tsui, D. C., and Allen, S. J., 1974, Phys. Rev. Lett., 32:1200.
Tsui, D. C., and Allen, S. J., 1975, Phys. Rev. Lett., 34:1293.
Wigner, E. P., 1939, Phys. Rev., 46:1002.
Zallen, R., and Scher, H., 1971, Phys. Rev. B, 4:4471.

SOME GENERAL INPUT-OUTPUT RULES GOVERNING HALL COEFFICIENT BEHAVIOR

R. S. Allgaier

Naval Surface Weapons Center
Silver Spring, Maryland 20910

ABSTRACT

Many calculations of the ordinary weak-field Hall coefficient have been carried out during the 100 years since its discovery. This paper examines a subset of those calculations which apply to transport models satisfying certain constraints. A set of general rules are developed which link the behavior of the Hall factor r ($R_o = r/ne$) to three basic features characterizing each of those models.

I. INTRODUCTION

The Hall coefficient R may be written as

$$R = r/ne \, , \tag{1}$$

where n is the carrier density, e is the charge per carrier, and r is a dimensionless factor which depends on the details of the transport model (i.e., band structure plus scattering function) under consideration. The main goal of this paper is to clarify the relationship between the factor r—sometimes called the Hall, anisotropy, mixing, or "fudge" factor—and the basic features of the transport model which significantly influence its value.

At first thought, it seems rather surprising that so little has been written about this relationship in any general sense during the past century. After all, the factor r is dimensionless, so that the Hall effect depends essentially on a single basic electronic parameter, charge density. This is surely a rare if not unique situation among the literally hundreds of distinct transport coefficients that have been defined (see, for example, Beer 1963, Jan 1957). But, on the other hand, this may make the behavior of r much more subtle. The problem may also be that many earlier articles, as well as the more elementary textbooks, have used models that are too simple for behavioral generalization, while more recent calculations (dealing with the bewildering variety of more realistic models which have appeared during the past 25 years or so) are too complex to provide much more than individual answers. Furthermore, as this article will point out, subtleties, "hidden" physics, and misconceptions are associated with the Hall effect even for the simplest transport model imaginable.

More specifically, this paper will identify three fundamental properties of transport models, and use them as the basis for developing a set of rules which predict the qualitative behavior of the Hall factor r. So far as the author is aware, these rules apply to all Hall effect calculations ever made, providing that they satisfy all of the following constraints:

(a) The Hall effect referred to is the ordinary, weak-magnetic-field coefficient (henceforth designated as R_0), calculated in the Ohm's Law (weak-electric-field) regime.

(b) The contributing carriers are contained in one band. Multivalley models, and those in which carriers originate from main and subsidiary extrema of a given band, are included. Multiband generalizations of eq. (1) are elementary and well-known (Chambers 1952), but including them would be a digression from the main goal of this paper.

(c) The calculation is based on the Jones-Zener (1934) solution of the Boltzmann equation, or uses another classical or quasi-classical approach, such as the kinetic, path-integral, hodograph, or trajectory methods (Shockley 1950, McClure 1956, Chambers 1957).

(d) The simple relaxation-time approximation is used for the collision-integral side of the Boltzmann equation, and corresponding assumptions about the existence of a scattering probability are employed in the other methods cited.

(e) The conducting medium is homogeneous but not necessarily isotropic.

(f) The only effect of sample boundaries (unless they are too far away to have any effect) is to determine the crystallographic orientation of the total, steady-state current flow.

This list of conditions may seem to limit the scope of this

paper to a rather small number of transport models of limited interest. But the period to be covered is 100 years, and a large number of Hall effect papers satisfying these constraints appeared during the first quarter and first half of the period (McKay 1906, Campbell 1923). After considering all of the papers which have appeared during the second 50 years, it still seems likely to the author that most weak-field Hall effect calculations which have been carried out up to the present fall within the category defined above.

In the centennial year of the discovery of the Hall effect, the results of these calculations, as a group, ought to be better understood than they seem to be. From the point of view of this paper, "better understood" means knowing in advance how the value of the weak-field Hall coefficient is going to be influenced by the nature of the model for which it is being computed.

II. SOME PERSONAL PERSPECTIVES

At this point, I want to shift this narrative to the first person, in order to emphasize that it reflects a very personal view of a particular aspect of Hall effect history. This viewpoint has evolved over a quarter century of my involvement with Hall effect measurements and analyses. Naturally, this centennial volume has a very special character. I hope that the flavor of my remarks will blend well with the personal comments and reminiscences made by a number of the speakers at the Symposium.

As stated in the introduction, I want to focus attention on the Hall factor r. That factor is "normally" positive (as in the free-electron model), as a consequence of the convention used to define the Hall coefficient. For a given one-band transport model, the sign of r does not depend on the sign of the charges present, since R and ne change sign together. One might say that the study of the factor r emphasizes the geometrical or spatial aspects of the transport models under examination.

I hardly need argue that the main characteristics of the transport model which determine the value of r are its energy dependence and its anisotropies. As "seen" from the origin of a coordinate system in momentum space, these features might be called the longitudinal and transverse aspects of the transport model. From the overall point of view of an entire energy band, transverse effects generally constitute the more important influence. After all, the Hall phenomenon involves a "rotatory power" (Thomson 1851), a turning away of the current flow from its original direction. It is only very near a band edge that truly isotropic transport models are to be found, and only in that region can energy dependence become the more significant aspect of the model.

In any event, the transverse characteristics are (to me at least) the more interesting ones. For example, anisotropy per se does not affect the value of r unless it leads to a "mixed response" of the carriers to a given set of applied fields. I will discuss this response of current carriers in terms of their mobilities, because it permits a more unified treatment of the input-output relation between transport model characteristics and the value of the factor r.

The essence of the Hall effect is diversion. It may be the direction of current which is altered, or it may be the total electric field which changes its orientation, depending on the boundary conditions present. The essence of magnetoresistance—the change of resistance in a magnetic field—is dispersion, i.e., the diversion of various current components through a range of angles spread about the overall direction of diverted current flow. Any proper discussion of the Hall effect should include more about magnetoresistance than space will allow in this paper.

Understanding the Hall effect is not made easier by the (in my view) widespread confusion about which carriers do and which do not contribute to charge transport. This subtle question is most easily discussed in the case of the zero-magnetic-field conductivity σ_o, written in the form

$$\sigma_o = ne\bar{\mu}, \tag{2}$$

where $\bar{\mu}$ is the average carrier mobility. I assert that if n refers to all of the current carriers, then $\bar{\mu}$ is the (admittedly complex) average value for all of those carriers. It is true that for degenerate statistics, σ_o for all of the carriers may be computed in terms of certain properties of the carriers at the Fermi surface only. It is not true that σ_o is solely due to the conductivity of the carriers at the Fermi surface. If the latter were a correct statement, then the total σ_o in a band would become negative as soon as the Fermi level rose high enough in the band that the average value of the effective mass became negative. Unfortunately, this matter becomes more complicated with respect to the Hall coefficient. I wish that my contribution to this centennial volume could have included a simple clear description of how the various carriers contribute their characteristics to the calculated value of the Hall coefficient.

The next section contains a brief overview of the mathematical and pictorial framework for the entire paper. It is followed by several sections in which I consider, one by one, a series of models of increasing complexity (but still, I hope, relatively simple).

The essential features of each model are identified, and their connection with the value of r is explored. Then, after focusing on an important missing link, I will combine those considerations into a set of "universal" (within the constraints defined in Section I) rules.

III. BASIC QUANTITIES, EQUATIONS, AND CONFIGURATIONS

In order to discuss the models in a unified fashion, I will relate them to a standard configuration in which the sample current density $\vec{i}$ lies in the 1-2 plane and the magnetic field $\vec{H}$ is parallel to axis 3 of an orthogonal coordinate system. What follows is only a summary of the basic formulas. Derivations and further details may be found in numerous texts (e.g., Beer 1963, Hurd 1972).

If the conductivity is not isotropic, the 1 and 2 axes will be those which diagonalize the conductivity tensor σ_{ii} in the absence of a magnetic field. Then the components of the current density are

$$i_i = \sigma_{ii} E_i \ , \tag{3}$$

with

$$\sigma_{ii} = ne\mu_i \tag{4}$$

and

$$\mu_i = e\tau_i / m_i \ , \tag{5}$$

where E_i, μ_i, τ_i, and m_i $(i = 1, 2)$ are the components of the total electric field, carrier mobility (an algebraic quantity determined by the signs and magnitudes of e and m_i), scattering probability, and effective mass.

When $|\vec{H}| = H_3 \neq 0$, eq.(3) becomes

$$i_1(H) = \sigma_{11}(H)E_1 + \sigma_{12}(H)E_2 \tag{6}$$

and

$$i_2(H) = \sigma_{21}(H)E_1 + \sigma_{22}(H)E_2 \ . \tag{7}$$

For weak electric and magnetic fields, none of the conductivity components depends on $\vec{E}$, the off-diagonal σ_{ij} are the negatives of each other and are proportional to H, and (for the purpose of calculating R_o) the magnetic field dependence of the diagonal components σ_{ii} (which, to lowest order, has the form $a_o + a_2 H^2$) may be neglected.

The Hall angle θ_H is the angle between the directions of $\vec{i}$ and $\vec{E}$ (the total electric field). Its sense is defined by the rotation from the former to the latter direction. The Hall angle may be expressed as

$$\tan \theta_H = \mu_H H, \tag{8}$$

where the Hall mobility μ_H is given by

$$\mu_H = R_o \sigma_o \quad , \tag{9}$$

and σ_o is the conductivity when $\vec{H} = 0$.

Two alternative versions of this basic configuration, labelled "A" and "B", are sketched in Fig. 1. The first corresponds to the standard "Hall bar" (a long narrow sample used to avoid end effects) configuration in which $\vec{i}$ is fixed in direction and (almost always) magnitude. The application of a transverse magnetic field along the +3 axis (out of the page) generally causes $\vec{E}_1$ to increase from

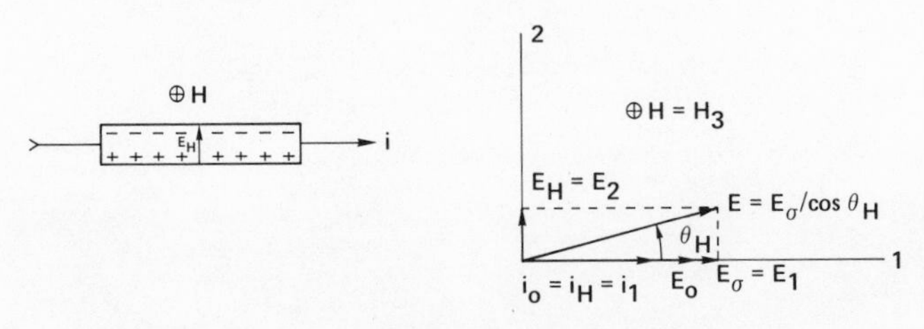

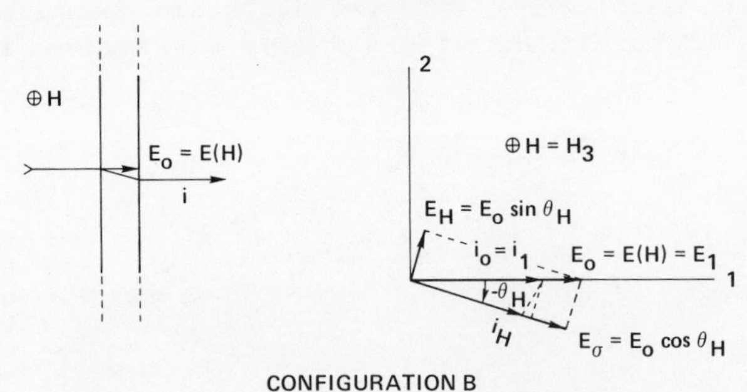

Fig. 1. Hall effect configurations (see Section III). A, long narrow sample, B, short wide sample. In both cases, current density $\vec{i}$ and electric field $\vec{E}$ lie in the 1-2 plane, the magnetic field $\vec{H}$ along the 3-axis (out of page). Positive charges are assumed. θ_H is the Hall angle.

$\vec{E}_o$ to $\vec{E}_\sigma$ (a positive magnetoresistance) as well as a buildup of the Hall field $\vec{E}_H = \vec{E}_2$. Thus the total field $\vec{E}$ becomes larger in magnitude than $\vec{E}_o$ and is rotated counterclockwise (in the positive carrier case) through the Hall angle θ_H.

Configuration B represents the opposite extreme of a thin layer across which current is flowing. The transverse boundaries are so far removed that their presence and effect may be neglected. In this case, $\vec{E}$ remains fixed, and for $\vec{H} \neq 0$, $\vec{i}$ rotates through the Hall angle θ_H, in the opposite sense from that occurring in configuration A. In this configuration, the presence of a positive magnetoresistance makes $i_H < i_o \cos\theta_H$.

Because of the neglect of magnetoresistance and the small angle approximations which are permissible when calculating the weak-field Hall coefficient, the appropriate formulas for R_o are

$$R_o = E_H/i_oH \tag{10}$$

in configuration A, and

$$R_o = (-i_2/i_o)E_o/i_oH \tag{11}$$

in configuration B.

There is a strong tradition in the literature to regard the Hall effect and magnetoresistance solutions obtained in the two configurations as somehow basically different from each other. For instance, one reads that configuration A generates magnetoconductivity while B produces magnetoresistance; in A, $\vec{E}_H$ builds up and leads to an average, or sometimes exact, cancellation of the Lorentz force, while in B this cannot occur because $\vec{E}_H$ is absent or is shorted out.

This is very misleading. A formal solution to a transport problem, such as that obtained using the Boltzmann equation, is usually carried out in configuration B. That is, $\vec{E}$ and $\vec{H}$ are the independent, input parameters, while $\vec{i}$ is the dependent output variable. In order to match the boundary conditions of the more commonly utilized configuration A, this solution is mathematically inverted, so that $\vec{i}$ and $\vec{H}$ are the independent, and $\vec{E}$ the dependent, variables. The effect is merely to rotate the configuration-B solution. The relative directions and magnitudes of $\vec{i}$, $\vec{E}$, and $\vec{H}$ are unchanged.

This last statement is exact if the conductivity is isotropic in the 1-2 plane. If it is not, then there will be an infinitesimal change in the solution, but for the infinitesimal rotation corresponding to the weak-field limit, this represents a negligible, higher-order correction to the calculated value of R_o.

There is one more basic equation that I wish to quote in this section (Tsuji 1958). When expressed in terms of k-space, it is

$$R_o = \frac{12\pi^3}{e} \frac{\int^{S_F} \tau^2 \; v_F^2 \; (\overline{1/\rho}) dS_F}{\left(\int^{S_F} \tau \; v_F dS_F \right)^2} , \tag{12}$$

where v_F is the magnitude of the carrier velocity, $\overline{1/\rho}$ is the mean value of the curvature, and dS_F is a differential element of Fermi surface area, all evaluated at a point on the Fermi surface. Had I known of this formula earlier and appreciated its significance, it would have been much easier to arrive at the general rules being reviewed in this paper.

IV. THE ONE-MOBILITY MODEL (THE SIMPLEST TRANSPORT MODEL)

Using the Boltzmann equation and Fermi-Dirac statistics to calculate R_o for an isotropic, parabolic band, assuming the same scattering probability τ for all carriers, leads to the result $R_o = 1/ne$, i.e., to

$$r = 1 \quad , \tag{13}$$

and also to a magnetoresistance which is zero for any angle between $\vec{i}$ and $\vec{H}$. In the absence of quantum effects, these results are exact for all values of $\vec{H}$.

In accordance with eq. (5), all carriers have one and the same mobility, since they all have the same charge, mass, and τ. All carriers are therefore rotated through the same Hall angle. Consequently, $\vec{H}$ has produced a diversion of the carriers, but no dispersion about the direction of total current flow. The latter effect must be present to produce magnetoresistance.

Many solid state texts obtain this standard result, using configuration A, by noting that one value of E_H can balance the Lorentz force v_1H_3 for all carriers, since $v_1 = \mu_1E_1$ and every carrier has the same mobility. It is frequently stated or implied that the magnetoresistance then vanishes because the carrier trajectories have been straightened out by the Hall field-Lorentz force cancellation.

As noted in Section III, the overall current rotates as $\vec{E}_H$ builds up. But regardless of the sources of the electric field, a charge in crossed electric and magnetic fields will follow a cycloidal trajectory. When subjected to a constant scattering probability, this trajectory becomes a sequence of cycloidal segments of random lengths.

It is clear that this kind of trajectory is longer than that for $\vec{H} = 0$. But it turns out that a charge traverses each cycloidal segment faster than it would along the straight line connecting its end points. For the one-mobility model, this effect is exactly balanced by the zigzag nature of the trajectory due to the random times between scattering events (Allgaier 1973).

From time to time, galvanomagnetic calculations have appeared which assumed that all carriers were always scattered after the same <u>time</u> had elapsed. For the one-mobility model, the result for R_o corresponds to

$$r = 2/3 \ . \tag{14}$$

This result (which is only valid in weak magnetic fields) has appeared in a number of books over a long period (Richardson 1914, Becker 1933, Page and Adams 1949 and 1969). This nonrandom model also leads to a negative magnetoresistance, since all of the faster cycloidal paths now have the same length and thus there is no compensating zigzag effect.

It is not obvious whether or not such nonrandom models might become physically reasonable under some circumstances, but it is clear that it can be dangerous to use such a model. Early in this century, a well-known physicist "proved" that magnetoresistance is positive using the simple nonrandom model (Thomson 1902). But he worked in configuration B, and actually proved that there was less current in the direction of $\vec{E}$ when the magnetic field was present. But this reduction will always occur as a simple consequence of carrier diversion. For zero magnetoresistance, the forward current component is reduced by the factor $\cos^2\theta_H$. Had he computed the true magnetoresistance (i.e., the ratio of $\vec{E}$ to $\vec{i}$ in the direction of the latter) for this model, he would have obtained a negative effect.

The simple one-mobility result, $r = 1$, did not attract much attention prior to the appearance of quantum theory and Fermi-Dirac statistics (Campbell 1923). It did not predict the correct magnitudes for R_o, except for some of the alkali metals and a few other coincidental cases, and of course it could not account for both signs of R_o in terms of charges of a single sign. Furthermore, the isotropic parabolic model did not predict $r = 1$ when Maxwell-Boltzmann statistics were used, unless it was assumed, without justification, that τ did not depend on carrier energy.

V. A SIMPLE ANISOTROPIC MODEL

If the diagonal components of the conductivity tensor in eq.(4) are not equal, the conducting medium becomes anisotropic, and is characterized by two carrier mobilities μ_1 and μ_2. This situation corresponds to a single-valley ellipsoidal energy-surface model, or more properly, to a "mobility ellipsoid," since τ as well as m may have tensor properties.

The galvanomagnetic properties of this simple model may surprise some readers. Figure 2 illustrates the pitfalls which may be encountered in determining R_o for this model. For concreteness, the figure assumes that $\mu_2 = 2\mu_1$, and to simplify the sketches, the units have been arbitrarily changed so that the magnitude of $\vec{i}_1 (= ne\mu_1\vec{E}_1)$ is the same as that of the field $\vec{E}_1$ which produces that current density.

If R_o is calculated for configuration A, only μ_1 is contained in eqs. (4) and (10) for $\vec{E}_H$ and $\vec{i}$, and the result is again r = 1, as in eq.(13) for the one-mobility model. This situation is shown in Fig. 2(a). Figure 2(b) corresponds to configuration B. The current component i_2 contains μ_2; using eq.(11) then leads to

$$r = \mu_2/\mu_1 , \tag{15}$$

which equals two for the case being considered. One might have guessed that some gradual variation in the value of r would appear, as the current direction was rotated in this anisotropic model. But in truly weak fields the angular rotation between configurations A and B is infinitesimal. Yet r has doubled, corresponding

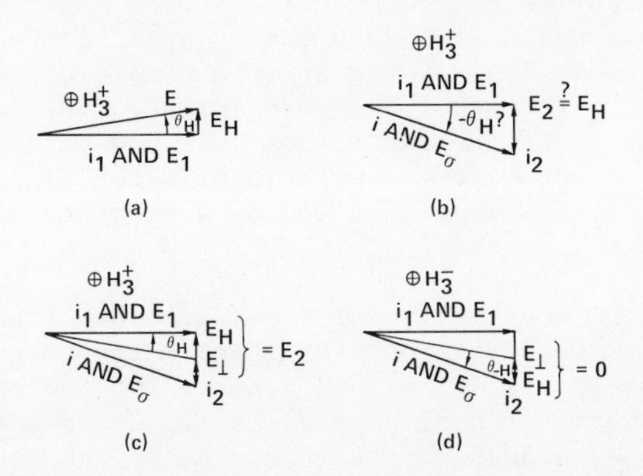

Fig. 2. A simple anisotropic model (see Section V). (a) Current along the 1-axis. (b) Total electric field along the 1-axis. (c) and (d) The true and spurious components of the Hall electric field, with the magnetic field $\vec{H}$ along the positive and negative directions of the 3-axis.

to the doubled values of $\vec{E}_H$ and θ_H suggested by Fig. 2(b).

Parts (c) and (d) of Fig. 2 explain the apparent paradox. If a conventional Hall sample were actually oriented in the direction taken by $\vec{i}$ in configuration B, and the magnetic field were zero, there would already be a perpendicular electric field component, $\vec{E}_\perp$, pointing in the positive 2-axis direction, because the conductivity is anisotropic and $\vec{i}$ does not lie along a symmetry axis. Fig. 2(c) shows the correct interpretation when $\vec{H}$ points along the positive 3-axis. Then the true $\vec{E}_H$, like $\vec{E}_\perp$, points in the positive 2-axis direction and (for the particular case $\mu_2 = 2\mu_1$) has the same magnitude as $\vec{E}_\perp$. What occurs when $\vec{H}$ is reversed is sketched in Fig. 2(d). The measured transverse fields in (c) and (d) will be $2\vec{E}_H$ and zero. The average gives the correct value, leading to $r = 1$, as in configuration A. The answer for *any* current direction in this anisotropic model is $r = 1$, and the magnetoresistance is zero, just as in the one-mobility model.

This result illustrates a fundamental point. The mobility in this model is a function of the direction of the applied fields, but for any given set of applied forces, *all* carriers respond with the *same* mobility. Thus the model, although anisotropic, remains a one-mobility model and leads to the same results as that simplest-of-all model. Anisotropic models usually do lead to $r \neq 1$. But the essential ingredient is not anisotropy *per se*, but a *mixed response* (i.e., the simultaneous response of carriers with different mobilities) to a single set of applied fields.

VI. THE TWO-MOBILITY MODEL

The two-mobility model discussed in this section might well be called the "next-simplest model." The mobilities are associated with two groups of contributing carriers, all carriers in each group having the same isotropic mobilities μ_1 and μ_2. Basically, this is a very old model, many versions of which are included in Campbell's book (1923). In almost all of those cases, one mobility is negative and the other positive, so that when weighted by carrier densities and other factors, the formulas are capable of being fitted to a range of both positive and negative values of R_0 obtained from measurements on a variety of materials.

Because of the one-band constraint cited earlier, the version of the two-mobility model considered in this section will consist of two groups of carriers having the same sign. Such a model, for example, might describe R_0 for an extrinsic semiconductor in which carriers from the main and subsidiary extrema of a given band participate. The Hall effect in such a case may be written

$$R_o = \frac{n_1 e_1 \mu_1{}^2 + n_2 e_2 \mu_2{}^2}{(n_1 e_1 \mu_1 + n_2 e_2 \mu_2)^2} \tag{16}$$

or, more compactly,

$$R_o = \left[\frac{1 - x + xb^2}{(1 - x + xb)^2}\right]\frac{1}{ne} = \frac{r}{ne}\ , \tag{17}$$

where $n = n_1 + n_2$, $x = n_2/n$, and $b = \mu_2/\mu_1$. Note that the sign of R_o, but not of r, is affected by the sign of e (as was noted in Section II).

Unless all of the carriers are in subband 1 or 2, there is a mixed, two-mobility response which always leads to

$$r > 1. \tag{18}$$

The proof is a matter of simple calculus. This same inequality is easily generalized to the case of N groups of carriers of the same sign, each group having its own single mobility. Here then is a tentative first rule: Mobility mixing in an isotropic transport model always leads to $r > 1$. Note that this model provides an example of mobility mixing without anisotropy, leading to $r \neq 1$. In contrast to this result, the previous section considered an <u>an</u>isotropic model without mixing, and that model led to $r = 1$.

VII. THE SIMPLE, MAXWELL-BOLTZMANN MODEL

The rule leading to eq.(18) can be broadened by referring to a very old calculation (Gans 1906). This paper appears to be the first detailed and complete derivation of R_o for a very precisely defined transport model. In more modern terms, it corresponds to an isotropic parabolic band model which assumes an isotropic $\tau \propto E^{-\frac{1}{2}}$ (E is carrier energy) and uses Maxwell-Boltzmann statistics.

The result, generalized to cover any dependence of τ on energy and to Fermi-Dirac statistics, appears in many solid state texts, and may be written as

$$r = \langle\tau^2\rangle \ \langle\tau\rangle^2, \tag{19}$$

where

$$\langle\tau^s\rangle = \frac{\int_0^\infty \tau^s \ E^{3/2}(\partial f_o/\partial E)dE}{\int_0^\infty E^{3/2} \ (\partial f_o/\partial E)dE}, \tag{20}$$

and f_o is the unperturbed Fermi-Dirac distribution function. It follows from a Schwarz inequality that $r \geq 1$ in eq.(19). There are two special cases for which $r = 1$, viz., when τ is a constant, or when the limiting case of "infinitely degenerate" statistics is assumed, so that the derivative $\partial f_o/\partial E$ becomes a delta function at the Fermi energy. In both cases, the model has again reduced to the one-mobility case.

Note that this model may be regarded as an infinite-mobility extension of the two- or N-mobility cases described in the previous section. The traditional notation for this standard semiconductor model does not focus attention on that kind of characterization of its nature. The model assigns the same e and the same m to all carriers, but since τ is a continuous function of E, all values of carrier mobility can contribute. In a similar way, an isotropic <u>non</u>parabolic model will generate an infinity of mobilities through the energy dependence of the effective mass (whether or not τ depends on energy), again leading to $r > 1$.

As noted above, $r \rightarrow 1$ as the statistics become degenerate, and the $\partial f_o/\partial E$ factor confines the τ-average to an ever-narrowing energy interval straddling the Fermi energy E_F. It is often stated in the literature that in the completely degenerate limit, only the carriers at E_F contribute to the transport coefficient being considered. Yet in this limit, $R_o = 1/ne$, where n is <u>all</u> of the carriers inside the Fermi surface. And this result has only changed modestly (typically, 10-20%) from the opposite extreme of Maxwell-Boltzmann statistics, despite the drastic change in the nature of the weighting function.

Surely it is not just a <u>mathematical</u> coincidence that $R_o = 1/ne$ in this limit. As noted in Section II, it is appropriate to say that in the degenerate limit, R_o is determined solely by <u>certain related properties</u> of the carriers at the Fermi surface. As will become evident in a later section, that related property, for the Hall coefficient, is Fermi-surface curvature, or, more loosely, Fermi-surface shape.

VIII. A SLIGHTLY ANISOTROPIC MODEL

The next model, originally treated in the metallic approximation, incorporates a Fermi surface which is slightly distorted from a spherical form, and a scattering probability which is slightly anisotropic (Davis 1939, Cooper and Raimes 1959). Thus the model hardly differs from the simple one-mobility model. The consequence of this slight deviation, however, was some complex mathematics which was so tedious and lengthy that it has never been published. The anisotropies in τ and in the E-p relationship (p = carrier momentum) near the Fermi surface were expressed as

$$p = \alpha_o(E) + \alpha_1(E)Y_4^c(\theta,\phi) \tag{21}$$

and

$$\tau = \tau_o(E) + \tau_1(E)Y_4^c(\theta,\phi), \tag{22}$$

where Y_4^c is a sum of fourth order cubically symmetric spherical harmonics and θ and ϕ are the usual angles used in spherical coordinates. The result is

$$r = 1 + \frac{4}{21}[- 9A^2 - 18A(B-C) + (B-C)^2], \tag{23}$$

where

$$A = \frac{\alpha_1}{\alpha_o}, \quad B = \frac{d\alpha_1/dE}{d\alpha_o/dE}, \quad \text{and} \quad C = \frac{\tau_1}{\tau_o}. \tag{24}$$

The quantities A, B, and C are to be evaluated at $E = E_F$, and are all much less than unity, because of the restriction to slight anisotropy. Equation (23) for r is exact to second order in A, B, and C and combinations thereof.

Despite its _slightly_ distorted character, this model is very important. For the first time, it provided an example which shows how anisotropies in both band structure and scattering affect the value of r. Moreover, the authors (Davis 1939, Cooper and Raimes 1959) examined these effects (input-output links) and came to some general conclusions. My own version of their discussions follows.

If the band model is isotropic, A = B = 0, and eq.(23) becomes

$$r = 1 + (4/21)C^2 . \tag{25}$$

Thus scattering anisotropy alone always makes $r > 1$. This result can be generalized for any kind of τ-anisotropy in a metallic model, not just that given by eq.(25) (Böning, 1970, and private communication). It seems obvious that combining this result with the two-, N-, or infinite-μ isotropic models must lead to the general result that mobility mixing, whatever its source, will make $r > 1$ for any _isotropic band_ model using the general form of the Fermi-Dirac distribution function.

On the other hand, if τ is isotropic, then choosing various non-zero values of A and B demonstrates that a weakly anisotropic band structure can lead to both $r > 1$ and $r < 1$. So it appears that band structure anisotropy alone can always reproduce the effect of τ-anisotropy alone, but not _vice versa_.

An important ingredient is missing from the above discussion. Before identifying it and its crucial role, however, two further models and their effects on r will be described.

IX. THE ELLIPSOIDAL MULTIVALLEY MODEL

So far, the models examined have been isotropic, slightly anisotropic, or very anisotropic. But the last-mentioned was the single-valley ellipsoidal model, and its anisotropy turned out to be completely ineffective—no mobility mixing was involved, so that it produced results which are identical to the isotropic single-mobility model.

The first model to introduce strong anisotropy and strong mixing effects without any mathematical complexities was one which consists of a cubically-symmetric array of ellipsoids of revolution (i.e., spheroids). This kind of model became well-known about 25 years ago (Abeles and Meiboom 1954, Shibuya 1954) when it was applied to the multivalley conduction bands of Ge and Si. In those two materials, the symmetry axes of the valleys are oriented along the <111> and <100> directions in the Brillouin zone, respectively. Actually, the <100> version originated much earlier (Frenkel and Kontorova 1935).

This type of model constitutes a very important conceptual breakthrough. Before its development, the constraints of cubic symmetry seemed to make enormous mathematical complications inevitable whenever the slightest anisotropies were incorporated. Now, a strongly anisotropic but mathematically simple band structure was proposed, and within a few years it became evident that this model is appropriate for many nonmetallic materials.

No mathematical complications appear in the multivalley model because, as noted in Section V, there is no mobility mixing from a single ellipsoid; hence the mixing occurs (and the resulting formulas differ from the one-mobility model) only because all of the ellipsoids are not oriented in the same direction. Consequently, each valley (or subset of valleys) responds to a given set of applied forces with its own characteristic single mobility value.

In the degenerate limit, the Hall factor for the ellipsoidal multivalley model is

$$r = 3K(K + 2)/(2K + 1)^2, \qquad (26)$$

where $K = \mu_t/\mu_\ell$, and μ_t and μ_ℓ are the carrier mobilities perpendicular and parallel to the symmetry axis of each ellipsoidal valley. For the more general case of Fermi-Dirac statistics, the above result must be multiplied by the scattering average defined in eq.(19).

Equation (26) has a maximum value of unity when K = 1. For that case, r = 1 is to be expected, since the ellipsoids become identical spheres, and the model reduces to the one-mobility case. At this point, it seems as though this special kind of anisotropic Fermi surface model cannot lead to $r > 1$, in contrast to the results of the previous section, where a very slight anisotropy sufficed to achieve that inequality. But as I will show below, eq.(26) lacks an essential ingredient, and this prevents the Hall factor from exceeding unity, no matter how strongly anisotropic the valleys become. That additional factor is already present in eq.(23) of the previous section.

X. PLANAR-FACED ENERGY SURFACE MODEL

The last section considered the first case of an anisotropic model which could produce strong mobility mixing, the essential ingredient for changing r from unity. The Hall factor r has also been determined for a group of anisotropic energy surfaces which are composed of planar faces. The first two such cases treated were a cube and an octahedron (Goldberg et al. 1957, Miyazawa 1962). These shapes led to

$$r = 1/2 \quad \text{and} \quad r = 2/3, \qquad (27)$$

respectively. Such values represent strong mixing effects, especially the first one. By comparison, prolate multivalley models (the type that almost always seems to occur in real materials) predict

$$1 > r > 3/4. \qquad (28)$$

The r values for the cube and octahedron were originally obtained using the long and rather complex kinetic technique (Shockley 1950, McClure 1956, Chambers 1957). I later showed (Allgaier 1967, 1968, 1970) that a short and simple method can be used to obtain these two results, and can be extended to more complicated and realistic cases without becoming too cumbersome.

The calculations are carried out in configuration B, assuming degenerate statistics. Since eq.(11) is used to determine r in this configuration, the current components i_x and i_y must be determined. Fig. 3(a) shows how these are obtained for the case of a cube. The x-component of the current density is obtained from the displaced portion of the carrier distribution (Δn) in momentum space, using the formula $i_x = (\Delta n)ev_F$, where v_F is the carrier velocity at the Fermi surface. The y-component is determined from that portion of the <u>displaced</u> carriers which drifts around the corners of the cube during the time τ, due to the Lorentz force. Many different planar-faced models have been worked out, and they always lead to $r < 1$, as in the case of the ellipsoidal multivalley model.

Part (b) of Fig. 3 indicates the kind of displaced distribution which would occur if the relaxation time τ increased in going from the edges of a cube face to its center. Introducing this, or any kind of, τ-anisotropy does not complicate matters significantly, since cubic symmetry requires that the Fermi velocity remain constant over each face of the cube. If the average scattering time is given by

$$\bar{\tau} = s\tau_{edge} , \tag{29}$$

then the Hall factor becomes

$$r = 1/2s^2 . \tag{30}$$

If the cube is regarded as an approximation to a Fermi surface which is more sharply curved near the edges of the cube and flatter away from them, then $s > 1$ increases the current contributions from the flatter portions of the surface, since τ and therefore μ become larger there. Thus the average curvature of the surface decreases, as though it corresponded to a larger carrier density. This would decrease R_o, and r as well, which is what eq.(30) predicts for $s > 1$.

Thus it begins to become evident that an important part of the model-dependent link between R_o and the carrier density n is the average curvature of the Fermi surface which encloses n. This connection is elegantly and compactly described by eq.(12) above (Tsuji 1958). In any event, an examination of all of the results for planar-faced energy surfaces with isotropic and anisotropic scattering [plus eq.(23) for the slightly anisotropic model] suggests the following rule: An anisotropic Fermi surface almost always leads to $r < 1$, and this value may be increased or decreased by an anisotropic τ which more strongly weights (i.e., increases the mobility on) the sharper or flatter portions of the Fermi surface (to which the planar-faced model is an approximation). This corresponds logically to an effectively decreased or increased value of n. The sole exception at this point seems to be in the slightly anisotropic model of Section VIII, which sometimes permits $r > 1$.

Parts (c) and (d) of Fig. 3 illustrate a second effect caused by an anisotropic τ (Allgaier 1968). Part (c) shows a planar-faced energy surface which is composed of three differently-oriented types of surfaces normal to the symmetry-equivalent sets of <100>,

<111>, and <110> directions. In this case, τ-anisotropy can be introduced by assigning a constant, but different, τ to each type of surface. A "smoothly varying" τ-anisotropy is achieved by defining a single anisotropy parameter T as

$$T = \frac{\tau_a}{\tau_b} = \frac{\tau_b}{\tau_c} , \tag{31}$$

where a, b, and c refer to the three types of faces, taken in different sequences. The three resulting curves for r versus T are shown in Fig. 3 (d). The different sensitivities of r versus T which result can be understood in terms of the curvature distribution of the energy surface to which the model is an approximation.

The new feature is the minimum in r which appears in all three curves. This leads to a second anisotropy-related rule: The initial effect of a τ-anisotropy which weights the flatter portions

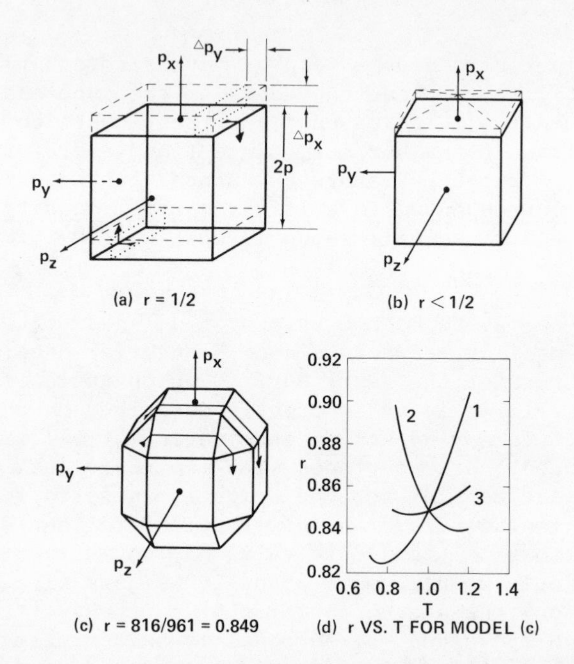

Fig. 3. Planar-faced energy surface models (see Section X). (a) Cubic surface. (b) Cubic surface with scattering anisotropy (scattering becomes weaker in going from edge to center of each face). (c) A surface with {100}, {111}, and {110}-oriented faces. (d) Hall factor r for model (c) with anisotropic scattering. Curves 1, 2, and 3 correspond to different sequences of scattering ratios T for the three types of faces.

of an energy surface more strongly will always be a decrease in r. But when that anisotropy becomes strong enough, r will begin to increase again. It is as though (this is a rather speculative interpretation) very strong weighting, in effect, causes some carriers to disappear, as in the isotropic two-mobility model when $\mu_2/\mu_1 \to 0$. Thus the influence of anisotropy is ultimately overcome by a more powerful mobility-weighting effect. It is the same kind of effect which increases r from the very beginning when the models do not include any shape anisotropy, viz., the two-, N-, and infinite-mobility isotropic models.

XI. THE MISSING INGREDIENT

Up to this point, two basic transport model ingredients—τ variation and energy surface shape—have been related to the behavior of r. The effect of the first ingredient on r can be described satisfactorily and consistently, but the influence of the second contains an element of uncertainty, although the predominant trend is clear enough. This shortcoming can be eliminated by a third ingredient that has not yet been mentioned. The oldest clue to its existence that I know of is contained in the papers dealing with the slightly anisotropic model described in Section VIII (Davis 1939, Cooper and Raimes 1959).

That model contains two parameters, A and B, relating to the Fermi surface, and a third, C, which specifies τ-anisotropy. The parameter B contains energy derivatives of the quantities which determine the shape anisotropy. Thus A describes the Fermi surface shape while B permits that shape to change with energy. Even though the calculation assumed completely degenerate statistics, <u>shape evolution</u> does have an important effect, since it can change the variation, or anisotropy, of v_F, the carrier velocity on the Fermi surface.

It is illuminating to reexamine eq.(23) for the case in which the energy surface shape does <u>not</u> change with energy. In that case, B = A, and when τ is isotropic (C = 0),

$$r = 1 - (104/21)A^2. \qquad (32)$$

Thus for the slightly anisotropic model, an <u>unchanging</u> anisotropic Fermi surface alone <u>always</u> makes $r < 1$. Note that none of the other models considered thus far contain any parameters which allow the Fermi surface shape to change with energy. The rule relating to anisotropic band models may now be strengthened and made more specific: For all band models with anisotropic Fermi surfaces of unchanging shape, $r < 1$.

Another important characteristic of the Hall factor r in the Davis-Cooper-Raimes model is that the effects of shape evolution and scattering anisotropy are completely interchangeable. As is easily seen from eq.(23), the condition is B = -C. The general equivalence between v_F and τ as weighting factors is obvious from the symmetrical way those variables enter eq.(12).

This equivalence also explains a characteristic of a model, not yet mentioned, which at one time also seemed to be an exceptional case. The weak-field galvanomagnetic coefficients have been calculated exactly for a cubically symmetric version (Allgaier 1966) of the nonparabolic, nonellipsoidal multivalley band model originally worked out for bismuth (Cohen 1961). When one of the two "extra" parameters is set equal to zero, the energy surfaces become ellipsoidal, but the model retains a direction-dependent nonparabolicity. Consequently, the elongation of the ellipsoids changes with energy, i.e., with carrier density. The results of the calculation for this case reveal (in contrast to the multivalley model treated in Section IX) that a range of carrier densities exists for which $r > 1$, even though no scattering anisotropy was included in the model. A closer look at the model shows that the shape evolution emphasizes the sharper, end regions of the prolate Fermi surfaces, causing the Hall factor r to increase.

XII. THE RULES

The synthesis of rules which I have extracted from the models considered in the previous sections may be summarized as follows:

1. For a one-mobility <u>transport</u> model, whether isotropic or anisotropic, $r = 1$.

2. For a multi-mobility (mobility-mixing) but isotropic <u>band</u> model, $r > 1$. This includes models with an energy-dependent mass, an energy-dependent τ, and a direction-dependent (anisotropic) τ.

3. For a <u>metallic</u> (degenerate limit) anisotropic <u>band</u> model with isotropic τ, $r < 1$, always, if that anisotropy does not change with energy. Here, anisotropy means the non-ellipsoidal, multi-mobility type.

4. When <u>additional</u> mobility mixing occurs under rule 3, due to an energy-dependent mass or τ (because the statistics are not completely degenerate), but without introducing <u>new</u> anisotropies, r will always increase, and may exceed unity if the mobility mixing is strong enough.

5. When the mobility mixing for the anisotropic metallic model in rule 3 (with its characteristic value $r < 1$) is altered by

introducing τ-anisotropy or energy-surface shape evolution (through altered v_F-anisotropy), r will initially increase or decrease according to whether these w ghting factors emphasize the sha flatter portions of the Fermi surface.

6. When the anisotropic w ghting becomes strong enough in the second alternative under rule 5, r will stop decreasing and begin to increase again. This means that the kind of mobility mixing described under rules 2 and 4, which always increases r, ultimately wins out over the mixing which involves anisotropic weighting of an anisotropic band model, even if the initial effect of the latter is to decrease r.

None of the models in the earlier sections discussed the case of an extremely distorted energy surface which includes regions of both positive and negative curvature. I must therefore add the caveat, under rule 6, that if the anisotropic τ- and v_F-weighting factors emphasize regions of opposite curvature strongly enough, r will change sign first before its magnitude increases again.

XIII. CONCLUSIONS

This paper has concentrated on one segment of the history of Hall effect analyses. I have long felt that there were shortcomings in the general understanding of the type of model discussed in this paper. To me, these shortcomings were most clearly revealed by the lack of any guidelines or rules which, for any transport model with a given set of basic characteristics, would predict even a qualitative value for the Hall factor r.

In this contribution to the Hall Effect Commemorative Symposium, I have summarized my proposal for a set of such rules. I believe that they apply to all models which fall within the constraints mentioned in the introduction.

Some of the other aspects of Hall effect behavior and other classes of models not discussed in this paper are still poorly understood, and a number of these remain topics of strong controversy. I hope that these complexities and controversies will not persist much longer, but will become as simple and straightforward as the subject of the present paper.

If the reader does not agree with this characterization of the present subject, please blame it on my inadequate explanation, not on the subject itself. I am willing to admit that almost any aspect of Hall effect behavior can be very difficult to explain clearly, despite the unique basic simplicity which I have ascribed to this fascinating transport coefficient.

ACKNOWLEDGEMENTS

The preparation of this paper was supported by the NSWC Independent Research Fund. I am deeply indebted to Dr. T. R. McGuire of the IBM Watson Research Center, Yorktown Heights, New York, and Professor C. L. Chien of Johns Hopkins University for making it possible for me to submit this contribution. I am also grateful for the understanding and support which my wife and children gave me during the Christmas holidays, when most of this paper was written.

REFERENCES

Abeles, B., and Meiboom, S., 1954, Phys. Rev., 95:31.
Allgaier, R. S., 1966, Phys. Rev., 152:808.
Allgaier, R. S., 1967, Phys. Rev., 158:699.
Allgaier, R. S., 1968, Phys. Rev., 165:775.
Allgaier, R. S., 1970, Phys. Rev. B, 2:3869.
Allgaier, R., 1973, Phys. Rev. B, 8:4470.
Becker, R., 1933, "Theorie der Electrizität, Band II," Teubner, Leipzig, p. 208.
Beer, A. C., 1963, "Solid State Physics, Suppl. 4," Academic Press, New York and London.
Böning, K., 1970, Phys. Kondens. Mat., 11:177.
Campbell, L. L., 1923, "Galvanomagnetic and Thermomagnetic Effects—The Hall and Allied Phenomena," Longmans, Green & Co., New York. Reprinted, 1960, Johnson Reprint Corporation, New York.
Chambers, R. G., 1952, Proc. Phys. Soc. (London), A65:903.
Chambers, R. G., 1957, Proc. Roy. Soc. (London), A238:344.
Cohen, M. H., 1961, Phys. Rev., 121:387.
Cooper, J. R. A., and Raimes, S., 1959, Phil. Mag., 4:145.
Davis, L., 1939, Phys. Rev., 56:93.
Frenkel, J., and Kontorova, T., 1935, Phys. Zeits. d. Sowjetunion, 7:452.
Gans, R., 1906, Ann. Physik, 20:293.
Goldberg, C., Adams, E., and Davis, R., 1957, Phys. Rev., 105:865.
Hurd, C. M., 1972, "The Hall Effect in Metals and Alloys," Plenum Press, New York-London.
Jan, J.-P., 1957, "Solid State Physics, Vol. 5," Academic Press, New York and London, p. 1.
Jones, H., and Zener, C., 1934, Proc. Roy. Soc. (London), A145:268.
McClure, J. W., 1956, Phys. Rev., 101:1642.
McKay, T. C., 1906, Proc. Am. Acad., 41:385.
Miyazawa, H., 1962, "Proceedings of the Conference on the Physics of Semiconductors," The Institute of Physics and the Physical Society, London, p. 636.
Page, L., and Adams, N. I., 1949, "Principles of Electricity," Van Nostrand, Princeton, 2nd ed., p. 287; 1969, 4th ed., p. 249.

Richardson, O. W., 1914, "The Electron Theory of Matter," University Press, Cambridge, p. 436.
Shibuya, M., 1954, Phys. Rev., 95:1385.
Shockley, W., 1950, Phys. Rev., 79:191.
Thomson, J. J., 1902, Phil. Mag., 3:353.
Thomson, W. (Lord Kelvin), 1851, Trans. Roy. Soc. (Edinburgh), March issue; see also, 1882, "Mathematical and Physical Papers," Volume 1, p. 273.
Tsuji, M., 1958, J. Phys. Soc. Japan, 13:979.

HALL CURRENTS IN THE AURORA

T. A. Potemra

Applied Physics Laboratory
The Johns Hopkins University
Laurel, Maryland 20810

ABSTRACT

In the eighteenth century, Anders Celsius noted that magnetic disturbances occurred during periods of auroral displays. At the end of the nineteenth century, Kristian Birkeland established magnetic observatories in northern Scandinavia, and determined that the "auroral" magnetic disturbances were caused by intense electric currents flowing horizontally above the earth's surface in the as yet undiscovered ionosphere. These currents are now known to be <u>Hall currents</u> that flow perpendicular to both the earth's magnetic field and intense electric fields which are generated at great distances from the earth. These Hall Auroral currents can be as large as one million amperes during periods when more power is deposited in the auroral zones than is generated in the entire United States. Numerous satellite, rocket, and ground-based observational programs are presently being directed toward an understanding of the auroral currents and how they are coupled to interplanetary space. Progress in understanding some of these phenomena is reviewed.

THE AURORA: A HISTORICAL INTRODUCTION*

The early immigrants from Asia, who crossed the Aleutian Island land bridge to North America thousands of years ago, may have been guided and inspired by the dramatic displays of light over their heads. These lights in the north and south polar regions are called

---

* For more complete review see Chapman and Bartels, 1940 or Chapman, 1968.

the aurora borealis and aurora australis, respectively, and have been the source of superstition, inspiration and scientific research for at least two thousand years.

The basic forms of the auroral are shown schematically in Figure 1 from Akasofu (1965) (see Potemra, 1977 for more examples). Modern studies of the aurora have confirmed that the dramatic light displays are caused by the excitation of atmospheric gases by stream of high energy (1 keV to 100 keV) electrons and protons which bombard the atmosphere. These particles are guided to the polar region by the earth's magnetic field, and produce "curtain-like" forms of light as shown in Figure 1.

Many other types of auroral forms exist, such as recorded by Seneca in his Naturales Questiones. "Sometimes flames are seen in the sky, either stationary or full of movement. Several kinds are known: the abysses, when beneath a luninous crown the heavenly Fire is wanting, forming as it were the circular entrance to a cavern; the turns, when a great rounded flame in the form of a barrel is seen to move from place to place, or to burn immovable; the gulfs, when the heaven seems to open and to vomit flames which before were hidden in its depths. Among these phenomena should be ranged these appearances as of the heavens on fire so often reported by historians; sometimes these fires are high enough to shine among the stars; at others, so low that they might be taken for the reflection of a distant burning homestead or city. This is what happened under Tiberius, when the cohorts hurried to the succour of the colony of Ostia, believing it to be on fire. During the greater part of the night the heaven appeared to be illuminated by a faint light resembling a thick smoke."

In the Pennsylvania Gazette of 1737 Benjamin Franklin wrote of an aurora so red and vivid "that People in the Southern Parts of the Town imagin'd there was some House on Fire near the North End: and some ran to assist in extinguishing it." Franklin believed the aurora was caused by the transport of humid, warm air from the lower latitudes to the polar regions where it condensed as snow and discharged through the upper atmosphere, thereby causing the rays of the aurora.

Edmond Halley published his observation of a great aurora in 1716 in the Royal Society's Philosophical Transactions and suggested that it was due to magnetic particles arising from within the earth. Other eighteenth century scientists who watched the aurora included Dalton who published his observations in his Meteorological Essays and Cavendish whos determined the height of an aurora in 1784. Celsius discovered in 1741 that the aurora is associated with disturbances in the earth's magnetic field and in 1770 Wilcke determined that the auroral rays were aligned with geomagnetic field lines. These facts prompted several scientists including Gauss in 1838 to suggest that "there is every appearance that electricity in motion plays a principal part."

By the beginning of the twentieth century several polar expeditions had already been conducted to establish magnetic observatories under the "auroral regions". The most notable explorers came from Norway and included Fridtjof Nansen and Kristian Birkeland.

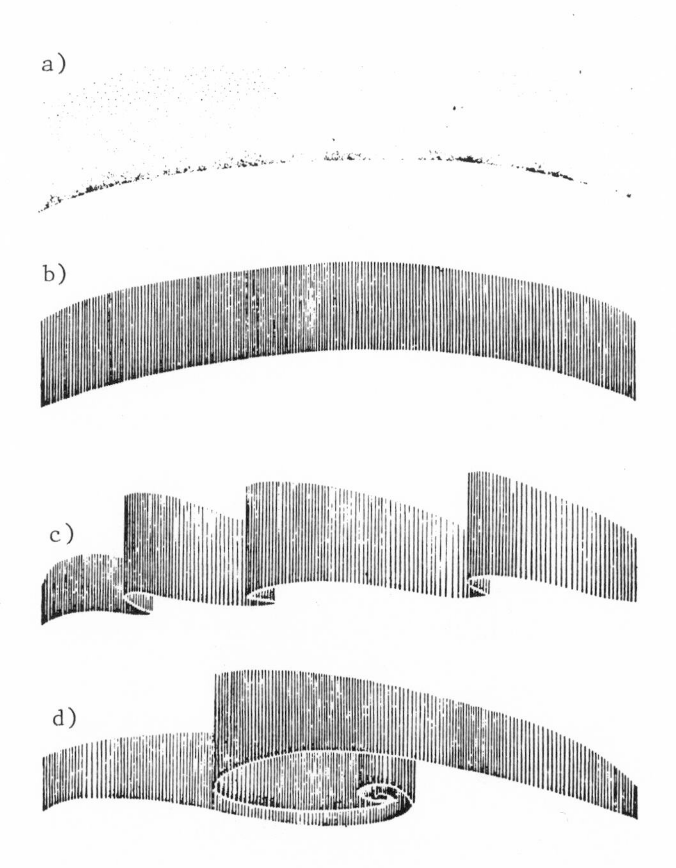

Fig. 1. The basic curtain-like form of the aurora (a) and its distorted forms (b), (c), (d). The form (a) is known as a homogeneous arc. When it is activated slightly, it develops fine folds (of wavelength of order of a few kilometers). Such a form is called a rayed arc. With more intense activation, larger scale folds are superimposed on the rayed arc form, and such an active form is called a rayed band. In the most activated form, a fold of scale of order a few hundred kilometers develops. Such a form is often called a drapery or a horseshow form. After Akasofu (1965).

Birkleland (1902, 1908) provided the following description of his work: "As leader of the expedition started by the Norwegian Government for the study of Earth-Magnetism, Polar Aurora and Cirrus clouds, I beg to inform you that during the time from August 1st 190 and June 30th 1903, four Norwegian Stations will be erected, viz. at Bossekop (Finmarken), at Dyrafjord (Iceland), at Axel Island (Spitzbergen) and Matotchkin-Schar (Novaja Zemlja). The abovementioned expedition has assumed the task of determining the connection existing between earth-magnetical perturbations, boreal lights and cirrus-clouds."

From his study of surface magnetic disturbances associated with the aurora, Berkleland came to the following important conclusion, "I demonstrated namely, that certain well-defined magnetic perturbations that occurred over large portions of the earth might be naturally explained as the effect of electric currents, which, it might be supposed, in the polar regions flowed approximately parallel with the surface of the earth at heights of several hundred kilo metres, and strengths of up to a million amperes, if they could be measured by their effect as galvanic currents. By such registrations, other important, unexplained phenomena that are very characteristically developed in the polar regions, might be excellently studied, e.g., the tremendous changes in the magnetic components, which often occur at short intervals, especially during an aurora."

These horizontally flowing currents deduced by Birkeland are now known to be Hall currents that flow perpendicular to both the earth's magnetic field and intense electric fields. A brief description is provided here of the characteristics of these Hall currents and their relationships to Solar-Terrestrial phenomena determined from surface, rocket and satellite observations.

AURORAL CURRENTS AND MAGNETIC DISTURBANCES

The geomagnetic field can be thought of as being produced by a bar magnet near the earth's center, with the axis of the magnet tilted 11.5° away from the earth's rotational axis. The poles of this magnet are located near Thule, Greenland, and Vostok, Antarctica, a Soviet research station. The fact that the earth possessed a magnetic field was first pointed out by Gilbert in his DeMagneta, published in 1600, nearly a century before Newton's Principia was published in 1687. Gauss established the first network of surface magnetic observatories to study the geomagnetic field and disturbance to it. Presently, a global network of surface magnetometers provides a fundamental and important set of observations of great importance to the exploration of the earth's environment and outer space. The network of surface magnetometers in the polar regions, such as shown in Figure 2 (Akasofu et al., 1965, their Figure 12), has been used to deduce the large-scale currents in the auroral

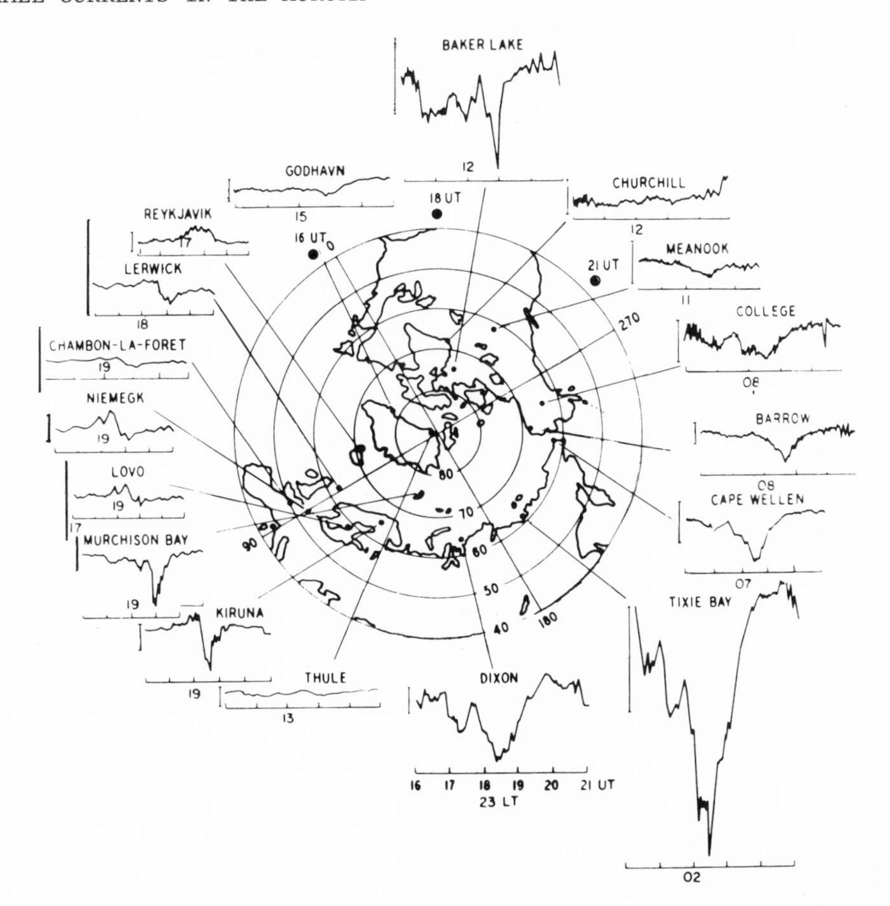

Fig. 2. Simultaneous magnetic records between 1600 and 2100 UT on 16 December 1957 from a number of stations in the northern hemisphere; view from above the magnetic dipole north pole; the direction of the sun at 1600, 1800 and 2100 UT is indicated, and also the local time at each station at 1800 UT; scale = 200 nT. After Afasofu et al. (1965), their Figure 13.

regions that were suggested by Birkeland.

Figure 3 is a schematic drawing of the ionospheric current and related magnetic disturbance. This figure is a "latitude slice" through which a slab of current (10 km thick by 500 km wide) flows toward the west (i.e., "into" this figure). The total current

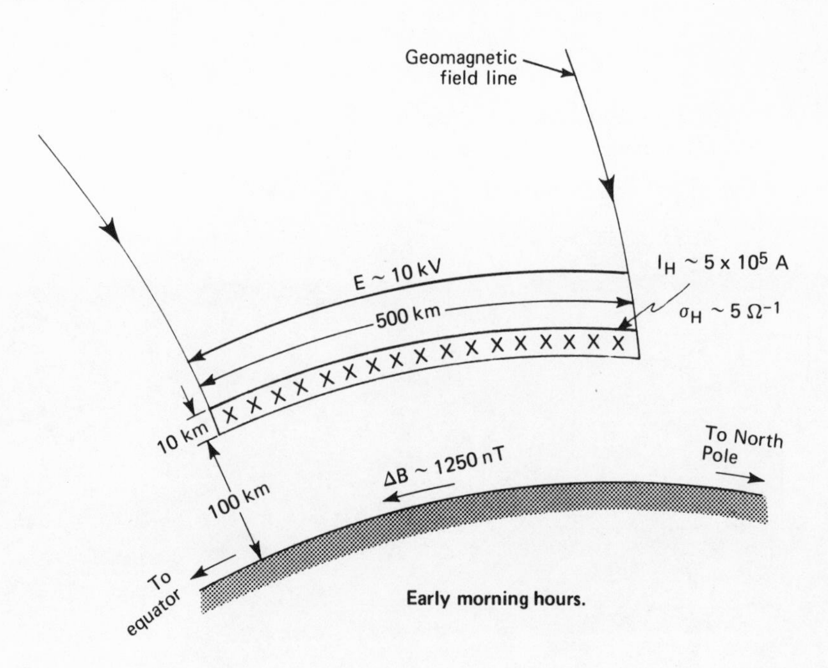

Fig. 3. A schematic and highly simplified diagram of the westward auroral electrojet. The view is toward the west from the early morning hours. The geomagnetic field is directed downward toward the earth, and the electric potential of 10 kV is directed toward the equator thereby driving a Hall current (comprised mainly of ionospheric electrons flowing in the opposite direction) toward the west. The magnetic disturbance underneath this current sheet is 1250 nT (computed with an infinite sheet approximation for the overhead current—the magnetic disturbance at higher or lower latitudes is smaller due to edge effects).

intensity of 5 x 10^5 A produces a magnetic field underneath at the earth's surface directed toward the south (in the equatorward direction) with a magnitude of 1250 nT (nT = "nano Tesla", where the surface geomagnetic field intensity is approximately 0.5 Gauss = 50,000 nT). From an examination of the direction and amplitude of the magnetic disturbances observed along a latitude chain of observatories, the height and latitude extent of the ionospheric current can be deduced. As will be described later, the westward ionospheric current depicted in Figure 3 is related to a southward directed electric potential of 10 kV by a Hall conductivity of 5 Ω^{-1}. Because of its association with a variety of auroral phenomena, the large-scale current in Figure 3 is often referred to as an "auroral electrojet current", or a "Hall auroral current".

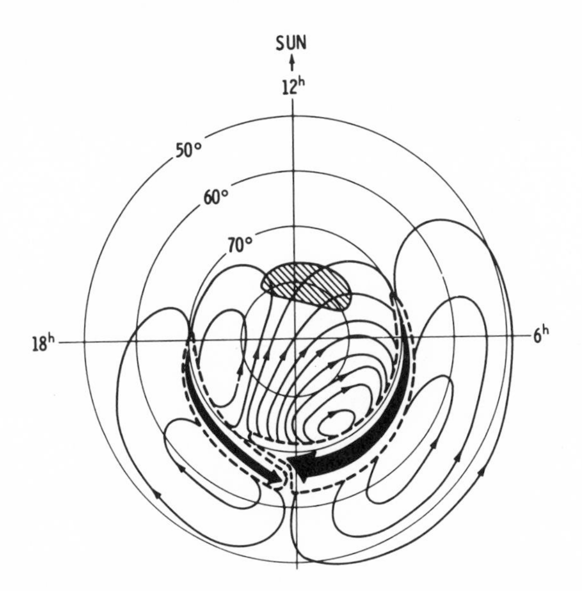

Fig. 4. Illustrative pattern of ionospheric electric currents causing magnetic bay disturbances in the auroral zone and associated currents at higher and lower altitudes. From Sugiura and Heppner (1965), their Figure 14.

The configuration of the auroral zone and its associated current system is fixed with respect to the position of the sun, and the earth appears to rotate underneath it. Therefore observatories often record magnetic disturbances directed equatorward in the dawn hours, and poleward in the evening hours (for example in Figure 2). These magnetic disturbances, sometimes referred to as "magnetic bays" because of their resemblance (on chart records) to a "bay" on a navigational map, are interpreted as being caused by a westward flowing Hall current in the dawn hours and an eastward flowing Hall current in the evening hours as depicted in Figure 4 (Sugiura and Heppner, 1965, their Figure 14). The intensities of the eastward and westward auroral electrojets, as deduced from surface magnetic measurements, have been used as index of auroral activity for many years. They are a more accurate index than, for example, optical observations which can be masked by poor weather or the full moon. Some auroral experimenters have joked that the best optical auroral forms invariably occur during periods of heavy cloud cover, thereby substantiating some claims of a close connection between auroral processes and the weather.

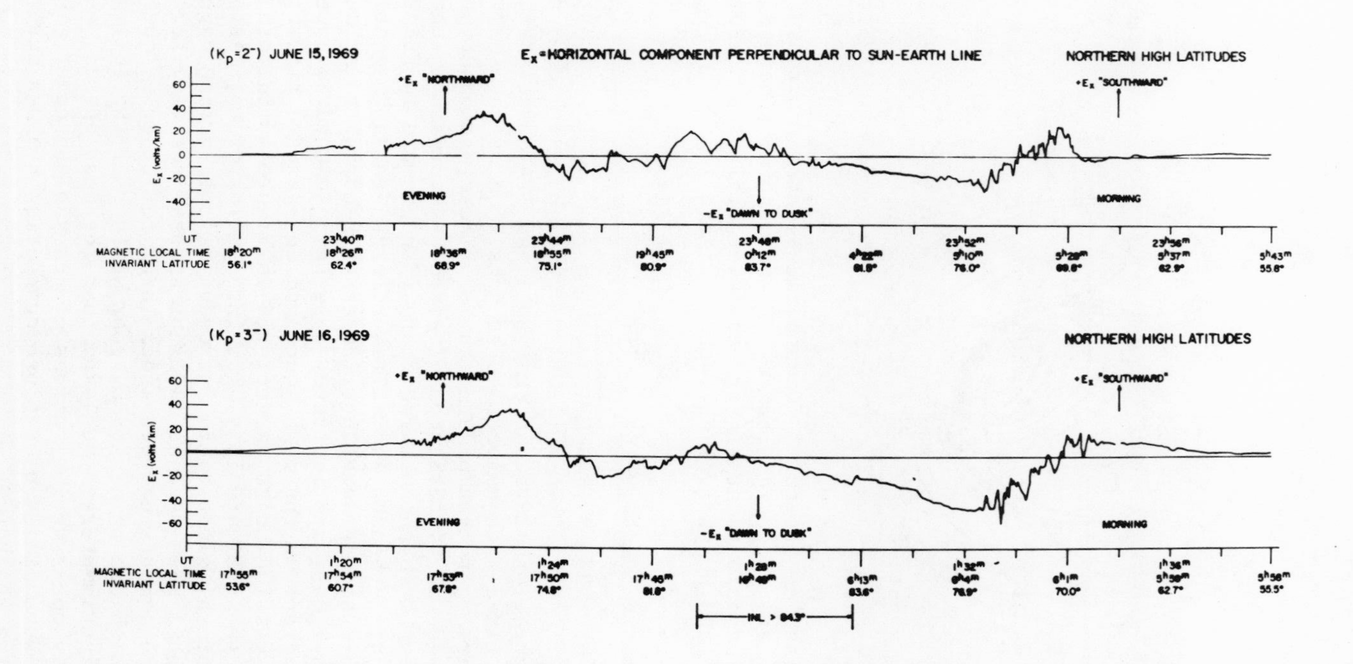

Fig. 5. OGO-6 electric field measurements from two successive northern high latitude passes. From Heppner (1973), his Figure 5.

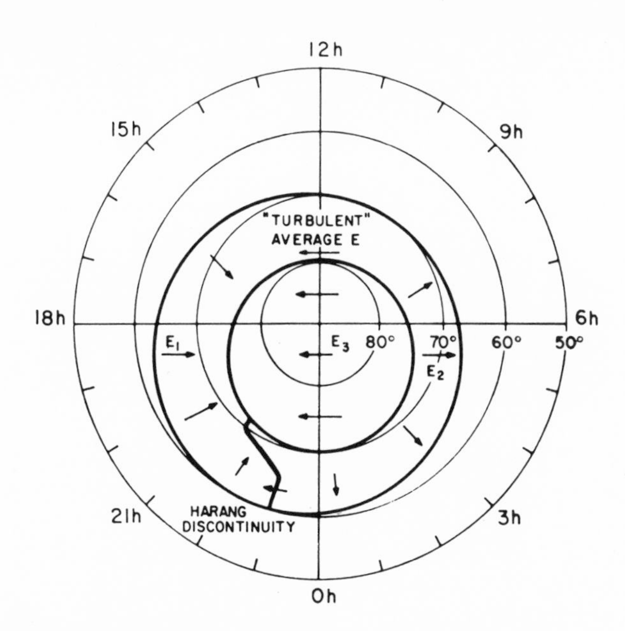

Fig. 6. Characteristic directions of the high-latitude electric field. From Heppner (1973), his Figure 3.

HIGH LATITUDE ELECTRIC FIELDS

Satellite experiments have confirmed the presence of large scale electric fields in the auroral and polar regions. Figure 5 shows two examples of the electric field data acquired by the OGO-6 satellite over the north pole (Heppner, 1973). Each of these orbits extends from the evening side of the auroral region, over the north pole, and down to the morning side of the auroral zone. The principal features in these data are the presence of electric fields directed toward the pole in the evening auroral region, directed toward the equator in the morning auroral region, and directed from dawn to dusk in the polar cap as schematically shown in Figure 6 (Heppner, 1973, his Figure 3). These electric fields reach intensities of 50 mV/m resulting in potential drops of tens of kilovolts over a few hundred km (see the example of Figure 3).

The auroral electric fields reverse direction near midnight where the eastward and westward electrojet currents meet (see Figures 4 and 6). This area has been referred to as the "Harang Discontinuity" in honor of the Norwegian scientist who made substantial contributions to the understanding of ionospheric currents from surface magnetic measurements (Heppner, 1972 and references therein). The auroral electric field orientation shown in Figure 6 is also of the proper sense to drive the eastward and westward Hall currents (since the geomagnetic field is directed downward in the northern polar region, an equatorward directed electric field would drive ionospheric electrons toward the east thereby producing a westward directed Hall current—see Figure 3).

CONDUCTIVITIES IN THE AURORAL REGION

Calculations of ionospheric conductivity parallel the development of modern Plasma Physics (e.g. Cowling, 1932; Chapman, 1956; Boström, 1964).

In general, the current density in the auroral ionosphere is given by:

$$\vec{J} = \sigma_{||}\vec{E}_{||} + \sigma_p \vec{E}_\perp + \sigma_H \frac{\vec{B} \times \vec{E}}{B}$$

where $\vec{B}$ is the geomagnetic field; $\vec{E}$ is the electric field with components parallel and perpendicular to $\vec{B}$ denoted by $\vec{E}_{||}$ and $\vec{E}_\perp$, respectively; and $\sigma_{||}$, σ_p, and σ_H are the field-aligned, Pedersen and Hall conductivities, respectively (e.g. Akasofu and Chapman, 1972, Chapter 4.4). The Pedersen conductivity is named in honor of the Danish scientist who provided some early contributions to an understanding of the ionospheric conductivity (see, for example Chapman, 1956). It has been suggested that the field-aligned component be named in honor of Birkeland, who originally proposed their existence (e.g. Potemra, 1978). These conductivities are given by the following formulas:

$$\sigma_{||} = q^2 n \left(\frac{1}{m_e \nu_{en}} + \frac{1}{m_i \nu_{in}} \right)$$

$$\sigma_p = q^2 n \left(\frac{1}{m_i} \frac{\nu_{in}}{\omega_i^2 + \nu_{in}^2} + \frac{1}{m_e} \frac{\nu_{en}}{\omega_e^2 + \nu_{en}^2} \right),$$

$$\sigma_H = q^2 n \left(\frac{1}{m_e} \frac{\omega_e}{\omega_e^2 + \nu_{en}^2} - \frac{1}{m_i} \frac{\omega_i}{\omega_i^2 + \nu_{in}^2} \right).$$

Here m_e and m_i, respectively are the mass of electrons and positive ions, q the charge, n the number density of electrons (and positive ions); ν_{en} and ν_{in}, respectively, the collision frequencies between electrons and neutral particles and between positive ions and neutral particles; and ω_i and ω_e, respectively, the Larmor frequencies (or gyrofrequencies) for the ions and electrons given by

$$\omega_i = \frac{qB}{m_i}$$

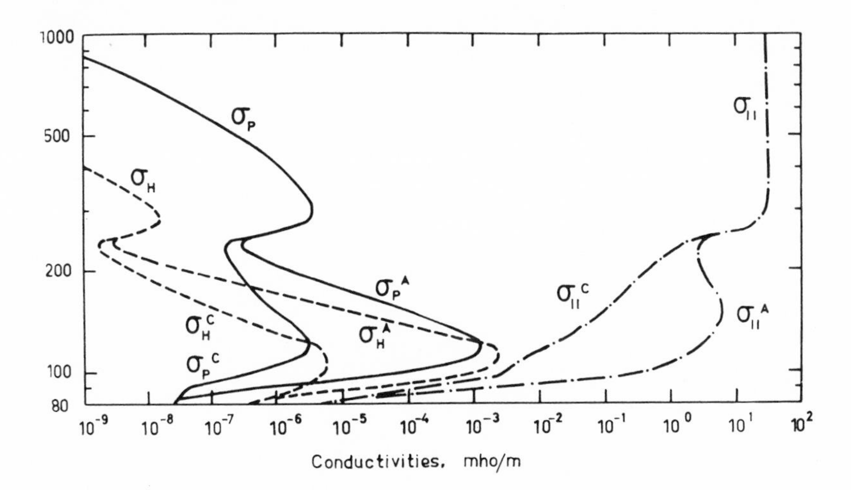

Fig. 7. Conductivities within and outside an auroral arc computed by Boström (1964), his Figure 3. The superscript "A" refers to the auroral arc, and "C" refers to the undisturbed ionosphere.

$$\omega_e = \frac{qB}{m_e}$$

Figure 7 shows height profiles of these conductivities computed by Boström (1964) for model electron densities appropriate for the undisturbed ionosphere, denoted by the "C" superscript, and for an auroral arc, denoted by the "A" superscript. The peak Hall and Pedersen conductivities occur just above 100 km altitude in the E-region of the ionosphere. In this region the electrons do not collide very often with the neutrals so that $\omega_e > \nu_e$, whereas the ions suffer many collisions, $\omega_i < \nu_i$. Consequently, the Hall current is carried mainly by the electrons which move with nearly the same velocity ($\vec{v}_e$), to a good approximation, as in the absence of collisions

$$\vec{v}_e = \vec{E} \times \vec{B}/B^2$$

The current density due to these electrons is given by

$$\vec{J} = n\, q\, \vec{v}_e = n\, q\, \vec{E} \times \vec{B}/B^2 = \sigma_H\, \vec{E} \times \vec{B}/B$$

where the Hall conductivity is

$$\sigma_H = n\, q/B$$

For electrons the sign of q is negative so that an electric field directed equatorward in the northern hemisphere (or southward), where the geomagnetic field is directed predominantly downward, will drive a Hall current toward the west (see Figures 3 and 4). For a geomagnetic field value of 0.5 gauss (= 5×10^{-5} Tesla) and a peak electron density of 10^6 cm^{-3} in an auroral arc, the corresponding Hall conductivity is about 3×10^{-3} Ω^{-1}/meter as shown in Figure 7. For the model height distributions shown in Figure 7, the height-integrated Hall and Pedersen conductivities in the auroral arc are

$$\Sigma_P = \int \sigma_P dh = 36\ \Omega^{-1}$$

$$\Sigma_H = \int \sigma_H dh = 56\ \Omega^{-1}$$

A combination of satellite and ground-based measurements indicate values of integrated Pedersen conductivities ranging from 12 to 20 Ω^{-1} for auroral arcs (Horwitz et al., 1978; Greenwald et al., 1980) and Hall to Pedersen conductivity ratios (Σ_H/Σ_P) ranging from 2 to 4 (Sulzbacher et al., 1979). The integrated conductivities outside the auroral arcs are somewhat lower with typical values of Σ_P = 1 to 2 Ω^{-1} and Σ_H = 5 Ω^{-1} as used in the example of Figure 3.

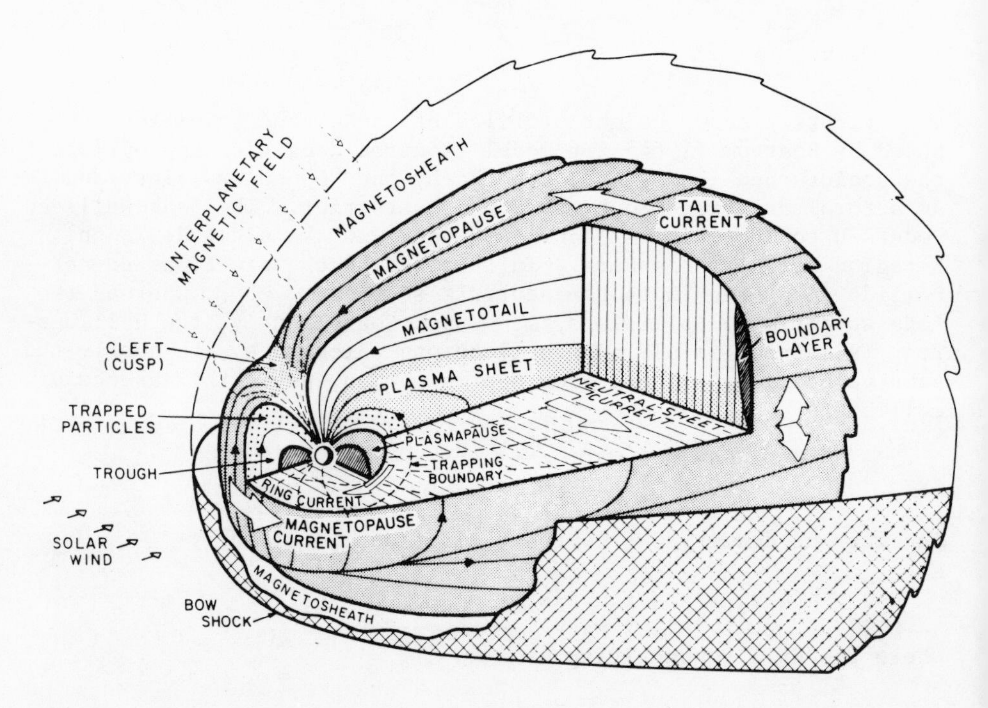

Fig. 8. The configuration of the magnetosphere produced by the solar wind flowing past the earth. From Heikkila (1973), his Figure 1.

THE SOLAR WIND AND MAGNETOSPHERE

From a study of type I (ionized) comet tails, Biermann (1951) proposed the existence of a particle flux flowing continously and radially outward from the sun. This flowing plasma, called the solar wind, is comprised of an equal number of low energy electrons and protons. When the solar wind arrives at the earth, being a good conductor, it cannot penetrate the geomagnetic field, but instead deforms it into a comet-like configuration shown in Figure 8 called the "magnetosphere" (Heikkila, 1973, his Figure 1). The magnetosphere extends up to about 10 R_e (R_e = earth radius = 6371 km) upstream toward the sun and to a great distance downstream away from the sun beyond the moon's orbit. As in the aerodynamic case of supersonic flow around an obstacle, a shock wave is formed upstream of the magnetosphere.

It has been proposed that "friction" or a "viscous-like" interaction between the solar wind and the magnetosphere causes a large-scale "convective motion" of plasma in the magnetosphere (Piddington, 1960; Axford and Hines, 1961). The convective flow consists basically of a two-cell pattern, with flow directed away from the sun in the polar cap of the earth, and directed toward the sun in the auroral zones as drawn schematically in the top of Figure 9. This pattern is similar to that produced in a tea cup when one blows gently across the surface of the tea. The flow of the plasma through the geomagnetic field produces a polarization electric field as shown in the bottom of Figure 9. The antisunward flow in the earth's polar cap will produce an electric field directed from the dawn to dusk, labelled "E_{PC}" in the top of Figure 9. The sunward convective flow in the auroral zones will produce electric fields in the opposite direction, labelled "E_{AZ}" in Figure 9. These auroral zone electric fields will drive Hall currents in an opposite direction to the convective flow, namely antisunward. These Hall currents are labelled "I_H" in Figure 9. Typical convective flow speeds of 1 km/sec will produce polar cap electric fields of 50 mV/m in the 50,000 nT (nT = nano Tesla) geomagnetic field. Consequently, variation in solar activity and the solar wind can be communicated to the polar cap and auroral zones of the earth by the chain of processes described above.

The distribution of field-aligned (Birkeland), Pedersen, and Hall currents associated with the auroral zones is shown in Figure 10 (Boström, 1975, his Figure 6). The flow pattern of Birkeland currents shown in this figure has been determined with satellite magnetic field observations (Iijima and Potemra, 1976; Potemra, 1978). The Birkeland currents may be connected in the auroral ionosphere by the Pedersen currents, as shown in Figure 10, which flow in the same direction as the convection electric fields (E_{AZ}) shown in Figure 9.

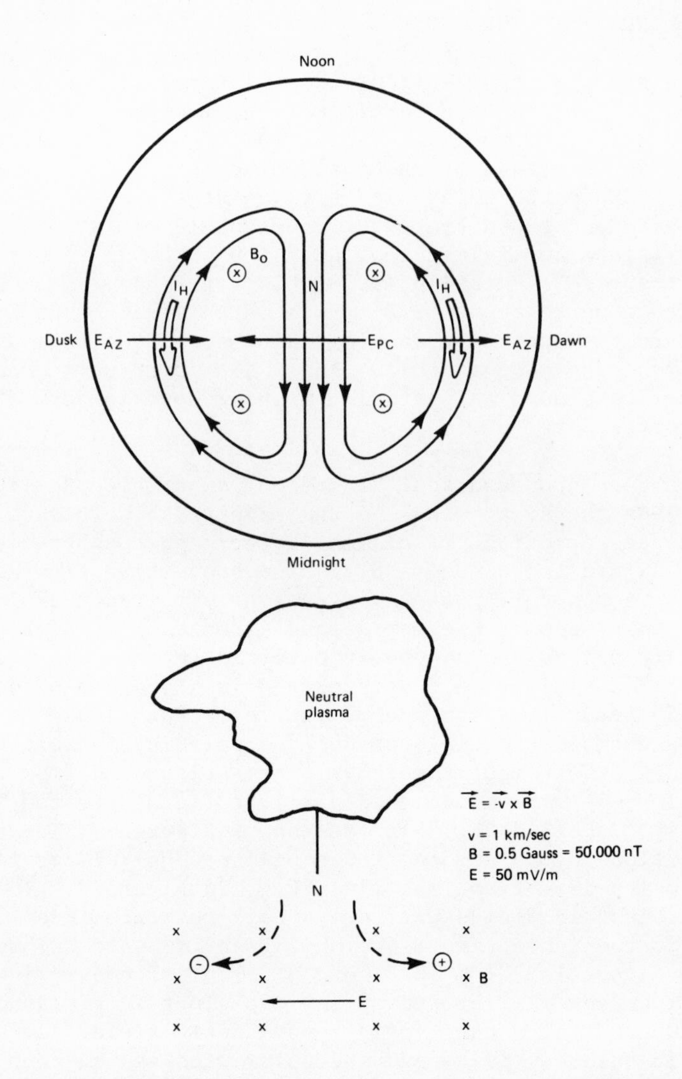

Fig. 9. A schematic view of the north polar region of the earth (in the top) and of the polarization electric field produced by a neutral plasma convecting through a magnetic field (at the bottom). In the top diagram, the two-cell convective flow pattern is denoted by the closed solid lines with arrows directed toward midnight (i.e. antisunward) in the polar cap, and directed toward noon (sunward) in the auroral regions. The electric fields produced by this convective flow are denoted by E_{PC} in the polar cap, and E_{AZ} in the auroral zones. The auroral Hall currents are labelled I_H.

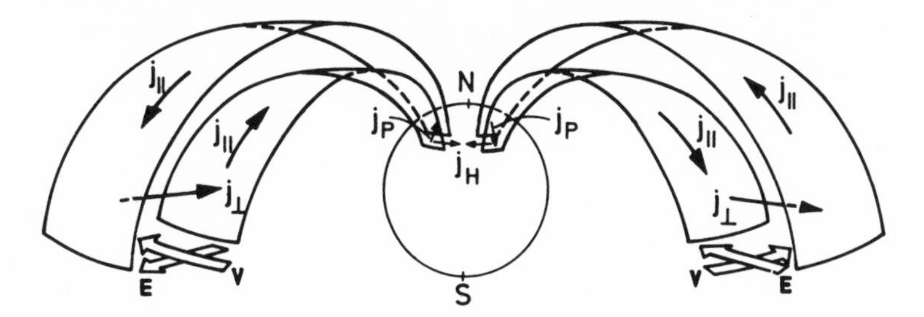

Fig. 10. Distribution of the field-aligned currents and the associated Pedersen (ionospheric) currents, plasma motions and electric fields in the equatorial plane. From Boström (1976), his Figure 6.

SUMMARY

The auroral zones are the focal points of the solar-terrestrial interaction. The spectacular visual forms are only a partial manifestation of the tremendous deposition of energy by electric currents which can dissipate more power in the auroral atmopshere than generated in the entire United States. A variety of effects are produced in the auroral ionosphere including infra-sonic waves that propagate to long distances, chemical changes, ionization, and Bremsstrahlung X-rays. Although these phenomena have been studied for many years, many puzzles remain unsolved. For example, how do solar particles enter the earth's magnetosphere and how are they stored? How are particles energized and directed toward the earth to form field-aligned currents and produce light emissions? Does the vast amount of energy deposited in the auroral atmosphere somehow trigger long-range weather effects? Do auroras affect biological processes? Most of the auroral energy is absorbed high in the atmosphere and therefore would not be expected to directly affect any life at the earth's surface. But is it possible that chemical changes in the auroral atmosphere somehow modify other wavelengths of solar energy to produce serious effects? A limited amount of data predicts that the earth's magnetic field is slowly decreasing and will ultimately vanish. What will happen to the aurora and the magnetosphere then?

Many mysteries of the aurora have been solved since the sightings of Seneca and Franklin. But as with many fields of research, more questions are raised than perhaps answered. Those new questions span a wide range of scientific disciplines including physics, chemistry, biology, and anthropology. The vast amount of energy available in the aurora may be impractical and impossible to tap to alleviate the energy crisis. But studies of this unique phenomenon can help us understand the delicate, complicated, and ever changing environment of our planet.

"...When we have time to spare we will set about devising a machine whereby the electricity of the aurora borealis may be harnessed, and made to do duty in a practical way. We will make it run the dynamos to supply our houses and streets with electric light; it shall propel our machinery, and thus take the place of steam; it shall be used for forcing our gardens, in the way that electricity is supposed to make plants grow; and it shall develop the brains of our statesmen and legislators, to make them wiser and better and of more practical use than they are at present. Hens shall lay more egges, cows must give cream in place of milk, trees shall bear fruit of gold or silver; tear-drops shall be diamonds, and the rocks fo the fields shall become alabaster or amber. Wonderful things will be done when we get the electricity of the aurora under our control."

"Yes," responded Fred, "babies shall be taken from the nursery and reared on electricity, which will be far more nutritious than their ordinary food. When the world is filled with giants nourished from the aurora, the ordinary mortal will tremble. We'll think it over, and see what we can do."*

REFERENCES

Akasofu, S.-I., December 1965, The aurora, Sci. Amer., 213:54-62.

Akasofu, S.-I., Chapman, S., and Meng, C.-I., 1965, The polar electrojet, J. Atmos. Terr. Phys., 27:1275.

Akasofu, S.-I., and Chapman, S., 1972, "Solar-Terrestrial Physics," Clarendon Press, Oxford.

Axford, W. I., and Hines, C. O., 1961, A unifying theory of high-latitude geophysical phenomena and geomagnetic storms, Can. J. Phys., 39:1433.

Biermann, L., 1951, Kometenschweife und solare korpuskularstrahlung, Z. Astrophys., 29:274.

Birkeland, Kr., June, 1902, The proposed magnetic researches of the Norwegian Polar Stations in 1902-1903, Terr. Magn. and Atm. Elec., 81.

Birkeland, Kr., 1908, On the cause of magnetic storms and the origin of terrestrial magnetism, in: "The Norwegian Aurora Polaris Expedition, Vol. 1," H. Aschehoug and Co. Pub., Oslo.

Boström, R., 1964, A model of the auroral electrojets, J. Geophys. Res., 69:4983.

Boström, R., 1975, Mechanisms for driving Birkeland currents, in: "Physics of the Hot Plasma in the Magnetosphere," ed. B. Jultqvist and L. Stenflo (Proc. Nobel Sym. April, 1975), 341.

* From The Voyage of the Vivian by Thomas W. Knox, 1884 (courtesy of S.-I. Akasofu).

Chapman, S., and Bartels, J., 1940, "Geomagnetism, Vols. I and II, Clarendon Press, Oxford.

Chapman, S., 1956, The electrical conductivity of the ionosphere: A Review, Nuovo Cimento, 4X:1385.

Chapman, S., 1968, Historical introduction to aurora and magnetic storms, Ann. Geophys., 24:497.

Cowling, T. G., Nov., 1932, Electrical conductivity of ionized gas in the presence of a magnetic field, Roy. Astron. Soc., M. N. 93:90.

Greenwald, R. A., Potemra, T. A., and Saflekos, N. A., 1980, STARE and TRIAD observations of field-aligned current closure and Joule heating in the vicinity of the Harang Discontinuity, J. Geophys. Res., 85:563.

Heikkila, W. J., 1973, Aurora, EOS (Transactions of the Am. Geophys. Union), 54:764.

Heppner, J. P., 1972, The Harang Discontinuity in auroral belt ionospheric currents, Geofysiske Publikasjoner, 29:105.

Heppner, J. P., 1973, High latitude electric fields and the modulations related to interplanetary magnetic field parameters, Radio Sci., 8:933.

Horwitz, J. L., Doupnik, J. R., and P. M. Banks, 1978, Chatanika radar observations of the latitudinal distributions of au-oral roral zone electric fields, conductivities, and currents, J. Geophys. Res., 83:1463.

Iijima, T., and Potemra, T. A., 1976, Field-aligned currents in the dayside cusp observed by TRIAD, J. Geophys. Res., 81:5971.

Piddington, J. H., 1960, Geomagnetic storm theory, J. Geophys. Res., 65:93.

Potemra, T. A., Feb., 1977, Aurora borealis: the greatest light show on earth, Smithsonian, 7:64.

Potemra, T. A., 1978, Observation of Birkeland Currents with the TRIAD satellite, Astrophys. Space Sci., 58:207.

Sugiura, M., and Heppner, J. P., 1965, The earth's magnetic field, in: "Introduction to Space Science," ed. W. N. Hess, Gordon and Breach Pub., Chap. 1.

Sulzbacher, H., Baumjohann, W., and Potemra, T. A., 1979, Coordinated magnetic observations of morning sector auroral zone currents with TRIAD and the Scandinavian magnetometer array: A case study, J. Geophys., (submitted).

HALL EFFECT FORMULAE AND UNITS

R. C. O'Handley

IBM Thomas J. Watson Research Center
Yorktown Heights, N. Y. 10598

The two most widely used units for the Hall coefficients are SI units, m^3/A-sec = m^3/C, and the hybrid unit Ohm-cm/G (which combines the practical quantities volt and amp with the cgs quantities centimeter and Gauss). Two numbers for the same quantity expressed in these two units are simply related by R(SI) = 100 x R(hybrid) because the unit m^3/C transforms to Ohm-m/Tesla which is one hundredth of the unit Ohm-cm/G.

The larger problem comes in the form for the Hall resistivity.

In SI units care must be taken to clarify the meaning of the magnetization. One recommended convention (Bennett et al. 1976) is:

$$B = \mu_o(H + M) \qquad \text{and } B = \mu_o H + J$$

where B and J are in V-sec/m^2 (Tesla) and H and M are in A/m. Then the total Hall resistivity $\rho_H = E_y^H/j_x$ is:

$$\rho_H = R_o B + R_s \mu_o M \qquad \text{(SI)} \quad 1)$$

or

$$\rho_H = R_o B + R_s J. \qquad \text{(SI)} \quad 2)$$

R_o and R_s are the ordinary and spontaneous Hall coefficients. The latter form seems to be preferred (although the symbol J is often written as M) possibly because of the habit of expressing induction and magnetization in the same units.

In Gaussian units of course no such distinction is made and we always have:

$$\rho_H = R_o B + R_s 4\pi M \qquad \text{(cgs)} \quad 3)$$

Consistent variations of these forms that often appear in the literature are $\rho_H = R_o \mu_o H_i + R_1 \mu_o M$ (SI) or $\rho_H = R_o \mu_o H_i + R_1 J$ (SI) and $\rho_H = R_o H_i + R_1 4\pi M$ (cgs) with $R_1 = R_o + R_s$. But these forms give the impression that H_i rather than B causes the ordinary Hall effect. The ordinary Hall effect arises from the Lorentz force on the charge carriers and the field experienced by carriers with typical Fermi velocities is B and not H (Panofsky and Phillips 1962). These forms are useful only as intermediate steps in expressing the Hall resistivity in terms of the applied field $H_a = H_i + NM$ (SI or cgs) or $H_a = H_i + NJ/\mu_o$ (SI) where N is the demagnetizing factor:

$$\rho_H = R_o \mu_o H_a + (R_1 - NR_o)\mu_o M \qquad \text{(SI)} \quad 4)$$

or

$$\rho_H = R_o \mu_o H_a + (R_1 - NR_o) J \qquad \text{(SI)} \quad 5)$$

and

$$\rho_H = R_o H_a + (R_1 - NR_o/4\pi) 4\pi M. \qquad \text{(cgs)} \quad 6)$$

In the usual Hall geometry, i.e. with H_a normal to a thin sample, N = 1 (SI) or N = 4π (cgs). Then below the Curie temperature T_C and well below saturation $H_a = (4\pi)M$ or $= J/\mu_o$ and Eqs. 4) - 6) become

$$\rho_H = R_1 \mu_o H_a \qquad \text{(SI)} \quad 4'),5')$$

and

$$\rho_H = R_1 H_a. \qquad \text{(cgs)} \quad 6')$$

If $R_s >> R_o$ then $(1/\mu_o) d\rho_H/dH_a\big]_{H_a=0}$ just give the spontaneous Hall coefficient. Above T_C and for $H_a \cong 0$ we have $\chi H_a = (4\pi)M$ or $= J/\mu_o$ and Eqs. 4) - 6) become

$$\rho_H = (R_o + R_s \chi)\mu_o H_a \qquad \text{(SI)} \quad 4''),5'')$$

and

$$\rho_H = (R_o + R_s 4\pi\chi) H_a. \qquad \text{(cgs)} \quad 6'')$$

Eqs. 4") - 6") may be further simplified depending on the relative magnitude of $(4\pi)\chi R_s/R_o$ which depends strongly on temperature. Note that $(1/\mu_o)d\rho_H/dH_a]_o$ does not give the same quantity above T_C as it does below T_C unless $R_s = 0$.

Clearly the potential for confusion is present in the multiplicity of forms and units used for the Hall effect. It is imperative that units and symbols be defined at the outset.

REFERENCES

Bennett, L. H., Page, C. H., and Schwartzendruber, L. J., 1976, AIP Conf. Proc., 29:xix.

Panofsky, W. K. H., and Phillips, M., 1962, "Classical Electricity and Magnetism," Addison Wesley, New York, p. 143.

THE HALL EFFECT IN SILICON CIRCUITS

J. T. Maupin
MICRO SWITCH Division, Honeywell
Freeport, Illinois 61032

M. L. Geske
Solid State Electronics Center, Honeywell
Bloomington, Minnesota

INTRODUCTION AND OVERVIEW

The purpose of this paper is to describe the technology of integration of Hall effect sensors with various signal conditioning circuits in monolithic silicon integrated circuits.

On the 100th anniversary of the discovery of the Hall effect, it is appropriate to note the industrial impact of this technology. Over 150 million devices have been manufactured and put into use by our company. By comparison the total United States production of microprocessors now stands at about 90 million devices. At present the most significant Hall effect device applications include solid state keyboards and a variety of position sensing and electric current sensing applications. Pictures of some of these products are shown in Figures 1 and 2. Figure 3 is a photograph the first chip to go into production, now obsolete.[1,2] Figure 4 shows the switching output chip that is used in highest volume for keyboard applications. Figure 5 shows a linear amplifier circuit. The developmental trend toward more circuit functions on a chip is apparent from these pictures. Performance will be described later.

The dominant forces in making major commitments to this technology have been low cost and intrinsic solid state reliability. Although some details are proprietary, acceptance in highly competitive markets shows good progress on both counts. Any semiconductor Hall sensor requires amplification and perhaps level detection (digitizing) to generate a signal that is directly usable in electronic systems. Integration is the most cost effective way of

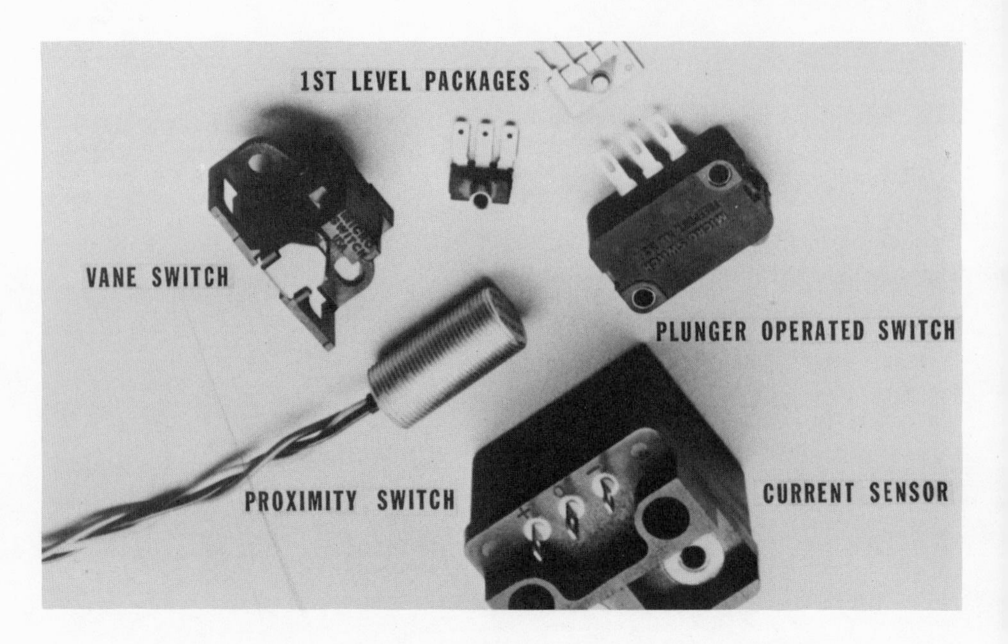

Fig. 1. Hall effect component products.

of achieving this result. This paper will give details on Hall parameters applicable to bipolar silicon IC processing, showing that the Hall effect is a remarkably well behaved and dependable element in such circuits, with predictable and producible properties. We

Fig. 2. Solid state keyboards. There is a silicon Hall effect switch under the plunger of each key position. A permanent magnet on the plunger operates the switch.

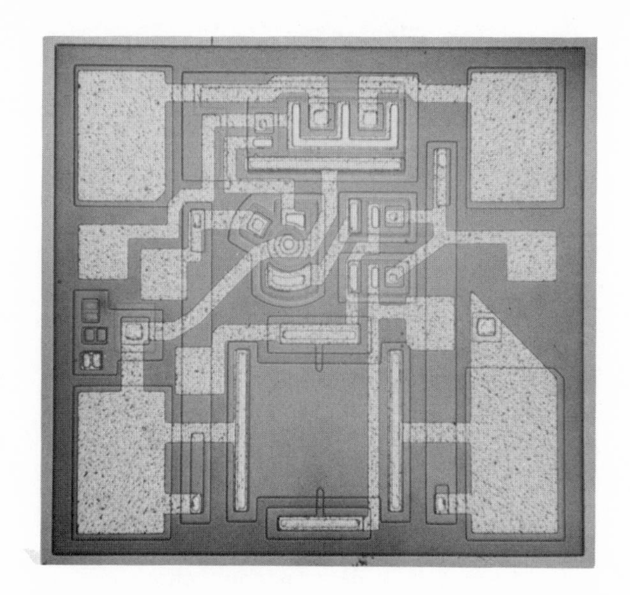

Fig. 3. First production chip (now obsolete). Note outline of rectangular Hall element.

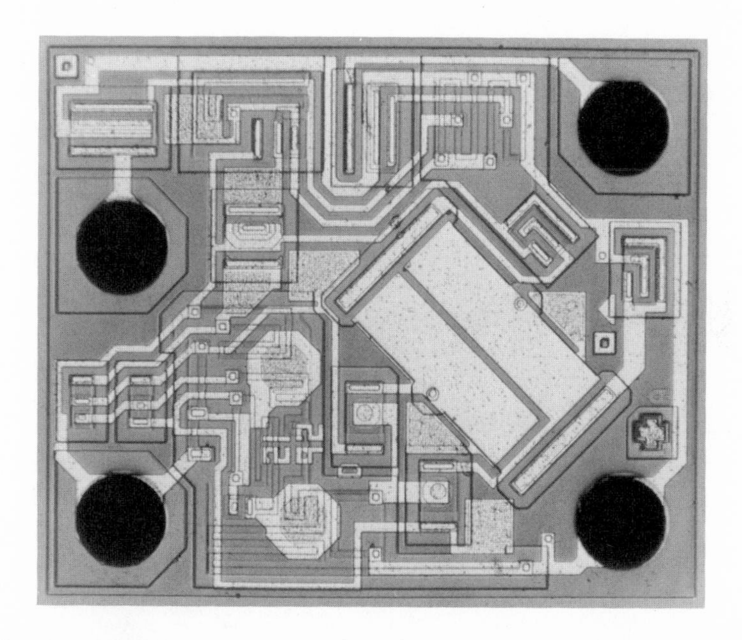

Fig. 4. Contemporary chip for keyboard applications. The dark circles are out-of-focus solder bumps.

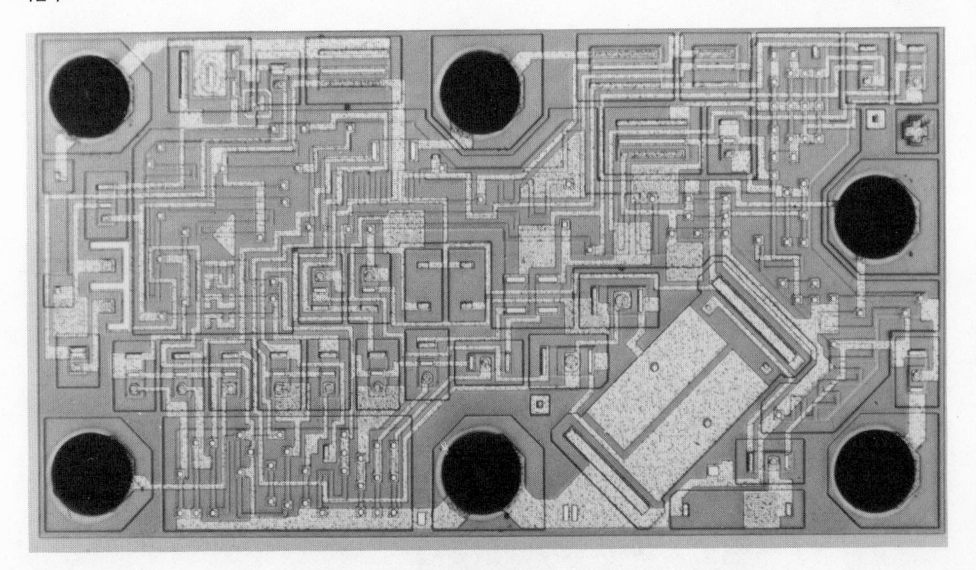

Fig. 5. Linear output chip.

will describe advances in the geometry of Hall elements, taking advantage of the unique non-reciprocal circuit property of the Hall effect. We will describe the synthesis of Hall elements and signal conditioning electronics using computer-aided design. Finally, some

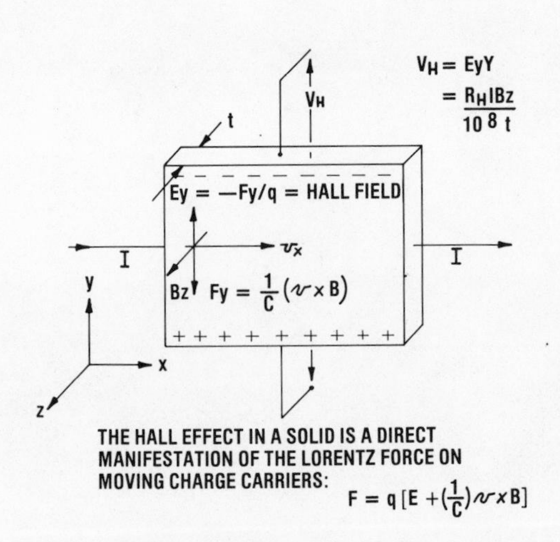

Fig. 6. Hall effect illustration.

speculations about future trends in this evolving field will be given.

HALL EFFECT PARAMETERS IN SILICON IC'S

The Hall effect in the usual discrete or unintegrated Hall generator is illustrated schematicall in Figure 6. As Beer[3] has noted, the Hall effect is a direct manifestation of the Lorentz force on moving charge carriers in a magnetic field. Since the charge carriers are constrained to remain within the conductor, the average velocity vector direction must be the same as the direction of current flow, under steady state conditions. This is realized by the generation of an equal and opposite force, cancelling the Lorentz force, in the only way that nature makes available, i.e. an electrostatic force resulting from internal charge redistribution. The electric field associated with the charge redistribution is called the Hall field, and the integration of the Hall field across the width of the conductor produces the Hall voltage that is picked up by conductive probes (or contacts) placed at opposite edges of the conductor.

Although many complications can alter this simple single particle force balance model, as noted in the literature, it has been found that it is quite accurate where there is predominantly one type of carrier present, the concentration is light, and for operation in the low field range of magnetization. These conditions are met in the silicon conducting layers of interest.

Performing the above noted integration of the Hall field results in this well-known expression for Hall voltage.

$$V_H = \frac{R_H I\ B}{10^8\ t} \tag{1}$$

where R_H = Hall coefficient in cm^3/coulomb

I = Bias current in amperes

B = Normal component of magnetization, in Gauss

t = Thickness in cm

V_H = Hall voltage in volts.

A somewhat more useful expression for Hall voltage results when (1) is converted to be a function of applied bias voltage rather than bias current. For an ideal rectangular geometry which is very long compared to its width, equation (1) converts to

$$V_H = V_S \mu \, B(\frac{W}{L}) \, 10^{-8} \tag{2}$$

where V_S = Bias voltage in volts

μ = Mobility in cm^2/volt-sec

W = Width

L = Length

Equation (2) shows the importance of the materials parameter, mobility. In this equation it is assumed that Hall mobility is equal to conductivity mobility. This relationship is generally complicated and of great interest in the study of conduction processes in solids. The research of Long and Tufte,[4] as well as our own empirical results, confirm this assumption. Practical Hall elements, especially in integrated circuits, cannot be long and skinny (W/L<<1) as required by (2). When this condition is not met a correction factor must be used due to the shorting effects of the end contacts on the Hall field. This has been analyzed by Kuhrt, et al.[5] The correction factor* is a constant for a given rectangular geometery, and is introduced as the factor K.

Equation 2 becomes:

$$V_H = V_S \mu \, B(K\frac{W}{L}) \tag{3}$$

where $K \leqq 1$.

Maximizing Hall sensitivity is, of course, important to most Hall devices. However, a point of diminishing return in respect to $(\frac{W}{L})$ is reached in the vicinity of unity. Optimum rectangular geometry IC Hall elements have $(\frac{W}{L})$ ratios slightly less than unity.

The most important physical difference between Hall elements in integrated circuits and discrete, or conventional, Hall elements is the method used to isolate the element from its surroundings and define its geometry in the X, Y, and Z directions. With discrete elements, the isolation is air or other dielectric material. In IC's, the isolation is achieved by the space charge layer associated with pn junctions. This is the typical method of isolating components in bipolar process IC's.

---

*If equation (1) is used for geometries other than the long skinny type, K or other suitable geometric constant must be factored in, due to end contact shorting.

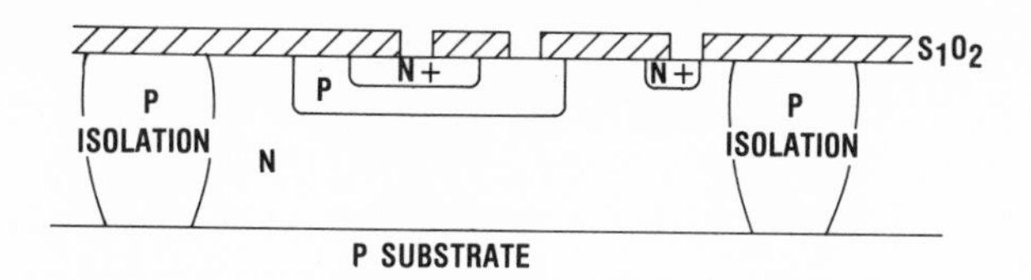

Fig. 7. Cross-section of typical bipolar IC showing NPN transistor region.

Figure 7 shows a cross-section of a typical bipolar integrated circuit, illustrating the conducting layers or regions that can be considered for Hall elements. These are the epitaxial or diffused layers associated with the emitter, base and collector regions of NPN transistors. Table I shows some approximate parameter values for these layers.

The collector layer has the most attractive properties for Hall elements. The Pinch collector layer varieties, with reduced thickness due to placing the base diffusion on top of the collector layer, also has potential. However, all practical Hall IC's have thus far used the unmodified epitaxial collector layer.

Figure 8 shows a Hall element embedded in silicon, illustrating the pn junction isolation method. As long as the junction is reverse biased or zero biased, and temperature is in the extrinsic range, this IC Hall element can be treated, in the main, as a discrete element. There are second order effects at very high frequencies due to distributed capacitance, and static characteristics are slightly affected by space charge layer widening near the end of the element that has the bias voltage connected to it. The isolation junction also contributes generation-recombination noise. All these are

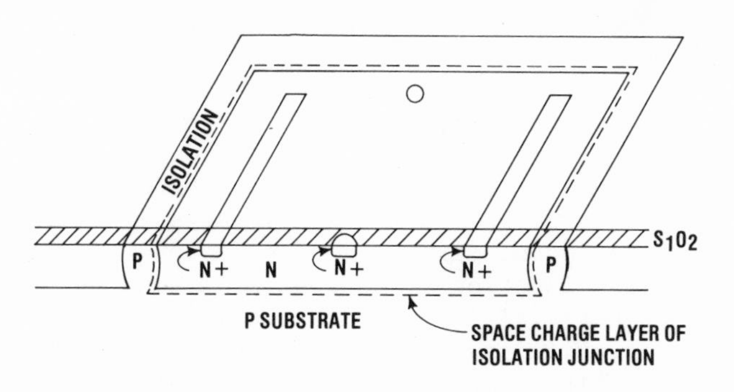

Fig. 8. Rectangular Hall element embedded in bipolar silicon IC.

TABLE I. BIPOLAR PROCESS PARAMETERS

Bipolar IC Conduction Layers	Approximate Carrier Concentration	Mobility	Sheet Resistance	Thickness
	cm^{-3}	cm^2/volt-sec	ohms/square	microns
Emitter (N+)	7×10^{19}	100	5	2.1
Base (P)	2×10^{18}	100	150	2.7
Collector (N)	4×10^{15}	1200	1500	10
Pinch Collector (N)	4×10^{15}	1100	4000	7

insignificant in the parameter ranges of interest. The top surface of the IC is covered with a stabilizing layer of SiO_2 thermally grown during the diffusion processing. This is, of course, standard for the planar process. The combination of high purity single crystalline starting material and excellent process control during epitaxy growth and the diffusion processes, results in conducting layers having exceptionally stable characteristics. Additional chip level protection for more demanding environments has been provided in more recent designs by a layer of silicon nitride (Si_3N_4) deposited over the planar process SiO_2.

Contact is made to the top of the Hall element rather than at its edges since these are not accessible. The epitaxial layer resistivity is too high for ohmic contact to be directly made by a metal. Therefore, additional openings are made in the surface oxide in the Hall element region to form n+ contacts simultaneously with the formation of n+ transistor emitters. The aluminum metalization is permitted to contact the n+ regions through small contact openings cut in the final oxide layer. The mask used to locate oxide openings for the n+ contact regions thus defines the Hall element geometry in two dimensions. No other technology can match the dimensional precision of the photolithography used in integrated circuit processing. The reproducibility of Hall element sensitivity with IC processing is good. This is especially so with the constant voltage mode of biasing where mobility is the only process dependent parameter. Figure 9 (after Sze and Irvine) shows mobility of N-type silicon as a function of carrier concentration. It is seen that mobility is a weak function of concentration in the range typically used in N-type epitaxial layers, 10^{15} to 10^{16} cm^{-3}. In this range, a concentration variation of $\pm 20\%$ produces a mobility variation of perhaps $\pm 2.5\%$. This is a most significant factor in the success of silicon Hall effect IC sensors.

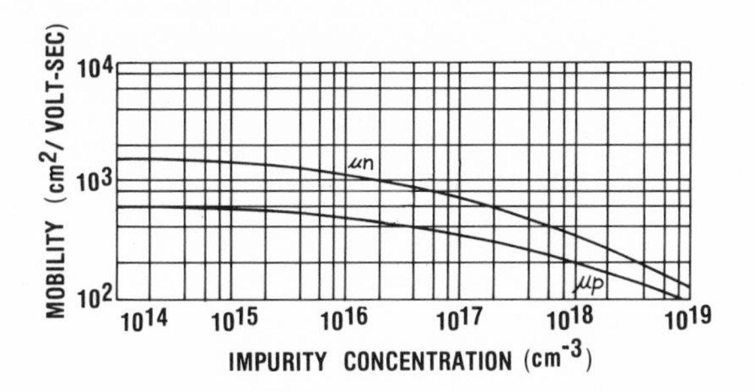

Fig. 9. Drift mobility of silicon at 300 K (after Sze and Irvine).

The chief temperature dependence of silicon Hall elements can be inferred from the familiar decline in mobility as temperature increases. Data for N-type silicon, after Gartner, is shown in Figure 10.

Figures 11 and 12 show measured temperature dependence of silicon Hall elements for constant voltage and constant current biasing, respectively, over the temperature range of greatest practical interest. The constant voltage bias data correlates well with mobility data, and, as expected, the data on various samples are bunched close together (good reproducibility). The small positive temperature coefficient exhibited when the biasing is constant current (Figure 12) is attributed to space change layer widening of the isolation junction, which slightly reduces element thickness (equation 1). Electrical resistance increases with temperature, thus increasing the voltage between Hall element and substrate. Figure 12 shows more spread in the data compared to Figure 11, reflecting

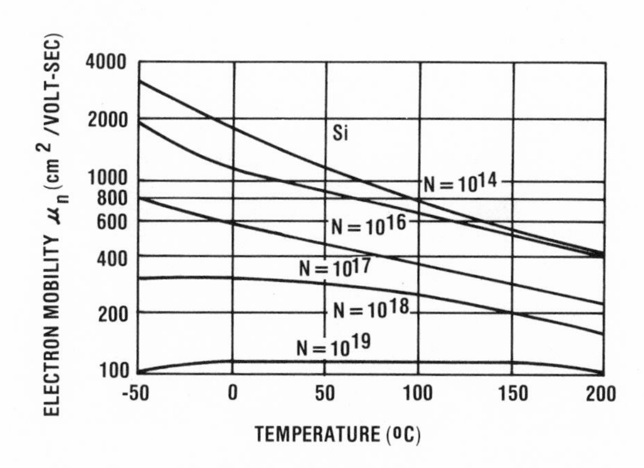

Fig. 10. Electron drift mobility in Si as a function of temperature and impurity concentration (after Gartner).

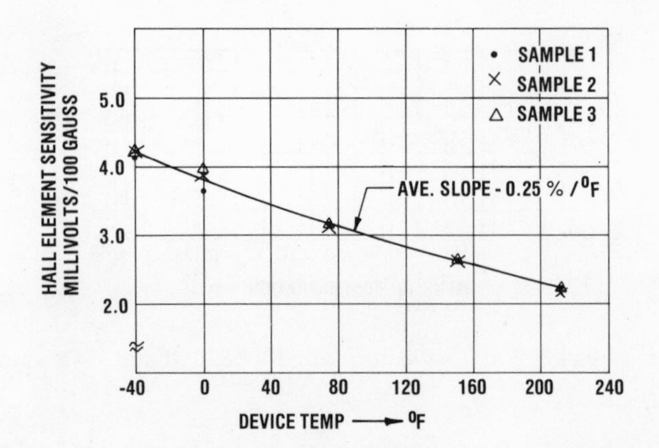

Fig. 11. Silicon Hall element sensitivity vs. temperature at constant voltage source of 5 volts and uniform field induction of 500 Gauss.

poorer process control over concentration and thickness (equation 1) than mobility (equation 2). The Hall coefficient is essentially constant in this temperature range, since donor impurities are completely ionized.

The magnitude of silicon Hall element sensitivity, shown in Figures 11 and 12, is very competitive with other Hall sensor materials considering the bias voltage and current levels. More recent devices use less bias current due to increases in epitaxial resistivity.

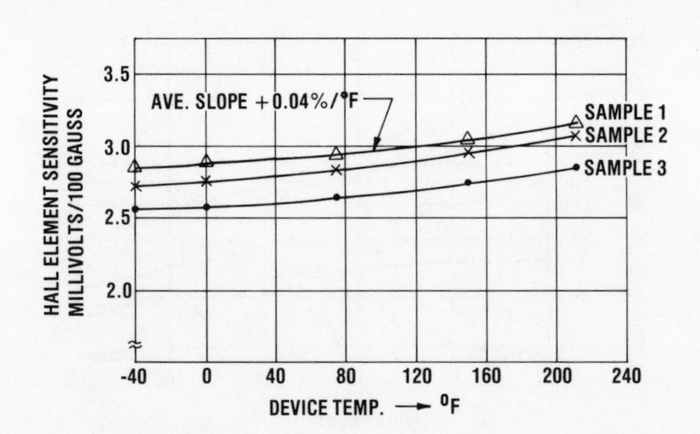

Fig. 12. Silicon Hall element sensitivity vs. temperature at constant supply current of 7 milliamp and uniform field induction 500 Gauss.

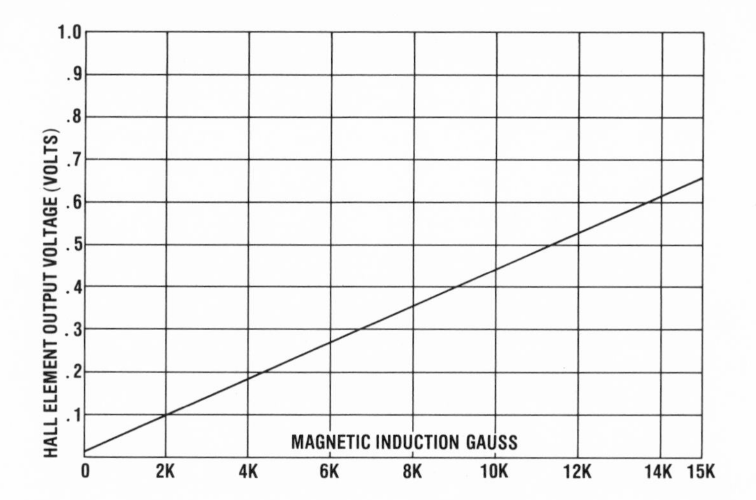

Fig. 13. Hall element output voltage vs. induction (constant current source).

The Hall effect in silicon is very linear, up to 10's of kilo-Gauss. We have data which shows excellent linearity up to 15 kilo-Gauss (Figure 13). In most applications of position sensing explored thus far, available magnetization is on the order of 200 to 800 Gauss, where linearity is assured since $\mu B << 1$.

PIEZORESISTIVE EFFECT IN SILICON INTEGRATED HALL ELEMENTS

One of the causes of variation in offset* voltage (sometimes called standoff voltage) in Hall elements in the absence of a magnetic field is the response of silicon to stress. This results in an undesirable output voltage and is perceived as an error in the Hall voltage. Silicon also has a strong piezeresistive effect (which is used effectively in force and pressure transducers). Difficulties arise due to changes in stress levels at wafer probe, between wafer form and after die separation, and in packaging. Changes in stress levels are, for example, set up by the differential in thermal expansions of the die and its mounting material.

The Hall element as a 4-terminal resistor responds to shear stress through the π_{44} piezoresistive coefficient. This effect can be minimized,[8] but not eliminated entirely, in the {100} face plane current direction to be <100>

* The term "offset voltage" refers to the static output of a Hall element in the absence of a magnetic field.

The analysis of the piezoresistive response in silicon Hall elements is mechanically complicated. In addition, most quantitative investigations[6,7] of piezoresistance in silicon have concentrated on P-type diffused layers with rather high impurity concentrations, and in two-terminal resistor configurations. These are the conditions of greatest practical interest for force and pressure transducers. Quantitative values for piezoresistance coefficients for lightly doped, N-type 4-terminal resistor configurations are not well established, addition to the uncertainty of analytical results. There is some indication in the literature[8] that the analysis of piezoresistance effects in Hall elements is getting more attention. Other than the directionality aspects previously noted, our efforts in minimizing piezoresistance effects has proceeded mostly in an empirical and conceptual vein.

OPTIMIZING CONVENTIONAL GEOMETRY HALL ELEMENTS

The rectangular Hall element has been the most common form used in Hall effect integrated circuit transducers. For a given bipolar process, optimization of Hall sensitivity and offset distributions is mostly through geometric considerations. Our objective is to get large signals, at low power, with minimum offset error due to process variables and stress. An additional requirement is usually added. We need to be able to set a nominal offset, or change it readily. This offset should be controlled by a mask late in the process sequence. Since the two masks of interest are the emitter and isolation, the best choice is the emitter.

Now, we will look at a figure of merit, the Hall sensitivity per unit power consumption. The Hall voltage is given by equation (3).

The power input is written

$$P_i = \frac{V_s^2}{R} \tag{4}$$

where the element resistance R can be described in terms of material properties.

$$R = \frac{L}{W} \cdot \frac{1}{nq\mu t} \tag{5}$$

Combining equations (3), (4) and (5) we have a figure of merit equation for integrated Hall elements.

$$\frac{V_H}{P_i B} = \frac{K}{V_S nqt} \tag{6}$$

where "n" is the carrier concentration and "q" is the electronic charge.

A second figure of merit is the ratio of the Hall voltage to the variation in offset voltage. This can be regarded as a dc signal to noise ratio. In integrated circuits we are interested in the standard deviation of the offset distribution for all products since with state of the art technology, no trims are made. Geometric shape, mask control, and Hall element orientation are all used to deal with offset errors.

Looking at Hall materials generally, the materials with much higher mobilities such as indium antimonide and indium arsenide, at first consideration, look better than silicon. However, the higher mobility comes at the expense of higher temperature coefficients of sensitivity, more restricted temperature range, and impedance level lower than desirable for direct coupling to semiconductor electronics. When most practical considerations are used, silicon is indeed a very competitive Hall effect sensor material in addition to integrability.

SYMMETRICAL HALL ELEMENTS

The trend in integrated circuits is continually toward smaller component sizes and more precise functions. Continuing efforts to improve Hall effect IC transducers have resulted in the invention and development of the symmetrical Hall element.[9]

The new geometry, shown in Figure 14, is a square with 4 identical contacts, one in each corner. Thus either pair of diagonally located contacts may be used for bias, the other pair serving as the Hall voltage terminals. The symmetrical design, with small input and output contacts, is typically of the size of two minimum geometry transistors which is much less than the area of a rectangular geometry Hall element.

Although small size is valuable in its own right, the combination of small size and symmetry of the terminals, allows the nonreciprocal property of the Hall effect to be utilized in practical designs. Let's explore this significant point in some detail.

A four-terminal network with passive components obeys the law of reciprocity. The two-part network characterized by the matrix

$$\begin{vmatrix} r_{11} & r_{12} \\ r_{21} & r_{22} \end{vmatrix}$$

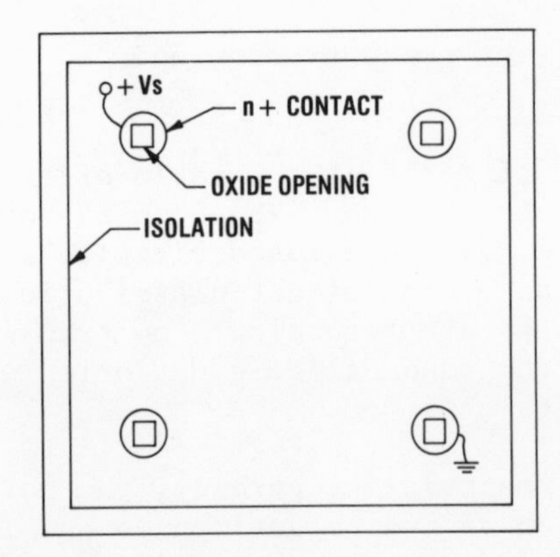

Fig. 14. Symmetrical Hall element. Note: input and output terminals are interchangeable.

is reciprocal if it is symmetrical about the main diagonal, i.e. $r_{12} = r_{21}$.

Networks which do not satisfy this criterion are called 'non-reciprocal'. The Hall element is non-reciprocal.

This can be demonstrated as follows. Referring to Figure 17, suppose that a small resistive unbalance appears between pins 1 and 4 of the Hall element. The device may be conceived as a four resistor bridge, which has one unbalanced element. An offset V_q is developed at the output as a result of the incremental resistant unbalance ΔR. Thus for the top connection it will be seen that the Hall voltage and offset voltage have the same polarity. The lower connection has the inputs and outputs interchanged. This reverses the offset polarity but <u>not</u> the Hall voltage polarity. Now the output is $V_{34} = V_h - V_q$.

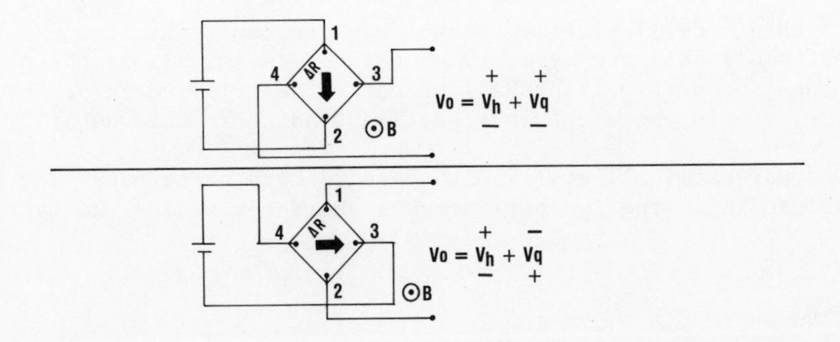

Fig. 15. Illustrating non-reciprocity in Hall elements.

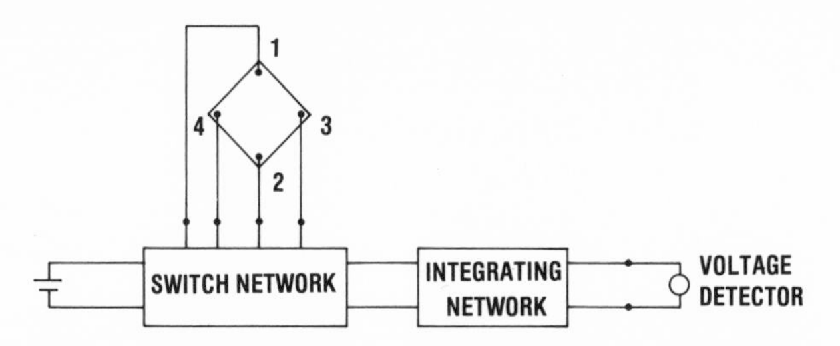

Fig. 16. Concept for using non-reciprocity with a single Hall element.

The Hall effect non-reciprocity can be qualitatively understood by recalling (Figure 6) that the Hall voltage is the result of a vector cross product. The Hall voltage rotates along with the control current in the illustrations in Figure 15. However, the incremental resistance, ΔR, is fixed in the solid space of the Hall element, and does not rotate with the control current when the connections are changed.

To exploit this property, a circuit as shown in Figure 16 could be implemented. The ideal switching network, operating with an equal duty cycle would alternately switch inputs and outputs. A second network would integrate these outputs canceling out the offset component. (Note that this requires a symmetrical Hall element with respect to interchangeability of inputs and outputs.) The ideal switching network is not practical now, but integrated circuit technology provides an easier way.

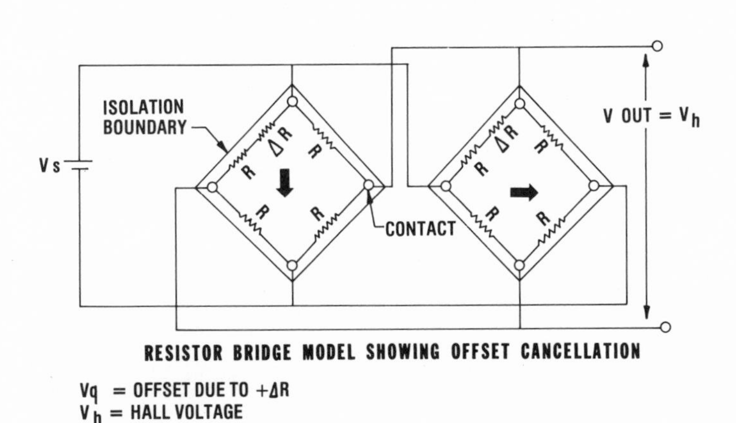

Fig. 17. Using non-reciprocity with dual matched symmetrical Hall elements.

We postulate that there should be a tendency of resistance type unbalance effects in IC Hall elements to be systematic rather than random within the confines of the chip; i.e., two Hall elements processed side by side should exhibit "matching" properties much as two transistors or resistors do. To explain how offset compensation and strain cancellation results from such matching we will again use a bridge analogy. See Figure 17. The symmetrical design ensures equal bridge elements and allows swapping input and outputs of one element. An incremental resistance ΔR is introduced in the same geometric position in each bridge (ΔR could have a number of systematic sources: a resistivity gradient, strain, misalignment of geometry defining mask, etc.). Considering each element independently as biased, it will be observed that equal and opposite offset voltages appear at the output terminals. When interconnected as shown, the offset terms cancel. The Hall voltage is preserved because this depends only on the direction of the bias current and the direction of magnetization.

The symmetrical square geometry also has a miniaturization advantage. Elements are of small size and can be nested close together. Matching works best in integrated circuits when the matched elements are as close together as possible. In contrast, conventional asymmetrical rectangular elements do not effectively match the "resistive inputs" because of their non-symmetrical geometries and wider separation. Thus, offsets do not fully cancel out. Chip area is also conserved and Hall sensitivity maintained using the square form instead of other symmetrical geometries such as the cruciform, where the additional series resistance reduces the voltage bias across the Hall region. The square elements fit well with other integrated components, achieve the orthogonal biasing requirement and have their principle direction of current in the minimum strain direction. These geometrical features are illustrated in Figure 18 which is a photo of a test chip containing two miniature matched symmetrical elements.

FOUR CONNECTION SCHEMES

The four non-redundant paralleling interaconnection schemes for dual matched elements are shown in Figure 19. These are

I. Orthogonal biasing, parallel output
II. Orthogonal biasing, anti-parallel output
III. Parallel biasing, parallel outputs
IV. Parellel biasing, anti-parallel outputs

Connection I is of greatest interest for integrated Hall technology: the Hall voltage is preserved (averaged) and the resistance generated outputs (including piezoresistance, or stress effects) cancel. Connection II is possibly of interest to the piezoresistance

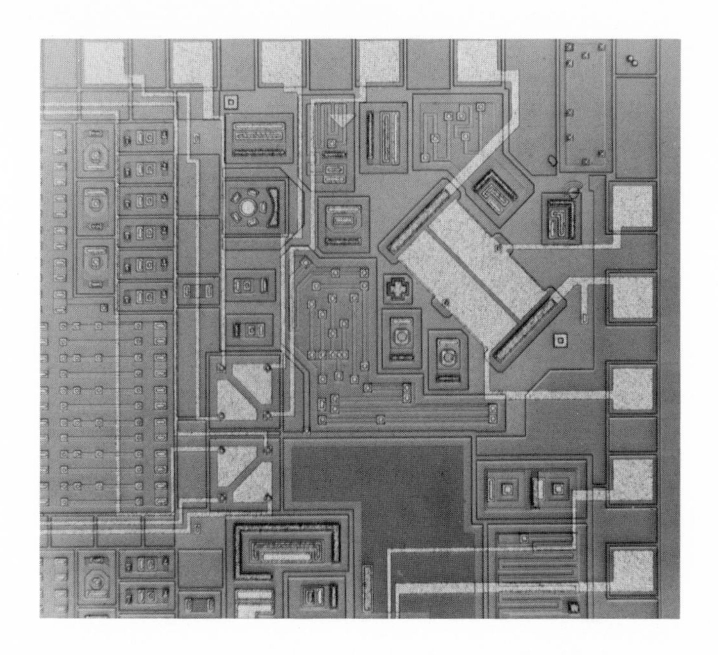

Fig. 18. Test chip. Two matched symmetrical Hall elements are in upper left corner.

technology, not previously done, without the concern over magnetically generated signals. Connections III and IV have bias currents in parallel and do not exhibit the selective cancellation of systematic inputs. Connection III, included mainly for completeness, has no advantage over one element. With Connection IV, all systematic inputs, magnetic or resistive, cancel. This is of interest to either the piezoresistive or Hall technology if the geometry can be arranged so that inputs due to the sensed variable are not systematic, i.e. the inputs are reversed or at least unequal for one element compared to the other, or one element is in a neutral design.

Table II shows experimental data on parallel connected symmetrical Hall elements from test chips shown in Figure 18. Average offset for the parallel devices was calculated, then the deviation from average was recorded for 3 cases; each Hall element separately and the parallel combination. It is clearly seen that the offset deviations tend to match and be of opposite sign and that the parallel combination produces a tighter distribution.

The data in the table were taken on chips packaged in 8 pin DIP's with the chip substrate bonded to the lead frame with hard epoxy. A simple strain rejection test was also conducted by clamping one end of the package and torquing the free end with a pair of pliers. In terms of actual applications, this is a drastic case of

NUMBER	CONNECTION SCHEMATIC	CONNECTION DESCRIPTION	Vh MAGNETIC RESPONSE	Vq RESISTIVE RESPONSE
I	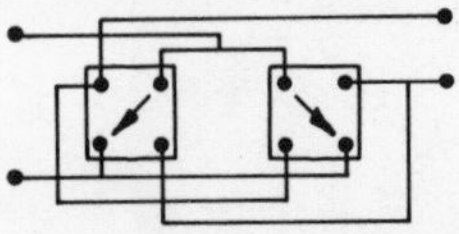	ORTHOGONAL CURRENT BIASING * PARALLEL OUTPUT	AVERAGES	SUBTRACTS (CANCELS)
II	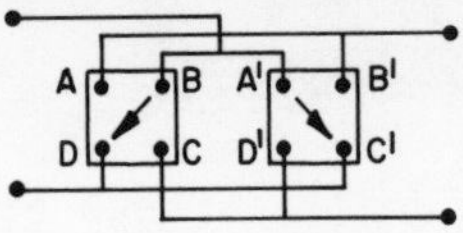 A B A' B' D C D' C'	ORTHOGONAL CURRENT BIASING ANTI-PARALLEL OUTPUT	SUBTRACTS	AVERAGES
III	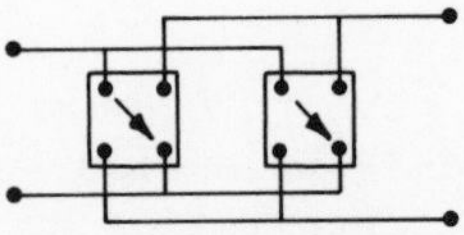	PARALLEL CURRENT BIASING PARALLEL OUTPUTS	AVERAGES	AVERAGES
IV	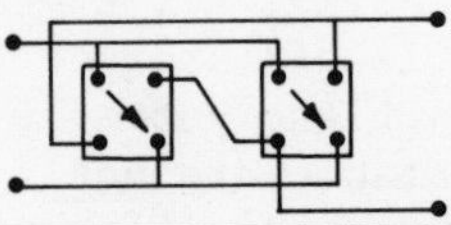	PARALLEL CURRENT BIASING ANTI-PARALLEL OUTPUT	SUBTRACTS	SUBTRACTS

* PARALLEL OUTPUT REFERS TO THE CONNECTION THAT PRESERVES THE HALL VOLTAGE

Fig. 19. Paralleling interconnections for two matched Hall elements.

TABLE II. DUAL HALL ELEMENT OFFSET VOLTAGE DATA. DATA REPRESENTS DIFFERENCE FROM THE AVERAGE IN MILLIVOLTS

Bias Voltage - 5 volts

Unit	2	5	6	7	8	9	10
H5	+6.25	+2.04	+1.41	-5.11	-1.16	-5.24	+1.83
H6	-7.58	-3.14	-1.61	+6.08	+1.8	+4.6	- .16
Parallel	- .08	- .25	- .08	- .14	- .08	- .65	+1.18

strain input. Offset voltages, singly and in parallel combination (Connection I of Figure 21), were monitored with 5 volts bias voltage. For a strain level that produced 5 to 6 millivolts of offset change in single units, the parallel connection showed offset changes of 0.2 to 0.7 millivolts, yielding strain rejection factors due to paralleling of 7 to 30.

Up to now we have seen that silicon Hall elements can be made in bipolar integrated circuits. Optimization is limited to geometric factors and that constant voltage is best. We have also seen that multiple symmetrical Hall elements offer some distinct advantages over conventional designs in that they can be connected to cancel strain and offsets. We would now like to move into combining the Hall sensor and integrated circuit electronics to realize a low cost fully integrated transducer.

SENSOR AND SIGNAL CONDITIONING ELECTRONICS ON ONE CHIP

I. C. technology has made possible economical Hall effect devices. Since the Hall element is fabricated without additional process steps, a low cost transducer can be made. Close proximity and simultaneous fabrication with the electronics eliminates interfacing problems between the sensor and electronics. As discussed previously, constant voltage biasing eliminates the temperature dependence of offset. This, however, leaves us with a temperature dependent Hall voltage (Figure 12), which must be compensated for in the electronic circuitry to result in a precision transducer. The temperature dependence of the transistor's V_{BE} is very predictable and reproducible; it can be and is used as a component part of temperature compensation.

The base resistor temperature dependence is of minor consequence since most circuits depend on resistor ratios, not absolute values. Matched transistors provide low offset amplifiers. A 1 mV offset on an amplifier translates to about 30 Gauss error in most

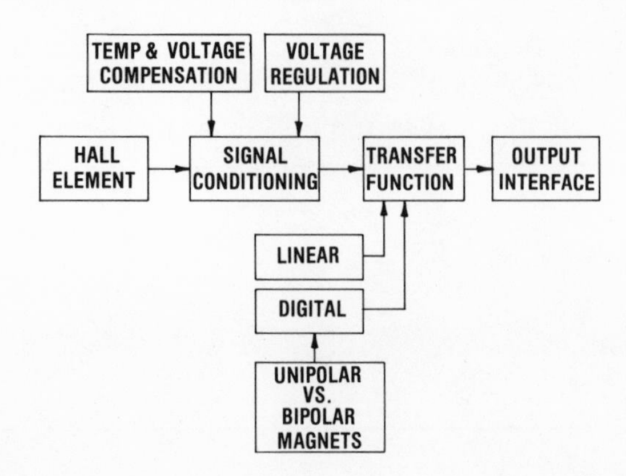

Fig. 20. Generalized Hall effect transducer.

applications.

Combining the sensor and signal conditioning electronics results in simplifying packaging and reducing magnetic circuit parameters. The approach also capitalizes on the inherent reliability of Integrated Circuits. Early devices were designed for benign environments, but now rugged versions work over -55 to +150°C ambients and for supply variations of 4 volts to 30 volts. On chip voltage regulators help increase the voltage range. Transient protection is readily available up to +50 volts with the present state of the art.

GENERALIZED HALL EFFECT TRANSDUCER

Putting it all together, the integrated Hall effect transducer can be diagrammed as in Figure 20. The Hall element properties have already been described and depend directly on the magnetic field, bias voltage and temperature. The signal conditioning electronics serve to alter in some manner the Hall voltage, and affect linearity, temperature and voltage compensation, voltage regulation, signal amplification, rectification, etc.

The Hall element is linear with field, therefore a linear output requires a linear amplifier. Other functions such as digital outputs require a switching function provided by a trigger circuit. Hysteresis is introduced to give a true 'snap action' switch. Snap action means that the switching transitions, once initiated, proceed in a regenerative manner without input control. The device 'operates' at a particular magnetic field level and 'releases' after return to a lower level. Most silicon IC Hall effect transducers manufactured to date have had static binary logic level outputs

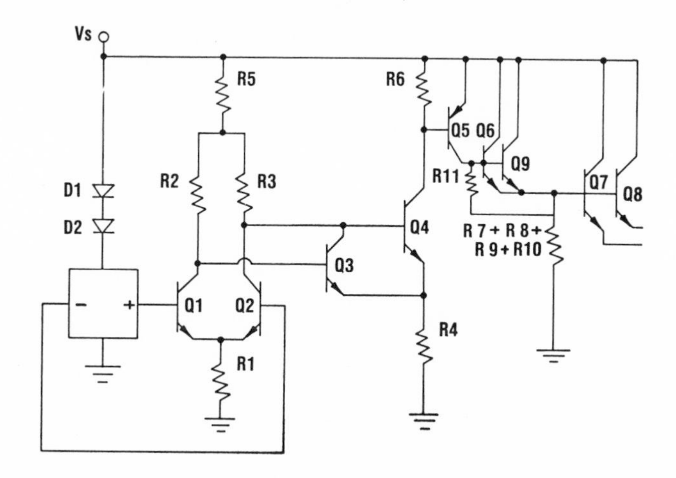

Fig. 21. Circuit diagram of a typical Hall switch.

(Figure 4 is a photograph of this type of chip). In keyboard applications, logic level I corresponds to the depressed condition of the key; logic level 0 the free position. A special variation of this type of chip has dynamic rather than static output coupling; a pulse appears at the output terminal(s) to indicate that a key has been depressed.

In Hall effect switches, the operate point is controlled by the design of the electronic circuits, and the nominal built-in offset in the Hall element. A statistical distribution around the nominal operate point arises becuase of the processing variables which influence Hall element offsets, transistor V_{BE} matching and resistor matching. The standard deviation, in continuous production, of the operate point of a chip like Figure 4 is about 100 Gauss. This chip has the conventional rectangular geometry Hall element and the standard deviation of offset voltage accounts for about 65 Gauss of the operate point deviation.

TYPICAL HALL SWITCH DESIGN APPROACH

The circuit of a typical Hall switch for keyboard applications is shown in Figure 21. It will be observed that the signal from the Hall element is first amplified by a difference amplifier stage, which is coupled to a Schmitt trigger stage, then to an output amplifier stage. The interaction between stages to achieve compensation for temperature and supply voltage variations is, at once, the beauty of this design and the source of difficulty in its analysis. The evolution of a circuit design is a creative interaction between analysis, synthesis and experimental testing. The following list shows the steps in this procedure.

Objective: derive a system of equations to predict magnetic operate and release points, and use it to optimize performance.

Steps to solution:

1. establish transistor and Hall element model parameters;
2. write Kirchoffs voltage equations;
3. rearrange equations to solve for unknown currents in transcendental form;
4. establish switching condition;
5. write equations to solve for switching points;
6. solve equations by iterative technique using a computer;
7. optimize performance by performing a sensitivity analysis.

The analysis of the circuit involves the solution of transcendental equations using a computer. The main objective of the analysis is to predict accurately the operate and release points in Gauss as a function of temperature, supply voltage and device parameters. This information is used iteratively to synthesize and optimize the design for maximum yield for the specified application conditions.

The diodes D1 and D2 shown in Fig. 21 provide some temperature compensation. Additional temperature compensation, and voltage compensation, is obtained through interactions of the characteristics of all components. This iterative method involving all components is made practical only through the use of computer aids. The mathematical models for computer abalysis used for the Hall element and transistor are quite simple. A 4-element transistor model is shown in Figure 22. The lumped Hall element model of Figure 23 is complete enough for analysis using Kirchoff's voltage equations.

Having established models, the next step is to write Kirchoff's voltage equations. These must be rearranged to solve for unknown currents in transcendental form. The criteria for switching is established and the switching point equations are generated. Since the equations are transcendental the computer converges on a solution through its own iterative process.

After a solution to the equations is achieved, a sensitivity analysis is performed. This is essential to optimizing the circuit design, over wide temperature and supply voltage ranges.

Figure 24 shows the operate point vs. temperature and supply voltage for a Hall switching IC optimized for a rather severe temperature environment. Both computed and measured results are shown.

The dynamic performance of Hall effect integrated transducers

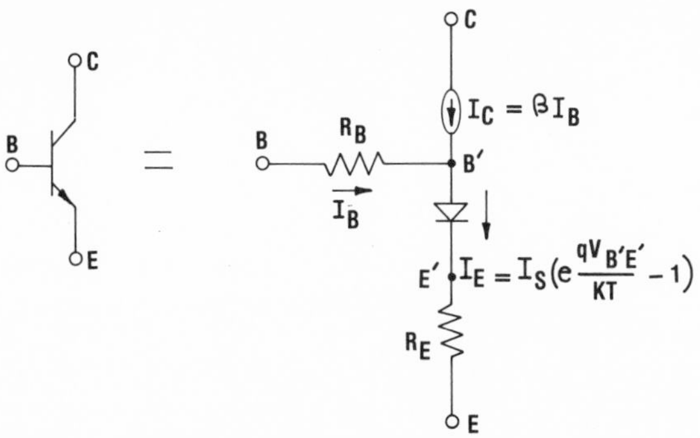

Fig. 22. Transistor model in the active role.

is excellent, as would be expected. We benefit from the fact that the Hall effect has no intrinsic limitation at either end of the frequency spectrum. At the high end of spectrum the usual gain bandwidth and charge storage parasitic effects of integrated active circuits applies. We have made no attempt to analyze or optimize these, since the typical electromechanical actuation does not require fast response. Operation rates of switching IC's of the type shown in Figures 3 and 4 have been measured to be in excess of 10^5 Hertz.

The yield, i.e. percentage of good chips per wafer, is a key economic consideration in any integrated circuit design. Maximizing yield permeates the entire development/design/process/applications cycle. We feel that our yield experiment with Hall effect IC's has been quite favorable.

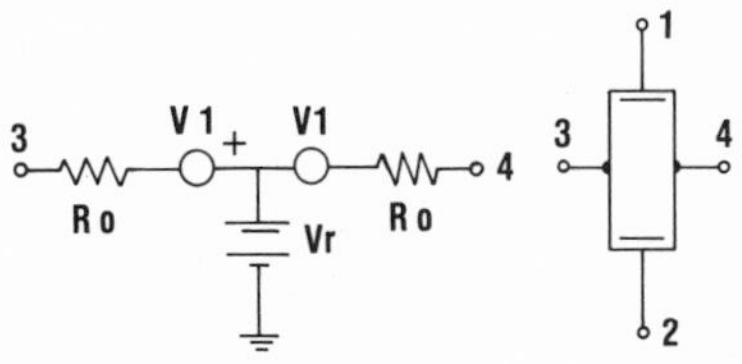

$$V1 = \frac{V_h + V_q}{2}$$

$$V_h = q\mu_h B \; V12$$

$$Ro = \left. \frac{\Delta V34}{I42} \right|_{V12 = CONST}$$

$$Vr = \left. \frac{V3 + V4}{2} \right|_{B = 0}$$

Fig. 23. Hall element model.

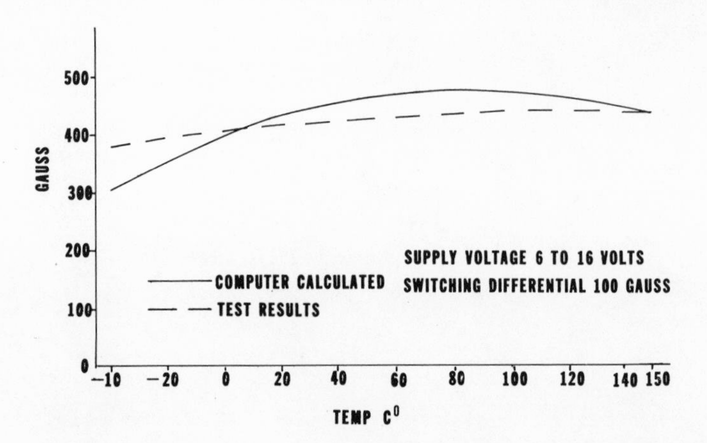

Fig. 24. Operate point vs. temperature of Hall effect integrated circuit switching chip for automotive applications.

FUTURE TRENDS

We expect the developmental progress in Hall effect sensor IC's to continue in all areas: process improvement, Hall element geometry, circuit design, environmental protection and packaging.

As process geometry control in the semiconductor industry improves toward micron-size dimensions, it will be possible to add more electronics without increasing size or cost. Most of the devices emphasized in this paper have binary or on-off switching outputs. A precision linear output chip was also shown in Figure 5. It will soon be possible to add A to D conversion on chip without much cost penalty.

New techniques will all but eliminate the pervasive matter of stress sensitivity in silicon Hall elements, allowing cheaper packaging, and operation in lower ranges of magnetization. Hall effect IC transducers will thus become more precise and the range of applications in position sensing and electric current sensing will increase.

All the work and progress reported here has been with the bipolar IC technology. Hall elements can be built with the MOS technology,[10] using surface inversion layers as the conductive layers. The surface inversion layers, thus far, have not been sufficiently reproducible, predictable, and stable for product use. This may change in the future as MOS processing improves. Hall effect transducers in MOS would be even smaller and require less power than those described here.

ACKNOWLEDGEMENT

Many people from all sectors of our company: Research, Development, Design, Marketing, and Production have contributed to technical and business success of the Hall effect IC technology. It would not be possible to cite everyone. Those omitted will not mind if we single out Dr. David E. Fulkerson, Richard W. Nelson and Everett A. Vorthmann.

REFERENCES

1. Maupin, Joseph T., and Vorthmann, Everett A., Solid State Keyboard, AFIPS Proceedings, 1969 Spring Joint Computer Conference, 34:149.
2. Ibid., July 27, 1971, United States Patent #3,596,144, Hall Effect Contactless Switch with Prebiased Schmitt Trigger.
3. Beer, A. C., May 1966, The Hall Effect and Related Phenomena, Solid State Electronics, 9:339.
4. Long, D., and Tufte, O. N., Honeywell Corporate Materials Science Center, private communication.
5. Kuhrt, F., and Hartel, W., 1969, Arch. Elektrotech. 43:1, cited in Weiss, H., "Structure and Application of Galvanomagnetic Devices," Pergamon Press, Ltd., p. 96.
6. Tufte, O. N., and Long, D., 1963, Recent Developments in Semiconductor Piezoresistive Devices, Solid State Electronics, 6:23.
7. Tufte, O. N., and Stelzer, E. L., February 1963, J. of Appl. Phys., 34:2, 34:313.
8. Kanda, Y., and Migitaku, M., 1976, Effect of Mechanical Stress on the Offset Voltages of Hall Devices in Si IC, Phys. Stat. Sol. (a), 35:K115.
9. Geske, M. L., United States Patent #3,994,010.
10. Gallagher, R. C., and Corak, W. G., May 1966, A Metal-Oxide-Semiconductor (MOS) Hall Element, Solid State Electronics, 9:571.

PACKAGING HALL EFFECT DEVICES

Everett A. Vorthmann

MICRO SWITCH Division
Honeywell
Freeport, Illinois 61032

In order to share in today's tributes to Doctor Edwin Hall, I would like to tell you about some practical applications that one company has derived from his discovery. The company is my employer, MICRO SWITCH, a Honeywell division.

It is intriguing to think that a discovery as basic as the Hall effect should gestate nearly ninety years before it began having a significant impact on the man in the street. As I think about this, it becomes apparent that it lay dormant waiting for the right technology to become practical. Certainly people have been using the Hall effect since its discovery one hundred years ago, but the cost of electronics to amplify the Hall voltage to a usable level has been too expensive for the creation of high volume production. Further, the Hall effect devices themselves had been expensive; large Hall sensitivity is associated with rather exotic materials such as indium antimonide and indium arsenide. With the advent of the transistor and silicon integrated circuit technology, it became practical for the Hall effect to become part of high volume products. Joe Maupin and I teamed up in 1965 to find a practical, low-cost solid state switch. We examined many different concepts: photoelectric, eddy current sensing, differential transformer, piezoresistance, capacity sensing and Hall effect.

We chose the Hall effect for one very basic reason; it was the only concept evaluated which could be entirely integrated on a silicon chip. In those days, the other approaches all required additional electrical or mechanical connections to the silicon chip, either for external sensors or light sources or to strain the chip in the case of the piezo device.

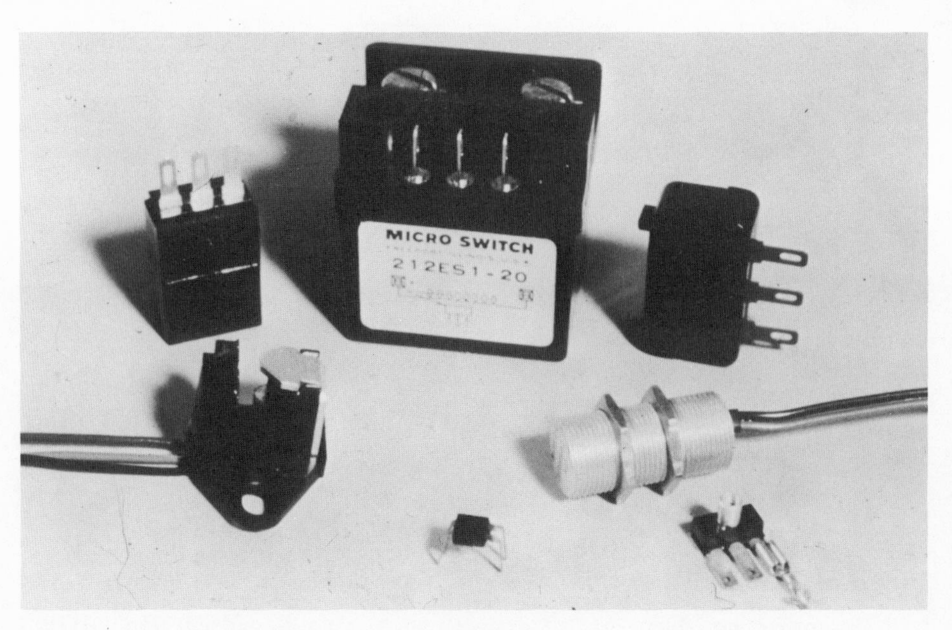

Fig. 1. Hall effect sensing devices for a variety of industrial and commercial applications.

In due order, we did develop the first Hall effect integrated sensor, or switch, and have helped it evolve over the years to accomplish many different tasks, as Joe Maupin has so ably pointed out in the previous paper.

I am sure that Doctor Hall never appreciated the significance of his discovery on mankind. Similarly, you in today's scientific community probably do not fully comprehend the impact that your work will have on future generations. However, you have an advantage over Doctor Hall; with our exponential growth in scientific understanding and applications, your discoveries may very well bear fruit in your lifetime. It might interest you to know that there are well over 1,000 people in Honeywell earning their livelihood as a result of Doctor Hall's discovery. These people may be found in Minneapolis and in Colorado Springs as part of Honeywell's Solid State Electronics Center; as well as in Freeport, Illinois, where MICRO SWITCH Division is located. Honeywell's Solid State Electronics Center fabricates the silicon integrated Hall chip, while we at MICRO SWITCH package the devices and combine the Hall sensors with magnetics. Development efforts are shared by both organizations.

Since our first product was released in 1968, a Hall effect keyboard, we have produced and delivered to our customers over

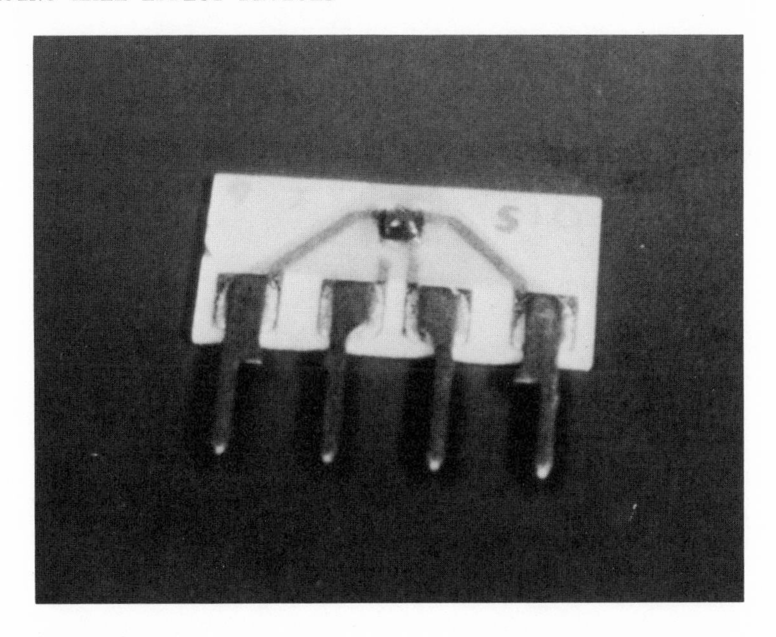

Fig. 2. Hall effect chip package utilizing "flip chip" manufacturing procedure.

150,000,000 Hall effect devices. If you prefer, that's 1.5 x 10^8.

Let's look at some of the packages and applications where these 150 million devices have been used.

In Figure 1 we see some of the Hall effect configurations which we produce in Freeport. Starting with the largest device and going counter-clockwise, we have a current sensor, a Hall sensor operated by an external magnet, a vane switch where the magnet is an integral part of the package, two small Hall effect chip packages, which we use as building blocks in some of our other devices, a 103SR which is a Hall device mounted in a metal bushing, and finally the XL, which is a plunger-operated Hall switch that is mechanically identical to one of our electromechanical basic switches.

In many of our packages, we use conventional wire bonding to make the interconnection from the Hall chip to the outside world. In our highest volume products, however, we use the "flip chip" technique to make this interconnection. These are seen as the white chip packages in the nutshell.

Figure 2 shows our highest volume chip package which uses the flip chip technique. To make this package we employ a thick film palladium silver screened onto a ceramic substrate. To this palladium silver pattern, we solder the leads, as well as the Hall

Fig. 3. Hall effect chip with four solder bumps. Positioning of bumps permits automated aligning of chips before entering reflow furnace.

Fig. 4. Scanning electron microscope photograph of the face-down chip after being soldered to the palladium silver conductor path.

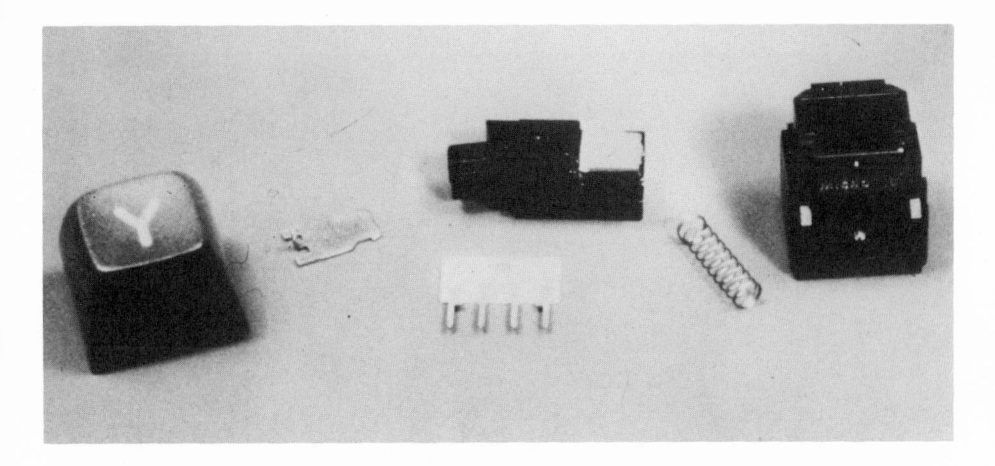

Fig. 5. (left to right) keytop, shunt, plunger, with molded-in magnet, Hall effect chip package with four leads, spring, housing.

effect chip.

Figure 3 shows the Hall effect chip with four solder bumps. Figure 4 is a scanning electron microscope view of the flip chip soldered to the palladium silver conductor path. This Hall effect flip chip package is the heart of our solid state keyboard, the world's first practical electronic keyboard. Figure 5 shows the package as part of the keyboard module. The magnet in this case is molded as part of the plunger, which is depressed as the operator pushes the key. Figure 6 shows an assembled module.

Figure 7 is a typical Hall effect keyboard in which some 70 Hall effect integrated circuits are used.

Figure 8 shows one of our favorite operational modes, a Hall vane switch. In this case, the magnet is an integrated part of the package. The Hall chip is located so that the flux from the magnet is constantly flowing through the chip. The device is operated when a ferro-magnetic object is placed in the gap, diverting the flux away from the chip. This is shown in Figures 9 and 10. The movement of the shunt can be either linear or rotary.

Probably the Hall vane switch application we are most proud of is found in the automotive industry. A vane switch is being used in automobiles to replace the ignition points.

Fig. 6. Assembled Hall effect keyboard module.

Fig. 7. Typical Hall effect keyboard. This model has a microprocessor mounted just above the top row of keys.

Fig. 8. Hall effect switch in a vane configuration. The switch operates when a ferro-magnetic object passes through the vane gap, diverting the flux away from the chip.

Figure 11 shows the vane switch used in the Dodge Omni and the Plymouth Horizon. In August of this year, Earl Myer, Jr., Chrysler's chief engineer for electrical engineering, noted that the low mass of the Hall effect distributor pickup was excellent for the higher vibration levels of four-cylinder engines, while its high output signal levels allow the use of a much simpler, smaller and less costly control unit. While Chrysler has built some 800,000 of these units so far, there will be many other ignition applications in the near future.

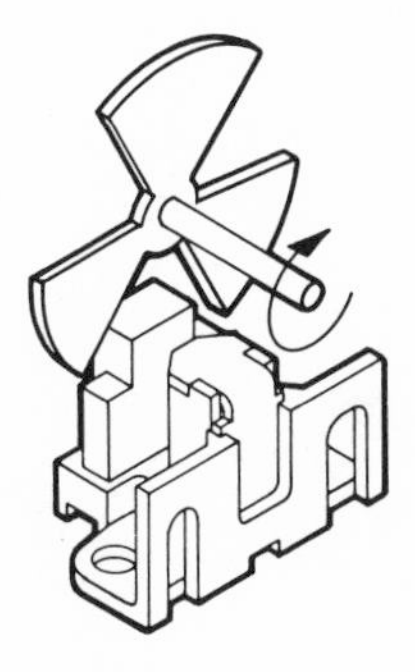

Fig. 9. Hall vane switch turned on

Fig. 10. Hall vane switch turned off.

As shown in Figure 12, another application of Doctor Hall's discovery is a current sensor. The magnetic field, in this case, is generated by an electromagnet. The Hall switch can be set to operate at a precise current level. A wide range of current levels is made possible by changing the structure of the coil in the electromagnet. Typical applications of these devices include solid state circuit breakers or as crowbars in power supplies. A product which we use in large quantities in our automatic assembly equipment in Freeport is the 103SR, which is a bushing-mounted Hall device.

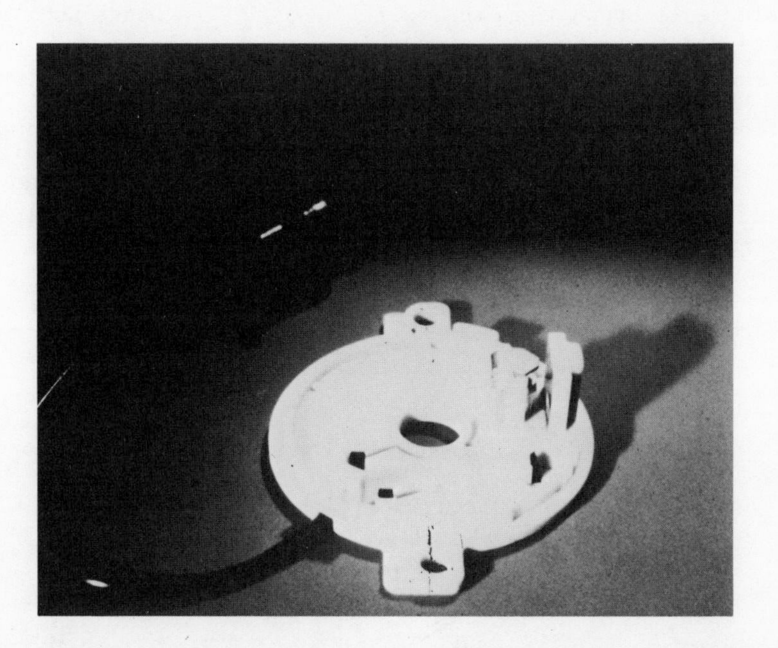

Fig. 11. Hall effect distributor pickup for the Dodge Omni and Plymouth Horizon. This solid state ignition eliminates points.

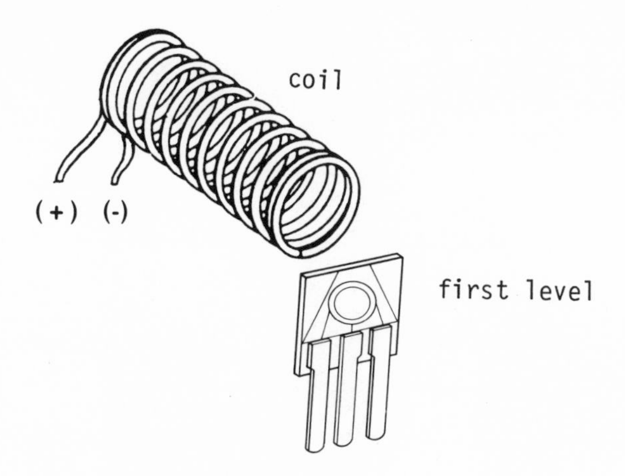

Fig. 12. A Hall effect current sensor utilizes an electromagnet to generate a magnetic field.

Another way we use Hall effect is through the conversion of some of our standard electromechanical switches to solid state switches by replacing the precious metal contacts with Hall effect transducers. In Figure 13 we see our Advanced Manual Line with the products on the left using metal contacts and the ones on the right using Hall effect devices.

Another example of Hall products replacing mechanical contacts is the XL device. Figure 14 shows the mechanical version of which we have made many million. Figure 15 shows this device with the Hall chip package and the plunger-operated magnet.

As you can see, our applications of Doctor Hall's discovery have been used to meet a very broad range of needs. These devices are inexpensive, rugged and reliable. One of the most demanding applications we have seen is this velocity measuring device (see Figure 16) using our Hall effect switch. This device was used by Draper Labs of MIT in 1970 to measure the current flow in the Gulf Stream at a depth of 2,000 feet.

So much for past applications of Doctor Hall's discovery. What of the future? Just as the invention of the transistor and the development of the integrated circuit were the ingredients needed to allow us to begin using the Hall effect, the further expansion of this technology in association with the microprocessor is going to cause an exponential demand for Hall sensors. These microprocessors, however, are of no use by themselves; they must have electrical imputs to allow them to do a useful function. At this point, the Hall sensor is probably the most developed approach to meet many of these sensor needs.

Fig. 13. Hall effect element (right) is used as a solid state version of a manual switch. There is no change in the outer dimensions of the switch.

Fig. 14. An electromechanical switch commonly found in vending machines and office equipment.

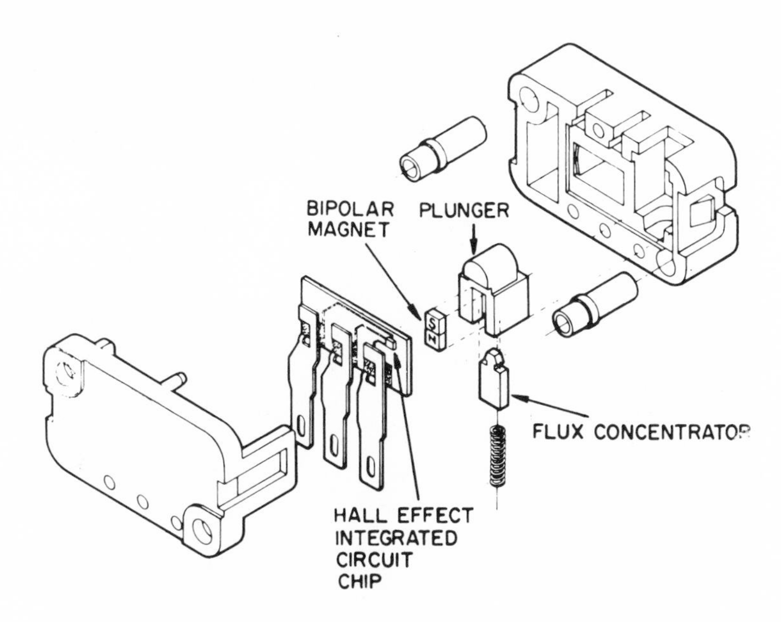

Fig. 15. The same switch in Hall effect design with a plunger operated magnet.

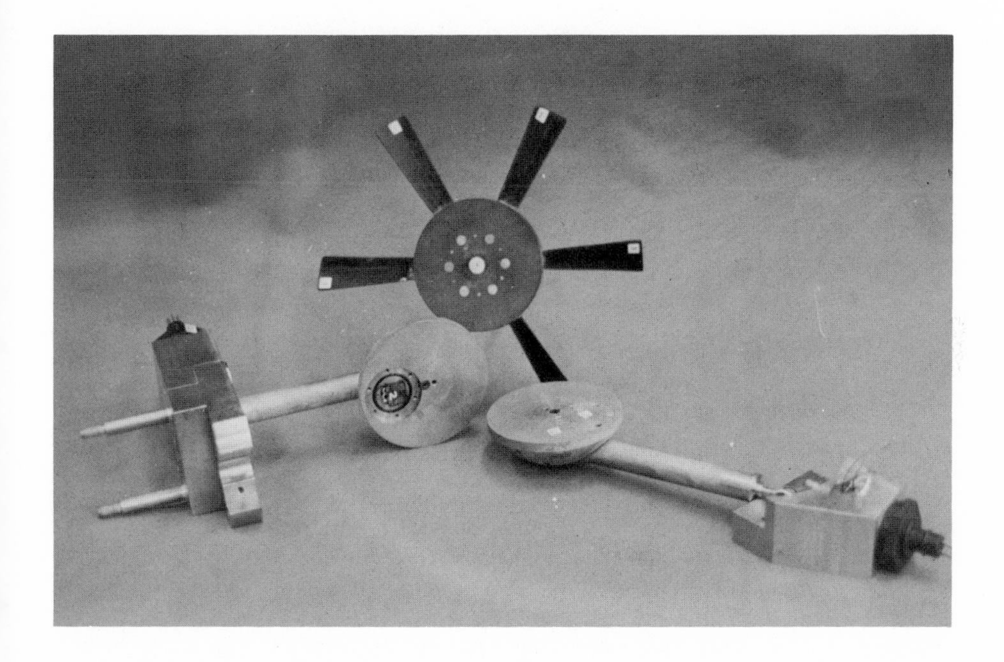

Fig. 16. A Hall device for measuring the flow of ocean currents.

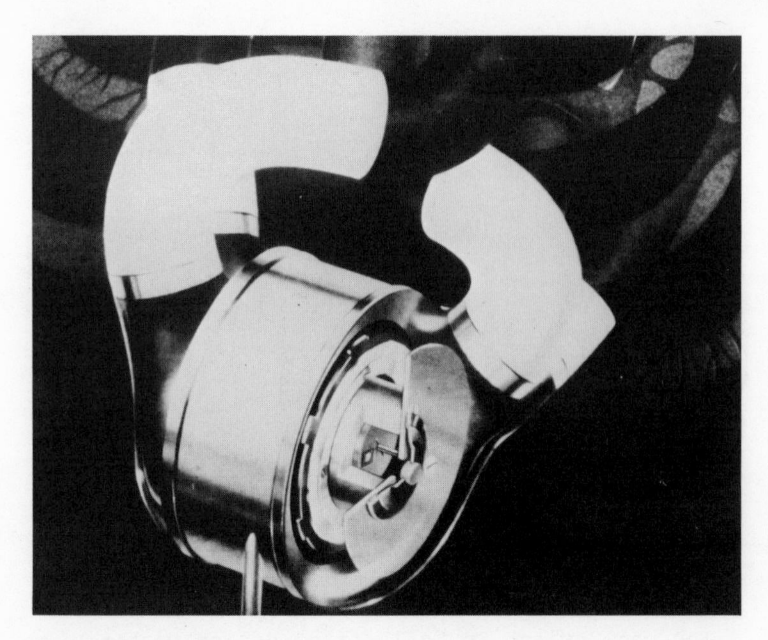

Fig. 17. A Hall effect switch controls the precise movement of this electrically-driven artificial heart under development at the Hershey Medical Center.

In mid-February of 1980 this artificial heart will begin to function for the first time as part of a left ventribular assist device for a calf at the Hershey Medical Center at Hershey, Pennsylvania. The Hall switch (see Figure 17) in this artificial heart signals the motor when it is at the end of its stroke and should begin reversing. The motion of this motor duplicates the pumping action of the human heart.

Last week in London, one of our engineers introduced a product at the World Automotive Engineering Conference (see Figure 18). We call this device a biased Hall gear-tooth sensor. This sensor is a Hall effect device with special circuits. The sensor is responsive to the rate of change of magnetic flux rather than absolute magnitude. Its function is crankshaft sensing as a better method of timing for automobile engines, since the distributor is becoming obsolete for this function. The goals are better mileage and better performance.

We can see in Figure 19 that the magnet is mounted behind the sensing chip. As a gear-tooth passes in front of the Hall integrated circuit, the tooth acts as a flux concentrator, increasing the field over the no-tooth condition. Since the device is sensitive to rate of change of flux, it has an operate and release point with

Fig. 18. A biased Hall gear tooth sensor. The magnet is mounted behind the sensor.

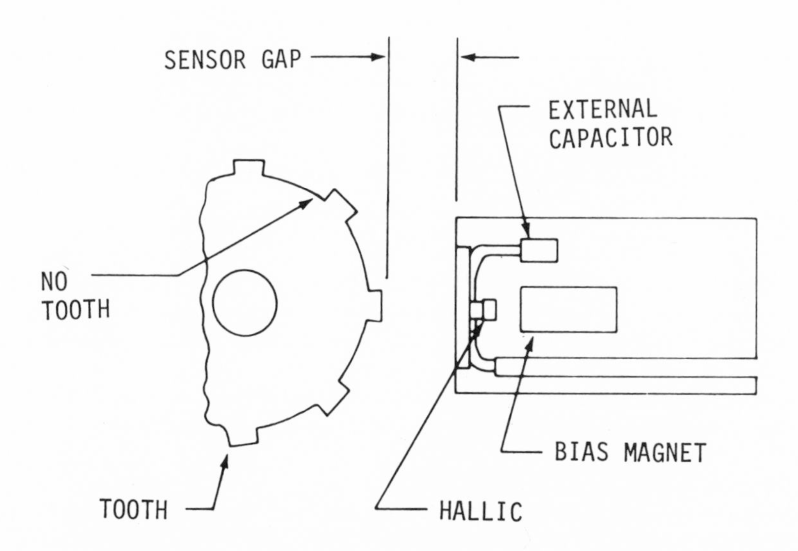

Fig. 19. Biased Hall gear-tooth sensor. The sensor is responsive to the rate of change of magnetic flux rather than to absolute magnitude.

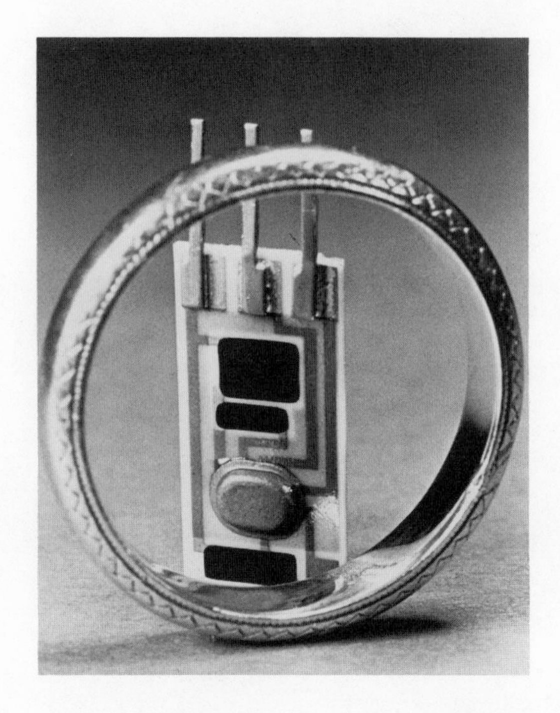

Fig. 20. Linear Output Hall Effect Transducer (LOHET), with laser-trimmed resistors to adjust nul voltage and sensitivity. Primary use is linear feedback in analog control systems.

respect to the leading and trailing edge of the gear-tooth. This product will be on the market by mid-1980.

In Figure 20 we see another new product, this time framed by a wedding ring that suggests a marriage of technologies. We call this device LOHET—Linear Output Hall Effect Transducer. The output of the Hall generator is amplified and temperature compensated right on the cip so that we have a stable output signal of several volts. Both ratiometric and regulated outputs are available. LOHET is a hybrid, as we can see, making use of laser trimmed resistors to adjust the nul voltage and sensitivity. The LOHET chip is located under the small brass cap shown in the figure. Its primary usage will be linear feedback in analog control systems.

One of the most intriguing uses for LOHET is in brushless D.C. motors. By making the rotor from permanent magnet material, and using coil windings for the stator, the LOHET can vary the drive current to the stator, depending on the position of the rotor in a continuously variable manner. A real advantage would be very smooth motor operation even at very low speeds, and there would be very little noise.

Throttle position feedback to an on-board microprocessor is another use. We are also looking into anti-skid application, as well as other safety-oriented projects for the automotive industry, both for standard and off-road vehicles.

Hall effect devices are just beginning to enter the major appliance field where low cost, long life and high reliability make them attractive. Remote meter reading is an application of very large potential. In this case, the meter mechanism is replaced with a ring magnet and a bipolar Hall chip to provide a pulse output. The pulses are electronically counted and stored in a transponder located in the home. Data would be fed to the utility computer via telephone lines. This system would allow variable pricing to encourage off-peak energy usage.

It has been a pleasure sharing a few of the practical aspects of Doctor Hall's discovery with you. For some in this room, the work of Hall has been of passing interest. For me, it has been a matter of total dedication for the past 14 years. I, for one, thoroughly appreciate the privilege of being able to pay homage to his brilliance.

SELECTED READINGS

Beer, A.C., May, 1966, The Hall Effect and Related Phenomena, Solid State Electronics, 9:339.

Brockman, David E., and Nelson, Richard W., October 16, 1975, Hall-Effect Sensors-Magnetic Switches That Have No Contacts, Machine Design, 123.

Dilatush, Earle, January 20, 1979, Keyboard Assemblies, EDN, 65.

Gaines, Donald J., March, 1978, When Are Solid State Position Sensors the Best Choice? Instruments & Control Systems.

Grossman, Morris, September 17, 1979, Low Cost Inspires New Uses of Linear Hall-Effect IC's, Electronic Design.

Herzenberg, Leonardo and Andres, Kent W., October, 1971, An LSI Approach to Keyboard Encoding, Computer Design, 81.

Howell, David, May, 1978, Solid State Keyboards, Electronic Products, 27.

McDermott, J., December 6, 1978, Keyboards for Microprocessor-Based Systems Grow in Versatility, Performance, Electronic Design.

Pople, DuWayne A., April 15, 1974, Keyboards Expand in Complexity, Electronic Products.

Ruhl, Roland L., Metz, L. Donald, and Hackman, Ronald, February, 1977, Reduction of Low Level Incidence of Solder Shorts in Wave Soldering, Electronic Packaging & Production.

Runyan, Stanley, Nov. 9, 1972, Focus on Keyboards, Electronic Design.

Vorthmann, E. A., and Maupin, Joseph T., Solid State Keyboard, in: "Proceedings of the 1969 Spring Joint Computer Conference," pp. 149-159.

Yates, Warren W., July, 1976, A Wide World of Solid-State Switches, Electronics in Industry.

HALL EFFECT MAGNETOMETERS FOR HIGH MAGNETIC FIELDS AND TEMPERATURES BETWEEN 1.5°K AND 300°K

L.G. Rubin
Francis Bitter National Magnet Laboratory
Massachusetts Institute of Technology
Cambridge, Massachusetts 02139

H.H. Sample
Physics Department
Tufts University
Medford, Massachusetts 02155

I. INTRODUCTION

Perhaps the most obvious of all of the applications of the Hall effect has been its use in the measurement of magnetic fields. Since the late forties,[1] the technology associated with the Hall effect magnetometer (HEM) has matured[2] to the point where off-the-shelf inexpensive instruments are available from many manufacturers. Metrologically speaking, there are very few unsolved problems relating to HEM materials, device fabrication, circuit design, and system characterization—assuming certain performance limits are observed. Thus, we can routinely measure a dc or low frequency ac magnetic field in the range ~1 mT to ~3 T (~10 G to ~30 kG) with a precision[3] of 0.5-2%, with spatial resolution of several millimeters, and in an environment maintained at or near room temperature.

The results of research and development efforts to extend HEM performance beyond such limits have been and continue to be published. High frequency (pulsed) magnetic fields can be measured with the aid of appropriate circuit design;[4,5] the measurement of very low fields benefits from advances in higher sensitivity materials and the use of concentrators;[6,7] thin film lithographic techniques have made it possible to fabricate sensors with a spatial resolution of several micrometers;[8,9] and a combination of improvements by instrument makers have increased the precision of measurement to the 0.1% range.[10] But it is the extension of HEM performance to the regime of high fields and low temperatures that is the main concern of this paper.

The magnetometric requirements of research programs which make use of Bitter water-cooled magnets and superconducting solenoids can often be most conveniently met by the physically small and simple to operate HEM that is also capable of operating at liquid helium temperature. This possibility was recognized by many of those initially involved in such programs, and so it was not a coincidence that the first publications on high field and/or low temperature HEM's appeared in the early sixties, when high-field magnets first became generally available. Then, as now, HEM users sought devices that were stable and reproducible with respect to thermal cycling. In addition to this necessary condition, other desirable operating characteristics that were simply convenient and useful, were also looked for: high sensitivity, low temperature dependence of sensitivity, small zero offset errors, and good linearity at high fields.

References 11-34 include the results of many of the efforts to investigate the properties of devices suitable for high field/low temperature HEM's. The majority of these studies were concentrated on the semiconductor materials InAs[11-21] and InSb,[15,22-30] but results were also reported for HgSe,[31] Bi,[32] Bi_2Se_3,[33] Ge,[29] and GaAs.[34] In many cases, studies were made of material parameters such as carrier concentration and mobility, dimensional and packaging parameters, and bulk vs thin film construction. Although most of the work was done on homemade sensors, several investigators preferred the convenience offered by commercially available HEM's.[11-14,20] The latter included devices manufactured by Siemens A.G. (Models FC33 and 34, SBV552, and RHY17 and 18) and by F. W. Bell (Models BH200 and 203).

Whereas all but two of the papers[19,22] include a discussion of the operation of HEM's at cryogenic temperatures, the coverage on the high field aspects has been less complete. Less than a third of the references include data and/or discussion of operation in fields $\geq$ 10 T,[16,19,20,27,29,33] and none of these offer details on the characteristics of commercially available devices. It was for this reason that we decided to undertake a program whose purpose was to discover and evaluate those HEM's which could satisfy the following criteria: 1) they were available from manufacturers as off-the-shelf items, 2) they had been designed for low temperature use and were so specified, and 3) they had either been specifically designed to have superior characteristics in magnetic fields as high as 15 T, or some evidence existed that they did.

We were able to locate three types of HEM's which seemed to satisfy the above requirements. These were an InAs Model BHT921 from F. W. Bell,[35] an InSb Model LMK from Electrotechnicky Ustav,[36] and a GaAs Model TC-8101 from Copal.[37] The fact that each of these devices was fabricated from a different semiconductor was a fortunate coincidence, in that we were able to make performance comparisons among the materials. Our study did not include any of the devices made by the pioneer manufacturer of HEM's, Siemens A.G. The reason was that the only models they had specifically designed for low temperature operation (RHY17 and 18) exhibited characteristics at 5 T which did not lead us to expect satisfactory performance at 15 T.[38]

The results of our investigation of the three devices were published in two papers,[39,40] from which most of the following material was drawn.

II. PROPERTIES OF THREE HALL EFFECT MAGNETOMETERS

All of the HEM's we studied were transverse models, having thicknesses ranging from 1-1.4 mm, lateral dimensions less than 16 mm, and active areas of about 1 mm^2 (see Ref. 40 for further information on probe configurations).

It is convenient to arrange the discussion into five general areas:

A. Zero Offset Errors
B. Sensitivity[41]
C. Temperature Dependence of Sensitivity
D. Non-Linearities
E. Reproducibility

These are not in order of importance, but rather in a logical sequence determined by our program of measurement. There are practical considerations associated with each of these areas, which must be appreciated for proper selection and/or use of HEM's in real, experimental situations. Each of these areas is discussed separately below.

A. Zero Offset Errors

As is well known, it is difficult to construct a Hall probe with perfectly aligned contacts. Hence, there is usually a zero offset voltage to be concerned with. Special measuring techniques are required to ensure that this zero offset error is properly accounted for as the magnetic field is increased.

We have used a zero offset correction circuit due to F. Kuhrt,[42] shown in Fig. 1, and found it to be quite satisfactory. The sensor leads are connected to the correction circuit box, as are the leads from a stable dc constant current source, and a high

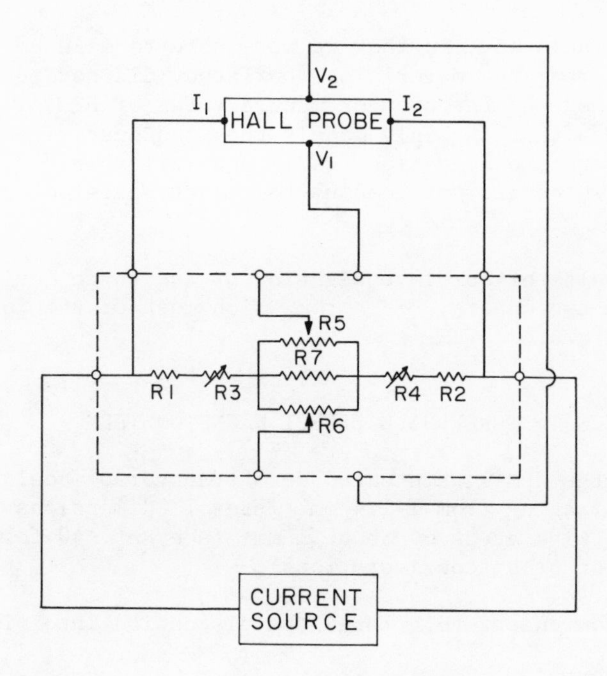

Fig. 1. Zeroing circuit. Typical values: R_1, R_2 - 1 kΩ, R_3, R_4 - 1 kΩ adjustable; R_5, R_6 - 100 Ω adjustable; R_7 - 2 Ω. The voltmeter is first connected between V_1 and the arm of R_5, and with the current on, R_3, R_4, and R_5 are adjusted to zero the voltmeter. V_1 may then be tied to the arm of R_5 at what represents an equipotential point (no current flows between them). The voltmeter is then connected to its normal measuring position between V_2 and the arm of R_6. Once the latter is adjusted for zero (at zero field, of course), the sensor-voltmeter system is capable of measurement over the full field range without the problem of a magnetoresistive zero shift. It should be noted that the shunting effect of the zeroing circuit on the ~1 Ω probe is < 0.05%.

precision dc voltmeter. It should be noted that no "linearizing resistor" was used across the output of any of the probes. See the caption of Fig. 1 for a detailed explanation of the zeroing circuit and its operation.

B. Sensitivity

While sensitivity alone is not the most important of the many parameters to be considered in choosing an HEM, it can become critical if its value becomes too low. In an operational sense, 'too low' might be defined as the point below which thermal emfs in the

Table I. Hall Effect Magnetometer Sensitivities

HEM Type	Range of Observed Sensitivities at I_m = 60 mA (mV/T)
FWB	4.2 - 5.2
ETU	7 - 17
CPL	1.3 - 1.5

measuring circuit become a factor. We suggest that a full scale output of ~10 mV is such a lower limit. Moreover, this sensitivity should be attainable at probe power dissipation levels of no more than tens of milliwatts, particularly if it is to be used at low temperatures.[43] These conditions were satisfied by all three of the HEM's studied. For comparison, we give a range of their room temperature sensitivities in Table I, normalized to the same measuring current, I_m, of 60 mA.

ETU devices were, on the average, more sensitive than the FWB or CPL models, but they also exhibited the greatest variation in sensitivity between nominally identical units.

C. Temperature Dependence of Sensitivity

Under this heading, one should make a distinction between what are, operationally, two kinds of dependence on temperature: narrow range and wide range. It is the former which often sets a limit on the output stability of an HEM when it is operating at a "fixed" temperature and that temperature is fluctuating by a small amount. It is also the former that should be significant when a manufacturer specifies a "temperature coefficient of Hall voltage" at a particular temperature. The "wide range" parameter is used to account for differences in sensitivity when an HEM is operated at widely separated temperatures, say, 4.2 K and 300 K.

The results of measurements made to evaluate the performance of the FWB, ETU, and CPL devices are summarized in Table II.

Some explanation of Table II, is in order. First of all, as is discussed in detail below, all three models eventually show changes in sensitivity brought about by repeated thermal cycling between room temperature and 4.2 K. Thus, the data in Table II were taken on "virgin" probes (i.e., not previously thermally cycled) measured at 300, 77, 4.2, and 1.5 K, in sequence, and

Table II. Percent change in sensitivity with temperature for CPL, FWB, and ETU Hall effect magnetometers

HEM	Sensitivity change from 4.2 K value (%)			
Type	T = 1.5 K	T = 4.2 K	T = 77 K	T = 300 K
FWB	0	0	+0.4	-0.2
ETU	0	0	+0.5	+0.2
CPL	-0.5	0	-0.3	+0.4

without warming the probe between measurements.

Secondly, it is in fact not possible to define HEM "sensitivity" in a completely unambiguous way, due to the existence of HEM non-linearities. That is to say, the measured Hall voltage, V, is not a strictly linear function of magnetic field, B. Deviations from linearity are displayed most easily by first fitting the Hall voltage data to the equation

$$V_{calc.} = A \cdot B + C, \tag{1}$$

where the slope A (the sensitivity) and intercept C are constants. Then, deviation voltages

$$\Delta V = V - V_{calc.}$$

are calculated for each measured data point, and plotted vs B. Results are shown for an FWB and an ETU device at 300, 77, and 4.2 K in Fig. 2. In this figure, the diagonal lines represent a deviation of $\pm$ 1% of V, so it is evident that except for the lowest fields with the ETU probe, both models have linearities of better than 1% at all temperatures.

By suitable choices of the constants A and C in Eq. (1), it is always possible to make the high-field deviation plot for any HEM oscillate about the $\Delta V = 0$ line. This procedure was always followed by us, and has been done for the probes in Fig. 2: A and C were arbitrarily varied, until the "best" deviation plot at high B was obtained. The value of A appropriate to this "best" deviation plot was then defined as the HEM sensitivity.

From Fig. 2, it is clear that the above procedure leads to an unambiguous A (within ~0.1%) at 300 and 77 K, but A is uncertain

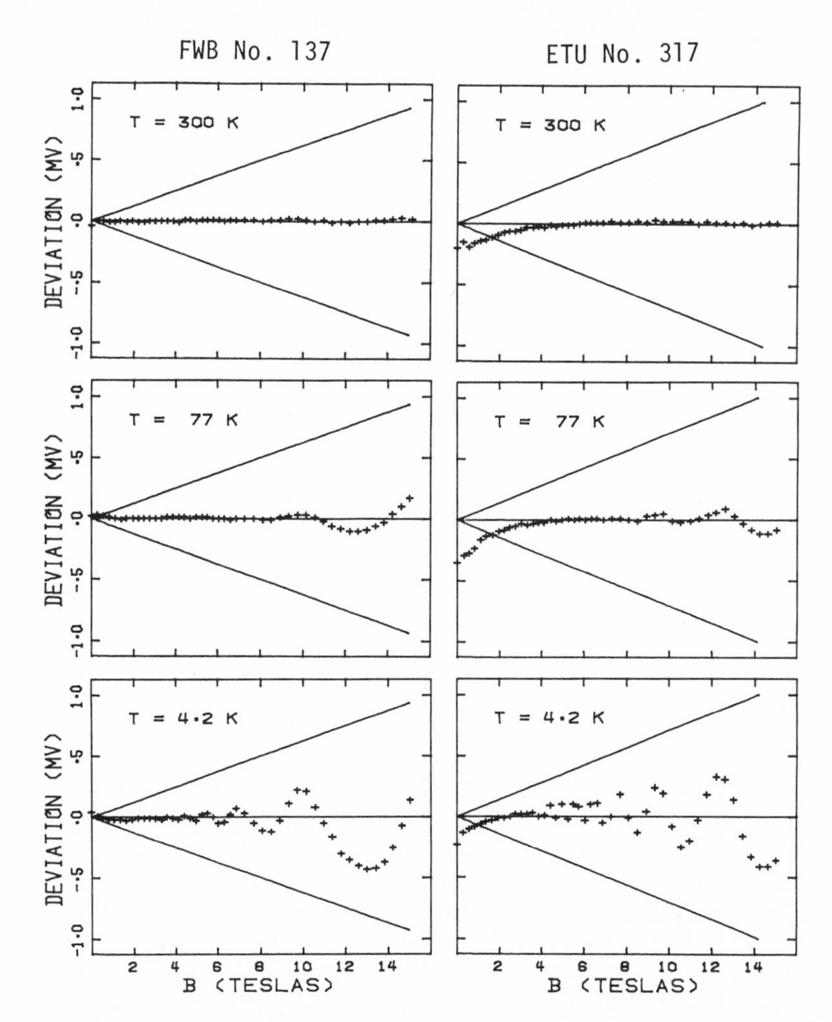

Fig. 2. Deviation of the Hall voltage from the linear relationship defined by Eq. (1) for an FWB and an ETU probe at 300, 77, and 4.2 K. The solid lines represent ± 1% of the total Hall voltage. For the FWB probe, the slopes used in calculating the deviation plots were 6.215, 6.250, and 6.225 mV/T at 300, 77, and 4.2 K, respectively, using a measuring current of 80 mA. For the ETU probe, the slopes were 7.070, 7.088, and 7.055 mV/T at 300, 77, and 4.2 K, respectively, at 60 mA.

by as much as 0.2% at 4.2 K. Deviation plots at 1.5 K are indistinguishable from those at 4.2 K for the same probe. Thus, the 1.5 K sensitivity is also uncertain by ~0.2%.

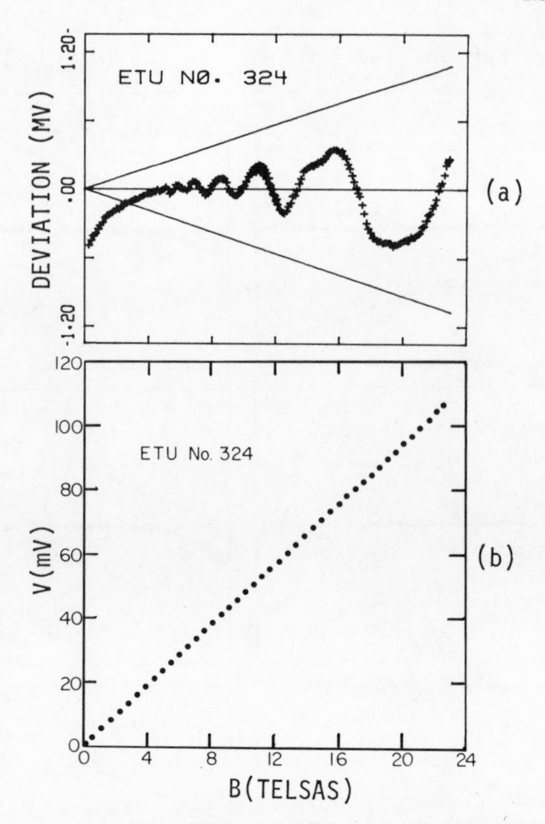

Fig. 3. Results for an ETU probe at T = 4.2 K, and B to 23 T. The measuring current is 20 mA. (a) Deviation, ΔV, of the Hall voltage from the linear relationship, Eq. (1). The solid lines represent $\pm$ 1% of the total Hall voltage, V. (b) The total Hall voltage, V. The high field oscillations are just perceptible, but the low field change in slope is not.

In light of the above discussion, the results given in Table II can be summarized as follows. There are real dependences of HEM sensitivity on temperature, and these "wide range" changes are most noticeable between 77 and 300 K. It is also evident that "A" is somewhat more sensitive to temperature for CPL devices than for FWB and ETU models, particularly in the 1.5-4.2 K range. Nevertheless, the sensitivity changes are all less than 1%, so that for all three types, a room temperature calibration is sufficient to determine the value of the slope to ~1% at <u>any</u> of the listed temperatures of Table II.

While we made no attempt to measure the "narrow range"

temperature coefficient at either 77 K or 300 K,[44] an estimate can be made of the 4.2 K value by using the 4.2 K and 1.5 K data.

D. Non-Linearities

Another aspect of HEM non-linearities is illustrated in Fig. 3. Fig. 3b is a plot of the total Hall voltage, V, vs B for an ETU probe, at T = 4.2 K and B to 23 T. On this plot, V appears to be quite linear except at the highest fields, where wiggles, due to quantum oscillations, become noticeable. For clarity, only about one-fourth of all the data points are plotted. Figure 3a is the deviation plot for this same probe, where the slope, A, and intercept, C, (in Eq.(1)) were chosen so as to force the high field data to oscillate about the $\Delta V = 0$ line. (From Fig. 3a, the value of C is about +0.5 mV.)

It is evident that there are two distinct types of non-linearities that may be present in the Hall voltage: quantum oscillations, and/or a change in slope at fields B ~3-4 T. But it should be noted that the low field change in slope is imperceptible in Fig. 3b, a plot of the total Hall voltage. For this device, the slope change is about 2.5%, and is centered at about 3 T. This may be seen by comparing the low field and high field deviation voltages with the solid diagonal lines in Fig. 3a, which represent ± 1% of the total Hall voltage. The quantum oscillations for this probe are less than about 0.5% of the Hall voltage for all fields above about 5 T, i.e., after the slope change non-linearity is complete. We wish to emphasize the significance of the differences between the results obtainable from a plot such as that of Fig. 3a, as compared to those from Fig. 3b. Without the increased resolution offered by the former, any evaluation of HEM performance would be severely limited.

Representative deviation plots for the three models of HEM's at T = 4.2 K are shown in Fig. 4. The results for all of the devices studied are summarized in Tables III and IV.

Table III characterizes the quantum oscillation type non-linearities at 4.2 K, and compares the height of the quantum oscillation peaks for a particular probe to the 15 T Hall voltage for that same probe. The 15 T output was chosen as a 'normalizing' factor for reasons of practicality, there being few laboratories in the world with higher fields available. Clearly, the average quantum oscillation non-linearities for the three HEM models are comparable.

The change-in-slope type of non-linearities have been approximately analyzed at 4.2 K and are given in Table IV. It is important to note that FWB InAs probes do not show this second type of

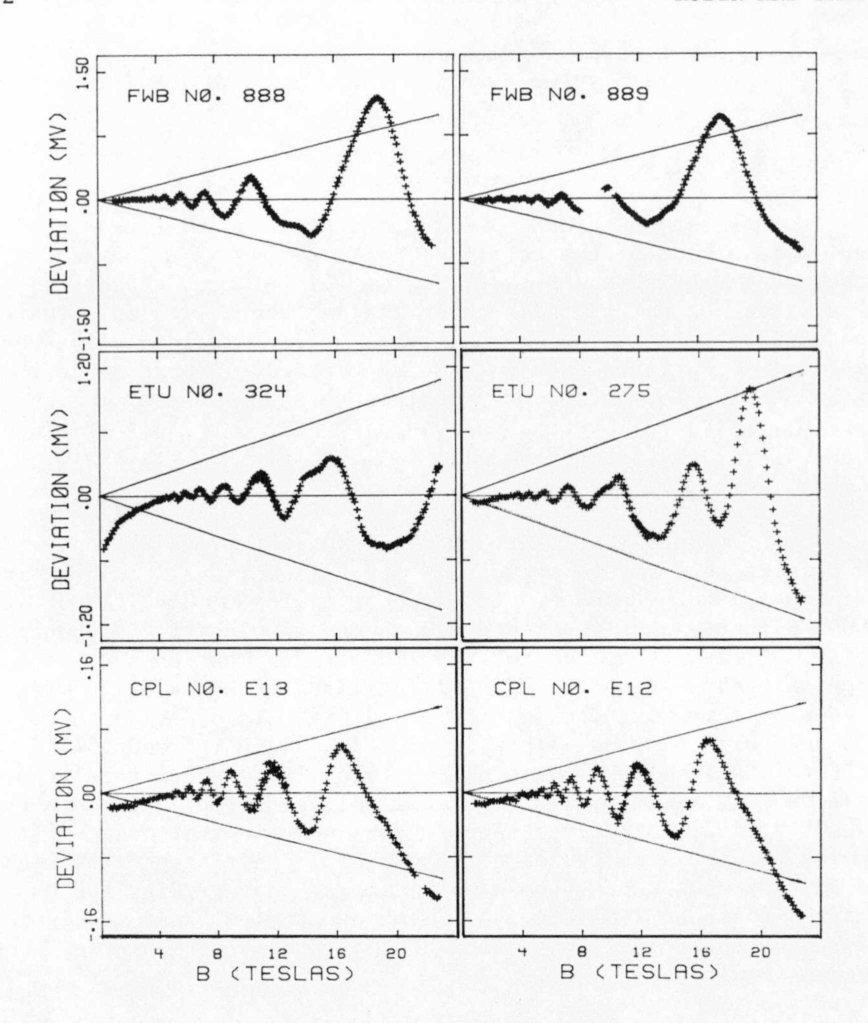

Fig. 4. Deviation ΔV of the Hall voltage from the linear relationship (1) for FWB, ETU, and CPL probes at 4.2 K. The solid lines represent $\pm$ 1% of the total Hall voltage. The measuring currents used were: 60 and 50 mA for FWB Nos. 888 and 889; 20 and 18 mA for ETU Nos. 324 and 275; 20 mA for CPL Nos. E13 and E12.

nonlinearity within the precision of the measurements. Both types of non-linearities are independent of temperature in the 1.5-4.2 K range, although they do change in the 4.2-300 K range. The development of the quantum oscillations as T is decreased from 300 K to 77 K to 4.2 K can be seen in Fig. 2 for FWB and ETU devices.

Table III. Quantum oscillations of Hall effect magnetometers at 4.2 K.

HEM Type		Maximum quantum oscillation correction to linear fit (as a percent of the 15 T Hall voltage) for B less than: 8 T	15 T	23 T
FWB (InAs)	best	0.1	0.5	0.5
	worst	0.2	0.8	1.9
	average	0.17*	0.6*	1.7+
ETU (InSb)	best	0.1	0.2	0.6
	worst	0.3	0.7	2.6
	average	0.12*	0.4*	1.6+
CPL (GaAs)	best	0.1	0.7	1.8
	worst	0.4	0.9	2.1
	average	0.24*	0.8*	2.0+

* Based on the results for six probes.
+ Based on the results for two probes.

E. Reproducibility

The question of HEM reproducibility becomes extremely important if it is to be shifted into and out of its measuring position, and/or is to be thermally cycled between room temperature and some low temperature, e.g., 4.2 K.

A mechanical movement or readjustment of the Hall probe almost certainly will lead to a change in probe orientation (and a large error), unless it is contained in a holder which accurately reproduces its orientation with respect to the B-field. For example, an 8° misorientation can result in a 1% change in sensitivity; the orientation must be reproduced to within 2.5°, if sensitivity changes of 0.1% are important.

We have made extensive measurements on reproducibility of the three HEM models under thermal cycling to 4.2 K. Most of these measurements were carried out only to B = 14 T. Complete details of the reproducibility test procedures are given elsewhere.[39] Very briefly, a set of probes was mounted in a holder which limited misorientation errors to less than 0.1%. The holder was then thermally cycled between room temperature and 4.2 K, with Hall

Table IV. Change-in-slope non-linearities of Hall effect magnetometers at 4.2 K.

HEM Type	Magnitude of slope change, %	Centered at B (T)
FWB (InAs)	< 0.2	-
ETU (InSb)	1-3	~ 3
CPL (GaAs)	0.2-1	~ 4

measurements taking place at 4.2 K periodically. A given set of probes remained untouched in the holder until all the measurements on that set were completed. Care was exercised to eliminate moisture from the sample holder before cooling to avoid strains on the probes due to ice formation.

The primary goal of these tests was to compare the sensitivity changes for the three types of probes. To do this, an initial "calibration" run was performed on a given probe the first time it was cooled to 4.2 K. These data were then fit to Eq. (1), and a deviation plot calculated, as described earlier. On this first calibration run, A and C were varied arbitrarily, until a "good" deviation plot was obtained (i.e., one which oscillated about $\Delta V = 0$ at higher fields).

Subsequent runs on the same probe were done in a similar manner, except that A and C were varied until the deviation plot reproduced that of the initial calibration run. With very few exceptions, this process could always be carried out to within the experimental error of the measurement, and produced values for the sensitivity (A in Eq. (1)) which were uncertain to about $\pm 0.1\%$. In this way changes of A for a given probe could be studied as a function of thermal cycling. After a sufficient number of thermal cycles, all of the 18 devices studied changed their sensitivities by at least 0.2%; two changed by more than 2.5%.

It is perhaps surprising, at first glance, that the deviation plots from different runs (with $|\Delta V| \leq 1\%$ of the total Hall voltage) can be matched even in cases where A has changed by 2.5%. Nevertheless, this was the case for 90% of the probes studied. The effect is indicated in Fig. 5 for an ETU probe. Evidently, percentage changes in the non-linearities themselves (i.e., the deviation plot) are less than, or on the same order as changes in the sensitivity.

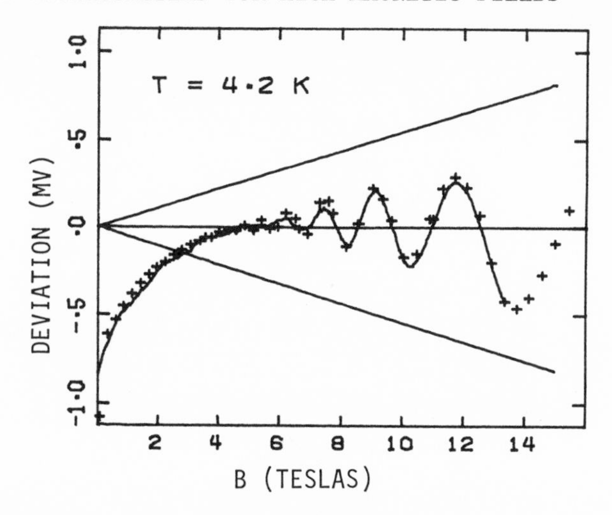

Fig. 5. Deviation of the Hall voltage from the linear relationship defined by Eq. (1) for an ETU probe. The points are from the initial run, where the slope was 5.430 mV/T at 30 mA. The drawn curve is the computer-smoothed result of a subsequent run where the slope was 5.320 mV/T at 30 mA, or a change of about 2%. The straight lines represent ± 1% of the total Hall voltage.

An "average" reproducibility under thermal cycling for the three HEM models is shown in Fig. 6. Three factors complicate the computation of the "average" reproducibility of the six probes of each type studied. First, not all probes were measured at the same thermal cycle number. Second, once a probe's slope starts to change (after a certain number of cycles) it can then change in either direction (i.e., towards or away from the initial value) on subsequent measurements. Third, after the slope starts to change initially, the probe often becomes "unstable" even while maintained at 4.2 K. That is, two measurements during the same run can give different values of A. Therefore, the maximum magnitude slope change during all previous thermal cycles was used for each probe in computing the average in Fig. 6.

The FWB devices are clearly the best choice as far as stability under thermal cycling is concerned. None of the six we tested changed sensitivity by more than 1%, even after 60 thermal cycles. Three out of six ETU probes changed by 2% or more after just 25 cycles. CPL probes seemed to be of intermediate stability.

III. SUMMARY AND CONCLUSIONS

For the purpose of summarizing the HEM results, we begin with

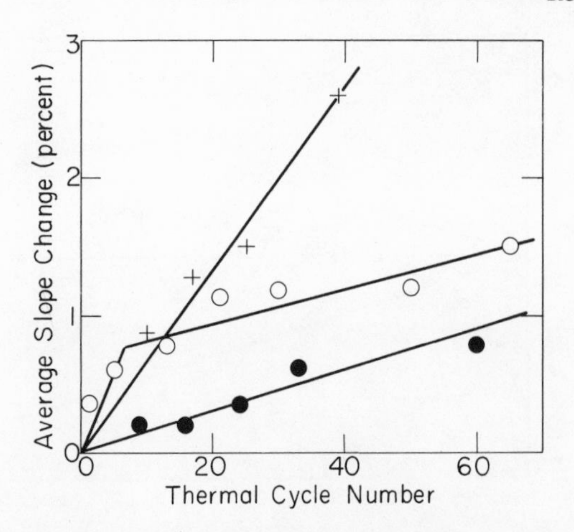

Fig. 6. Average change in slope with thermal cycling for CPL probes (O), FWB probes (●) and ETU probes (+). See the text for the method used to calculate the average slope change.

an assumption that a user has calibrated a probe in situ against a magnetometric standard and has determined the deviation plot to the required accuracy. What are the errors to be expected in subsequent measurements and how may they be guarded against?

1. A mechanical movement or readjustment of the probe or its holder almost certainly will lead to some change in probe orientation with respect to the field. This can produce a large error unless proper precautions are used. We are of the opinion that the ultimate limit on HEM accuracy is set by the orientation error.

2. There will be a slope change whose magnitude is a function of thermal cycling and the type of probe. This error can range from tenths of a percent to several percent. The recommended corrective action (for both 1. and 2.) is a periodic in situ recalibration at two or more field points. Note, the original deviation plot can continue to be used. The HEM should probably be replaced once its slope has been observed to change by more than the desired accuracy.

3. If a direct readout of the HEM is desired, e.g., a DVM used with a probe adjusted for 1.000 or 10.000 mVT^{-1} sensitivity, then the only error (after recalibration) is strictly a function of the quantum oscillation effects above ~6 T and/or the change-in-slope non-linearity below ~4 T (for some probe types). Since both of these non-linearities appear to reproduce within ± 0.1% FS, it is simple to correct for this error by using the deviation plot. While

on the subject of quantum oscillations, it is worth noting that their effect on probe linearities was the same (within a factor of two) for all of the materials (InAs, InSb, and GaAs).

4. In the 'ideal' case of an HEM mounted in a fixed position in a magnet, and maintained at a constant temperature for long periods of time, there is good reason to believe that a magnetometric accuracy of 0.2% is achievable. For an application such as control of a fixed or slightly variable field, HEM stability is almost completely determined by its narrow range temperature dependence. Thus, a properly thermostatted probe, used in conjunction with a carefully designed ac feedback circuit, should provide controllability to 10 ppm or better.

A century after the discovery of the Hall effect and about 30 years after the development of the first practical Hall effect magnetometer, there is available to users of high magnetic fields a mature technology, complete with sources of off-the-shelf devices and instruments.

REFERENCES

1. It is generally agreed that the first publication dealing with a practical HEM was that of G. L. Pearson, Rev. Sci. Instrum. 19:263 (1948).
2. C. Germain, in his review of methods of measuring magnetic fields [Nucl. Instrum. Methods 21:17 (1963)], lists 44 papers on the Hall effect published between 1953 and 1960.
3. Precision, rather than accuracy, is specified so as to bypass considerations of calibration traceability, sensor orientation errors, etc.
4. H. H. Wieder, J. Appl. Phys. 33:(S)1278 (1962).
5. R. L. Baty and P. B. Weiss, IEEE Trans. Instrum. Meas. IM-22:18 (1973).
6. I. M. Ross and E. W. Saker, J. Electronics 1:233 (1955).
7. N. P. Milligan and J. P. Burgess, Solid State Electron. 7:323 (1964).
8. R. N. Goren and M. Tinkham, J. Low Temp. Phys. 5:465 (1971).
9. H. T. Minden and M. F. Leonard, J. Appl. Phys. 50:2945 (1979).
10. There are numerous examples of carefully designed systems where HEM performance has been extended even beyond this. Typical among these have been the Varian "Fieldial"; J. R. Mulady, IEEE Trans. Instrum. Meas. IM-13:343 (1965); G. W. Trott et al., J. Phys. E 12:979 (1979).
11. J. P. McEvoy and R. F. Decell, Rev. Sci. Instrum. 34:914 (1963).
12. M. S. Lubell and B. S. Chandrasekhar, Rev. Sci. Instrum. 35:906 (1964).

13. D. A. Hill and C. Hwang, J. Sci. Instrum. 43:581 (1966).
14. K. G. Günther and H. Freller, Cryogenics 7:49 (1976).
15. J. E. Simpkins, Rev. Sci. Instrum. 39:570 (1968).
16. B. G. Lazarev, L. S. Lazereva, S. I. Goridov and S. G. Shul'man, Sov. Phys. Dokl. 13:696 (1969).
17. S. G. Shul'man, Instrum. Exp. Tech. 184 (1969).
18. Y. G. Agbalyan, Y. I. Kazantsev, E. N. Lysenko, N. M. Noginova, and G. K. Yagola, Measurement Tech. 1723 (1969).
19. V. G. Veselago, M. V. Glushkov, V. M. Ivanov, and S. G. Shul'man, Instrum. Exp. Tech. 1580 (1970).
20. G. W. Donaldson, Advances in Cryogenic Engineering 18:416 (1973).
21. G. K. Yagola, Y. I. Kazantsev, and E. N. Lysenko, Measurement Tech. 1535 (1973).
22. H. Hieronymous and H. Weiss, Solid State Electron. 5:71 (1962) [in German].
23. S. S. Shalyt, R. V. Parfen'ev, and M. S. Bresler, Sov. Phys.-JETP 48:1212 (1965).
24. G. N. Harding, W. H. Mitchell, and E. H. Putley, Solid State Electron. 9:464 (1966).
25. I. Hlásnik, F. Chovanec, and M. Polák, Cryogenics 6:89 (1966).
26. M. Polák and I. Hlásnik, Solid State Electron. 13:219 (1970).
27. A. A. Kolmakov, R. V. Parfen'ev, V. I. Pogodin, and G. A. Yur'eva, Instrum. Exp. Tech. 289 (1971).
28. M. Polák, Rev. Sci. Instrum. 44:1794 (1973).
29. V. G. Veselago, V. M. Ivanov, V. I. Pogodin, V. I. Tikhonov, and G. A. Yur'eva, Measurement Tech. 424 (1975).
30. H. W. Weber, G. P. Westphal, and I. Adaktylos, Cryogenics 16:39 (1976).
31. S. V. Odenov, G. A. Udzulashvili, V. E. Khuedelidze, D. G. Chigvinadze, and V. A. Skukhman, Cryogenics 6:46 (1966).
32. W. C. Hubbell, A. D. Haubold, and C. J. Bergeron, Cryogenics 7:373 (1967).
33. J. A. Woollam, H. A. Beale, and I. L. Spain, Rev. Sci. Instrum. 44:432 (1973).
34. P. Kordos, L. Jansak, and V. Benc, Cryogenics 13:313 (1973).
35. F. W. Bell, Inc., 6120 Hanging Moss Rd., Orlando, Florida 32807, U.S.A. Designation: FWB.
36. Electrotechnicky Ustav, Bratislava, Czechoslovakia (the Institute of Electrical Engineering of the Slovak Academy of Sciences in Bratislava). Designation: ETU. It is our understanding that HEM's are no longer available from this supplier.
37. Copal Co., Ltd. Shimura 2-16-70 Itabashi-ku Tokyo, Japan 174. Designation: CPL.
38. From the data reported in the manufacturer's specifications [Instrument Systems Corporation, "Siemens Galvanomagnetic Devices" (1968)], and also in Ref. 14, it was apparent that the RHY17/18 was strongly non-linear. At 5 T, its output was down ~25% from a linear extrapolation of the 0-1 T data,

and from the shape of the curve, it appeared likely that there would be an even greater droop at 15 T.

39. L. G. Rubin, D. R. Nelson, and H. H. Sample, Rev. Sci. Instrum. 46:1624 (1975).
40. H. H. Sample and L. G. Rubin, IEEE Trans. Magn. MAG-12:810 (1976).
41. We define sensitivity in Sec. II. C.
42. F. Kuhrt, Siemens Z. 28:370 (1956).
43. An example of an HEM with otherwise superior performance, but with only marginally useful sensitivity, can be found in Ref. 33.
44. The manufacturer specifies a value of 0.007%/°C for the FWB Model BHT921 at 300 K.

APPLICATIONS OF HALL EFFECT DEVICES IN SPACE TECHNOLOGY

Frederick F. Mobley

Applied Physics Laboratory
Johns Hopkins Road
Laurel, Maryland 20810

ABSTRACT

Hall effect devices have found important applications in space technology. Artificial magnetic hysteresis damping of the attitude motion of the DODGE and RAE-A satellites was achieved with Hall effect devices. They are used to measure the magnet strength in the trim magnet systems for NIMBUS, LANDSAT, SAS-A, and SAS-B satellites. They are used to provide commutation of DC power for brushless DC motors for SATCOM, DMSP, and the Atmosphere Explorer satellites.

The drag-free sensing system (called DISCOS) for the NOVA satellites requires a very slender magnet for damping proof-mass motions. The uniformity of magnetization of this magnet is critical. A Hall effect detector was used to measure the magnet strength and confirm the required uniformity on the magnets for flight.

INTRODUCTION

The unique technical problems presented by space systems presents many challenges to the scientist and engineer which are distinctly different from those met in earth-bound applications. It is not too surprising to find that Hall effect devices have found important places in space technology where their peculiar and unique capabilities make them ideally suited to solve our problems. Some, but by no means all of these applications will be described here.

ARTIFICIAL HYSTERESIS DAMPING

One of the unique features of the TRANSIT navigation satellites has been the use of magnetic hysteresis as a means of removing spin and tumble from the satellite and providing other damping functions. Figure 1 shows the first such application—on the TRANSIT IB satellite.[1] Four slender rods of a highly permeable nickel-iron alloy are arranged in a plane perpendicular to the satellite spin axis. These rods are magnetized by induction from the earth's magnetic field. As the satellite rotates the sense of the magnetization is reversed and a magnetic hysteresis loop is traversed in the rod material. Figure 2 shows typical hysteresis loops. Each complete hysteresis loop causes energy dissipation of $(\mathrm{Vol}/4\pi)\oint BdH$ ergs (with B in gauss and H in oersted). This dissipation causes a gradual decay in satellite spin rate. The hysteresis loops shown in Figure 2 are for MuMetal and AEM4750 alloys. The latter alloy of nickel

Fig. 1. Internal view of the TRANSIT 1B satellite.

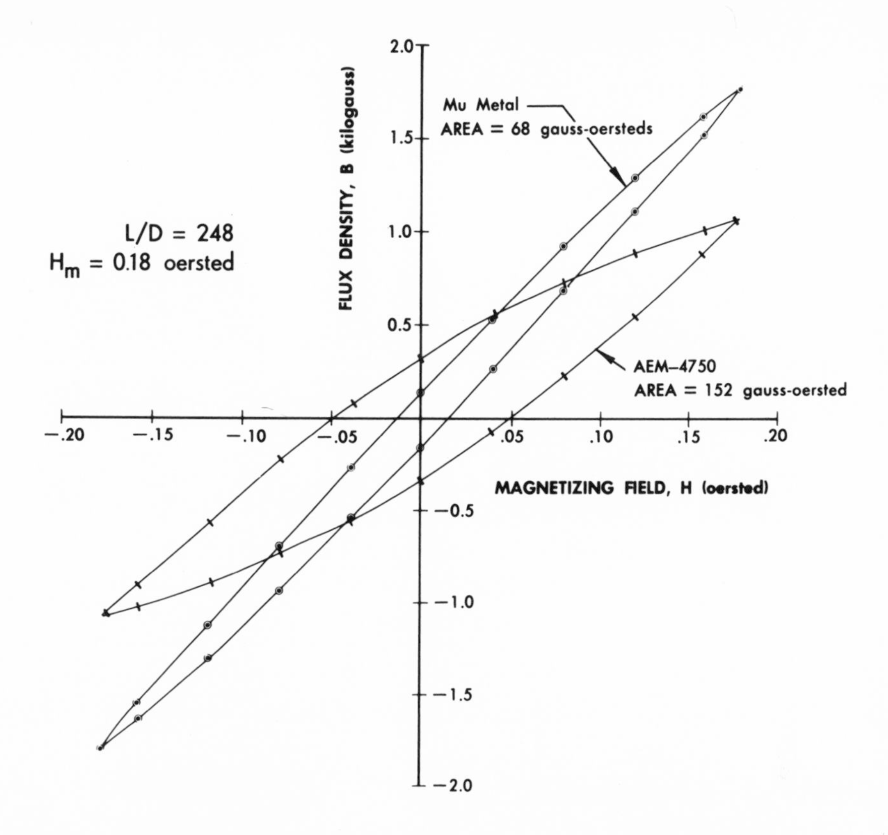

Fig. 2. Hysteresis loops for two different magnetic materials.

and iron was chosen for the TRANSIT satellite applications.

Figure 3 shows the results of spin decay versus time observed with the TRANSIT 2A satellite in 1960. This entire despin process is a result of energy dissipation in the magnetic hysteresis rods.

Magnetic hysteresis damping is also used to damp the oscillations of the gravity-gradient stabilized TRANSIT 5A[2] and subsequent TRANSIT satellites. Figure 4 shows an artist's concept of TRANSIT 5A-3 in orbit. A 100 ft long boom is extended to increase the gravity-gradient torques to the level required for stabilization. A helical damping spring is attached to the end of the boom, with a flashing light device for length measurement. The damping spring was found to be unnecessary and dropped from the operational TRANSIT satellites.

It is evident that magnetic hysteresis damping works well for

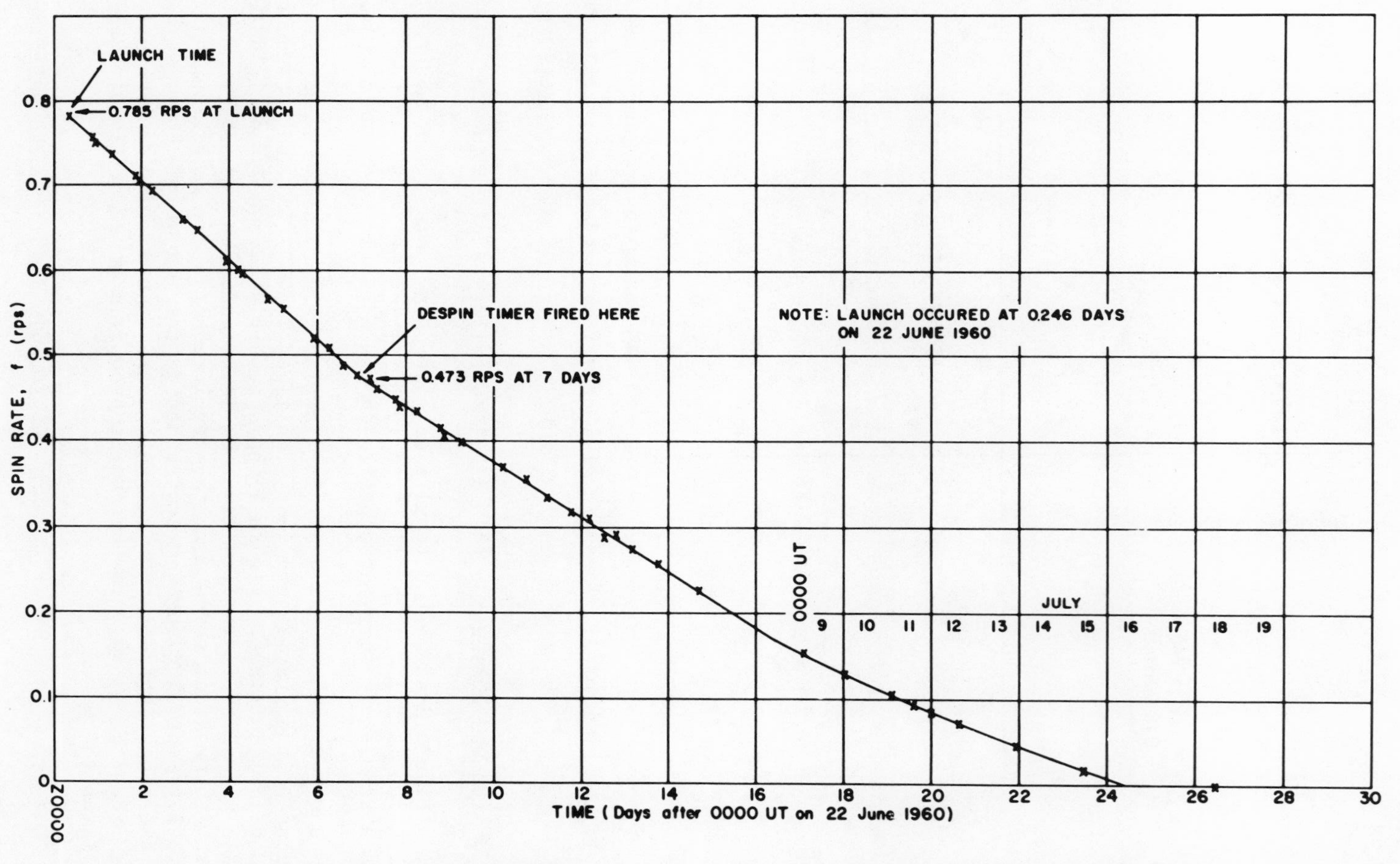

Fig. 3. Despin of the TRANSIT 2A satellite by magnetic hysteresis damping.

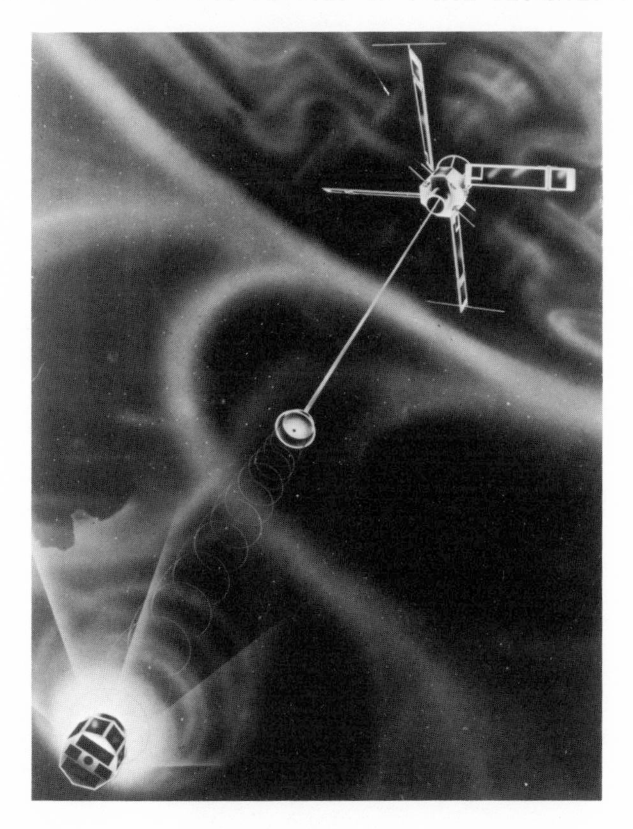

Fig. 4. The TRANSIT 5A-3 satellite, gravity-gradient stabilized.

the TRANSIT satellites. In 1965 we were working on a satellite called "DODGE" (Department of Defense Gravity Experiment) to prove the feasibility of gravity-gradient attitude stabilization at near synchronous-satellite altitude, ~18,000 n. miles. We wanted to use magnetic hysteresis damping, but the magnetic field is only 1/200th as strong as it is at the TRANSIT altitudes of 500 n. miles.

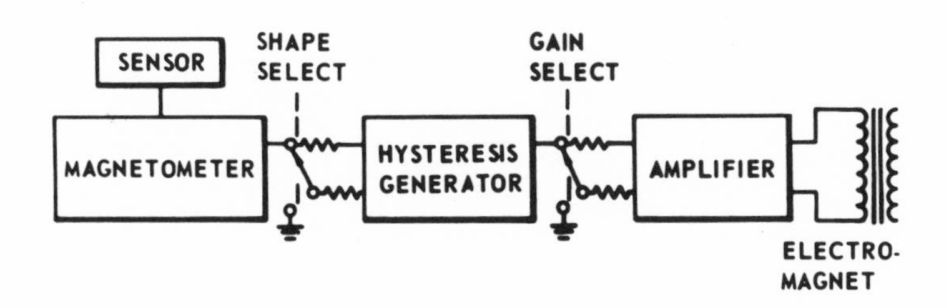

Fig. 5. Single channel of enhanced magnetic damping.

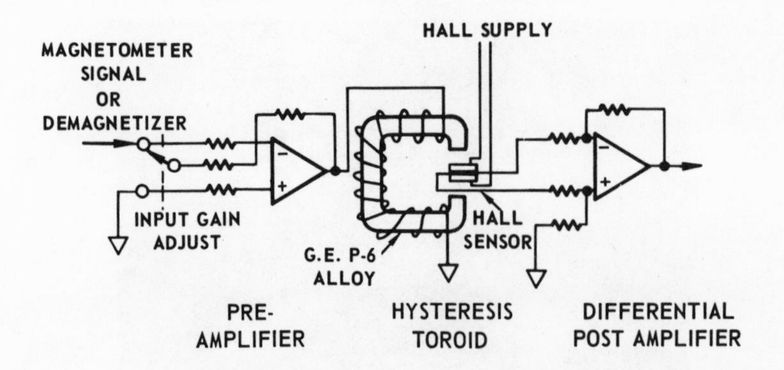

Fig. 6. Hysteresis generator functional design

Passive magnetic damping would be too weak. Some form of enhanced or artificial magnetic damping was needed, and the concept shown in Figure 5 was adopted.[3] Here a vector magnetometer sensor is aligned with a satellite axis and detects the component of the earth's magnetic field along that axis. The output voltage is proportional to the earth 's field. The signal goes into a "hysteresis generator" whose output depends on the input in a hysteretic manner. That output is amplified and used to energize a linear electromagnet that interacts with the earth's field just as a hysteresis rod would. Figure 6 shows the "hysteresis generator". A small toroid

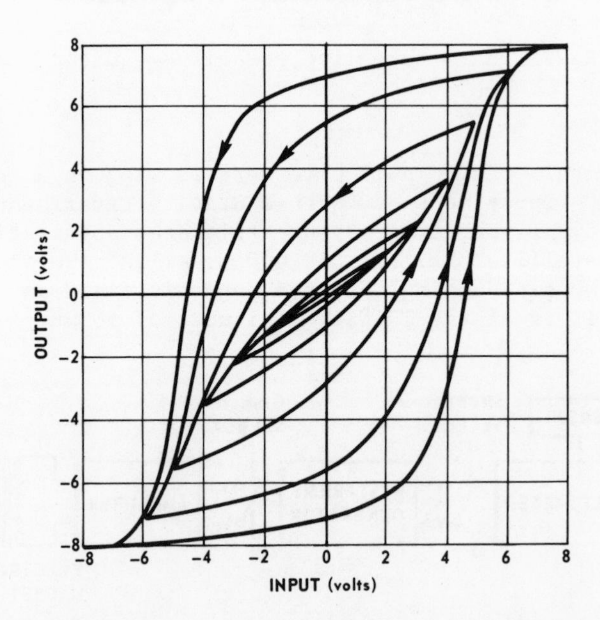

Fig. 7. Hysteresis generator signal characteristics.

Fig. 8. The DODGE satellite.

of a semi-hard magnetic material, GE P-6, has a small slot cut in it, and a Hall effect detector inserted. The magnetometer output

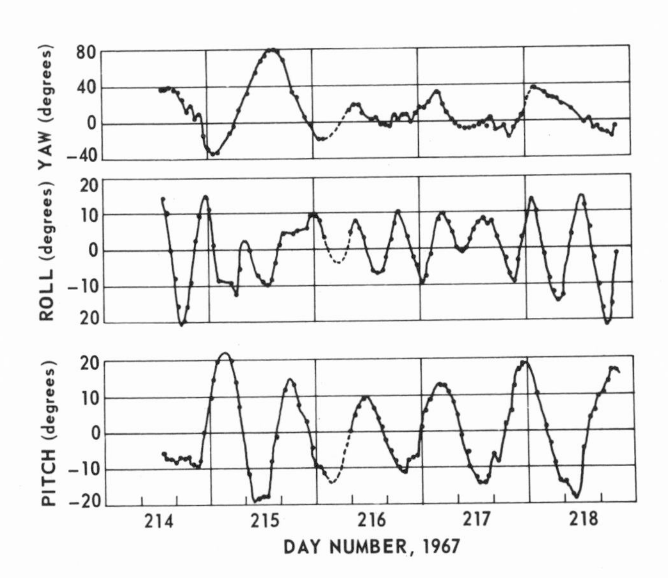

Fig. 9. Gravity-gradient stabilization of DODGE.

Fig. 10. Earth as seen from the DODGE satellite.

is used to energize a winding around the toroid so that the toroid becomes partially magnetized.

The Hall sensor detects the magnetic flux in the toroid and generates a signal proportional to the magnetization of the toroid. Thus the output depends on the input with the required hysteresis. Figure 7 shows the typical hysteresis loops obtained with this system.

This concept of enhanced (or artificial) magnetic damping was incorporated in the DODGE satellite, shown in Figure 8. This satellite was launched on July 1, 1967. Four booms were extended ~150 ft from the satellite to achieve the gravity-gradient stabilization torque. Figure 9 shows the roll, pitch and yaw angles achieved during one period of stabilization. Roll and pitch angles are within ± 20 degrees, which is "fair" stabilization. Yaw angles approach 80 degrees indicating that yaw is not well stabilized, as was expected. DODGE was successful in proving the feasibility of gravity-gradient stabilization at synchronous altitude—it also showed that the accuracy was not sufficient for most communication satellite applications.

We were pleased to get the first color picture of the full earth sphere with DODGE. A b&w version is shown here in Figure 10.

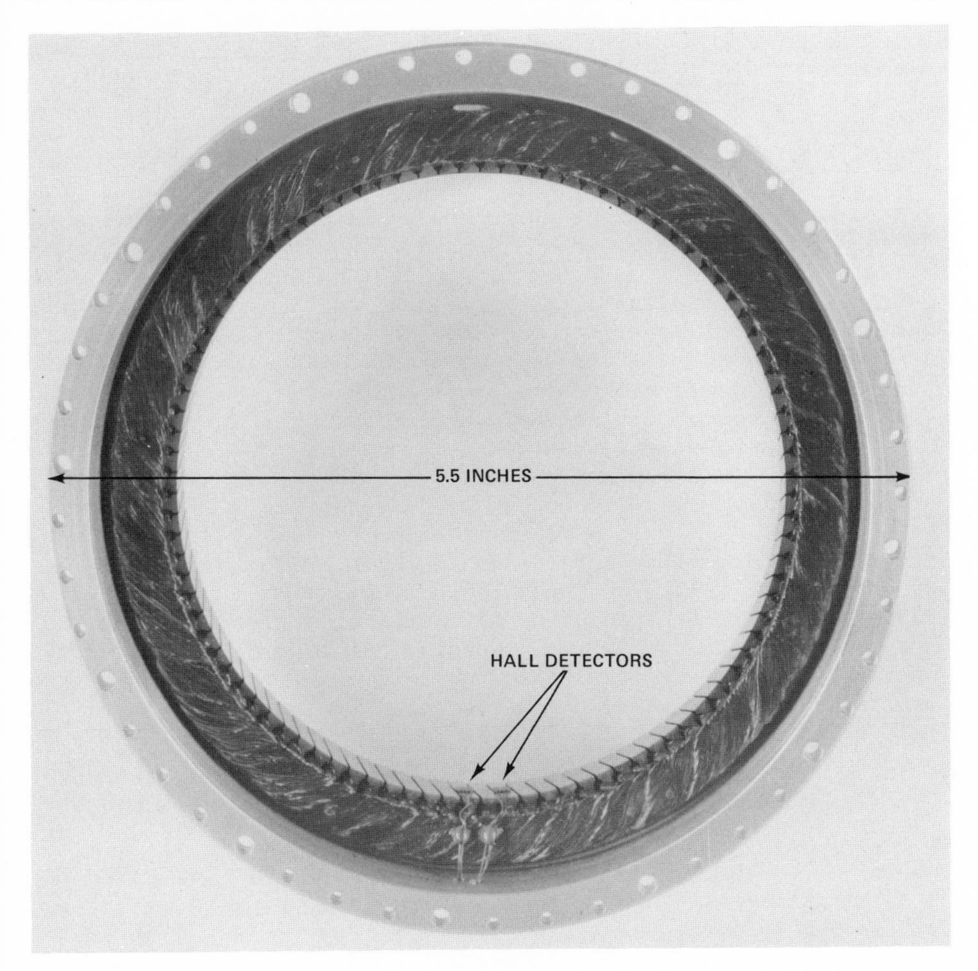

Fig. 11. Stator of the AE-C satellite wheel showing Hall detectors.

The small circle in the upper portion of the picture is a sphere on the end of a short boom which can be seen in Figure 8. The color picture was printed in National Geographic, November 1967.

Enhanced magnetic damping with Hall effect detectors was also used on the RAE-A satellite.[4]

THE USE OF HALL DETECTORS IN BRUSHLESS DC MOTORS

DC electric motors for atmospheric applications typically use graphite brushes and a segmented commutator to alternate the input current to the separate electrical windings of the rotor. Graphite is a good electrical conductor and is inherently self-lubricating so it makes a good choice for brush material but only in the environment of the earth's atmosphere. The lubricating properties of graphite depend on absorbed water vapor—in a vacuum environment, such as a satellite orbit, the water vapor is gradually lost, the graphite brushes loose their lubricating properties and rapidly wear away resulting in motor failures. One solution to this problem for space applications is to use a magnet on the rotor and Hall detectors on the stator which respond to the magnet as it passes by and electronically switch stator windings on and off in response to the Hall detector signal. These are called "brushless" DC motors and are very important in space applications. RCA-Astronautics has used Hall effect detectors on brushless DC motors in a number of spacecraft, including the Atmosphere Explorer-C, D, and E satellites the first of which was launched in December 1973. A photo of the stator of the AE wheels is shown in Figure 11. The two Hall detectors respond to the permanent magnetization of the rotor and the stator windings are switched on the basis of the Hall detector outputs. This wheel is used in attitude control. Another application of brushless DC motors is in the RCA "SATCOM" satellite—a synchronous satellite designed for communication purposes. Figure 12 shows a cross-section of a SATCOM reaction wheel which uses 3 Hall detectors to switch the 3 phases of this motor. The reaction wheels in SATCOM are used in a closed-loop system of attitude control. Wheels of this type are also used in the Defense Meteorological Satellite Program (DMSP), TIROS-M, and ANIK-B satellites. They are capable of speeds up to 10,000 rpm.[5]

MAGNETIC TRIM SYSTEM OF THE SAS-A AND B SATELLITES

SAS-A was launched in 1970 with soft X-ray detectors to scan the sky for new X-ray sources (see Fig. 13). It was christened "Uhuru" which is Swahili for freedom because it was launched from the coast of Kenya on Independence Day. SAS-B carried a gamma ray telescope—it was launched in 1972. Both satellites required stable spin axis orientation in space. APL designed and developed the "magnetic trim system" to permit in-orbit change in the satellite net magnetic dipole moment to counteract spurious magnetic dipoles which would cause the satellite attitude to drift. Three small Alnico magnets are used—these can be magnetized by capacitor discharge as shown in the diagram of Figure 14. Small Hall detectors are glued to the end of each magnet and the outputs are telemetered to the ground to confirm the magnet strength. The Hall detectors were the FWBell BH702 model with active area of .080 x .180 inches and thickness of .023 inches. Figure 15 shows the assembled flight package for SAS-A with magnets, and electronics in a foamed module. Figure 16 is a typical calibration curve for the X axis magnet as obtained by a sequence of pulse dischagres. Note that there are two different calibration curves depending on the direction of change from a saturated condition.

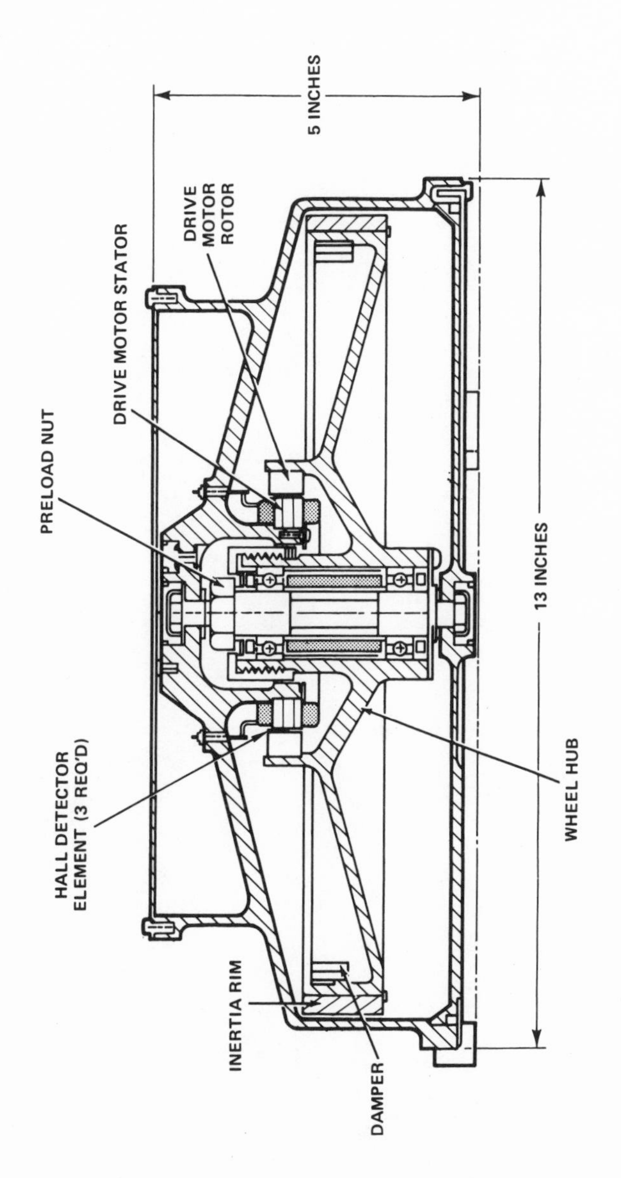

Fig. 12. Cross-section of the SATCOM wheel showing Hall detector placement.

Fig. 13. The SAS-A satellite.

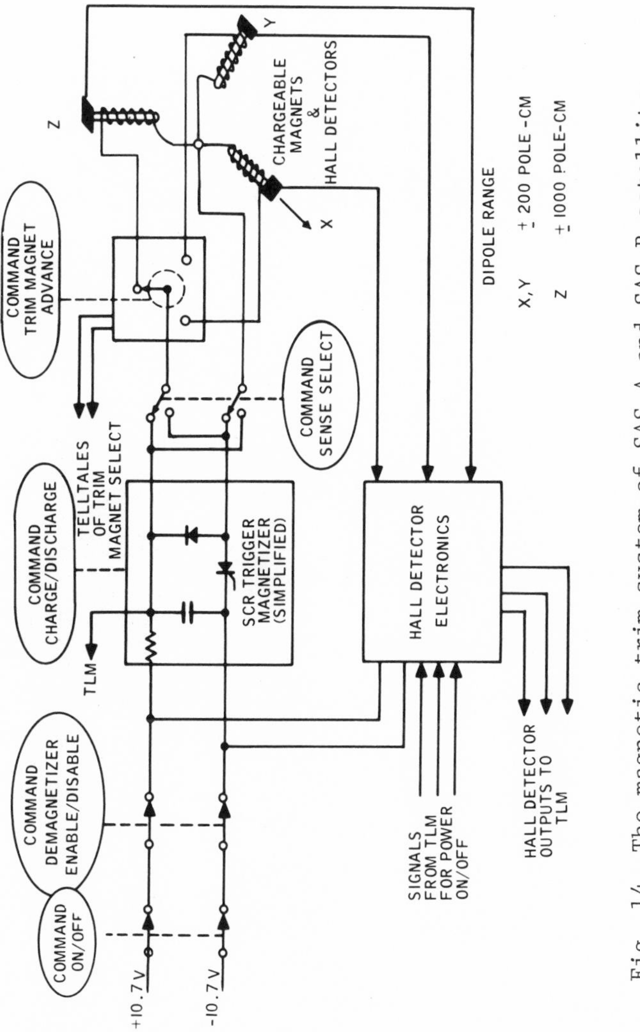

Fig. 14. The magnetic trim system of SAS-A and SAS-B satellites.

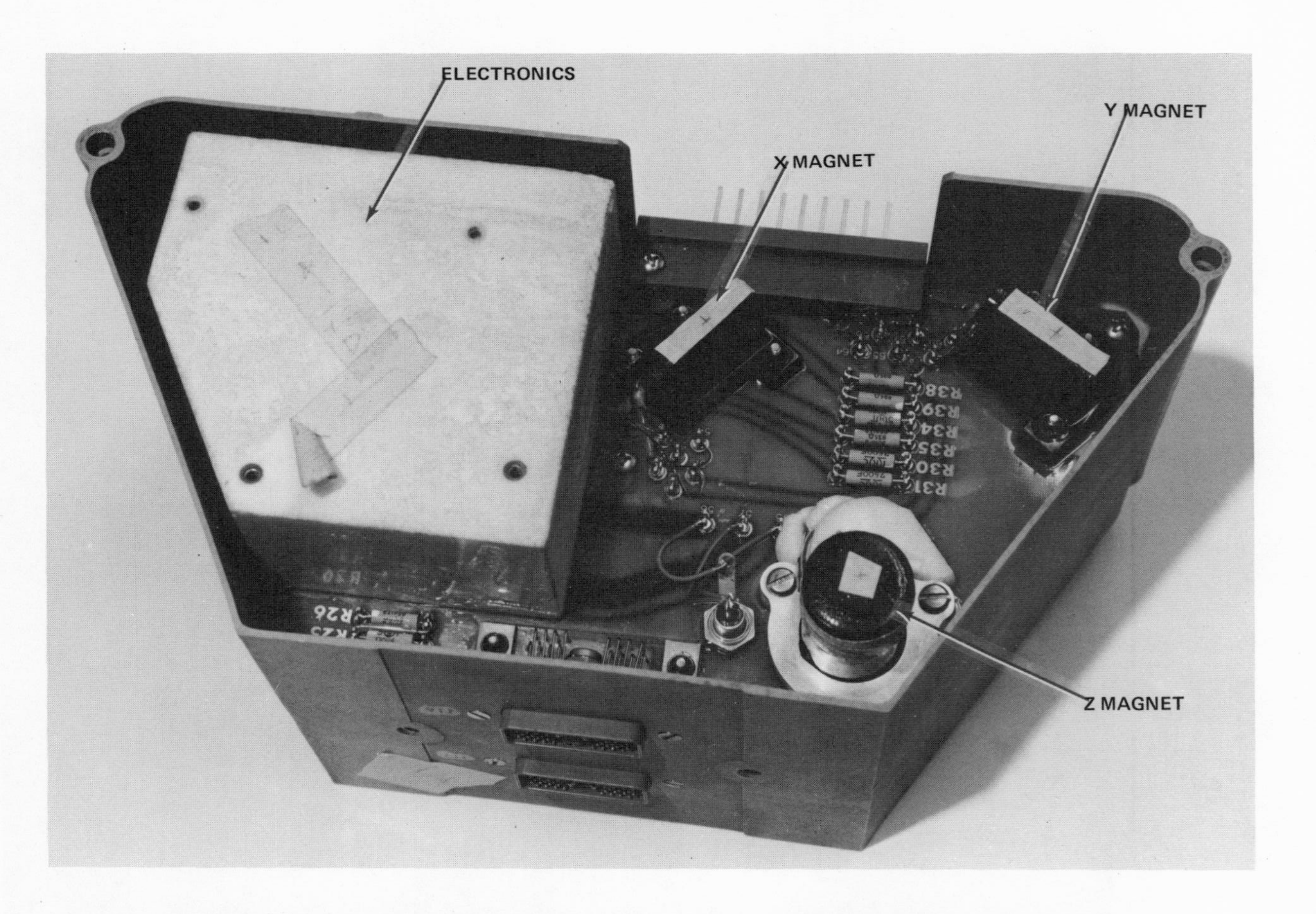

Fig. 15. Magnetic trim system for the SAS-A and SAS-B satellites.

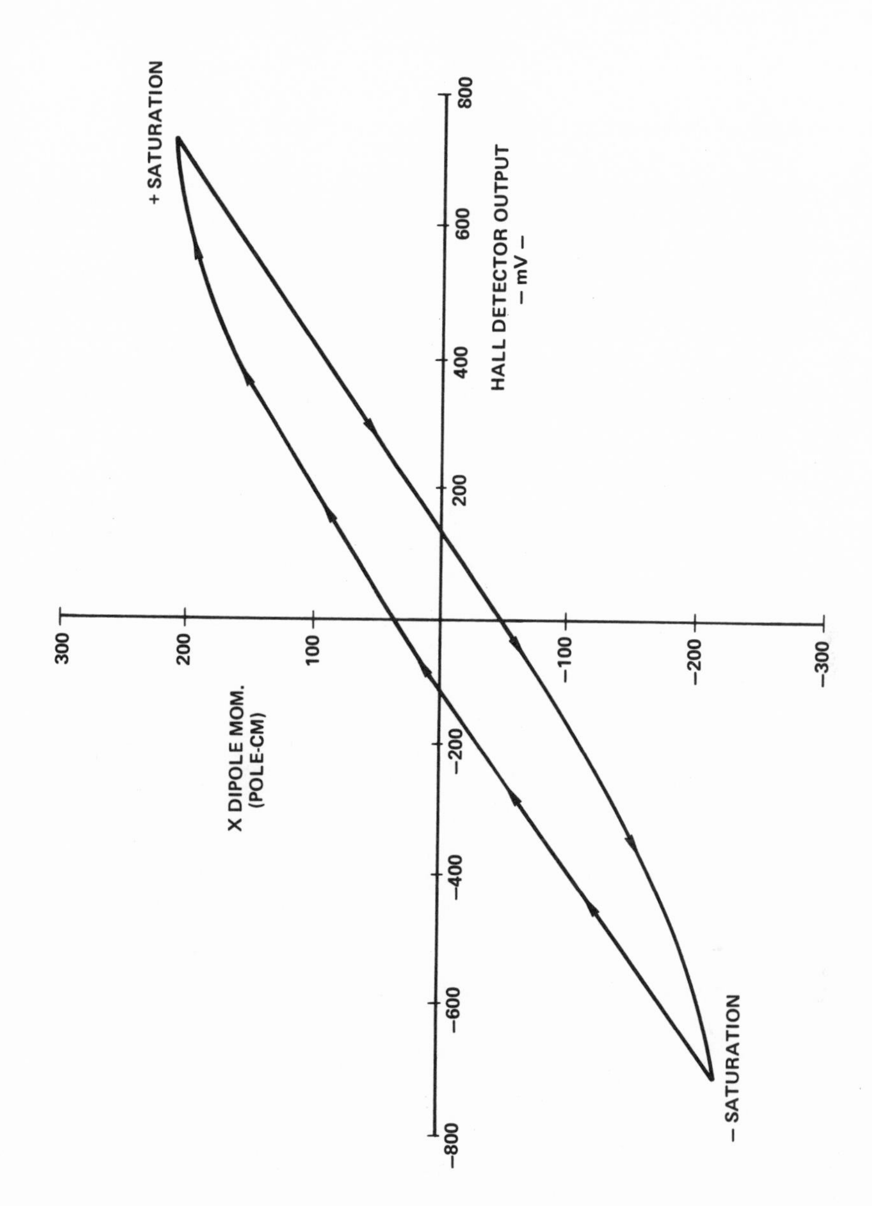

Fig. 16. Hall detector output vs magnet strength for the SAS-A satellite.

The X and Y magnets have maximum strength of 200 pole-cm and the Z magnet is 900 pole-cm magnetic dipole moment at full magnetization. One of the attractions of this scheme for dipole compensation is that no further electric power is required once the desired magnet strength has been reached.

MAGNETIC MOMENT COMPENSATOR

This system has been developed by Ithaco, Inc. for trimming large satellite residual magnetic dipole moments. The latest version is shown in Figure 17. It was used on the MEAMA satellite (P72-2) for the U. S. Air Force to give improved yaw accuracy. Three chargeable magnets, each 1 cm x 30 cm are arranged orthogonally in the satellite. The alloy for the core is chosen for its permanent magnet characteristics and its ease of magnetization by electric current in solenoid windings around each core. GE P-6 alloy (vanadium-cobalt-iron alloy) or Bell Labs "Remendur" can be used. Hall effect detectors (from FWBell) are attached to an end of each core. The satellite operator sends a command to the system to change the magnetic dipole. A change of ~300 pole-cm is accomplished by driving the solenoid until the Hall detector output changes by a pre-determined value—then the current is switched off automatically. The polarity of the current can be reversed by command. Each magnet can be driven over a range of ±14,000 pole-cm in this fashion. The Hall detector output is also telemetered to the ground for determination of the magnet strength.

Using the Hall detector at the end of the magnet does result in a hysteresis loop in its response characteristic as noted above in a similar application in the SAS-A and SAS-B satellites. This makes the precise value of the dipole somewhat in doubt, however, in the present application this is not critical since the ground controlled is more interested in adjusting the magnet strength until the disturbances in satellite attitude are minimized.

USE OF HALL DETECTORS IN THE DISCOS DAMPING MAGNET DEVELOPMENT

One of the unique and exotic devices developed for satellite use is the "drag-free" control system pioneered by Dr. Dan DeBra of Stanford University and built and flown by APL for the first time on the TRIAD satellite in 1972.[6] A small sphere of a gold-platinum alloy is allowed to float freely in a chamber inside the satellite. As the sphere approaches one side of the cavity its proximity is detected by capacitance change, and thrusters are fired to push the satellite away from the sphere. In this way the satellite is made to follow the sphere (or "proof-mass"). Since the sphere is inside the satellite it is protected from external drag forces and solar radiation pressure, therefore it would proceed

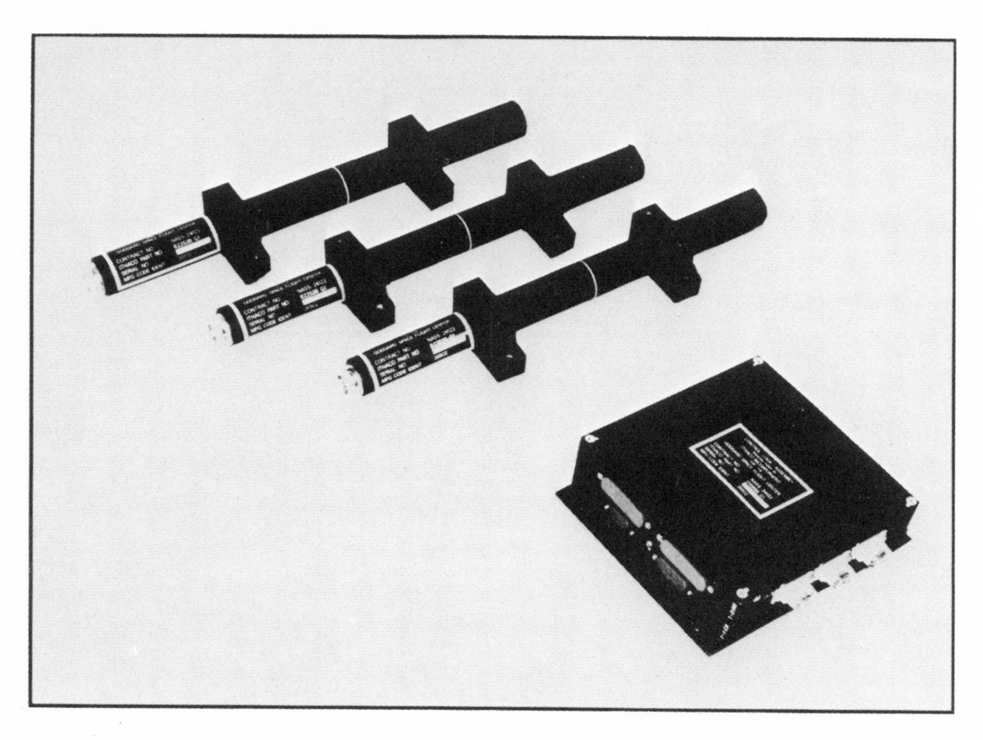

Fig. 17. Magnetic dipole moment compensation system.

in a "drag-free" orbit, governed only by gravitational forces. The objective here is to achieve a satellite orbit which is highly predictable—the atmospheric drag and its variability with latitude, time-of-day, and solar activity is the major cause of the very small but important unpredictability in the satellite orbit—which limits the accuracy of navigation by satellite.

This system is known at APL by the acronym "DISCOS" for Disturbance Compensation System. For the TIP-II and III satellites (the successors to TRIAD) we designed a "single-axis" DISCOS system to replace the 3 degree of freedom system of TRIAD.[7] The single axis DISCOS proof-mass is free to move only along one axis, and that axis aligned with the flight path. The two transverse motions are constrained by a frictionless suspension. Figure 18 shows a cross-section of the single-axis DISCOS sensor. The proof-mass is a hollow cylinder of aluminum about 1" in diameter, suspended from a slender rod by eddy-current repulsion due to AC electric currents in the rod at a frequency of 2 kHz. The proof mass is free to move along the rod axis without restraint. The axial position of the proof mass is measured with a light beam and detector scheme, and teflon thrusters are fired to move the satellite to keep the proof-mass centered.

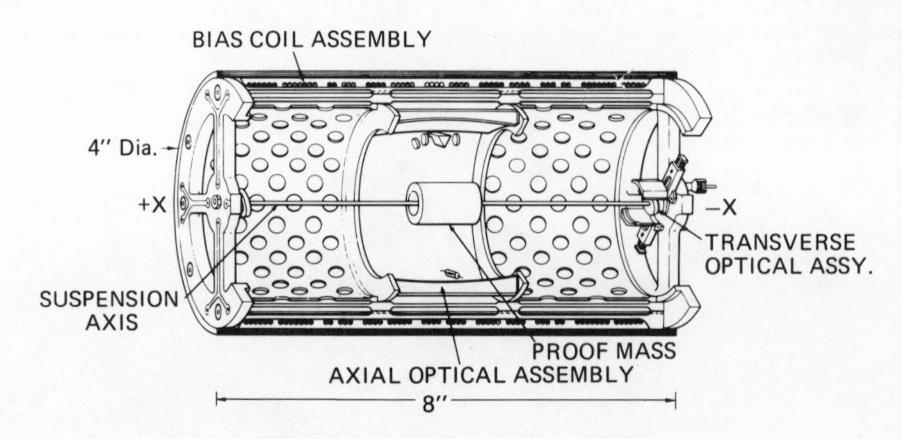

Fig. 18. Single axis DISCOS.

The configuration of the TIP-II and III satellites is shown in Figure 19. TIP-II was launched in 1975 and TIP-III in 1976. Our flight results with TIP-II and TIP-III showed that the proof-mass would occasionally begin to move in the transverse direction, overcome the eddy-current suspension, and make contact with the wire. After extensive analysis we concluded that increased damping of transverse oscillations was required.

Damping of the transverse oscillations will be improved by inserting a slender permanent magnet inside the center rod. This magnet is magnetized transverse to its long axis. Transverse motion of the proof-mass will cause eddy-currents to be developed in the proof-mass, the energy of the oscillation will be dissipated in heat, and the motion damped.

The transverse magnetization of this damping magnet must be quite uniform so as not to affect the freedom of the proof-mass to move without internal forces along the suspension axis. The magnet is 9.6 inches long x .026 inches in diameter of a platinum-cobalt alloy. It must be magnetized to a strength of 1 pole-cm/cm with a uniformity of magnetization of ~1% or better.

Since the diameter of the magnet is quite small we require a detector which is comparably small to measure the uniformity of magnetization. The Hall detector was the ideal choice. We used the FWBell model FH-301-40 detector shown in Figure 20. It is made of a 1 micron thick layer of indium arsenide, in a rectangle of .04 x .08 inches.

Figure 21 shows how the Hall detector was mounted within a few thousandths of an inch of a beryllium-copper tube with the damping

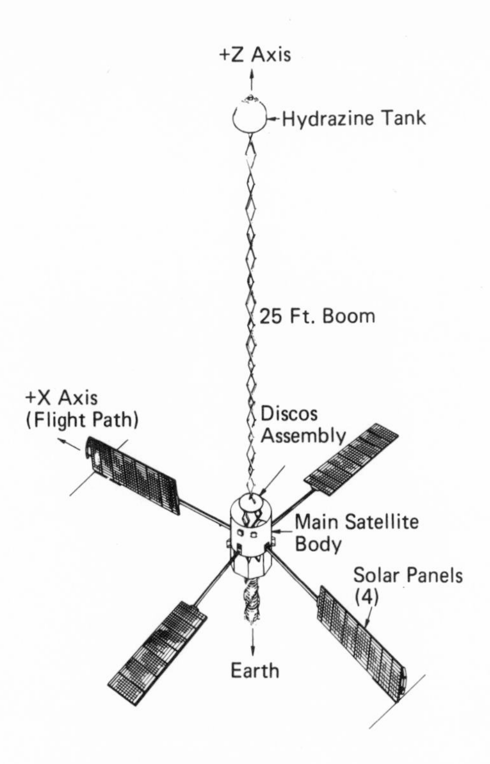

Fig. 19. The TIP-II satellite.

magnet inside. The tube is mounted in a lathe and the Hall detector is driven by the tool holder to scan along the rod axis as the rod is rotated in the lathe. Figure 22 is a photograph of the lathe set-up. Figure 23 shows a typical output of the Hall detector during a complete scan along one prototype magnet. The transverse magnetization of the magnet produces a periodic variation in the Hall detector output as the magnet rotates in the lathe. An interesting phenomenon occurs near the ends of this magnet. The average value of the output deviates from zero, then returns to zero as the detector gets well off the end of the magnet. We ascribe this deviation to some partial axial magnetization of the rod—the flux from the axial magnetization does not vary as the magnet rotates, it simply biases the oscillation to one side or the other. Axial magnetization is also undesirable in this application.

Very uniform magnetization was achieved by magnetizing the magnets with the discharge from a 60 μfd capacitor charged to 50 kilovolts into aluminum rods that were placed side by side against the magnet. This concept is shown in Figure 24.[8] The intense magnetic flux is forced through the narrow gap containing the magnet.

INDIUM-ARSENIDE, 1 MICRON THICK

HALL PLATE
0.040 x 0.080

0.125

INPUT RESISTANCE, R_{in}	40-80 Ω
OUTPUT RESISTANCE, R_{out}	2.2 R_{in} approx.
MAGNETIC SENSITIVITY γ_B, MIN. @ I_{cn}	12.0 mV/kG
PRODUCT SENSITIVITY, γ_{1B}, MIN	0.8 V/A · kG
RESISTIVE RESIDUAL VOLTAGE, V_M @ I_{cn}, B = 0	6 mV max.
NOMINAL CONTROL CURRENT, I_{cn}	15 mA
MAXIMUM CONTINUOUS CONTROL CURRENT, I_{cmos}	30 mA
MEAN TEMPERATURE COEFFICIENT OF V_H (−20°C to +80°C), β_T	−0.1%/°C max.

Fig. 20. Hall detector model FH-301-040.

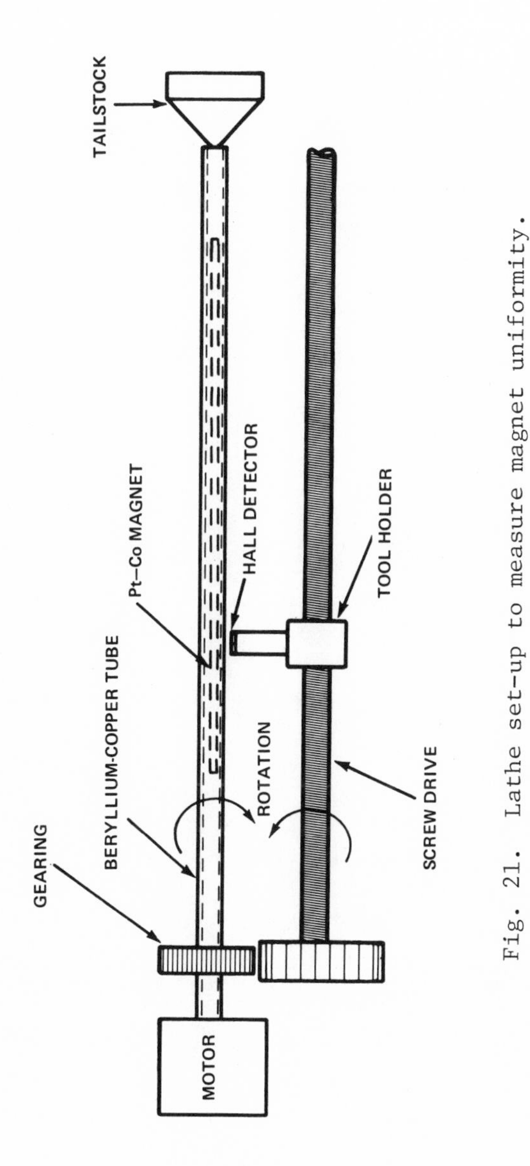

Fig. 21. Lathe set-up to measure magnet uniformity.

Fig. 22. Lathe set-up for measuring magnet uniformity.

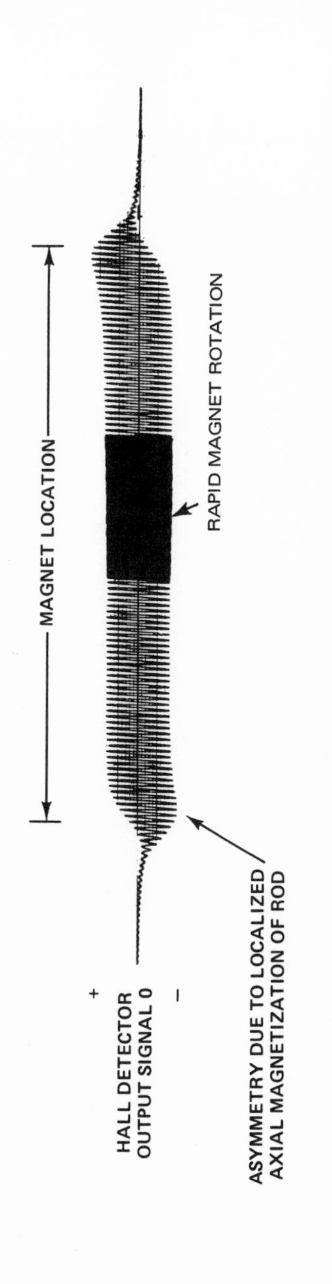

Fig. 23. Hall output signal while rotating the magnet and moving the detector horizontally.

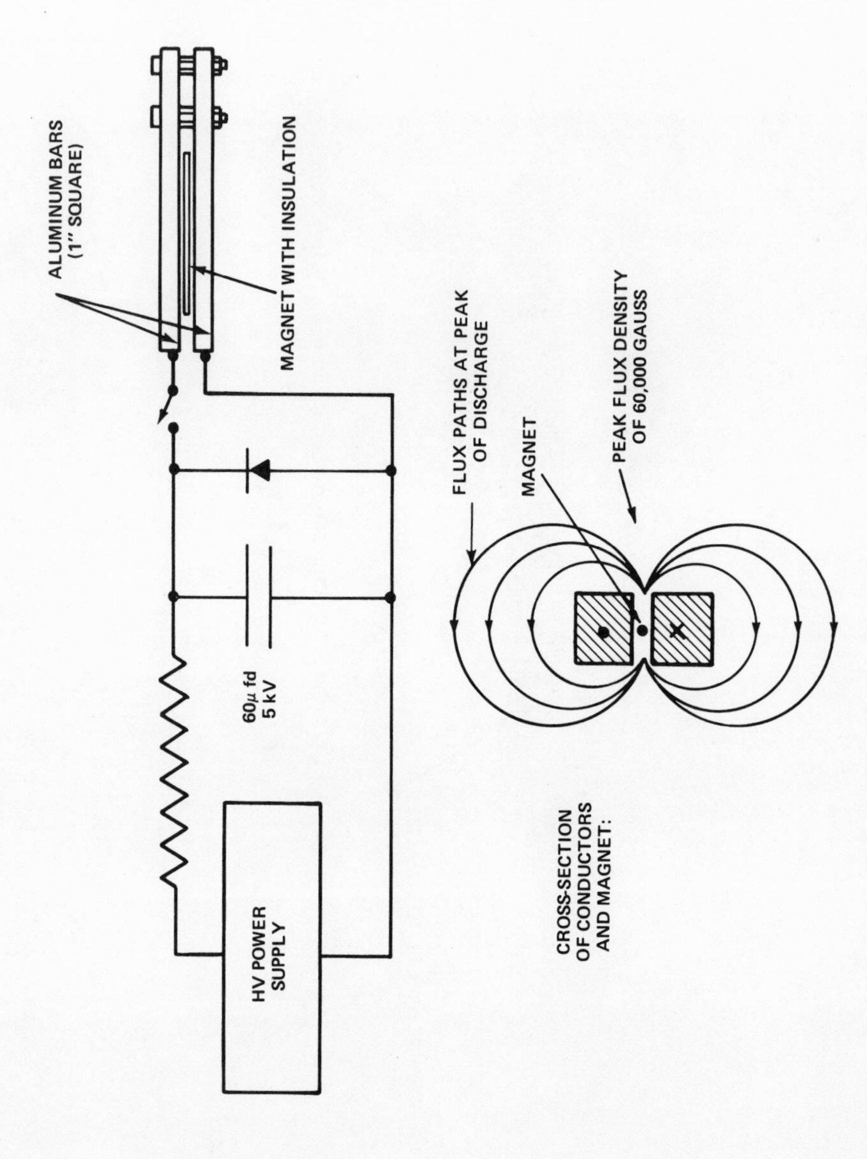

Fig. 24. System for intense magnetization of magnet to achieve best uniformity.

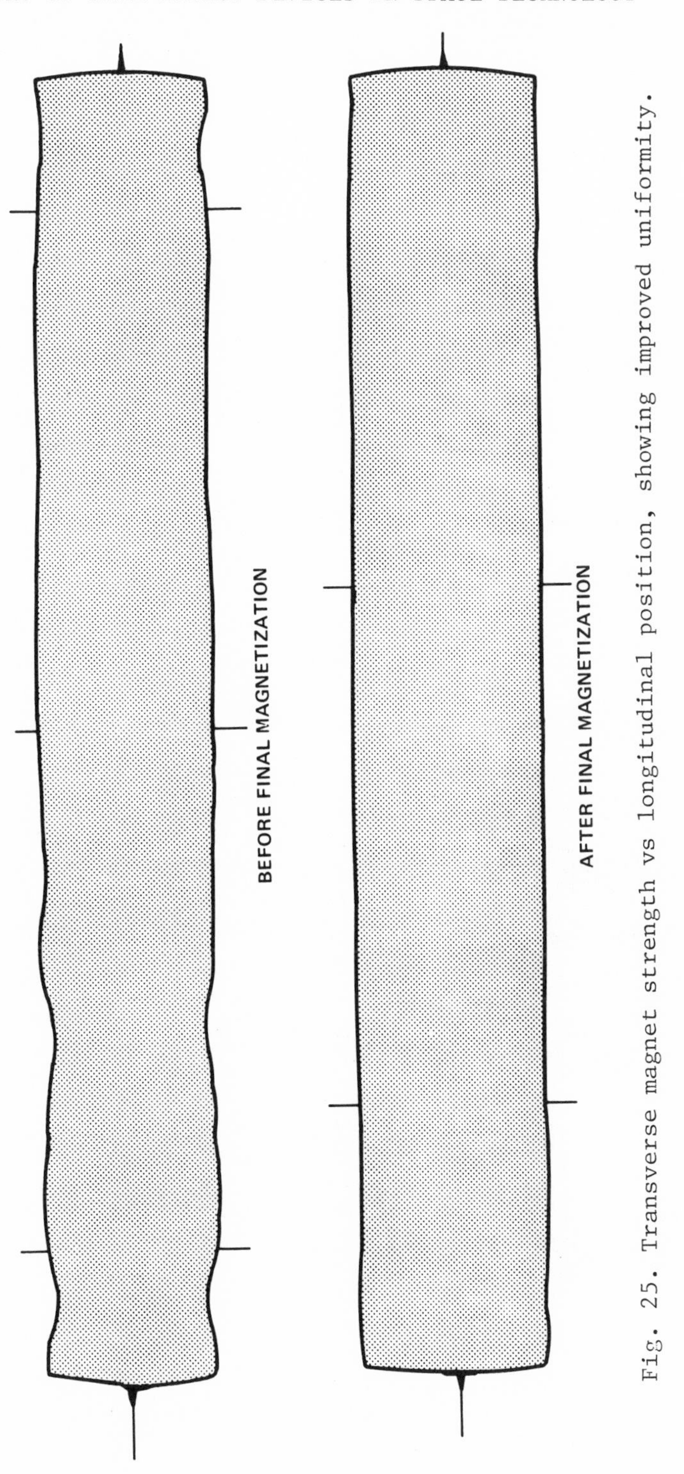

Fig. 25. Transverse magnet strength vs longitudinal position, showing improved uniformity.

Fig. 26. The MAGSAT satellite.

Figure 25 shows a comparison of the uniformity, before and after final magnetization, achieved with this technique. Note the significant improvement in uniformity.

The damping magnets are now being incorporated in the successors to the TIP satellites, called "NOVA"; the first is to be launched in 1980.

In summary it is clear that Hall effect detectors have found important applications in space technology. Perhaps the greatest number of such applications has been in brushless DC motors.

The applications of Hall effect devices has been only one aspect of the interest in magnetic fields in space technology. Figure 26 shows a new satellite for magnetic field research—the MAGSAT satellite, which was launched on October 30, 1979. This satellite has two very accurate magnetometer sensors at the end of a 20 ft scissors boom. One sensor is a three axis vector magnetometer, the other is a cesium vapor scalar magnetometer. Each sensor has an accuracy on the order of 10^{-5} gauss. This satellite was built by APL for the National Aeronautics and Space Administration. The vector magnetometer was built by Goddard Space Flight Center and the scalar magnetometer by Ball Bros and Varian of Canada. The purpose of this satellite is to make the most accurate satellite survey ever obtained of the earth's magnetic field for field mapping, geological prospecting, and refinement of mathematical models of the field.

REFERENCES

1. R. E. Fischell, "Magnetic and Gravity Attitude Stabilization of Earth Satellites," Applied Physics Laboratory Report CM-996, May 1961.
2. R. E. Fischell and F. F. Mobley, A System for Passive Gravity-Gradient Stabilization of Earth Satellites, in: "Progress in Astronautics and Aeronautics," Academic Press, Inc., N. Y., (1964), Vol. 13.
3. F. F. Mobley, "Gradient-Gradient Stabilization Results from the DODGE Satellite," AIAA 2nd Communications Satellite Systems Conference, April 1968.
4. B. E. Tossman, "Magnetic Attitude Control System for the Radio Astronomy Explorer-A Satellite," Paper No. 68-855, AIAA Guidance, Control and Flight Dynamics Conference, August 1968.
5. Private communication from L. Muhlfelder, RCA-Astronautics, Hightstown, New Jersey.
6. J. Dassoulas, The TRIAD Satellite, APL Technical Digest, 12:2 (April-June 1973), Applied Physics Laboratory, Laurel, Maryland.

7. F. F. Mobley, G. H. Fountain, A. C. Sadilek, P. W. Worden, Jr., and R. VanPatten, Electromagnetic Suspension for the TIP-II Satellite, IEEE Transactions on Magnetics, MAG-11:6 (November 1975).
8. Designed by W. R. Powell of APL.

E. H. HALL AND PHYSICS AT HOPKINS: THE BACKGROUND TO DISCOVERY

O. Hannaway

History of Science Department
The Johns Hopkins University
Baltimore, Maryland 21218

Edwin Hall's experiment, performed almost exactly one hundred years ago, had an elegant simplicity to it. A current from a carbon-zinc battery was passed through a strip of gold foil (2 cm. x 9 cm.) fixed firmly on a glass plate by means of brass clamps. The plate bearing the gold leaf was placed between the poles of an electromagnet is such a way that the lines of magnetic force passed perpendicularly through the horizontal plane of the foil. Opposite edges at the mid-point of the gold foil strip were tapped to a high-resistance galvanometer in order to detect any influence of the strong magnetic field on the current flowing through the gold foil. The results indicated the existence of an electromotive force at right angles to the direction of the primary current and perpendicular to the magnetic field. It was this transverse potential, produced by the action of an external magnetic field on a permanent current, which became known as "the Hall effect."

Hall, then a second-year graduate student in physics at the Johns Hopkins University, quickly established priority in a paper entitled: "On a New Action of the Magnet on Electric Currents," published before the end of the year (1879) in the *American Journal of Mathematics*.[1] This journal was one of the new organs of science and scholarship established at Johns Hopkins as outlets for the work of the new research-oriented university. It was edited by the university's professor of mathematics, J. J. Sylvester. In the course of 1880, Hall worked up his initial investigations into a Ph.D. thesis which appeared in published form towards the end of the year in the *American Journal of Science*,[2] and in the British scientific periodical the *Philosophical Magazine*.[3] This was the second Ph.D. awarded by the Physics Department at Johns Hopkins.[4]

But notice of Hall's research was not confined to the scientific press. The leading New York weekly magazine, The Nation, carried a paragraph on it in its issue of Christmas Day, 1879.[5] Here the discovery was hailed as one of the most important made in the field of electricity in the past fifty years, and stress was laid on the fact that it was not the result of serendipity but the designed outcome of elaborate and delicate experiments carried out under the direction of Edwin Hall's professor Henry Augustus Rowland. This early effort at scientific reportage concluded with the sentence: "The new force is exceedingly feeble, so that we cannot predict any practical applications for it."

In what follows I would like to set the background of Hall's discovery with particular reference to the character of the early Physics Department at Hopkins and the work of its founder Rowland. In this I would like to acknowledge my debt to John David Miller, whose 1970 Oregon State dissertation on Henry Rowland is indispensable for the study of the first phase of Hopkins physics,[6] and to Robert Rosenberg, a graduate student in our own department of the History of Science, who is currently exploiting some new sources to throw further light on electrical studies at Hopkins in this period.[7]

Edwin Hall came to Baltimore as a graduate student from Maine, where his family's roots stretched back into the seventeenth century.[8] He had attended college at Bowdoin where he graduated at the top of the class in 1875. Two years of schoolteaching followed, before Hall's interests turned to science. He looked first to Harvard, but John Trowbridge, professor of physics there, directed his footsteps south to Baltimore where just the year before Daniel Coit Gilman had launched his great experiment in higher education at The Johns Hopkins University. Trowbridge was well informed on this venture, for he had been approached by President Gilman to join the faculty at Hopkins, but had declined for family reasons.[9] Trowbridge's initial deference to Hopkins in the matter of Hall was well rewarded; after the triumphs of Hall's graduate career, it was Trowbridge who successfully negotiated with Gilman to bring him to the faculty at Harvard.[10]

Of the transition from schoolmaster to physics student, Hall has been reported as saying: "I turned to science, after two years of schoolteaching, because it was progressive and satisfied my standards of intellectual and moral integrity, not because I had any passionate love of it or felt myself especially gifted for scientific undertakings."[11] Both the sentiment and the candor of this statement ring a little oddly to twentieth-century ears; but science as progress with moral integrity was very much part of the ethos of early Hopkins. Gilman, Rowland and the first professor of chemistry, Ira Remsen, all expressed this ideal frequently in words that echoed the long heritage of Protestant dissent which they shared with Edwin Hall. The explicit mission of Hopkins, as stated by its leaders, was to ennoble American society by introducing to it the measured, rational advance of research, thereby rescuing it from the helter-skelter scramble for material goods and advantage as represented by Tammany in the north, carpetbagging in the south, the railroad pushing to the west and unrestrained entrepreneurship everywhere.[12]

But what, more specifically, of the physics department which Hall joined in 1877? Here the formative and controlling influence was that of Henry Augustus Rowland, whose education as a physicist was in sharp contrast to that which would later be offered to Hall.[13] Rowland was formally trained as a civil engineer at Rensselaer Technological Institute where he graduated in 1870. In his sophomore year, however, he had determined to devote himself not to practical technology, but to "pure science" (a favorite expression of Rowland),[14] and henceforth his real education took place in his lodgings and not in the classroom. Beginning in 1868, Rowland steadily built a personal library in physics, took detailed notes on his reading and experimented ingeniously within the modest limits of his circumstances. Much of his work centered around a detailed study of Faraday's text, _Experimental Researches in Electricity_.[15] This independent program of study and research continued past graduation and into fitful periods of employment at Wooster College (Ohio) and back at Rensselaer, until Rowland finally brought his mathematical analysis of Faraday's lines of magnetic force to such a level that he could submit his results directly to James Clerk Maxwell. The year was 1873: Maxwell's _A Treatise on Electricity and Magnetism_[16] had just appeared. Prompt publication at the urging of Maxwell of two papers on magnetic permeability in the _Philosophical Magazine_[17] set the seal on this amazing autodidactic transformation from engineering sophomore to theoretical physicist of international standing.

It was this growing reputation that prompted Gilman to recruit Rowland to the Hopkins faculty in 1875, one year in advance of the planned opening of the university. Gilman took his new "find" to Europe with him in the summer of 1875 to strengthen Rowland's contacts with the leading European physicists and to allow him to study the lay-out, equipment and organization of the major laboratories.[18] Rowland also took the opportunity to assess his own future as a physicist. It lay, he believed, "midway between the purely mathematical and [the] purely experimental, and in a place where few are working."[19] Exactly where Rowland demonstrated that winter, when having stayed over to work in Helmholtz's laboratory in Berlin, he performed there an experiment conceived and designed in his period at Rensselaer. This was his famous experiment of rotating a gilded disc carrying an electrostatic charge in an effort to determine whether or not a moving charge produced a magnetic effect similar to that produced by a current moving in a conductor. His results were positive, although they were subsequently contested. Rowland would refine and repeat this experiment with graduate students several times in the course of his career at Hopkins.[20]

During that same winter in Europe, Rowland began choosing and ordering the instruments and equipment for the Hopkins laboratory. This was a matter on which Rowland was insistent in two respects: first, that the instruments be of the highest quality and designed for research, not teaching or demonstration; and second, that the Trustees foot the bills, no matter what. Already by Hall's time in Baltimore, it was widely conceded that Rowland's laboratory was as well equipped as any, especially in the area of electricity and magnetism. The place where this excellent equipment was housed, however, left something to be desired. For almost ten years the Physics Department occupied several rooms in an annex extending from the rear kitchen of one of the two converted boarding houses which were the first buildings of The Johns Hopkins University. It was in these facilities that Hall worked as a graduate student.[21]

We know something of Hall's course preparation before he began his research. He took Rowland's advanced courses in the theory of heat conduction, and on electricity and magnetism: he studied the calculus with Thomas Craig, the young Pennsylvanian who worked under Sylvester in the Mathematics department; and there was also significant course and laboratory work in chemistry, reflecting the close relationship of the two departments in those early years. As an advanced student, Hall was expected to spend the balance of his time in the Physics laboratory and to attend the journal club which Rowland conducted personally.[22] A careful reading of the first account of Hall's discovery reveals how the experiments which brought him fame and his Ph.D. stemmed naturally from this preparation. As Professor Judd explains elsewhere in this volume, the inspiration for Hall's experiment was a passage in Maxwell's _A Treatise on Electricity and Magnetism_, the text of Rowland's advanced course.[23] It was the apparent conflict of this passage with views expressed by the Swedish physicist Erik Edlund in a paper in the _Philosophical Magazine_ of 1878,[24] which suggested the idea of the experiment to Hall. Who can doubt that he read this paper in the context of the journal club?

But if we examine Hall's experiment itself, we can see the continuity with Rowland's own researches on electromagnetism. Here again, Edwin Hall's remarkable candor is our guide. He tells us that his successful experiment was the third procedure he adopted to investigate the problem as he conceived it; that is, whether or not a current flowing in a fixed conductor was acted upon by a magnetic field independently of the effect on the conductor itself.

The first procedure, suggested by Hall himself, involved an attempt to displace a current flowing in a triangularly-drawn silver wire; the second was a modification of an experiment developed by Rowland in which it was planned to investigate the effect of a magnetic field on an electric current moving across the surface of a gilded disc; and the third procedure, suggested, as Hall tells us, by Rowland himself, was the one described at the beginning of this paper.[25] It is the second unsuccessful experiment--the one with the disc--which gives the clue to the conceptual and experimental antecedents to the discovery of the Hall effect, for in it we can see clearly the link with Rowland's Berlin experiment of 1875. That earlier experiment had sought to detect the magnetic effect of a static charge located on a moving conductor, namely a <u>rotating</u> disc covered with gold leaf: the "Hall experiment" reversed that procedure by attempting to detect the influence of a magnetic field on an electric charge flowing through a <u>fixed</u> conductor, in this case a strip of gold leaf clamped to a glass plate.[26] At the back of both experiments lay a fundamental question--still problematic in the era before the discovery of the electron--namely, what was the fundamental nature of electricity itself? Despite the formal triumph of Maxwell's electromagnetic theory, the phenomena of electricity still lack conceptual unity; at this level questions remained as to the relationship between a moving electric charge, a current flowing in a fixed conductor such as a wire, and the chemical processes taking place in a battery which produced such a current.

In emphasizing Hall's debt to Rowland, both in terms of institutional facilities as well as theoretical and experimental orientation, it has not been my aim to weigh their individual contributions and claims to an important discovery; but rather to point up the novelty of the enterprise of which they were both an indispensable part. We have so come to accept the process whereby a student is led to the frontier of knowledge through advanced course work, initiated into the techniques of investigation in the laboratory, and made part of an ongoing research program whose object is to publish new findings, that we forget that once it was all quite radically new. In America its genesis was here at Hopkins. In commemorating the centennial of the Hall effect we do more than mark a notable discovery; we celebrate the first product of an educational revolution which transformed the practice and prestige of physics in the United States.

REFERENCES

1 E. H. Hall, On a New Action of the Magnet on Electric Current, <u>Amer. J. Math.</u> 2:287 (1879). The account of Hall's experiment given above is from this paper, p. 290.

2. E. H. Hall, On the New Action of Magnetism on a permanent Electric Current, <u>Amer. J. Sci.</u> ser. 3, 20:161 (1880).

3. idem, Phil. Mag. ser. 5, 10:301 (1880).
4. The original version of Hall's thesis does not appear to have survived at Hopkins. The first Ph.D. in physics at Hopkins was awarded in 1879 to William White Jacques (1855-1932) for a thesis published as "distribution of Heat in the Spectra of Various Sources of Radiation," John Wilson & Son, University Press, Cambridge (1879); see also, Proc. Amer. Acad. Arts and Scis. 14:142 (1879).
5. The Nation, Dec. 25, 1879, p. 44.
6. John David Miller, "Henry Augustus Rowland and his Electromagnetic Researches," Ph.D. thesis, Oregon State University (1970). See also the following articles by Miller, based upon his thesis: Rowland and the Nature of Electric Currents, Isis 63:5 (1972); Rowland's Magnetic Analogy to Ohm's Law, Isis 66:230 (1975); and Rowland's Physics, Physics Today 29:39 (July, 1976).
7. Robert Rosenberg is making a study of electrical engineering at Hopkins and its relationship to the Physics Department. His knowledge of the newly accessioned Presidential Papers in the Archives of the Milton S. Eisenhower Library at Johns Hopkins was valuable to me on a number of points.
8. The fullest and best account of Hall's life remains P. W. Bridgman's obituary notice, Edwin Herbert Hall 1855-1938, in "Nat. Acad. Biogr. Mem.," vol. XXI (1939-40).
9. Miller, Thesis (1970), pp. 189-190. See also, Hugh Hawkins, "Pioneer: A History of The Johns Hopkins University," Cornell University Press, Ithaca (1960), pp. 45-46.
10. Miller, Thesis (1970), p. 282.
11. Quoted by Bridgman (ref. 8), pp. 74-75.
12. The best indications of these sentiments are the occasional addresses of the individuals involved. For Gilman see D. C. Gilman, "University Problems in the United States," New York (1898), for Rowland see the nontechnical papers in "The Physical Papers of Henry Augustus Rowland," The Johns Hopkins University Press, Baltimore (1902). No such collection exists for Remsen, but see O. Hannaway, The German Model of Chemical Education in America: Ira Remsen at Johns Hopkins (1876-1913), Ambix 23:145 (1976).
13. The following account of Rowland's education as a physicist is based largely on Miller, Thesis (1970), pp. 6-75.
14. See, for instance, Rowland's famous address before the AAAS in 1883, A plea for pure science, Proc. AAAS 32:105 (1883). The significance of this address is discussed in Daniel J. Kevles, "The Physicists: the History of a Scientific Community in Modern America," Alfred A. Knopf, Inc., New York (1977), pp. 43-44.
15. Michael Faraday, "Experimental Researches in Electricity," 3 vols., London, (1839-1855).
16. J. Clerk Maxwell, "A Treatise on Electricity and Magnetism,"

2 vols., Oxford (1873).
17. H. A. Rowland, On Magnetic permeability, and the maximum of magnetism of iron, steel, and nickel, Phil. Mag. ser. 4, 46:140 (1873); and idem, On the magnetic permeability and maximum of magnetism of nickel and cobalt, Phil. Mag. ser. 4, 48:321 (1874).
18. Rowland's trip to Europe is discussed in Miller, Thesis (1970), pp. 90-102.
19. Quoted in Miller, Thesis (1970), p. 97.
20. The "Berlin" experiment itself is described in H. A. Rowland, On the Magnetic Effect of Electric Convection, Amer. J. Sci. and Arts 15:30 (1878). The context and subsequent history of the experiment is fully discussed in Miller, Thesis (1970), pp. 114-182, and also in the same author's article in Isis 63:6 (1972).
21. Miller, Thesis (1970), pp. 183-205. For testimony on the instrumentation at Hopkins, see ibid., pp. 293-295.
22. This information can be gleaned from the class lists published in the University Circulars. See The Johns Hopkins University Circulars December 1879 - September 1882 (1882).
23. Hall, Amer. J. Math. 2:287 (1879). The passage from Maxwell is to be found in "A Treatise on Electricity and Magnetism," vol. 2 (1873), pp. 144-145.
24. E. Edlund, Unipolar Induction, Phil. Mag., ser. 5, 6:289 (1878).
25. All three experimental procedures are described in Hall, Amer. J. Math. 2:287 (1879). More detail concerning the second unsuccessful procedure using the gilded disc is given in Hall, Phil. Mag. ser. 5, 10:301 (1880), where no mention is made of the silver wire experiment. The fact that the silver wire was drawn through a triangular die is a detail provided by Miller, Thesis (1970), p. 253, who obtained it from Hall's laboratory notebook of the period which survives in the Houghton Library at Harvard University.
26. Miller has also drawn attention to the relationship of Hall's experiments to the "Berlin" experiment of Rowland. See esp. his article in Isis 63:19 (1972). He overlooks, however, the important clue provided by the second unsuccessful experimental procedure described by Hall.

THE HALL EFFECT IN THE CONTEXT OF NINETEENTH-CENTURY PHYSICS

B. R. Judd

Physics Department
The Johns Hopkins University
Baltimore, Maryland 21218

1. MAXWELL'S ASSERTION

Hall's motivation for carrying out the experiment whose result bears his name attests to the value of a close reading of an established text. Writing in a style that to our eyes seems remarkably open and unaffected, he reported[1] his surprise at reading Maxwell's statement[2] that "If the current itself be free to choose any path through a fixed solid conductor or a network of wires, then, when a constant magnetic force is made to act on the system, the path of the current through the conductors is not permanently altered, but after certain transient phenomena, called induction currents, have subsided, the distribution of the current will be found to be the same as if no magnetic force were in action." To Hall, this appeared "contrary to the most natural supposition," and, after consultation with Rowland, who had apparently already made some preliminary but unsuccessful attempts to detect the effects of a magnetic field on currents in conductors, he was able so to arrange matters that a positive effect was seen.

We can perhaps understand why Maxwell's erroneous assertion is not at first sight unreasonable by considering a linear superposition of two independent solutions to Maxwell's equations. In Fig. 1a, a metal conductor of rectangular cross section is shown carrying a current $\vec{J}$. The line joining the points s and t is perpendicular to a straight edge of the conductor, and it is obvious that the electric potentials at s and t are the same. In Fig. 1b, the same conductor is shown in the presence of a constant magnetic field $\vec{H}$ perpendicular to a plane face of the conductor. Again, it is obvious that the potentials at u and v are identical. Since Maxwell's equations are linear in all the fields, we can superpose

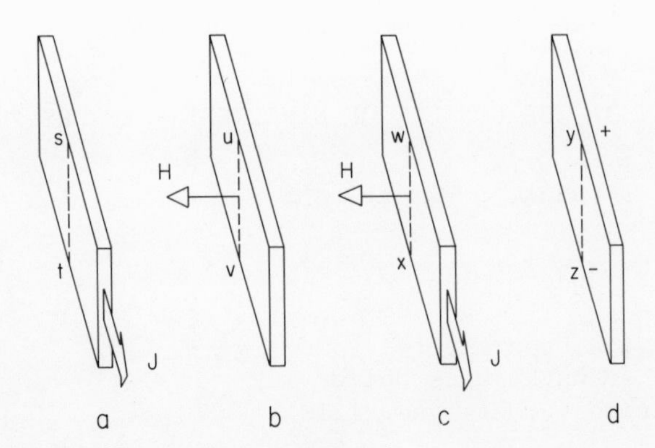

Fig. 1 An identical conductor subject to four electromagnetic environments. (c) represents the superposition of the two independent solutions (a) and (b); no Hall charges (indicated in (d)), appear in (c).

the two solutions. We thus obtain a situation where Maxwell's equations as well as all the boundary conditions are satisfied, as shown in Fig. 1c. Yet the Hall effect is clearly null, since the potentials at w and x, being given by the sums of the corresponding potentials at s and u, and at t and v, are the same. We cannot add a fourth solution, as illustrated in Fig. 1d, where electric charges of opposing signs are placed on opposite edges of the conductor, since this situation is unstable and would quickly decay.

2. NON-LINEAR SOLUTIONS

The resolution of the paradox lies in first noting that any solution to Maxwell's equations requires some assumption concerning the connection of $\vec{J}$ to $\vec{E}$, the electric-field intensity. Normally one takes Ohm's law in the form $\vec{J} = \sigma\vec{E}$, where σ is the conductivity, and, indeed, Maxwell states[2] that "The only force which acts on electric currents is electromotive force." But in the Hall effect, a transverse current proportional to $\vec{E}\times\vec{H}$ is developed so that the constitutive equation relating $\vec{J}$ to $\vec{E}$ becomes

$$\vec{J} = \sigma\vec{E} + a\vec{E}\times\vec{H}, \tag{1}$$

where a depends on the nature of the carriers of the current. Such an equation is automatically excluded when we restrict ourselves to the linear superposition of solutions in which $\vec{E}$ and $\vec{H}$ separately appear, since there is no way in which their product can be formed. Of course, the solution represented in Fig. 1c is a genuine one, and corresponds to a current formed from equal positive and negative charge densities moving in opposite directions. The correction that Rowland[3] envisaged is required, not in the four equations that bear Maxwell's name today, but instead in one of the constitutive equations.

Although Hall was stimulated by Maxwell's statement that the current through a conductor should not be affected by a constant magnetic field, it is a remarkable fact, pointed out by Hopkinson,[4] that Maxwell had considered[5] the modification to Ohm's law represented by Eq. (1): but instead of the vector $\vec{H}$ he used an undefined vector $\vec{T}$ for which "we have reason to believe that it does not exist in any known substance. It should be found, if anywhere, in magnets." Evidently Maxwell thought of $\vec{T}$ as being an intrinsic property of a solid body rather than an external magnetic field. He thus anticipated the existence of the so-called anomalous Hall effect that exists in ferromagnets.[6]

3. AXIAL CHARACTER OF H

A few years after Hall's discovery, it was recognized by Koláček[7] that the mere existence of an effect (irrespective of its sign) provides information about the nature of the vectors $\vec{E}$ and $\vec{H}$. A vector that reverses direction under inversion with respect to the origin of coordinates is said to be polar: one that maintains its direction (as well as its magnitude) is said to be axial. Classical mechanics provides examples of both types. As illustrated in Fig. 2, the momentum $\vec{p}$ of a particle is a polar vector, since an inversion reverses the motion of the particle. However, angular momentum is an axial vector, since the inversion of a circulating motion produces a new circulation in the same sense as the old. Mathematically, $\vec{r}\times\vec{p}$ is invariant to the substitutions $\vec{r}\to-\vec{r}$ and $\vec{p}\to-\vec{p}$.

If we suppose both $\vec{E}$ and $\vec{H}$ are polar vectors, then the effect of inversion on Hall's experiment as illustrated in Fig. 3a is to produce the arrangement of Fig. 3b. However, it is at once seen that the second pattern of vectors is incompatible with the first, since a simple rotation of the conductor by 180° about an axis

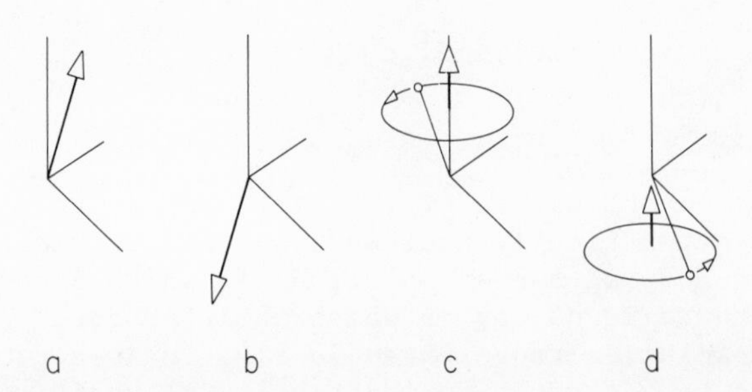

Fig. 2 Under inversion, the polar vector (a) changes direction (b); the axial vector (c), however, maintains its direction (d).

parallel to $\vec{H}\times\vec{J}$ restores $\vec{H}$ and $\vec{J}$ to their previous directions but leaves the charges on the edges with reversed signs. If we argue that $\vec{E}$ is an axial vector rather than a polar one, we must conclude that neither $\vec{J}$ (which is driven by $\vec{E}$) nor the charges on the edges of the conductor (whose signs determine the direction of the transverse $\vec{E}$ field) change sign. We again arrive at an incompatible arrangement of vectors. The only possible conclusion is that $\vec{H}$ is an axial vector.

This result goes beyond what could be deduced from Maxwell's

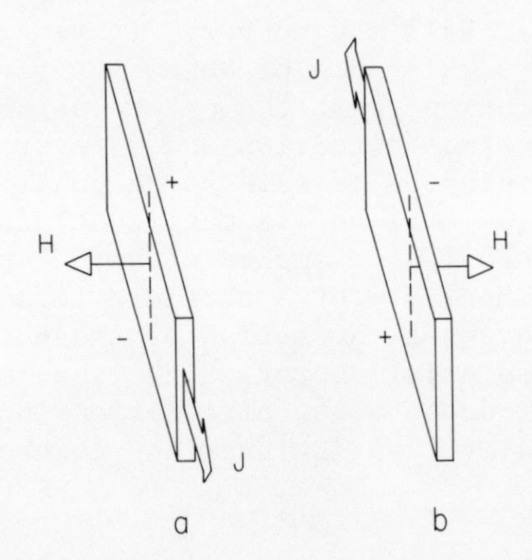

Fig. 3 The inversion of all particles and fields appearing in Hall's experiment (a) leads to (b) if $\vec{E}$ and $\vec{H}$ are assumed (erroneously) to be both polar vectors. The situation (b) is physically incompatible with (a).

four equations. The operator $\vec{\nabla}$, whether used as a curl or a divergence, changes sign under inversion: it thus converts a polar vector to an axial vector and vice-versa. Since the curls of $\vec{E}$ and $\vec{H}$ are related to $\vec{H}$ and $\vec{E}$ respectively, we cannot, on the basis of Maxwell's equations alone, make any statement as to the vectorial character of $\vec{E}$: but once $\vec{H}$ is established as an axial vector it can be inferred that $\vec{E}$ is a polar vector. The substitutions $\vec{H} \rightarrow \vec{H}$, $\vec{E} \rightarrow -\vec{E}$ leave Eq. (1) invariant, as we expect.

The implications of Hall's non-null result need not stop here. From Poisson's equation, we can deduce that electric charge is a true scalar: it is unaffected by inversion. Magnetic charge, were it to exist, would necessarily be pseudoscalar in nature: that is, it would change sign under inversion. Of course, the arguments that lead to these results depend on the presumed invariance of Maxwell's equations to inversion. For our purposes, this presumption is still valid; but the discovery in the 1950's by Lee and Yang of the non-conservation of parity in the weak interactions shows that the assumption of invariance with respect to inversion is not universally valid.

4. CONCLUDING REMARKS

It would be out of place here to go into the most unsettling result of Hall's discovery: that the current in some conductors appears to be carried not by negatively charged electrons but rather by positive ions. This difficulty had to wait until quantum theory and the Pauli principle has been assimilated before a satisfactory resolution could be achieved. Nineteenth-century physics had no inkling of the surprises that were to come.

REFERENCES

1. E. H. Hall, Amer. J. Math. 2:287 (1879).
2. J. C. Maxwell, "A Treatise on Electricity and Magnetism," (Third Edition), Constable and Co. Ltd., London (1891); Reprinted by Dover Publications, New York (1954), Vol. 2, p. 157.
3. H. A. Rowland, Amer. J. Math. 2:354 (1879).
4. J. Hopkinson, Phil. Mag. 10:430 (1880).
5. J. C. Maxwell, Ref. 2, Vol. 1, pp. 420-423.
6. See, for example G. Bergmann, Physics Today, 32-8:25 (August 1979).
7. F. Koláček, Ann. Phys. 55:503 (1895). See also M. Abraham, "Theorie der Elektrizität," Teubner, Leipzig (1907), pp. 247-250.

THE DISCOVERY OF THE HALL EFFECT:

EDWIN HALL'S HITHERTO UNPUBLISHED ACCOUNT

Katherine Russell Sopka

Harvard University
Cambridge, Massachusetts 02138

ABSTRACT

An annotated transcript of Edwin Hall's notebook account of his 1879 laboratory investigations of the transverse effect of a magnetic field applied at right angle to the direction of electric current in a conductor.

INTRODUCTORY NOTE

The public account of Edwin Hall's laboratory investigations associated with the discovery of what has since become known as the Hall Effect has been widely available for a century. In late 1879, when Hall was a 24 year old graduate student, he published "On a New Action of the Magnet on Electric Currents" in the *American Journal of Mathematics Pure and Applied*,[1] a quarterly journal published under the auspices of the Johns Hopkins University, edited by J. J. Sylvester and William Story of the Mathematics Department with the cooperation of other faculty members, including Hall's thesis supervisor, Henry A. Rowland. There he gave the preliminary notice of that new effect. In addition, Hall's early findings were brought to wider international attention through the publication a few months later of exactly the same text in *The London, Edinburgh and Dublin Philosophical Magazine and Journal of Science*.[2] Less than a year after these preliminary notices, accounts of Hall's more complete investigations that formed his doctoral dissertation at the Johns Hopkins University were published in the *American Journal of Science* and in the *Philosophical Magazine*.[3]

The hitherto unpublished account of the early part of Edwin Hall's investigations into the effect of imposing a magnetic field perpendicular to a current carried by a conductor which is presented in the following pages comes from his handwritten Notebook of Physics. This item is one of the earliest of some 400 papers that were deposited in the Houghton Library of Harvard University by his daughter, Miss Constance Hall in 1964, more than a quarter century after his death. Hall had joined the faculty of the Harvard Physics Department in 1881, serving actively until his retirement in 1921. As an emeritus professor he continued to work in his laboratory at Harvard until shortly before his death in 1938.

The Notebook itself is in remarkably good condition. The paper is substantial and only slightly discolored. The ink has turned brown but is still dark and easily legible, thanks principally to Hall's good penmanship.

The notebook's 124 ruled pages, approximately 7" by 8", are sewn together and covered with a firm binding of a dark, mottled coloring. The early pages are devoted to data taken by Hall on about a dozen experiments ranging from "Use of Large Spherometer" to "On Sp[ecific] Gr[avity] of Carbonic Anhydride compared with Air". While these experiments are not dated they were presumably done before January 14, 1878, the first date recorded (on page 56). The next several pages contain a narrative description of Hall's studies in the areas of spectrometry, polarized light and photography, followed by a List of Experiments in Electricity and Magnetism done during the calendar year 1878.[4]

Beginning on page 73 and continuing for the remainder of the notebook is Hall's personal account of how he came to perform those experiments which revealed the effect, now bearing his name, that is recognized today as being of great practical as well as theoretical importance in solid-state physics.

For the historian of science a notebook such as this is a valuable document, revealing private aspects of doing science that are not found in the public record of scientific results. While Hall's first publication mentioned above[1] is similar in many ways to his notebook account which follows, it seems worthwhile to make the full text from the notebook available now to scientists and historians of science in this centennial year of his discovery of the Hall Effect. Especially noteworthy are the insights the reader can gain into such areas as Hall's student/mentor relationship with Henry A. Rowland and his association with others in the Johns Hopkins physics department, and, perhaps more importantly, the state of knowledge and understanding in physics a century ago through Hall's candid personal account of how he approached his laboratory investigations.

The page numbers referred to below, and found in the text of the transcript, correspond to those used by Hall, those in [ ] having been added to fill in gaps in the sequence written by him.

As we read these pages today some of the ideas expressed by Hall strike us as decidedly naive or just plain wrong. We must, however, keep in mind that in 1879 scientists had little understanding of the nature of electricity. The electron, for example, was not identified until more than a decade after Hall conducted the experiments he described in this notebook. Furthermore, Hall's expectation that a wire with triangular cross section would be better than a round wire in testing for the proposed magnetic effect on resistance was apparently shared by a physicist of no less stature than Rowland himself. Hall's use of the now long outmoded water stream analogy for electric current nevertheless provided him with a line of reasoning that did put him on the path to making his discovery. On the other hand, his expectation that continuous work could be done by a permanent magnet is clearly at odds with the principle of the conservation of energy which had been established well before 1879.

Hall's account transcribed below has some curious aspects of which the reader should be aware. Despite the existence of dated entries (June 13, 14, November 3, 10, 21 and December 15) it appears that the entire piece was not written sequentially, page by page. The first clue to this comes in the opening sentence on page 73 with his reference to the "history" of his experiment. The next two pages are blank and his narrative continues on page 76. Again, pages 88 and 89 are found to be blank also.

It is my conjecture that Hall purposely left several blank pages after his earlier entries in this notebook and then did in fact record the data on pages 82, 84 and 86 on June 13 and 14, 1879 after discussion with his advisor Rowland and at least one other member of the department. It is likely that he then wrote at least some of the accompanying narrative before leaving for the summer with the belief that he had found an increased resistance in his test coil when it was subjected to a magnetic field, but somewhat perplexed by the <u>smaller</u> effect observed when he used a <u>more</u> sensitive galvanometer (see page 85).

By the time Hall made his next dated entry, November 3 on page 90, he had been back at the Johns Hopkins University several weeks. During this time he had redone the initial experiment more carefully, recording his data elsewhere, and come to the conclusion, stated on page 91, that there was no measurable increase in resistance. In addition, however, he had changed his experimental setup to one proposed by Rowland (see page 95) and looked for a transverse electromotive force in a piece of gold foil carrying an electric current and subjected to a perpendicular magnetic field.

The fact that with this arrangement he did find clear evidence of such an emf is not recorded in this notebook until November 10 (on page 98) although it had been observed about two weeks earlier.

Hall's results were, of course, of considerable interest to Rowland and also to other younger members of the physics department, some of whom were present while Hall was conducting his experiment and assisted him in carrying it out.

Obviously further work was needed and the remainder of Hall's account discusses such variations as changing the location of the tapping points for detecting the emf and using metals other than gold. Each such variation shed further light on the nature of the effect, but also raised further questions that are not adequately resolved in these pages. The interested reader will find a more thorough probing of such questions in "The Experiments of Edwin Hall 1879 - 1881", Chapter IX of "Henry Augustus Rowland and his Electromagnetic Researches", the doctoral dissertation of John D. Miller submitted to Oregon State University in 1970, an account which draws not only on this notebook of Hall but also on other resources.

In Hall's notebook narrative he has been primarily concerned with describing <u>his</u> conception of the phenomena being studied and explaining the routes by which he arrived at his ideas. While it is regrettable that the associated "Experiment Book" has apparently not survived, we rejoice that we do have such a revealing personal account as is contained in the pages transcribed below.

TRANSCRIPT FROM EDWIN H. HALL'S NOTEBOOK OF PHYSICS OF THOSE ENTRIES RELATED TO THE DISCOVERY OF THE HALL EFFECT

page 73

For the last two or three weeks of the year have been engaged in an experiment of which the history is as follows.

I was surprised to read some months ago a statement in Maxwell,[5] Vol II page [144] that the Electricity itself flowing in a conducting wire was not at all affected by the proximity of a magnet or another current. This seems different from what one would naturally suppose, taking into account the fact that the wire alone was certainly not affected and also the fact that in Static Electricity it is plainly the Electricity itself that is attracted by Electricity.

(Note: page 74 and 75 are blank)

[page 76]

Soon after reading the above statement in Maxwell I read an article by Prof. Edlund entitled Unipolar Induction (Phil. Mag and Journal de Chemie et Physique) in which the author evidentally assumed that an electric current was acted on by a magnet in the same way as a wire bearing a current.

As these two authorities seemed to disagree I asked Prof. Rowland about the matter. He told me he doubted the truth of Maxwell's statement and had been thinking of testing it by some experiment though no good method of so doing had yet presented itself to him. I now began to give more attention to the matter and thought of a plan which seemed to promise well. As Prof. Rowland was too much occupied with other matters to undertake this investigation at present I proposed my scheme

page 77

to him and asked whether he had any objection to my making the experiment. He approved of my method in the main, though suggesting some very important changes.

My method was formulated on the following theory. If the Current of Electricity is itself attracted by a magnet while the wire bearing it remains fixed in position, the current should be drawn to one side of the wire and therefore the resistance experienced should be increased. I thought this effect might be magnified if the wire were made of this (∇) section and so wound that the tendency of the magnetic force would be to draw the current into the thin part of the wire. This seemed plausible to Prof. Rowland and he put

[page 78]

me in the way of carrying out the experiment. Dr. Hastings[6] urged that there seemed to be no reason for expecting any advantage in using wire of the propose[d] section and thought round wire ought to show the effect, if there were any, as well. I could not indeed give any very definite reason for using the wire of triangular section, though it seemed to me that the effect would certainly not be diminished and might possibly be increased. Prof. Rowland, however, said that some effect of this kind had already been observed with round wire though not explained on above theory. He thought there would be an advantage in using the triangular wire.

The wire selected was of German

page 79

Silver about 1/50 in. thick. Short pieces of this wire were drawn through a triangular hole punched in a steel ribbon and the pieces were then soldered together making a length of about three feet and having a resistance probably of about two ohms. This wire was wound in a spiral groove turned in the flat surface of a disk of hard rubber. It was held in place as wound by a soft cement and then covered by another disk of hard rubber laid upon the first. The coil thus prepared was pressed between the poles of an electromagnet and made one arm of a Wheatstone Bridge.

The magnet was operated by a battery of 20 Bunsen's cells[7] placed 4 in series, 5 broad.

My plan was simply to send

[page 80]

a current through the wire and observe whether the resistance was varied by operating the magnet.

The disturbing cause might be
1st Current in wires from the magnet battery might directly affect the Galvanometer.
2nd Magnet might directly affect Galvanometer.
3rd Induction currents would probably arise on closing circuit though Galvanometer.
4th Thermo-Electric currents might exist.

In order to avoid the second difficulty magnet was placed some thirty feet distant from the Galvanometer. It was found that the first and second of the above causes affected the Galv. only slightly and in the way of a permanent

page 81

deflection whenever the magnet circuit was closed. These causes then do not affect the experiment.

The combined effect of the 3^{d} and 4th causes was observed as follows. I replaced the small battery used with the Bridge by a short wire; then with the magnet circuit alternately open and closed, I would complete the circuit through the Galvanometer pressing down the key until the needle reached its maximum deviation. I usually found slight deviations in this way and have recorded them when about to try the main experiment.

Observations made June 13th and 14th are recorded on succeeding pages.

[page 82]

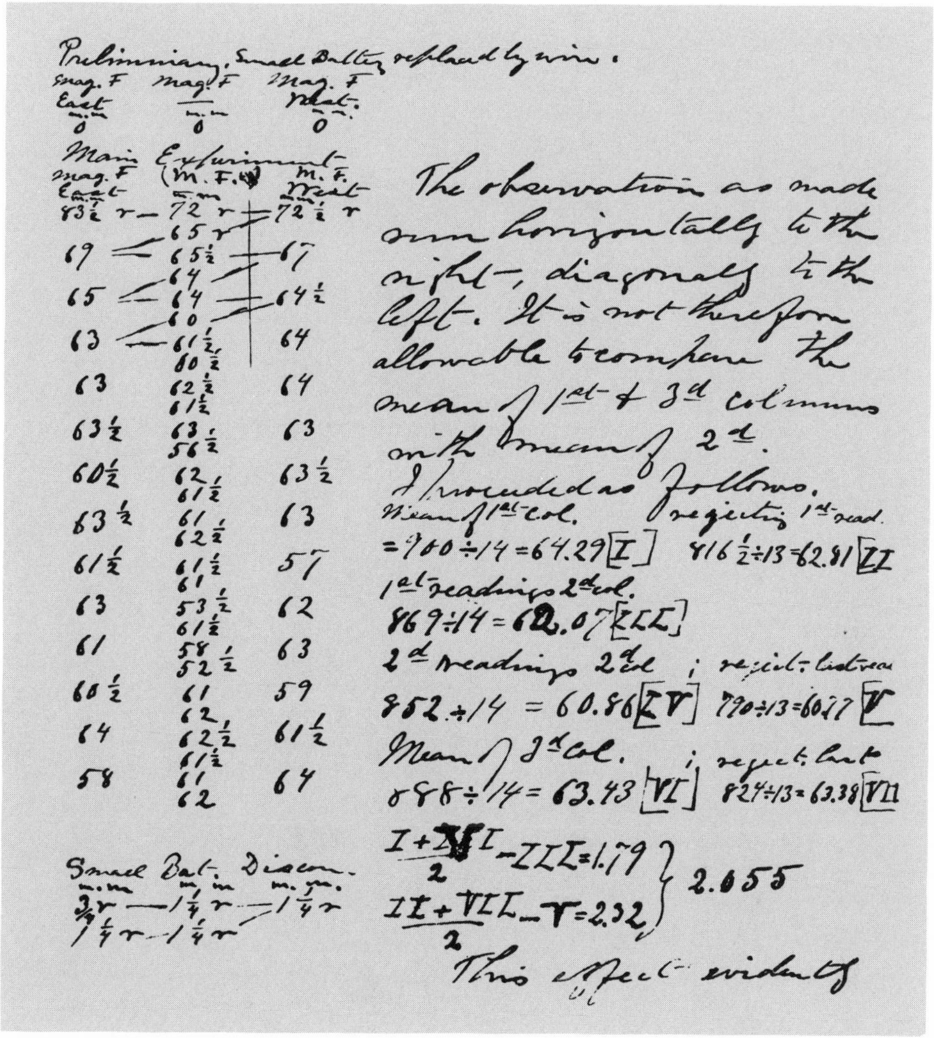

Preliminary, Small Battery replaced by wire.

mag. F East	mag. F —	Mag. F West
0	0	0

Main Experiment

mag. F East	(M. F.) —	M. F. West
83½ r	72 r	72½ r
	65 r	
69	65½	67
	64	
65	64	64½
	60	
63	61½	64
	60½	
63	62½	64
	61½	
63½	63	63
	56½	
60½	62	63½
	61½	
63½	61	63
	62½	
61½	61½	57
	61	
63	53½	62
	61½	
61	58½	63
	52½	
60½	61	59
	62	
64	62½	61½
	61½	
58	61	64
	62	

Small Bat. Discon.

m.m	m. m	m. m.
¾ r	1¼ r	1¼ r
1¼ r	1¼ r	

The observation as made seem horizontally to the right, diagonally to the left. It is not therefore allowable to compare the mean of 1st & 3d columns with mean of 2d.

I proceeded as follows.

Mean of 1st col. ; rejecting 1st read.

= 900 ÷ 14 = 64.29 [I]   816½ ÷ 13 = 62.81 [II]

1st readings 2d col.

869 ÷ 14 = 62.07 [III]

2d readings 2d col ; reject. last read

852 ÷ 14 = 60.86 [IV]   790 ÷ 13 = 60.77 [V]

Mean of 3d col. ; reject. last

888 ÷ 14 = 63.43 [VI]   824 ÷ 13 = 63.38 [VII]

$\frac{I + VI}{2} - III = 1.79$ }
$\frac{II + VII}{2} - V = 2.32$ } 2.055

This effect evidently

page 83

This effect evidentally cannot be accounted for by thermal currents or induction. It is such an effect as would be caused by a slight increase of resistance in the wire tested.

[page 84]

Saturday June 14th.

Test for Thermal Current etc.

M.F. East	M.F. —	M.F. West
3 r	4 r 3½ r	3½ r
3 r	3¾ r 3½ r	3¾ r
3¾ r	3½ r 3¾	3¾ r
3 r	3½ 3⅓ r	4 r
3½ r	3½ r	4 r

1st col
16.25 ÷ 5 = 3.25 ; reject first read 13.25 ÷ 4 = 3.31

1st readings 2d col.
18.6 ÷ 5 = 3.72 ; 2d reading 2d col 14.4 ÷ 4 = 3.60

3d col reject last read.
19 ÷ 5 = 3.80 15 ÷ 4 = 3.75

$\frac{3.25+3.80}{2} - 3.72 = -.20$
$\frac{3.31+3.75}{2} - 3.60 = -.07$ } −.135 m.m

Main Experiment

18½ r	15½ r 16¼	17¾ r
18.7	17.1 19.4	20.0
20.8	20.8 23.8	23.7
26.4	29.4 38.2	34.6
41.3	42.8 47.8	46.6
53.3	54.9 64.6	60.6

179.0 ÷ 6 = 29.83 160.5 ÷ 5 = 32.10
180.5 ÷ 6 = 30.08 145.45 ÷ 5 = 29.09
203.25 ÷ 6 = 33.88 142.65 ÷ 5 = 28.53

$\frac{29.83+33.88}{2} - 30.08 = 1.77$
$\frac{32.10+28.53}{2} - 29.09 = 1.22$ } 1.49

After readjustment.

6.4	8.0 12.0	12.0
15.4	16.6 19.4	20.3
21.9	23.9 28.0	27.0
29.6	30.4 33.5	32.9
35.0	36.4 41.5	41.4
41.3	46.1 52.7(?)	51.8(?)
55.2		

149.6 ÷ 6 = 24.93. 198.4 ÷ 6 = 33.07
161.4 ÷ 6 = 26.90 187.1 ÷ 6 = 31.18
185.4 ÷ 6 = 30.90

$\frac{24.93+30.90}{2} - 26.90 = 1.02$
$\frac{33.07+30.90}{2} - 31.18 = .80$ } .91

[There was a slight irregularity in way of taking these two observations, but I do not think they need be rejected.

page 85

In making observations recorded on the preceding page and the following a more sensitive Galvanometer has been substituted for the one previously employed. It is to be noticed however that the effect seems to be less marked with the sensitive Galvanometer than with the other. The continued and rapid increase in the deflection is probably due to the heating of some part of the circuit by the repeated currents.

[page 86]

Saturday June 14th continued,

M.F. East m.m	M.F. m.m	M.F. West m.m
17.l	14.l 13.9l	15.3l
15.8l	14.9l 15.0	15.5l
17.0	15.1 15.3	16.0
17.4	15.4 16.5	17.5
17.7	17.4 18.8	19.0
22.0	21.4 23.2	22.4
27.1	26.8 30.0	28.8
33.8		

134.0 ÷ 7 = 19.14   150.8 ÷ 7 = 21.54

125.0 ÷ 7 = 17.86   132.2 ÷ " = 18.89

134.5 ÷ 7 = 19.21

$\frac{19.14 + 19.21}{2} - 17.86 = 1.31$

$\frac{21.54 + 19.21}{2} - 18.89 = 1.48$

} 1.40

Test of Thermal & Induction Currents.

m.m		
.6l	.3l .1l	.2r
.6l	.2l .0	.2r
.2l	.2r .6r	.6r
.2r	.7r 1.0r	1.3r
.8r		

1.2 ÷ 4 = .3l   .2 ÷ 4 = .05r

.4 ... = .1r   1.5 " " = .37r

2.3 " " = .57r

$\frac{.3^l + .57^r}{2} - .1r = +.03$

$\frac{.05r + .57r}{2} - .37r = -.06$

} −.015

page 87

Deflections recorded on preceding page are mainly to the left while those previously recorded were to the right.

This was caused by reversing current rhrough the Bridge so that the current from the zinc pole entered the test coil by the circumference instead of the center as previously it had done. The last arrangement made current in the test coil flow from north to south above

(Note: pages 88 and 89 are blank)

[page 90]

Record for '79 and '80

Nov. 3 '79

In the preceding pages I have described some experiments made in the last part of last year, whereby it appeared that the resistance of a wire carrying an electric current was slightly increased by placing it in a strong magnetic field.
On my return to the University this fall I resumed the consideration of this subject.
It occurred to me one day that the effect observed might be due to the heating caused by the pressure exerted by the poles of the electromagnet on the spiral of wire between them. I therefore repeated the experiment taking precautions against any such heating effect

page 91

and the observed increase of resistance disappeared totally. Particulars of the experiment are recorded in my Experiment Book and it is there shown that with a galvanometer capable of indicating, I think, a change of one part in a million in the wire's resistance, no such effect appeared after some days' trial, but rather on the average a very slight decrease was observed, too slight to be considered anything more than an error of the experiment. It seemed therefore very probable that no measurable change in resistance was caused by the magnet. This might have seemed to settle the question of the action or non action of the magnet on the electric current as such, but the whole

[page 92]

ground was not yet covered. I reasoned thus:

If electricity were an incompressible fluid it might be acted on in a particular direction without moving in that direction. I took an example about like this. Suppose a stream of water flowing in a perfectly smooth pipe which is however loosely filled with gravel. The water will meet with resistance from the gravel but none from the pipe, at least no frictional resistance. Suppose now somebody brought near the pipe [something] which has the power of attracting a stream of water. The water would evidently be pressed against the side of the pipe but being incompressible

page 93

and, with gravel completely filling the pipe, it could not move in the direction of the pressure and the result would simply be a state of stress without any actual change of course by the stream. It is evident however that if a hole were made traversing through the pipe in the direction of the pressure and the two orifices thus made were connected by a second pipe water would flow out toward the attracting object and in at the opposite orifice. This suppose[s] of course that the attracting object acts upon the

[page 94]

current flowing in one direction without acting, equally at least, upon current in the other direction.

Nov. 4th '79 I mean by this that the attracting object is supposed to act, not upon the water at rest and under all circumstances, but only when the water is flowing and flowing in a certain direction or the opposite. In this way I arrived at the conclusion that in order to show conclusively that the magnet does not affect the current at all I must show, not merely that there was no actual deflection of the current, which seemed to be already shown by my experiments on resistance, but further that there was no tendency of the current to move.

page 95

In order to do this I concluded to repeat an experiment which Prof. Rowland had once tried without any positive results.
This was to determine whether the equipotential lines in a disk of metal carrying a current would be affected by placing the disk between the poles of the electro-magnet. I set something of the above reasoning before Prof. Rowland and he advised me to try the

experiment as he had not made the trial very carefully himself.

He advised me to use gold leaf mounted on a glass plate and I did so. The plan was this

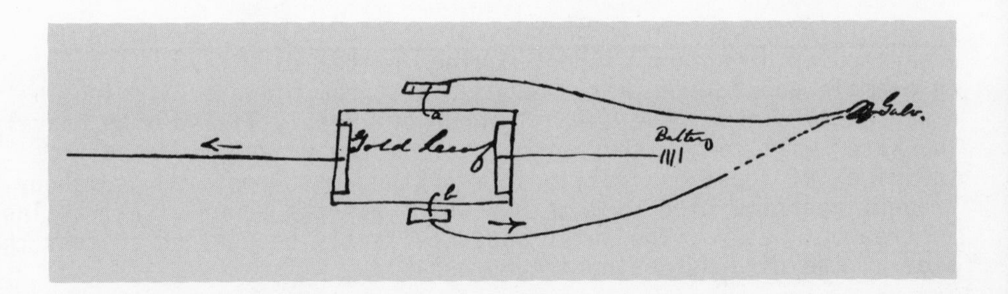

[page 96]

The theory was this

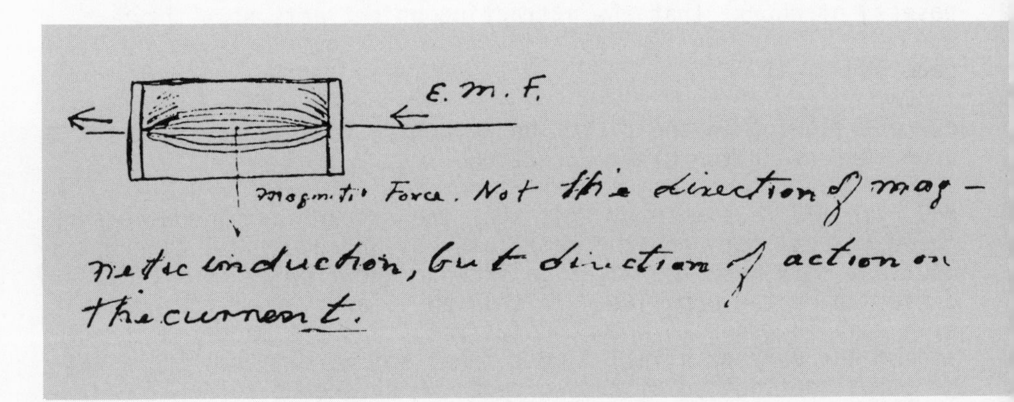

Prof. Rowland advised me to place my tapping points (a,b) near the end of the disk on the ground that the equipotential lines crossing the disk in the center would not be deflected by a deflection of the current. His theory was

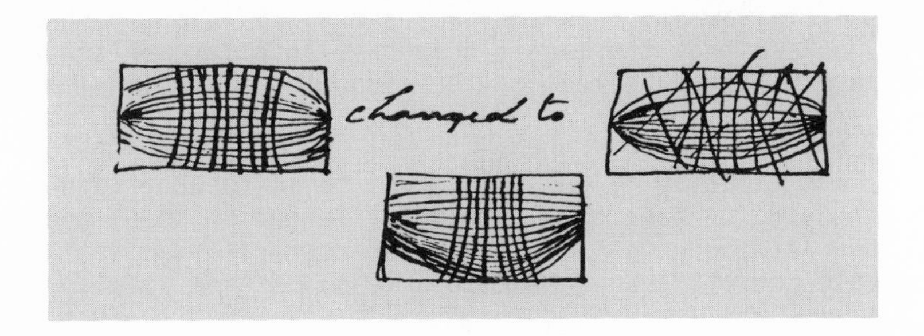

I followed his advice in placing my points but I pointed out to

page 97

him before trying the experiment that in case there was a tendency to deflect the current without any actual deflection, the current would be in a state of stress and the equipotential lines would not be simply perpendicular to the lines of flow but would be oblique to them and that the whole length of the disk, in the middle as elsewhere. Experiment afterward proved the correctness of my theory.

Nov. 10th '79
In trying this experiment I used a Thomson galvanometer which was perceptibly affected by an EMF of 1/1,000,000 of a Bunsen cell, the resistance of the galvanometer being 6491 ohms. With this instrument, having in its circuit an additional resistance of 100,000 ohms,

[page 98]

there was detected on the evening of, I believe, Friday the 24th of October, a somewhat doubtful indication of the action of the magnet. A trial the next day however seemed to fail in confirming this indication. Means were taken however to make the test more delicate and finally we were able to do away with all additional resistance in the galvanometer circuit. On the evening of, I believe, Tuesday Oct. 28th, very marked and seemingly unmistakable evidence of the looked for effect was observed. Mr. Freeman[8] was observing the galvanometer. The deflection observed was a permanent one of two or three centimeters. I was myself at the magnet. It seemed hardly safe even then to believe that a new phenomenon had

page 99

been discovered but now after nearly a fortnight has elasped and the experiment has been many times and under various conditions successfully repeated, meeting at the same time without harm the criticism of fellow students and professors, it is not perhaps too early to declare that the magnet does have an effect on the electric current or at least an effect on the circuit never before expressly observed or proved.

It now became a matter of great interest to me to ascertain the effect of moving my tapping points to different parts of the gold leaf plate. If there was an actual displacement or deflection of the electric current through the gold leaf, as Prof. Rowland had

[page 100]

thought possible, the effect should be reversed on moving the tapping points to the other end of the gold leaf and in the transverse diameter the effect should disappear. If the effect of the magnet was simply to cause a tending of the current to change its course without effecting any deflection, which I thought to be the only view consistent with my failure to discover an increase of resistance in the spiral of wire previously tried, then the observed effect would not be reversed in direction when the points were moved to the other end of the gold leaf and furthermore the effect on the galvanometer should not disappear when the points were placed

page 101

opposite the center of the gold foil. I stated both these views to Mr. Jacques[9] and Mr. Ayres[10] immediately before making the test and then asked Mr. Ayres to assist me in the trial telling him just what I expected to find, viz. that the effect would not be reversed.

The points were first moved to the opposite end of the gold leaf and then to the middle.

The result was what I expected. Prof. Rowland came in while the experiment was in progress. On learning the result he after a little thought proposed an explanation of the phenomenon which seemed to me at the time very like my own already given. It was discovered in these experiments just described or soon after that the effect was

[page 102]

in one respect the reverse of what had been anticipated by both Prof. Rowland and myself. That is, the current seemed to tend to move in the opposite direction to that in which the disk itself would tend to move under the action of the magnet.

At first this seemed to be easily accounted for on the supposition that the electric current really flows from negative to positive instead of from positive to negative as usually assumed for convenience. A little reflection has however convinced me that this explanation is not sufficient. If we make the suggested change in our conception of the direction of the current through the gold leaf,

page 103

we must make the same change in our conception of the direction of the current in the electromagnet and the two changes will annul each other leaving us still face-to-face with [a point] of difficulty. To me at present it seems probable that two parallel currents of electricity flowing in the same direction tend to repel each other, just as two quantities of static electricity do. Further experiments will probably be necessary to ascertain the truth of the matter.

Note by EHH: see p. 114

I have been for a week past making ready, by Prof. Rowland's advice and under his direction to make a series of observations which will be capable of reduction to absolute measure, thus giving the quantitative laws of the phe

[page 104]

nomenon observed. This is for the purpose of getting something definite to publish.

More than a week ago I suggested an experiment which Prof. Rowland, not getting my idea fully I think, did not regard favorably. I have since been able to discuss the matter more fully with him and think it altogether probable that he will ultimately approve of it. The experiment proposed is suggested by one described by Edlund in his article on "Unipolar Induction" (Journal de Chemie et Physique).

Edlund states that if a magnet be placed vertical and surrounded with a metal cylinder, this cylinder tends to revolve about the magnet when a current of electricity is

page 105

made to flow from one end to the middle of the cylinder, the wires from a battery being applied at these points. Some months ago as I was considering this phenomenon I wondered what could make the cylinder move if, as Edlund seemed to suppose, an electric current could be acted upon and moved by a magnet just as a wire bearing a current is acted upon and moved. Prof. Rowland to whom I mentioned my difficulty suggested that the current might be made by the magnet to move around the cylinder and by the metallic resistance drag the cylinder about. Some time after this however I approached Prof. Rowland again on this subject.

If, I argued, the current is ac

[page 106]

tually made to pursue a spiral course along the cylinder more work must be done in the cylinder than if the current flowed in its natural course. This must be true even when the cylinder is at rest. Now this additional work must be done either by the battery maintaining its current, in which case the work must show itself in an increased resistance of the cylinder, or the extra work must be done by the magnet. On one supposition we have the resistance of a circuit increased by the presence of a magnet, a thing hitherto unobserved; on the other supposition we nave a permanent magnet doing continuous

page 107

work, a thing equally improbable.

Prof. Rowland admitted the full force of my reasoning and remarked at the time that he didn't see why this argument was not conclusive. Conclusive against Edlund's theory I suppose he meant. He observed however at the same time that if there was no action at all on the current itself he didn't see how his experiment on Electric Convection[11] could succeed.

When I began the experiments with the gold leaf I expected to obtain a negative result as I had in the experiment on resistance and I expected to publish those negative results together with a criticism of Edlund's theory as applied to this phe

[page 108]

nomenon of the revolving cylinder. Naturally after discovering the action of the magnet on the gold leaf I recurred to this phenomenon and asked myself how I could reconcile it with the results of my own experiments. So I took up again the suggestion of Prof. Rowland that the current went in a spiral course along the cylinder and concluded to boldly face the two horns of the dilemma and test them in turn.

Who knows that the proximity of a magnet may not increase the resistance of an electric current? Who knows that a permanent magnet cannot do continuous work?

At present I reason thus. The deflection of the current in the cylinder cannot be the cause of the

page 109

cylinder's motion, for according to my experiment thus far the current would have a tendency to drag the cylinder in a direction contrary to the one it follows. Moreover the effect of the deflected current would be, so far as I can judge now, altogether inadequate to produce such a result as I presume the motion of of the cylinder is affected in one way while the current flowing in it is affected in the opposite way. It is not difficult to suppose this possible, though I have not now any clear conception of the way in which it is possible.

Note by EHH: see p. 114

The explanation, as I conceive, of the

[page 110]

non deflection of the current in a disk is that the circuit in the transverse direction is not completed. When we complete it through the galvanometer a current is set up. Suppose now we bend our disk into a cylinder and so dispose it in the magnetic field that many lines of force pass out through the wall of the cylinder thus giving the current flowing therein a tendency to turn or slide around the cylinder. Why will it not pursue a spiral course? If it does take a spiral course, and I think we can ascertain whether it does so or not, we will try the two horns of our dilemma in succession.

My opinion is that the experiment will succeed and that the magnet will be found to do the

page 111

extra work.

Nov. 21st '79

Since writing the preceding notes I have extended my experiments on the newly discovered phenomenon and the numerical results of my work are given in my experiment notebook where the observations made are recorded. It was my idea before performing my later experiments, and I expressed my view to Dr. Nichols[12] at least before making the trial, that with the apparatus used the effect on the Thomson galvanometer would be directly proportional to the product of the strength of the magnetic field

page 112

and the strength of the current through the gold leaf.

That this is the law is made extremely probable by the experiments made Nov. 12th and recorded in the notebook above referred to.

I had formed and expressed moreover the further opinion that the new phenomenon would not with our instruments appear when a strip of moderately thick copper was substituted for the gold leaf.

Prof. Rowland however thought the copper would serve quite as well as the gold leaf.

He happened to come in just as I was about to try the strip of copper which was of dimensions somewhat similar to those of the gold leaf except that the

page 113

thickness was perhaps 1/4 mm.

Immediately before making the trial which he witnessed, I told Prof. Rowland I thought the copper would not show the effect.

He declared his opinion that it would show it and further remarked that we would now see who had the right idea of the phenomenon. Upon making the trial he saw and at once admitted that I was right.[13]

In this experiment with the copper the magnetic field, as well as the current through the strip, was considerably stronger than had been the case when the measurements with the gold leaf were made as mentioned above. By Prof. Rowland's advice

[page 114]

I have written a short abstract of what has been recorded in the preceding pages, adding thereto the numerical results arrived at, and this article will, I presume, appear in the next number of the Mathematical Journal.[14]

Nov. 25th '79

On pages 102 and 103 I have expressed my opinion that there was no advantage to be gained by supposing the current of electricity to flow from the negative to the positive pole. Prof. Rowland however was inclined to think there would be an advantage, or at least a difference [in] conception of the direction of the newly discovered effect, by assuming that the current was from negative

page 115

to positive. Upon further consideration I have come to the same conclusion. I have said above that reversing our conception of the current in the gold leaf made it necessary to revse our conception of the current through the magnet circuit, and consequently these two changes would neutralize each other. I forgot however that we must also change our ideas of the course of the current in the Thomson galvanometer. It seems then that if we conceive of the current as flowing from the zinc pole of the battery through the gold leaf to the carbon pole of the battery, we find that the new phenomena are in accordance with the supposition that two parallel currents attract each other. I tried yesterday the

[page 116]

experiment of placing between the poles of the magnet a strip of gold leaf about 1 cm. wide lying in a horizontal position, flat side up and testing it for a change of relative potential of points on opposite edges of the strip when the magnet was put in operation. No effect was discerned beyond that ordinarily caused by the direct action of the magnet and its circuit on the galvanometer needle.

The current in [the] gold leaf and the strength of the magnetic field were in this case probably not so large as in some previous experiments, but were sufficient to have shown a marked effect had the gold leaf strip been placed in the usual position.

page 117

Dec. 15th '79

Since making my last entry in this book I have been at work quite steadily upon my experiments but have not reached any very important results as yet. I made a silver strip by depositing the metal on glass. This showed the same effect as the gold leaf though no absolute measurements have as yet been [made?] with the silver. I have a strip of tin foil prepared but have not yet tried this. Prof. Rowland wants me to try iron and thinks the effect in this case will be reversed.

I referred some time ago to an experiment I wanted to make with a cylinder. Prof. Rowland did not approve of this experiment at first but finally admitted

[page 118]

it would be worth trying. He suggested using instead of the cylinder a disk of metal with one pole at the centre. The radiating currents would of course be affected by the magnetic force. This was an obvious improvement over the cylinder and I adopted his plan.

Neither of us saw at first though Prof. R. has since pointed out that any increase of resistance would be very small indeed. Thus where I had found a transverse electromotive force equal to about 1/3000 of the direct electromotive force in the strip, we could look for a change of resistance equal to the square of 1/3000 or 1/9000000. This slight change is of course

page 119

very difficult to detect and I doubt whether it can be discovered with our instruments even if it exists which I somewhat doubt.

I have heretofore expressed my opinion that the magnet would be found to do the extra work, but here too the change of resistance (in the magnet circuit) would be extremely small; too small probably to detect even if it exists. At present however I see no absurdity in the way of expecting continuous work from a magnet. It is quite evident that a magnet does work while it is attracting any body toward itself. The difficulty appears to be that ordinarily in removing the attracted body just as much work has to be as it were returned to the magnet as has been done

[page 120]

by the magnet in attracting the body. That is, the body attracted has a Potential with respect to the magnet. I conceive the magnet to act like a stretched spring which is capable of doing work and losing energy by contraction but which recovers the same amount of energy when it is again stretched.

If now we could allow a body to approach a magnet in such a way as to be attracted during the approach and then remove the body by some path in which it was no longer affected by the magnet, it seems to me that we could have the magnet doing work and losing energy. Obviously this is impossible in ordinary cases. But suppose now that the magnet

[page 121]

acts upon the electric currents radiating from the center of the disk mentioned above and continues to act upon them until they reach the ring surrounding the disk. Beyond this there seems to be no reason for attributing to the magnet any considerable action on the current. Now if in this case a new electromotive force is set up causing a current around the disk at right angles to the original radiating currents, these radiating currents remaining unchanged meanwhile, it seems to me that in this particular part of the circuit the magnet does work and I do no see how the energy thus lost to the magnet can be made up in any other part of the circuit.

A week or two ago I expressed those views to Dr. Nichols who him

[page 122]

self had formed somewhat similar ones, though hardly so definite as mine. He immediately set about trying to find some way of making a magnet do work. A few days afterward he suggested the experiment of allowing a magnet to attract a stream of iron filings which after being attracted to the vicinity of the magnet were to be dissolved or otherwise changed in such a way as to lose their magnetic property. I suggested, though, he may have thought of it before, that in this case there might be a retardation or enfeeblement of the chemical reactions owing to the influence of the magnet on the filings. It appears that this must be true or that the magnet must be able to do con

[page 123]

tinuous work. I have been myself thinking of attacking the problem in a different way.

It seems to me that a magnet ought to do work when as in Faraday's experiment (Ganot p. 712)[15] a part of an electric circuit revolves about a magnet.

The difficulty is that if the magnet does do work in this case it ought to have been discovered long ago. There may be some consideration which I have overlooked and which will show the absurdity of my ideas at once. Nichols and I are thinking of making some experiments on the thing some time, in his way or mine or both, unless someone shows us our folly before we have a chance to test out theories practically.

NOTE, added by transcriber: There were no more entries after the above although 3 more pages were available before the end of the notebook.

ACKNOWLEDGEMENTS

I gratefully acknowledge the encouragement and helpful comments made to me during the preparation of this manuscript by Professors Gerald Holton and Edward M. Purcell of the Harvard Physics Department. I am indebted to Ms. Julia Morgan of the Ferdinand Hamburger Jr. Archives of the Johns Hopkins University for information on the identities of persons mentioned by Edwin H. Hall in his Notebook of Physics. The above pages were transcribed by permission of the Houghton Library, Harvard University.

NOTES

1. Amer. Jour. Math. 2:287 (1879). The exact date when this publication reached the hands of readers is not clear since it was in the "September" issue, but Hall's paper and appended note are dated November 19 and 22, respectively.
2. Phil. Mag.(Series 5) 9:225 (1880). This was in the March issue and bore the notation "From a separate impression from the 'American Journal of Mathematics' 1879, communicated by the Author."
3. On the new action of magnetism on a permanent electric current, Amer. Jour. Sci. 20:161 (1880) and Phil. Mag.(Series 5) 10:301 (1880).
4. The notebook described here is the only one presently known to exist from among those used by Hall during his years at the Johns Hopkins University. It is clear, however, from references made by Hall in this notebook that he did keep at least one other "Experiment Book".
5. J. Clerk Maxwell Treatise on Electricity and Magnetism was published for the first time in 1873 by the Clarendon Press of Oxford University.
6. Charles Sheldon Hastings, Associate in Physics at the Johns Hopkins University 1876-1883.
7. Bunsen's cells were zinc carbon batteries, invented by Robert Wilhelm Bunsen in 1843. These double fluid cells (sulfuric and nitric acids) yield an emf of 1.9 volts.
8. Spencer Hedden Freeman was a fellow graduate student of Hall.
9. William White Jacques received his Ph.D. in physics at the Johns Hopkins University in 1879 and was a "Fellow by Courtesy" in the Physics Department during the academic year 1879-80.
10. Brown Ayres was a Fellow in Mathematics, 1879-80, who also studied physics with H. A. Rowland.
11. While working in Helmholtz' laboratory in Berlin during the year 1875-6 H. A. Rowland successfully demonstrated the magnetic effect of a rotating charged disk. For further discussion of this work see J. D. Miller, op. cit. pp. 114-135.
12. Edward Leamington Nichols was a Fellow in the Physics Department in 1879-81. Later, 1887-1919, he taught at Cornell University and was a founding editor of The Physical Review.
13. The significance of the thickness of the metallic strip is now understood with the recognition that a crucial element in the Hall Effect is the current density rather than the total current.
14. See Note 1.
15. "Ganot" is Hall's shortened version of the title Elementary Treatise on Physics Experimental and Applied: For the Use of Colleges and Schools, translated and edited from Ganot's

Elements de Physique by E. Atkinson and published in New York by William Wood and Company. This work went through many editions in French and English and was widely used as a college text in the United States. Presumably the Faraday experiment referred to by Hall is one in which a metallic conductor shaped in a double loop and carrying a current can be mounted in such a way that it rotates in the presence of a suitably oriented magnetic field. Such an apparatus is pictured and described on page 817 of the 1886 edition of Ganot's Physics.

INDEX